Birkhäuser

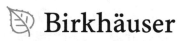

Progress in Mathematics
Volume 300

Daniel Bump • Solomon Friedberg
Dorian Goldfeld
Editors

Multiple Dirichlet Series, L-functions and Automorphic Forms

Birkhäuser

Editors
Daniel Bump
Department of Mathematics
Stanford University
Stanford, CA, USA

Solomon Friedberg
Department of Mathematics
Boston College
Chestnut Hill, MA, USA

Dorian Goldfeld
Department of Mathematics
Columbia University
New York, NY, USA

ISBN 978-0-8176-8333-7 ISBN 978-0-8176-8334-4 (eBook)
DOI 10.1007/978-0-8176-8334-4
Springer New York Heidelberg Dordrecht London

Library of Congress Control Number: 2012939728

Mathematics Subject Classification (2010): 11-06, 11F68, 11F70, 11M32, 11N05, 16T25, 22E50

Printed on acid-free paper

Springer is part of Springer Science+Business Media (www.birkhauser-science.com)

Preface

The theory of L-functions of one complex variable has a long and colorful history going back to Euler, Dirichlet, and Riemann, among others. L-functions associated to representations of Galois groups (Artin L-functions) as well as geometric L-functions associated to algebraic varieties (Hasse-Weil L-functions) have been extensively studied in the last 60 years. There is now a grand unification theory of L-functions of one complex variable, loosely called the Langlands program, which conjecturally associates to each L-function satisfying certain basic properties such as:

- Meromorphic continuation (with finitely many poles)
- Moderate growth
- Functional equation
- Euler product

an irreducible automorphic representation of a certain reductive group.

By comparison, the theory of multiple Dirichlet series (L-functions of several complex variables) is much less developed. It is natural to try to extend the Langlands program to L-functions in several complex variables satisfying the bulleted properties above. One quickly sees, however, that it may not be natural to assume the existence of Euler products in the theory of multiple Dirichlet series; a notion of twisted multiplicativity has been introduced instead. Some important progress in developing a theory of multiple Dirichlet series has been made in the last two decades. The main achievements so far include the construction of Weyl group multiple Dirichlet series, applications of multiple Dirichlet series to the problem of nonvanishing of twists of L-functions, to convexity breaking, to moments of zeta and L-functions, and to the connections to multiple zeta function values. In order to get interested researchers together and outline the latest advances in the subject, a workshop, Multiple Dirichlet Series and Applications to Automorphic Forms, was held in Edinburgh from August 4 to 8, 2008. It was hoped that this workshop would stimulate further research and lead to deeper insights into what a theory of multiple Dirichlet series should be. This has, in fact, happened and is the main motivation for publishing this volume which is focused on the more recent developments in the theory. The chapter "Introduction: Multiple Dirichlet Series," which opens this volume, gives an introduction to many of these developments. We would like to take this opportunity to thank especially Nikos Diamantis, Ivan Fesenko, and Jeffrey Hoffstein for their hard work and efforts in organizing the Edinburgh workshop and all the participants for their enthusiastic support and interesting talks. In particular, we thank Ben Brubaker, Alina Bucur, Gautam Chinta, Adrian Diaconu, Nikolaos Diamantis, Ivan Fesenko, Paul Garrett, Paul Gunnells, and Jeffrey Hoffstein for organizing and/or lecturing in the two minicourses at the workshop. We are extremely grateful to ICMS for hosting the workshop and making the stay in Edinburgh so enjoyable. We thank the EPSRC, the NSF, and the LMS for financial support. Finally, we would like to thank all the authors who contributed to this volume and also the referees for their timely and insightful contributions.

Stanford, CA, USA — Daniel Bump
Chestnut Hill, MA, USA — Solomon Friedberg
New York, NY, USA — Dorian Goldfeld

Contents

Chapter 1
Introduction: Multiple Dirichlet Series

Daniel Bump

Abstract This introductory article aims to provide a roadmap to many of the interrelated papers in this volume and to a portion of the field of multiple Dirichlet series, particularly emerging new ideas. It is both a survey of the recent literature, and an introduction to the combinatorial aspects of Weyl group multiple Dirichlet series, a class of multiple Dirichlet series that are not Euler products, but which may nevertheless be reconstructed from their p-parts. These p-parts are combinatorially interesting, and may often be identified with p-adic Whittaker functions.

Keywords Weyl group multiple Dirichlet series • Crystal graph • Solvable lattice model • Whittaker function • Metaplectic group • Yang–Baxter equation

This survey article is intended to help orient the reader to certain topics in multiple Dirichlet series. There are several other expository articles that the reader might also want to consult, though we do not assume any familiarity with them. The article [20] which appeared in 1996 contained many of the ideas in an early, undeveloped form. The articles [9, 26], which appeared in 2006, also survey the field from different points of view, and it is hoped that these papers will be complementary to this one. Further expository material may be found in some of the chapters of [14].

Since approximately 2003, there has been an intensive development of the subject into areas related to combinatorics, representation theory, statistical mechanics, and other areas. These are scarcely touched on in [9, 20, 26] and indeed are topics that have largely developed during the last few years. However, these combinatorial developments are discussed in [14] as well as this introductory paper and other papers in this volume.

D. Bump (✉)
Department of Mathematics, Stanford University, Stanford, CA 94305-2125, USA
e-mail: bump@math.stanford.edu

D. Bump et al. (eds.), *Multiple Dirichlet Series, L-functions and Automorphic Forms*, Progress in Mathematics 300, DOI 10.1007/978-0-8176-8334-4_1,

Preparation of this paper was supported in part by NSF grant DMS-1001079. We would like to thank Ben Brubaker, Solomon Friedberg, and Kohji Matsumoto for their helpful comments.

1.1 Moments of L-Functions

The subject of multiple Dirichlet series originated in analytic number theory. If $\{a_n\}$ is a sequence of real or complex numbers, then a typical Tauberian theorem draws conclusions about the a_n from the behavior of the Dirichlet series $\sum_n a_n n^{-s}$. If the a_n are themselves L-functions or other Dirichlet series, this is then a multiple Dirichlet series.

One may try to study moments of L-functions this way. For example, Goldfeld and Hoffstein [37] considered a pair of Dirichlet series whose coefficients are

$$Z_{\pm}(w,s) = \sum_{\pm d>0} A_d(s)|d|^{\frac{1}{2}-2w}, \tag{1}$$

where the coefficients $A_d(s)$ are essentially quadratic L-functions. More precisely, if d is squarefree

$$A_d(s) = \frac{L_2\left(2s-\frac{1}{2},\chi_d\right)}{\zeta_2(4s-1)},$$

and

$$\frac{A_{dk^2}(s)}{A_d(s)} = \sum_{\substack{d_1d_2d_3=k\\ d_2,\,d_3 \text{ odd}}} \chi_d(d_3)\mu(d_3)d_2^{-4s+3/2}d_3^{-2s+1/2},$$

where μ is the Möbius function and χ_d is the quadratic character $\chi_d(c) = \left(\frac{d}{c}\right)$ in terms of the Kronecker symbol. The subscript 2 applied to the L-function and zeta function ζ means that the two parts have been removed.

Goldfeld and Hoffstein applied the theory of Eisenstein series of half-integral weight to obtain the meromorphic continuation and functional equations of $Z_{\pm}$. They showed that there are poles at $w=\frac{3}{4}$ and $\frac{5}{4}-s$, then used a Tauberian argument to obtain estimates for the mean values of L-functions. For example, they showed

$$\sum_{\substack{1<\pm d<x\\ d \text{ squarefree}}} L\left(\frac{1}{2},\chi_d\right) = c_1 x\log(x) + c_2 x + O\left(x^{\frac{19}{32}+\varepsilon}\right)$$

with known constants c_1 and c_2.

Note that $Z_{\pm}$ is a double Dirichlet series (in s, w) since if we substitute the expression for the L-function $L(w,\chi_c)$, we have "essentially"

$$Z_{\pm}(s,w) = \sum_{d} L\left(2s - \frac{1}{2}, \chi_d\right) |d|^{\frac{1}{2}-2w} = \sum_{c,d} \left(\frac{d}{c}\right) |c|^{\frac{1}{2}-2s} |d|^{\frac{1}{2}-2w}. \quad (2)$$

Equation (2) gives two heuristic expressions representing the multiple Dirichlet series with the intention of explaining as simply as possible what we expect to be true and what form the generalizations must be. Such a heuristic form ignores a number of details, such as the fact that the coefficients are only described correctly if d is squarefree (both expressions) and that c and d are coprime (second expression). Later, we will first generalize the heuristic form by attaching a multiple Dirichlet series to an arbitrary root system. The heuristic version will have predictive value, but will still ignore important details, so we will then have to consider how to make a rigorous definition.

To give an immediate heuristic generalization of (2), let us consider, with complex parameters $s_1, \ldots, s_k$, and w a multiple Dirichlet series

$$\sum_{c} L\left(2s_1 - \frac{1}{2}, \chi_d\right) \cdots L\left(2s_k - \frac{1}{2}, \chi_d\right) |d|^{\frac{1}{2}-2w} \quad (3)$$

for $k = 0, 1, 2, 3, \ldots$. If one could prove meromorphic continuation of this Dirichlet series to all s with $w_i = \frac{1}{2}$, the Lindelöf hypothesis in the quadratic aspect would follow from Tauberian arguments. A similar approach to the Lindelöf hypothesis in the t aspect would consider instead

$$\int_1^{\infty} \zeta(\sigma_1 \pm it) \cdots \zeta(\sigma_k \pm it)\, t^{-2w}\, \mathrm{d}t, \quad (4)$$

where for each zeta function, we choose a sign $\pm$; if k is even, we may choose half of them positive and the other half negative. This is equivalent to the usual moments

$$\int_0^{T} \zeta(\sigma_1 \pm it) \cdots \zeta(\sigma_k \pm it)\, \mathrm{d}t,$$

which have been studied since the work in the 1920s of Hardy and Littlewood, Ingham, Titchmarsh, and others. It is possible to regard (4) as a multiple Dirichlet series, and indeed both (3) and (4) are treated together in Diaconu, Goldfeld, and Hoffstein [32]. See [31] in this volume for a discussion of the sixth integral moment and its connection with the spectral theory of Eisenstein series on GL_3.

Returning to (3), there are two problems: to make a correct definition of the multiple Dirichlet series, and to determine its analytic properties. If $k = 1, 2$, or 3, these can both be solved, and the multiple Dirichlet series has global analytic continuation. In these cases, the multiple Dirichlet series was initially constructed by applying a Rankin–Selberg construction to Eisenstein series ("of half-integral weight") on the metaplectic double covers of the groups $\mathrm{Sp}(2k)$ for

$k = 1, 2,$ and 3. No corresponding constructions could be found for $k > 3$, but since Rankin–Selberg constructions are often tricky, which did not constitute a proof that such constructions may not exist undiscovered.

In [20], a different approach was taken. If $k > 3$, then it may be possible to write down a correct definition of the multiple Dirichlet series, and indeed this has essentially been done in the very interesting special case $k = 4$. See Bucur and Diaconu [19]. Nevertheless, the approach taken in [20], which we will next explain, shows that the multiple Dirichlet series cannot have meromorphic continuation to all s_i and w if $k > 3$.

The analog for (3) of the second expression in (2) would have the form

$$\sum_{d, c_1, \ldots, c_k} \left(\frac{d}{c_1}\right) \cdots \left(\frac{d}{c_k}\right) |c_1|^{\frac{1}{2}-2s_1} \cdots |c_k|^{\frac{1}{2}-2s_k} |d|^{\frac{1}{2}-2w}. \tag{5}$$

It will be helpful to associate with this Dirichlet series a graph whose vertices are the variables $d, c_1, \ldots, c_k$. We connect two vertices if a quadratic symbol is attached to them. Our point of view (which is justified when rigorous foundations are supplied) is that due to the quadratic reciprocity law, we do not have to distinguish between $\left(\frac{d}{c}\right)$ and $\left(\frac{c}{d}\right)$. Thus, for heuristic purposes, the graph determines the Dirichlet series. If $k = 1, 2,$ or 3, the graph looks like this:

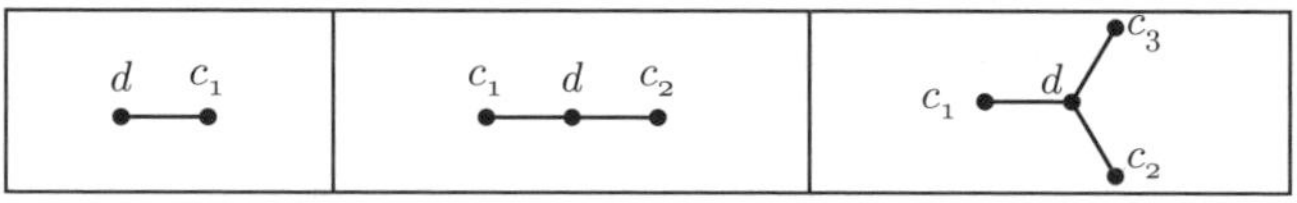

(6)

We could clearly associate a multiple Dirichlet series with a more general graph, at least in this imprecise heuristic form. The interesting cases will be when the diagram is a Dynkin diagram.

Here, we will only consider cases where the diagram is a "simply laced" Dynkin diagram, that is, the diagram of a root system of Cartan type A, D, or E. A *simply laced* root system is one in which all roots have the same length, and these are their Cartan types. More general Dynkin diagrams are also associated with multiple Dirichlet series, and we will come to these below.

We recall that a *Coxeter group* is a group with generators $\sigma_1, \ldots, \sigma_r$, each of order two, such that the relations

$$\sigma_i^2 = 1, \qquad (\sigma_i \sigma_j)^{n(i,j)} = 1$$

give a presentation of the group, where $n(i, j)$ is the order of $\sigma_i \sigma_j$. We may associate with the Coxeter group a graph, which consists of one node for each generator σ_i, with the following conditions. If $n(i, j) = 2$, so that σ_i and σ_j commute, there is no edge connecting i and j. Otherwise, there is an edge. If $n(i, j) = 3$, it is not necessary to label the edge, but if $n(i, j) > 3$, it is labeled with $n(i, j)$. (In Dynkin diagrams, it is usual to interpret these labels as double or triple bonds.) We will consider only the case where $n(i,j) = 2$ or 3. In these cases, the Coxeter group is finite if and only if the diagram is the finite union of the Dynkin diagrams of finite Weyl groups of types A_r, D_r, or E_r.

As we will now explain, the group of functional equations of a multiple Dirichlet series such as (5) is expected to be the Coxeter group of its Dynkin diagram. For example, consider (5) when $k = 3$. We collect the coefficients of c_1:

$$\sum_{d,c_1,c_2,c_3} \left(\frac{d}{c_1}\right)\left(\frac{d}{c_2}\right)\left(\frac{d}{c_3}\right) |c_1|^{\frac{1}{2}-2s_1}|c_2|^{\frac{1}{2}-2s_3}|c_3|^{\frac{1}{2}-2s_3}|d|^{\frac{1}{2}-2w}$$

$$= \sum_{d,c_2,c_3} \left(\frac{d}{c_2}\right)\left(\frac{d}{c_3}\right) |c_2|^{\frac{1}{2}-2s_2}|c_3|^{\frac{1}{2}-2s_3}|d|^{\frac{1}{2}-2w}\left[\sum_{c_1}\left(\frac{d}{c_1}\right)|c_1|^{\frac{1}{2}-2s_1}\right].$$

This equals

$$\sum_{d,c_2,c_3} \left(\frac{d}{c_2}\right)\left(\frac{d}{c_3}\right) |c_2|^{\frac{1}{2}-2s_2}|c_3|^{\frac{1}{2}-2s_3}|d|^{\frac{1}{2}-2w} L\left(2s_1 - \frac{1}{2}, \chi_d\right).$$

The functional equation for the Dirichlet L-function has the form

$$L\left(2s_1 - \frac{1}{2}, \chi_d\right) = (*)\,|d|^{1-2s_1} L\left(2(1-s_1) - \frac{1}{2}, \chi_d\right),$$

where $(*)$ is a ratio of Gamma functions and powers of π. This factor is independent of d, so we see that the functional equation has the effect

$$(s_1, s_2, s_3, w) \mapsto \left(1 - s_1, s_2, s_3, w + s_1 - \frac{1}{2}\right).$$

In the general case, let the variables be $s_1, \dots, s_r$. Thus, if we are considering (5), then $r = k + 1$ and $s_r = w$, but we now have in mind a general graph such as a Dynkin diagram, and r is the number of nodes. We have a functional equation which sends s_i to $1 - s_i$. If $j \neq i$, then

$$s_j \longmapsto \begin{cases} s_j & \text{if } i, j \text{ are not connected by an edge;} \\ s_i + s_j - \dfrac{1}{2} & \text{if } i, j \text{ are connected by an edge.} \end{cases} \tag{7}$$

These functional equations generate the Coxeter group associated with the diagram, and its group of functional equations is the geometric realization of that group as a group generated by reflections.

We may now see why the Dirichlet series (3) is expected to have meromorphic continuation to all s_i and w when $k \leqslant 3$ but not in general. These are the cases where the graph is the diagram of a finite Weyl group, of types A_2, A_3, or D_4. If $k = 4$, the graph is the diagram of the affine Weyl group $D_4^{(1)}$, and the corresponding Coxeter group is infinite. The meromorphic continuation in (s_1, s_2, s_3, s_4, w) cannot be to all of $\mathbb{C}^5$ since the known set of polar hyperplanes will have accumulation points.

We have now given heuristically a large family of multiple Dirichlet series, one for each simply laced Dynkin diagram. (The simply laced assumption may be eliminated, as we will explain later.) Only three of them, for Cartan types A_2, A_3, and D_4, are related to moments of L-functions. The case $k = 3$, related to D_4, was applied in [32] to the third moment after the combinatorics needed to precisely define the Dirichlet series were established in [21].

Although only these three examples are related to quadratic moments of L-functions, others in this family have applications to analytic number theory. Chinta gave a remarkable example in [25], where the A_5 multiple Dirichlet series is used to study the distribution of central values of biquadratic L-functions. The distribution of nth-order twists of an L-function was studied by Friedberg, Hoffstein, and Lieman [36], and it was shown in Brubaker and Bump [7] that these could be related to Weyl group multiple Dirichlet series of order n. (In this survey, the Dirichlet series we have considered in this section correspond to $n = 2$, but we will come to general n below.)

One may also consider the Dirichlet series that are (heuristically) of the form $L(w, \pi, \chi_d)\,|d|^{-w}$, where π is an automorphic representation of GL_k. If $k = 2$, then there is considerable literature of the case, where $n = 2$; see, for example, [6] and the references therein. If $k = 2$ and $n = 3$, there is a remarkable theory in [17]; there is a finite group of functional equations, which transform the Dirichlet series into various different ones. If $k = 3$ and $n = 2$, then there is also a finite group of functional equations; see [21]. The papers cited in this paragraph predate the recent development of the combinatorial theory, but the combinatorics of multiple Dirichlet series involving GL_k cusp forms is under investigation by Brubaker and Friedberg.

A double Dirichlet similar to that in [36] was considered by Reznikov [53]. This is the Dirichlet series $\sum L(s, \chi^n)|n|^{-w}$, where χ is a Hecke character of infinite order for $\mathbb{Q}(i)$. Using a method of Bernstein, he proved the meromorphic continuation of this multiple Dirichlet series and determined the poles. Despite the similarity of this multiple Dirichlet series to that of [36], this series does not fit the same way in the theory of Weyl group multiple Dirichlet series.

While the origins of our subject are in analytic number theory, our emphasis in this paper will not be such applications, but rather on emerging connections with areas of combinatorics, including quantum groups and mathematical physics, and the theory of Whittaker functions. As we will see, the problem of giving precise definitions of the multiple Dirichlet series, even when the general nature of the Dirichlet series is known, is a daunting combinatorial one. Early investigations, such as [21, 25], took an *ad hoc* approach substituting computer algebra or brute force computation for real insight. This is sufficient for applications on a case-by-case basis but also unsatisfactory. In recent years, the combinatorial theory has been examined more closely, and its study may turn out to be as interesting as the original problem.

1.2 A Method of Analytic Continuation

Let us consider a double Dirichlet series which might be written

$$Z^*_\Psi(s_1, s_2) = (*) Z_\Psi(s_1, s_2), \qquad Z_\Psi(s_1, s_2) = \sum_{n,m} A_\Psi(n,m) n^{-s_1} m^{-s_2}.$$

Here, $(*)$ denotes some Gamma functions and powers of π. The Dirichlet series is allowed to depend on a parameter Ψ drawn from a finite-dimensional vector space Ω. It is assumed convergent in some region $\mathcal{C}$ such as the one in the following figure, which shows the region for (2). We have graphed the projection onto $\mathbb{R}^2$ obtained by taking the real parts of s_1 and s_2.

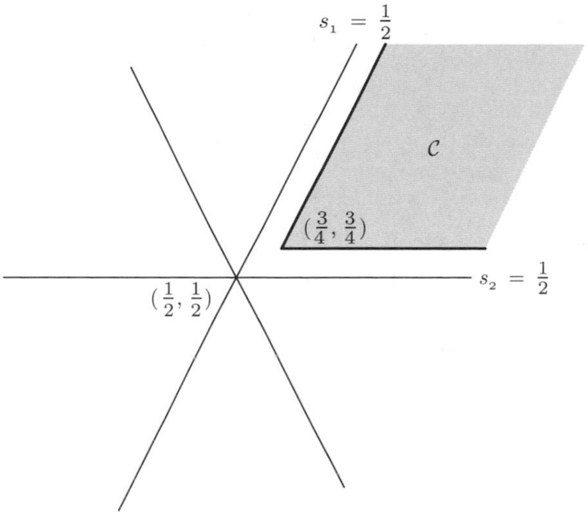

Collecting the coefficients of m^{-s_2} for each m gives a collection of Dirichlet series in one variable s_1 which have functional equations. These may be with respect to some transformation such as the following, which is a functional equation of (2):

$$\sigma_1 : (s_1, s_2) \longmapsto \left(1 - s_1, s_1 + s_2 - \frac{1}{2}\right).$$

More precisely, there may be an action of σ_1 on Ω, or more properly on $\mathcal{M} \otimes \Omega$, where $\mathcal{M}$ is the field of meromorphic functions in s_1 and s_2, such that the functional equation has the form

$$Z^*_{\sigma_1\Psi}\left(1 - s_1, s_1 + s_2 - \frac{1}{2}\right) = Z^*_\Psi(s_1, s_2).$$

Thus, $\Psi \longmapsto \sigma_1\Psi$ is a linear transformation of the vector space Ω which, when written out as a matrix, could involve meromorphic functions of s_1 and s_2. This is the *scattering matrix*. In some cases, these meromorphic functions are holomorphic, or even just Dirichlet polynomials in a finite number of integers. For example (2), we would take polynomials in 2^{-s_1} and 2^{-s_2}.

This gives the meromorphic continuation to the convex hull of $\mathcal{C} \cup \sigma_1 \mathcal{C}$.

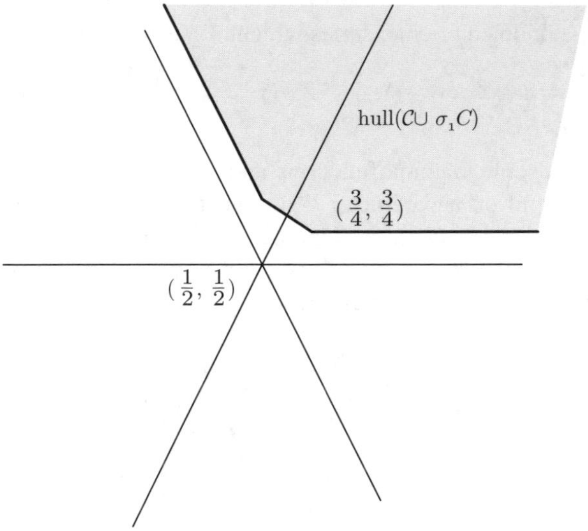

Similarly, we assume that collecting the coefficients of the other variable gives another functional equation, which in example (2) is

$$\sigma_2 : (s_1, s_2) \longmapsto \left(s_1 + s_2 - \frac{1}{2}, 1 - s_2\right).$$

The functional equations may be iterated, so we get analytic continuation to the the union of hull($\mathcal{C} \cup \sigma_1 \mathcal{C}$) with hull($\mathcal{C} \cup \sigma_2 \mathcal{C}$) and σ_1 hull($\mathcal{C} \cup \sigma_2 \mathcal{C}$):

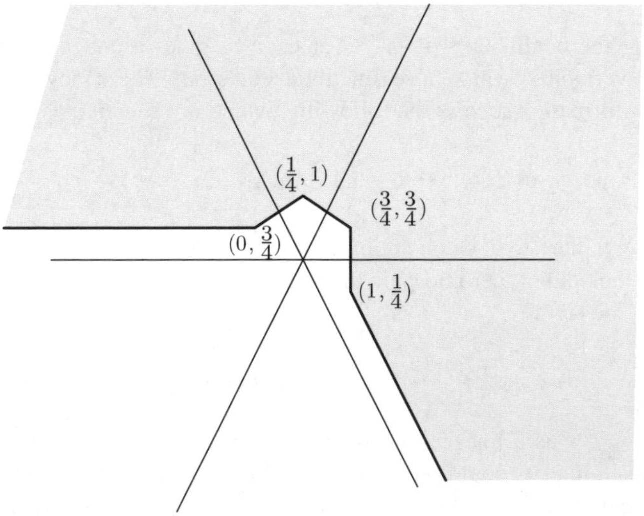

(8)

At this point, there are two ways of proceeding, one better than the other. We could continue to iterate the functional equations until we obtained meromorphic continuation to a region such as this:

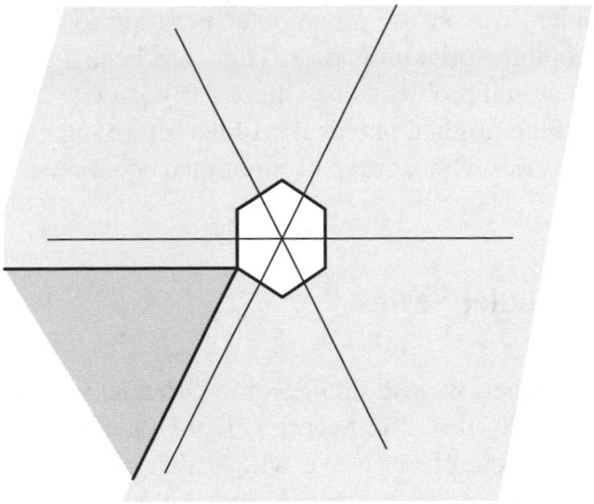

There are two problems with this. One is that we have not obtained the meromorphic continuation in the area near the origin. The other is that we have obtained two different meromorphic continuations to the region $\sigma_1\sigma_2\sigma_1\mathcal{C} = \sigma_2\sigma_1\sigma_2\mathcal{C}$ that is darkly shaded. We do not know that these two meromorphic continuations agree. This agreement, the *braid relation*, should be true in a suitable sense, but in fact since there is a scattering matrix involved, we must be careful in formulating it. We want an action of W on Ψ extending the one already mentioned for σ_1 such that $Z^*_\Psi(s_1, s_2)$ satisfies

$$Z^*_{w(\Psi)}(w(s_1, s_2)) = Z^*_\Psi(s_1, s_2), \tag{9}$$

and the braid relation means that $\sigma_1\sigma_2\sigma_1(\Psi) = \sigma_2\sigma_1\sigma_2(\Psi)$.

A better procedure is to use a theorem in complex variables, Bochner's convexity theorem [5], to assert meromorphic continuation once one has obtained meromorphic continuation to a region such as (8) whose convex hull is $\mathbb{C}^2$. Bochner's theorem is as follows: let U be an open subset of $\mathbb{C}^r$, where $r \geqslant 2$ that is the preimage of an open subset of $\mathbb{R}^r$ under the projection map; such a set is called a *tube domain*. Then, any holomorphic function on a tube domain has analytic continuation to its convex hull. In our case, we have a meromorphic function, but the polar divisor is a set of hyperplanes, and the theorem is easily extended to this case. Hence, once we have meromorphic continuation to (8), we obtain meromorphicity on $\mathbb{C}^2$. The braid relation $\sigma_1\sigma_2\sigma_1(\Psi) = \sigma_2\sigma_1\sigma_2(\Psi)$ is then a consequence. See [9, 11] for further details.

Now, we come to the fundamental combinatorial question. Once one has decided *roughly* what the Dirichlet series is to look like, the exact coefficients are still not precisely defined. How can the coefficients be determined in such a way that the functional equations (9) are true for both σ_1 and σ_2? For the Dirichlet series (3), this is not too hard when $k = 1$, but when $k = 3$, the combinatorics are rather daunting. They were treated in [21] using difficult manipulations that were the only way before the combinatorial properties of Weyl group multiple Dirichlet series began to be established. Similarly, in the example of Chinta [25], the method of solving the combinatorial problem was to use a computer program to find a Dirichlet series with very special combinatorial properties. There has been a great deal of progress in the basic combinatorial problem since these early papers, and this progress has implications beyond the original practical problem of giving a proper definition of a multiple Dirichlet series with a group of functional equations.

1.3 Kubota Dirichlet Series

Let n be a positive integer: we will define some Dirichlet series related to the n-th power reciprocity law, so $n = 2$ in Sect. 1.1. Let F be a number field containing the group μ_n of n-th roots of unity. We will further assume that F contains the group μ_{2n} of $2n$-th roots of unity, that is, that -1 is an n-th power in F. The assumption that $\mu_n \subset F$ is essential; the assumption that $\mu_{2n} \subset F$ is only a matter of convenience. We will make use of the n-th power reciprocity law.

We will define a family of Dirichlet series with analytic continuation and functional equations, called *Kubota Dirichlet series*. If $n = 2$, these are the quadratic L-functions $L(2s - \frac{1}{2}, \chi_d)$. If $n = 1$, these Dirichlet series are divisor sums, actually finite Dirichlet polynomials. For general n, they are generating functions of nth-order Gauss sums.

Let S be a finite set of places of F, containing all places dividing n and all archimedean ones. If v is a place of F, let F_v be the completion at v. If v is nonarchimedean, let $\mathfrak{o}_v$ be the ring of integers in F_v. Let $\mathfrak{o}_S$ be the ring of S-integers in F, that is, those elements $x \in F$ such that $x \in \mathfrak{o}_v$ for all $v \notin S$. Let $F_S = \prod_{v \in S} F_v$. We may embed $\mathfrak{o}_S$ in F_S diagonally. It is a discrete, cocompact subgroup. We may choose S so large that $\mathfrak{o}_S$ is a principal ideal domain. If $a \in F_S$, let $|a|$ denote $\prod_{v \in S} |a|_v$. It is the Jacobian of the map $x \longmapsto ax$. If $a \in \mathfrak{o}_S$, then $|a|$ is a nonnegative rational integer.

We recall the nth-order reciprocity law and nth-order Gauss sums. See Neukirch [51] for proofs. Properties of the reciprocity symbol and Gauss sums are more systematically summarized in [11].

The nth-order Hilbert symbol $(,)_v$ is a skew-symmetric pairing of $F_v^\times \times F_v^\times$ into μ_n. Define a pairing $(\, , \,)$ on $F_S^\times$ by

$$(x, y) = \prod_{v \in S} (x_v, y_v)_v, \qquad x, y \in F_S^\times.$$

Then, the nth power residues symbol $\left(\frac{d}{c}\right)$, defined for nonzero elements $c, d \in \mathfrak{o}$, satisfies the nth power reciprocity law

$$\left(\frac{c}{d}\right) = (d,c)\left(\frac{d}{c}\right). \tag{10}$$

Let ψ be an additive character of F_S that is trivial on $\mathfrak{o}_S$ but no larger fractional ideal. Let

$$g(m,d) = \sum_{c \bmod d} \left(\frac{c}{d}\right) \psi\left(\frac{mc}{d}\right).$$

The sum is well defined since both factors only depend on c modulo d. It has the *twisted multiplicativity* properties:

$$g(m,dd') = \left(\frac{d}{d'}\right)\left(\frac{d'}{d}\right) g(m,d)\, g(m,d') \text{ if } \gcd(d,d') = 1,$$

$$g(cm,d) = \left(\frac{c}{d}\right)^{-1} g(m,d) \qquad \text{if } c, d \text{ are coprime}$$

and the absolute value for p prime in $\mathfrak{o}_S$:

$$|g(m,p)| = \sqrt{|p|} \text{ if } \gcd(m,p) = 1. \tag{11}$$

Let Ψ be a function on $F_S^\times$ such that $\Psi(\varepsilon c) = (\varepsilon, c)\Psi(c)$ when $\varepsilon \in \mathfrak{o}_S^\times (F_S^\times)^n$. The vector space of such functions is nonzero but finite dimensional. Let

$$\mathcal{D}_\Psi(s;m) = \sum_c \Psi(c) g(m,c)\, |c|^{-2s}.$$

This has a functional equation under $s \longmapsto 1-s$. To state it, let

$$\boldsymbol{G}_n(s) = (2\pi)^{-(n-1)(2s-1)} \frac{\Gamma(n(2s-1))}{\Gamma(2s-1)}.$$

Define

$$\mathcal{D}_\Psi^*(s,m) = \boldsymbol{G}(s)^N \zeta_F(2ns - n + 1)\, \mathcal{D}_\Psi(s,m),$$

where N is the number of archimedean places (all complex) and ζ_F is the Dedekind zeta function of F. Then, Kubota [46] proved a functional equation for this, as a consequence of the functional equations of Eisenstein series on the n-fold metaplectic covers of SL_2, which he developed for this purpose. In the form we need it, this is proved by the same method in Brubaker and Bump [18], and a similar result is in Patterson and Eckhardt [34].

To state these functional equations, there exists a family of Dirichlet polynomials P_η indexed by η in $F_S^\times/(F_S^\times)^n$ such that

$$\mathcal{D}_\Psi^*(s,m) = \sum_{\eta \in F_S^\times/(F_S^\times)^n} |m|^{1-2s} P_{m\eta}(s)\, \mathcal{D}_{\tilde{\Psi}_\eta}^*(1-s,m), \tag{12}$$

where

$$\widetilde{\Psi}_\eta(c) = (\eta, c)\Psi\left(c^{-1}\eta^{-1}\right).$$

The polynomial P_η is actually a polynomial in q_v^{-s} where v runs through the finite places in S and q_v is the cardinality of the residue field. It is important for applications that P_η is independent of m.

1.4 A More General Heuristic Form

If $n = 2$ and m and c are coprime, then $g(m, c)$ equals $\left(\frac{m}{c}\right)^{-1}\sqrt{|c|}$ times a factor which may be combined with Ψ and ignored for heuristic purposes. Thus, $\mathcal{D}_\Psi(s; m)$ is essentially $L(2s-\frac{1}{2}, \chi_m)$. We may now give the following heuristic generalization of the Dynkin diagram multiple Dirichlet series described in Sect. 1.1. Let us start with a Dynkin diagram, which we will at first assume is simply laced (type A, D, or E). As in Sect. 1.1, for purely heuristic purposes, it is not necessary to distinguish between $\left(\frac{c}{d}\right)$ and $\left(\frac{d}{c}\right)$ since by the reciprocity law they differ by a factor (d, c) which may also be combined with Ψ and may be ignored for heuristic purposes. Ultimately, such factors must eventually be kept track of, but at the moment, they are unimportant.

The nodes $i = 1, \ldots, r$ of the Dynkin diagram are in bijection with the simple roots of some root system Φ. We choose one complex parameter s_i for each i, and one "twisting parameter" m_i, which is a nonzero integer in $\mathfrak{o}_S$. The multiple Dirichlet series then has the heuristic form

$$\sum_{d_1,\ldots,d_r}\left[\prod_{i,j \text{ adjacent}}\left(\frac{d_i}{d_j}\right)^{-1}\right] g(m_i, d_i)|d_i|^{-2s_i}.$$

The form of the coefficient is only correct if d_i are squarefree and coprime, and even then there is a caveat, but this heuristic form is sufficient for extrapolating the expected properties of the multiple Dirichlet series. Whereas before, on expanding in powers of one of the s_i parameters, we obtained a quadratic L-function, now we obtain a Kubota Dirichlet series.

If the Dynkin diagram is not simply laced, there are long roots and short roots. In this case, there is also a heuristic form, which we will not discuss here. For each district pair of simple roots α_i and α_j, let $r(\alpha_i, \alpha_j)$ be the number of bonds connecting the nodes connecting α_i and α_j in the Dynkin diagram. Thus, if θ is the angle between α_i and α_j, let

$$r(\alpha_i, \alpha_j) = \begin{cases} 0 & \text{if } \alpha_i, \alpha_j \quad \text{are orthogonal,} \\ 1 & \text{if } \theta = \frac{2\pi}{3}, \\ 2 & \text{if } \theta = \frac{3\pi}{4}, \\ 3 & \text{if } \theta = \frac{5\pi}{6}. \end{cases}$$

Normalize the roots so short roots have length 1; thus, every long root α has $||\alpha||^2 = 1, 2$, or 3, the last case occurring only with G_2. Let

$$g_\alpha(m,d) = \sum_{c \bmod d} \left(\frac{c}{d}\right)^t \psi\left(\frac{mc}{d}\right), \qquad t = \|\alpha\|^2.$$

Then, the heuristic form of the multiple Dirichlet series is

$$\sum_{d_1,\cdots,d_r} \left[\prod_{i,j} \left(\frac{d_i}{d_j}\right)^{-r(\alpha_i,\alpha_j)}\right] \left[\prod_i g_{\alpha_i}(m_i,d_i)|d_i|^{-2s_i}\right]. \tag{13}$$

1.5 Foundations and the Combinatorial Problem

The first set of foundations for Weyl group multiple Dirichlet series was given by Fisher and Friedberg [35], and these were used in all earlier papers. Another set of foundations were explained in [9, 11], and these have been used for the most part in subsequent papers. We recall them in this section.

Let V be the ambient vector space of Φ. Let $\langle\ ,\ \rangle$ be a W-invariant inner product on V such that the short roots have length 1. Let $B : V \otimes \mathbb{C}^r \longrightarrow \mathbb{C}$ be the bilinear map that sends $(\alpha_i, \mathbf{s})$ to s_i, where $\mathbf{s} = (s_1, \ldots, s_r)$ to $\sum k_i s_i$ and α_i is the ith simple root. Let $\rho^\vee$ denote the vector $(1, \ldots, 1) \in \mathbb{C}^r$. The reason for this notation is explained in [11]. The Weyl group action on $\mathbf{s}$, corresponding to the group of functional equations, may be expressed in terms of B: we require that

$$B\left(w\alpha, w(\mathbf{s}) - \frac{1}{2}\rho^\vee\right) = B\left(\alpha, \mathbf{s} - \frac{1}{2}\rho^\vee\right)$$

for $w \in W$.

We fix an ordering of simple roots of Φ, so that in order they are $\{\alpha_1, \ldots, \alpha_r\}$. Some of the formulas depend on this ordering, but in an inessential way. Following [11], let us define $\mathcal{M}$ to be the nonzero but finite-dimensional space of functions on $\Psi : (F_S^\times)^r \longrightarrow \mathbb{C}$ that satisfy

$$\Psi(\varepsilon_1 C_1, \ldots, \varepsilon_r C_r) = \prod_{i=1}^r (\varepsilon_i, C_i)_S^{\|\alpha_i\|^2} \left\{\prod_{i<j} (\varepsilon_i, C_j)_S^{2\langle\alpha_i,\alpha_j\rangle}\right\} \Psi(C_1, \ldots, C_r) \tag{14}$$

when $\varepsilon_1, \ldots, \varepsilon_r \in \mathfrak{o}_S^\times (F_S^\times)^n$ and $C_i \in F_S^\times$.

We seek a function $H(C_1, \ldots, C_r; m_1, \ldots, m_r)$ defined if the C_i and m_i are nonzero elements of $\mathfrak{o}_S$ with the following properties. There is the multiplicativity condition

$$\frac{H(C_1C_1',\ldots,C_rC_r';m_1,\ldots,m_r)}{H(C_1,\ldots,C_r;m_1,\ldots,m_r)\;H(C_1',\ldots,C_r';m_1,\ldots,m_r)}$$
$$=\prod_{i=1}^{r}\left(\frac{C_i}{C_i'}\right)^{\|\alpha_i\|^2}\left(\frac{C_i'}{C_i}\right)^{\|\alpha_i\|^2}\prod_{i<j}\left(\frac{C_i}{C_j'}\right)^{2\langle\alpha_i,\alpha_j\rangle}\left(\frac{C_i'}{C_j}\right)^{2\langle\alpha_i,\alpha_j\rangle}. \tag{15}$$

There is another multiplicativity condition which, unlike (15), *does* involve the m_i. If $\gcd(m_1'\cdots m_r', C_1\cdots C_r)=1$, we require:

$$H(C_1,\ldots,C_r;m_1m_1',\ldots,m_rm_r')$$
$$=\left(\frac{m_1'}{C_1}\right)^{-||\alpha_1||^2}\cdots\left(\frac{m_r'}{C_r}\right)^{-||\alpha_r||^2}H(C_1,\ldots,C_r;m_1,\ldots,m_r). \tag{16}$$

The conditions (14) and (15) together imply that if $C_1,\ldots,C_r$ is each multiplied by a unit, then the value of $\Psi H(C_1,\ldots,C_r)$ is unchanged. Since $\mathfrak{o}_S$ is a principal ideal ring, we see that $\Psi H(C_1,\ldots,C_r)$ is really a function of ideals. Let

$$Z_\Psi(s_1,\ldots,s_r;m_1,\ldots,m_r)$$
$$=\sum\Psi(C_1,\ldots,C_r)H(C_1,\ldots,C_r;m_1,\ldots,m_r)|C_1|^{-2s_1}\cdots|C_r|^{-2s_r} \tag{17}$$

where the summation is over ideals (C_i). Also, let

$$Z_\Psi^*(s_1,\ldots,s_r;m_1,\ldots,m_r)=\left[\prod_{\alpha\in\Phi+}\zeta_\alpha(s)G_\alpha(s)\right]Z_\Psi(s_1,\ldots,s_r;m_1,\ldots,m_r), \tag{18}$$

where, if α is a positive root

$$\zeta_\alpha(\boldsymbol{s})=\zeta_F\left(1+2n(\alpha)\left\langle\alpha,\boldsymbol{s}-\frac{1}{2}\rho^\vee\right\rangle\right),$$
$$G_\alpha(\boldsymbol{s})=\boldsymbol{G}_{n(\alpha)}\left(\frac{1}{2}+\left\langle\alpha,\boldsymbol{s}-\frac{1}{2}\rho^\vee\right\rangle\right) \tag{19}$$

with

$$n(\alpha)=\begin{cases} n \text{ if } \alpha \text{ is a short root,}\\ n \text{ if } \alpha \text{ is a long root and } \Phi\neq G_2\text{, and } n \text{ is odd}\\ \frac{n}{2} \text{ if } \alpha \text{ is a long root and } \Phi\neq G_2\text{, and } n \text{ is even}\\ n \text{ if } \alpha \text{ is a long root and } \Phi=G_2\text{, and } 3\nmid n\\ \frac{n}{3} \text{ if } \alpha \text{ is a long root and } \Phi=G_2\text{, and } 3|n.\end{cases}$$

The Kubota Dirichlet series $\mathcal{D}_\Psi(s;m)$ is the special case if Z_Ψ where the root system is of type A_1.

We still have not fully described H, so we have not given a proper definition of Z_Ψ. The multiplicativities (15) and (16) together imply that the function H is determined by its values on prime powers. In other words, if we specify $H(p^{k_1}, \ldots, p^{k_r}; p^{l_1}, \ldots, p^{l_r})$ for prime elements p, the function is determined.

The fundamental combinatorial problem is this: given a global field F in which -1 is an nth power and a root system, give a correct definition of the multiple Dirichlet series extrapolating the heuristic one, such that expanding in powers of every s_i gives a sum of Kubota Dirichlet series all having the same functional equations. Naturally, this must be made more precise. We will write $\boldsymbol{s} = (s_1, \ldots, s_r)$ and $\boldsymbol{m} = (m_1, \ldots, m_r)$.

Fundamental Combinatorial Problem. *Define $H(p^{k_1}, \ldots, p^{k_r}; p^{l_1}, \ldots, p^{l_r})$ in such a way that for each index i, the series $Z_\Psi(\boldsymbol{s}; \boldsymbol{m})$ has an expansion*

$$\sum_M \mathcal{D}_{\Psi_i}(s_i, M)\, P_M(\boldsymbol{s}) \tag{20}$$

for some Ψ_i, where P_M is a Dirichlet polynomial, such that for each i we have

$$P_M(\sigma_i \boldsymbol{s}) = |M|^{1-2s_i} P_M(\boldsymbol{s}). \tag{21}$$

If this can be done, then we have a functional equation

$$Z_\Psi(\boldsymbol{s}; \boldsymbol{m}) = Z_{\Psi'}(\sigma_i \boldsymbol{s}; \boldsymbol{m}) \tag{22}$$

for some Ψ'. Here, σ_i is the simple reflection in the Weyl group action on the parameters $\boldsymbol{s}$; if the root system is simply laced, it is the action (7), or see [11] for the general case. The method of analytic continuation described in Sect. 1.2 is applicable. This yields both the meromorphic continuation and the scattering matrix, which we recall from Sect. 1.2 amounts to an action of W on Ψ such that in (22) we have $\Psi' = \sigma_i \Psi$ and more generally

$$Z^*_{w\Psi}(w\boldsymbol{s}; \boldsymbol{m}) = Z^*_\Psi(\boldsymbol{s}; \boldsymbol{m}).$$

The normalizing factor in (18) works out as follows: the factor $\zeta_\alpha G_\alpha$ with $\alpha = \alpha_i$ is needed to normalize the Kubota Dirichlet series in (20). The other such factors are simply permuted by $\boldsymbol{s} \longmapsto \sigma_i(\boldsymbol{s})$.

Let us consider briefly how this works in the case of type A_2. See [9] for a complete discussion and detailed proof for this case. We have noted above that specifying the coefficients $H(p^{k_1}, p^{k_2}; p^{l_1}, p^{l_2})$ completely determines the function H. In this example, let us take $m_1 = m_2 = 1$ so $l_1 = l_2 = 0$ for all p. The coefficients to be described are given by the following table.

Let the nonzero values of $H(p^{k_1}, p^{k_2}; 1, 1)$ be given by the following table.

k_2 \ k_1	0	1	2
0	1	$g(1,p)$	
1	$g(1,p)$	$g(p,p)g(1,p)$	$g(p,p^2)g(1,p)$
2		$g(p,p^2)g(1,p)$	$g(p,p^2)g(1,p)^2$

Then, collecting terms with equal powers of $|p|^{-s_2}$, we have a decomposition (20) where the summation includes terms of the following type:

$$\mathcal{D}_{\Psi'}(s_1;1), \quad g(1,p)|p|^{-2s_2}\mathcal{D}_{\Psi''}(s_1;p), \quad g(1,p)g\left(p,p^2\right)|p|^{-2s_1-4s_2}\mathcal{D}_{\Psi'''}(s_1;1),$$

for suitable Ψ', Ψ'', and Ψ'''. We recognize the p-parts of these Kubota Dirichlet series from the tabulated values by collecting the terms in each column of the table.

Early papers in this subject gave *ad hoc* solutions to the combinatorial problem. Such direct verifications become fairly difficult, for example, in [21, 25].

1.6 p-Parts

Let us define the *p-part* of Z to be the Dirichlet series

$$\sum_{k_i=0}^{\infty} H\left(p^{k_1},\ldots,p^{k_r};p^{l_1},\ldots,p^{l_r}\right)|p|^{-2k_1s_1-\cdots-2k_rs_r}. \tag{23}$$

We fix the representative p of the prime and ignore Ψ. By twisted multiplicativity, if the p-parts are known for all p, the multiple Dirichlet series is determined.

Returning to (17), let us consider the effect of the parameters $m_1,\ldots,m_r$. These are called *twisting* parameters, and the term "twisting" is supposed to evoke the usual twisting of L-functions: if $L(s,f)=\sum a_n n^{-s}$ is some L-function and χ is a Dirichlet character, then $L(s,f,\chi)=\sum \chi(n)a_n n^{-s}$. The term "twisting" in the present context is both apt and in a way misleading, as we will now explain.

First, suppose that $m_1,\ldots,m_r$ are coprime to $C_1,\ldots,C_r$. Then, by (16), we have

$$H(C_1,\ldots,C_r;m_1,\ldots,m_r)$$
$$=\left(\frac{m_1}{C_1}\right)^{-\|\alpha_1\|^2}\cdots\left(\frac{m_r}{C_r}\right)^{-\|\alpha_r\|^2} H(C_1,\ldots,C_r;1,\ldots,1). \tag{24}$$

Thus, these coefficients are indeed simply multiplied by an nth-order character, as the term "twisting" suggests.

On the other hand, if the m_i are not coprime to the C_i, then the effect of the m_i is much more profound. For example, in $H(p^{k_1},\ldots,p^{k_r};p^{l_1},\ldots,p^{l_r})$, it is important to think of $(l_1,\ldots,l_r)$ as indexing a weight, $\sum l_i\varpi_i$, where $\varpi_1,\ldots,\varpi_r$ are the fundamental dominant weights of the root system Φ. Then, we may think of the p-part (23) as being something related to the character of an irreducible representation of the associated Lie group, times a deformation of the Weyl denominator, but with the weight multiplicities replaced by sums of products of Gauss sums. In particular, varying $m_i = p^{l_i}$ affects the p-part in a profound way, no simple twisting.

With m_i general, their meaning may be explained as follows: specifying $m_1,\ldots,m_r$ is equivalent to specifying, for each p, dominant weight λ_p such that $\lambda_p = 0$ for almost all p. Indeed, factor $m_i = p^{l_i}m_i'$ where $p\nmid m_i'$ and take $\lambda_p=\sum l_i\varpi_i$.

Now, let the prime p and the exponents $l_1,\ldots,l_r$ be fixed, and let $\lambda = \lambda_p = \sum l_i\varpi_i$ be the corresponding dominant weight. Also, let

$$\rho=\sum_{i=1}^{r}\varpi_i=\frac{1}{2}\sum_{\alpha\in\Phi+}\alpha$$

be the Weyl vector. Let W be the Weyl group of Φ. If $w\in W$, let $\boldsymbol{k}(w)$ be the r-tuple of nonnegative integers $(k_1,\ldots,k_r)$ such that $\rho+\lambda-w(\rho+\lambda)=\sum k_i\alpha_i$.

The coefficients $H(p^{k_1},\ldots,p^{k_r};p^{l_1},\ldots,p^{l_r})$ in general do not admit an easy description, but if $(k_1,\ldots,k_r) = \boldsymbol{k}(w)$ for some w, then it is a product of $l(w)$ Gauss sums, where $l(w)$ is the length of w. To make this explicit, let Φ_w be the set of all positive roots α such that $w(\alpha)$ is a negative root, so $|\Phi_w| = l(w)$. Then (see [12]), we have

$$H\left(p^{k_1},\ldots,p^{k_r};p^{l_1},\ldots,p^{l_r}\right)=\prod_{\alpha\in\Phi_w} g_{\|\alpha\|^2}\left(p^{\langle\lambda+\rho,\alpha\rangle-1},p^{\langle\lambda+\rho,\alpha\rangle}\right). \qquad (25)$$

Let Supp(λ) be the *support* of $H(p^{k_1},\ldots,p^{k_r};p^{l_1},\ldots,p^{l_r})$, that is, the set of $\boldsymbol{k} = (k_1,\ldots,k_r)$ such that $H(p^{k_1},\ldots,p^{k_r};p^{l_1},\ldots,p^{l_r})\neq 0$. Then, by (25), $\{\boldsymbol{k}(w)|w\in W\}$ is contained in Supp(λ).

Most importantly, the $|W|$ points $\boldsymbol{k}(w)$ are the extremal values of the support. That is, Supp(λ) is contained in the convex hull of $\boldsymbol{k}(w)$. These $|W|$ extremal points are called *stable* in [11,12] for the following reason. If n is sufficiently large, then Supp(λ) $=\{\boldsymbol{k}(w)|w\in W\}$, and in this case, the values (25) are the *only* nonzero values of $H(p^{k_1},\ldots,p^{k_r};p^{l_1},\ldots,p^{l_r})$. So these values are "stable," and the combinatorial problem is solved by (25).

For arbitrary n, Supp(λ) is at least contained in the convex hull of $\{\boldsymbol{k}(w)|w\in W\}$. But for interior points of this convex polytope, the description of $H(p^{k_1},\ldots,p^{k_r};p^{l_1},\ldots,p^{l_r})$ is much more difficult. We will next look at the various approaches.

1.7 Multiple Dirichlet Series and Combinatorics

In this section, we will introduce the modern combinatorial theory of the p-parts of Weyl group multiple Dirichlet series. There are several different methods of representing the p-parts of multiple Dirichlet series to be considered, each with its own individual combinatorial flavor. The combinatorial theory has only taken shape in the last few years. We will state things most fully in the "nonmetaplectic" case $n = 1$, leaving the reader hopefully oriented and ready to explore the general cases in the literature.

We see that correctly specifying the p-part of the function H produces a Dirichlet series Z_Ψ with global meromorphic continuation. These functions turn out to be extremely interesting. Several definitions of H emerged, and proving their equivalence proved to be nontrivial. Moreover, as the functions H were intensively studied, various clues seemed to suggest connections with the theory of quantum groups. We will discuss these points in this section.

The following main classes of definitions were found:

- Definition by the "averaging method," sometimes known as the Chinta-Gunnells method
- Definition as spherical p-adic Whittaker functions
- Definition as sums over crystal bases
- Definition as partition functions of statistical-mechanical lattice models

The first and second definitions give uniform descriptions for all root systems and all n. The third and fourth definitions are on a case-by-case basis and have not been carried out for all n. Nevertheless, they are very interesting, and it is the latter two approaches that suggest connections with quantum groups.

The equivalence of these different definitions is by no means clear or easy. However, it is now mostly proved by the following scheme.

Chinta-Gunnells method	$\longleftrightarrow$	Whittaker function	$\longleftrightarrow$	Crystal description	$\longleftrightarrow$	Statistical model

The equivalence of the averaging method with the Whittaker definition was proved by Chinta and Offen [18] for type A by generalizing the original proof of Casselman and Shalika. This was extended by McNamara [49] to arbitrary Cartan types. Two results which both assert that the Whittaker definition is equivalent to the crystal definition (in type A) are Brubaker, Bump, and Friedberg [13] and McNamara [50]. The first paper directly computes the Whittaker coefficients of Eisenstein series, and the second paper proceeds locally by partitioning the unipotent integration into cells that contribute the individual terms in the sum over the crystal. The relationship between the statistical model scheme and the crystal description must be done on a case-by-case basis, but we will discuss these below.

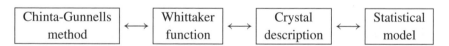

Yet another possibility has appeared on the horizon within the last few months:

- Approach p-adic Whittaker functions by means of Demazure-Lusztig operators and "metaplectic" generalizations of them.

The above remarks concern mainly what is, in the language of Whittaker models, the *spherical Whittaker function*. However, it is useful to consider a larger class of Whittaker functions, namely, the Iwahori-fixed vectors in the Whittaker model. When these are considered, the Demazure-Lusztig operators and their metaplectic analogs appear.

There are other objects in mathematics that may be related to these:

- Some examples of zeta functions of prehomogeneous vector spaces seem to be specializations of Weyl group multiple Dirichlet series. These connections are under investigation by Chinta and Taniguchi.
- Jacquet conjectured that an $O(r)$ period of an automorphic form on GL_r is related to a Whittaker coefficent of the Shimura correspondent on the double cover of GL_r. Applying this to Eisenstein series, this would mean that orthogonal periods of Eisenstein series on GL_r are related to type A_{r-1}. When $r - 1 = 2$, this was investigated by Chinta and Offen [30].
- Zeta functions of prehomogeneous vector spaces as well as the Witten zeta functions studied by Komori, Matsumoto, and Tsumura [45] in this volume are both special cases of Shintani zeta functions. It is by no means clear that the Witten zeta functions can be related to Weyl group multiple Dirichlet series, but potentially there are undiscovered connections.

Let us begin with p-adic Whittaker functions. Casselman and Shalika [24] showed that the values of the spherical Whittaker function are expressible as values of the characters of irreducible representations of the L-group times a deformation of the Weyl denominator. We begin by reviewing this important formula.

Let G be a split Chevalley group or more generally a split reductive group defined over $\mathbb{Z}$. Let F be a nonarchimedean local field with residue field $\mathfrak{o}/\mathfrak{p} = \mathbb{F}_q$, where $\mathfrak{o}$ is the ring of integers and $\mathfrak{p}$ its maximal ideal. Let $B = TN$ be a Borel subgroup, where T is a maximal split torus, and N is the unipotent radical. The root system lives in the group $X^*(T)$ of rational characters of T and the roots so that N is the subgroup generated by the root groups of the positive roots.

We may take the algebraic groups $G, T, B,$ *and* N to be defined over $\mathfrak{o}$. Then, $G(\mathfrak{o})$ is a special maximal compact subgroup. If w is an element of the Weyl group W, we will choose a representative for it in $G(\mathfrak{o})$, which, by abuse of notation, we will also denote as w.

Let $\hat{G}$ be the (connected) Langlands L-group. It is an algebraic group defined over $\mathbb{C}$. Then, G and $\hat{G}$ contain split maximal tori T and $\hat{T}$, respectively; T we have already chosen. Then, $\hat{T}(\mathbb{C})$ is isomorphic to the group of unramified characters of $T(F)$, that is, the characters that are trivial on $T(\mathfrak{o})$. If $z \in \hat{T}(\mathbb{C})$, let τ_z denote the corresponding unramified character.

Let Λ be the weight lattice of $\hat{T}$, that is, the group of rational characters. Then, Λ is isomorphic to $T(F)/T(\mathfrak{o})$. The isomorphism may be chosen so that if λ is a weight and a_λ is a representative of the corresponding coset in $T(F)/T(\mathfrak{o})$, then

$$\tau_z(a_\lambda) = z^\lambda. \tag{26}$$

There are now two root systems to be considered: the root system of G relative to T and the root system of $\hat{G}$ relative to $\hat{T}$. The latter is more important for us, so we will denote it as Φ. Thus, Φ is contained in the Euclidean vector space $\mathbb{R} \otimes \Lambda$. If $\alpha \in \Phi$, then the corresponding coroot $\alpha^\vee$ is a root of G with respect to T, and we will denote by $i_\alpha : \mathrm{SL}_2 \longrightarrow G$ the corresponding Chevalley embedding.

For example, let $G = \mathrm{GL}_{r+1}$. Then, $\hat{G} = \mathrm{GL}_{r+1}$. We take T and $\hat{T}$ to be the diagonal tori. We may identify the weight lattice Λ of $\hat{T}$ with $\mathbb{Z}^{r+1}$ in such a way that $\lambda = (\lambda_1, \dots, \lambda_{r+1}) \in \mathbb{Z}^{r+1}$ corresponds to the rational character

$$z = \begin{pmatrix} z_1 & & \\ & \ddots & \\ & & z_{r+1} \end{pmatrix} \longmapsto \prod_i z_i^{\lambda_i}.$$

If p is a generator of $\mathfrak{p}$, we may take

$$a_\lambda = \begin{pmatrix} p^{\lambda_1} & & \\ & \ddots & \\ & & p^{\lambda_{r+1}} \end{pmatrix}.$$

Then, (26) is satisfied with

$$\tau_z \begin{pmatrix} t_1 & & \\ & \ddots & \\ & & t_{r+1} \end{pmatrix} = \prod z_i^{\mathrm{ord}_p(t_i)}.$$

Returning to the general case, let $z \in \hat{T}(\mathbb{C})$. We may induce τ_z to $G(F)$ by considering the vector space V_z of functions $f : G(F) \longrightarrow \mathbb{C}$ that satisfy

$$f(bg) = (\delta^{1/2}\tau_z)(b) f(g), \qquad b \in B(F). \tag{27}$$

The group $G(F)$ acts on V_z by right translation. If z is in general position, this representation is irreducible and unchanged if z is replaced by any conjugate by an element of the Weyl group. If it is not irreducible, at least its set of irreducible constituents are unchanged if z is conjugated.

Let $\psi : N(F) \longrightarrow \mathbb{C}$ be a character. We will assume that if α is a simple root, then the character $x \longmapsto i_\alpha\left(\begin{smallmatrix} 1 & x \\ & 1 \end{smallmatrix}\right)$ of F is trivial on $\mathfrak{o}$ but no larger fractional ideal. If $f \in V_z$ and $g \in G(F)$, define the Whittaker function on $G(F)$ associated with f by

$$W_f(g) = \int_{N(F)} f(w_0 n g)\, \psi(n)\, \mathrm{d}n, \tag{28}$$

where w_0 is a representative in $G(\mathfrak{o})$ of the long Weyl group element. The integral is convergent if $|z^\alpha| < 1$ for positive roots α; for other z, it may be extended by analytic continuation. The space V_z has a distinguished *spherical vector* f° characterized by the assumption that $f^\circ(g) = 1$ for $g \in G(\mathfrak{o})$. Let $W^\circ = W_{f^\circ}$.

Theorem 1 (Casselman–Shalika [24]). *Let $\lambda \in \Lambda$. Then,*

$$\delta^{-1/2}(a_\lambda)\, W^\circ(a_\lambda) = \begin{cases} \left[\prod_{\alpha\in\Phi^+}(1 - q^{-1}z^\alpha)\right] \chi_\lambda(z) & \text{if } \lambda \text{ is dominant,} \\ 0 & \text{otherwise.} \end{cases} \tag{29}$$

Here, with λ dominant, χ_λ is the character of the finite-dimensional irreducible representation of $\hat{G}$ having highest weight λ. Note that the product on the right-hand side is a deformation of the Weyl denominator. Thus, if ρ is half the sum of the positive roots, on specializing q^{-1} to 1, the right-hand side of (29) becomes

$$\left[\prod_{\alpha\in\Phi+}(1 - z^\alpha)\right] \chi_\lambda(z) = z^\rho \sum_{w\in W} (-1)^{l(w)} z^{w(\lambda+\rho)}, \tag{30}$$

where we have used the Weyl character formula to rewrite the specialization as a sum over the Weyl group.

We next consider how expressions such as the character χ_λ may be interpreted as the p-parts of multiple Dirichlet series.

Let q be a power of a rational prime. Let λ be a dominant weight. We consider an expression $E = \sum_\mu m(\mu)\, z^\mu$, where the sum is over weights μ and $m(\mu)$ is a complex number that is nonzero for only finitely many μ. More precisely, we assume $m(\mu) = 0$ unless μ is in the convex hull of the polytope spanned by the W-orbit of λ, and moreover $\lambda - \mu$ is in the root lattice, which is the lattice in Λ spanned by Φ.[1] We will also assume that m does not vanish on the W-orbit of λ, though it may vanish for roots in the interior of the polytope. We will call E a *λ-expression*. For example, χ_λ is a λ-expression, and the *numerator in the Weyl character formula*, in other words (30), is a $(\lambda + \rho)$-expression.

Given a λ-expression E, let us show how to obtain a Dirichlet polynomial, that is, a polynomial in $q^{-2s_1}, \ldots, q^{-2s_r}$, where r is the rank of $\hat{G}$. Given μ, there exist nonnegative integers $(k_1, \ldots, k_r) = (k_1(\mu), \ldots, k_r(\mu))$ such that $\sum k_i \alpha_i = \mu - w_0(\lambda)$, where w_0 is the long Weyl group element. Then, we call

$$\sum_\mu m(\mu) q^{-2k_1(\mu)s_1 - \ldots - 2k_r(\mu)s_r}$$

[1] If $\hat{G}$ is semisimple, then the root lattice has finite index in Λ.

the *Dirichlet polynomial associated with the λ-expression E*. The p-parts of the multiple Dirichlet series that we are considering are all of this type. If $n = 1$, the $(\lambda + \rho)$-expression producing the p-part is

$$\left[\prod_{\alpha\in\Phi+}(1-q^{-1}z^{\alpha})\right]\chi_{\lambda}(z) \tag{31}$$

which, we observe, differs from (30) by the insertion of q^{-1}. Thus, the p-part is a *deformation* of (30). Comparing with the Casselman–Shalika formula (29), we see that this is essentially a value of the spherical Whittaker function.

Similarly, the p-part (23) with $q = |p|$ is derived from a certain $(\lambda + \rho)$-expression, and these $(\lambda + \rho)$-expressions turn out to be values of spherical Whittaker functions on metaplectic covers of G. The integer l_i in (23) is the inner product of λ with the coroot $\alpha_i^\vee$. These $(\lambda + \rho)$-expressions might be regarded as analogs of (31) in which the integers $m(\mu)$ have been replaced by sums of products of Gauss sums. As we will explain, they are extremely interesting objects from a purely combinatorial point of view.

The averaging method of Chinta-Gunnells expresses the p-part of the multiple Dirichlet series as a ratio in which the numerator is a sum over the Weyl group, and in the case $n = 1$, it reduces to the right-hand side of (30). When $n > 1$, the Weyl group action on functions is nonobvious; the simple reflections involve Gauss sums and congruence conditions, and verifying the braid relations is not a simple matter. See Chinta and Gunnells [27] for this action and Patterson [52] for a meditation on the relationship between the method and the intertwining operators for principal series representations.

We turn next to the crystal description. Crystals arose from the representation theory of quantum groups, that is, quantized enveloping algebras. Let $\hat{\mathfrak{g}}$ be a complex Lie algebra, which for us will be the Lie algebra of $\hat{G}$. Then, the quantized enveloping algebra $U_q(\hat{\mathfrak{g}})$ is a Hopf algebra that is a deformation of the usual universal enveloping algebra $U(\hat{\mathfrak{g}})$ to which it reduces when $q = 1$. The representations of $\hat{G}(\mathbb{C})$ correspond bijectively to characters of $\hat{\mathfrak{g}}$, and hence to $U(\hat{\mathfrak{g}})$; they extend naturally to representations of $U_q(\hat{\mathfrak{g}})$.

Suppose that λ is a dominant weight, which is the highest weight vector of an irreducible representation of $\hat{G}$ and hence of $U_q(\hat{\mathfrak{g}})$. This module $U_q(\hat{\mathfrak{g}})$ has a distinguished basis, Kashiwara's global crystal basis, which is closely related to Lusztig's canonical basis. Let us denote it as $\mathcal{B}_\lambda$. Let 0 be the zero element of the module. Let $\alpha_1, \ldots, \alpha_r$ be the simple roots. If $\alpha = \alpha_i$, then $di_\alpha\left(\begin{smallmatrix}0&1\\0&0\end{smallmatrix}\right)$ and $di_\alpha\left(\begin{smallmatrix}0&0\\1&0\end{smallmatrix}\right)$ are in a certain sense approximated by maps e_i and f_i from $\mathcal{B}_\lambda$ to $\mathcal{B}_\lambda \cup \{0\}$, the *Kashiwara operators*; each such operator applied to $v \in \mathcal{B}_\lambda$ either gives 0 or another element of the basis.

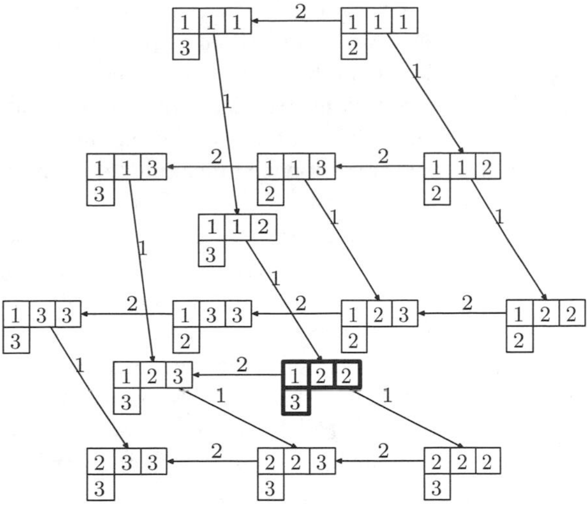

Fig. 1.1 The crystal $\mathcal{B}_{(3,1,0)}$ with an element of weight $(1,2,1)$ highlighted

Also, every element of the crystal basis lies in a well-defined weight space, so there is a map $\mathrm{wt} : \mathcal{B}_\lambda \longrightarrow \Lambda$, mapping each element to its weight. We have

$$\chi_\lambda(z) = \sum_{v \in \mathcal{B}_\lambda} z^\lambda. \tag{32}$$

The maps e_i and f_i shift the weights: if $x, y \in \mathcal{B}_\lambda$, then $f_i(x) = y$ if and only if $e_i(y) = x$, and in this case, $\mathrm{wt}(y) = \mathrm{wt}(x) - \alpha$. When this is true, we draw an arrow $x \xrightarrow{i} y$, and the resulting directed graph, with edges labeled by indices i, is the *crystal graph*.

Let us consider an example. Take $G = \mathrm{GL}_3$. Its Cartan type is A_2. In this case, the weight lattice Λ may be identified with $\mathbb{Z}^3$. If λ is a weight, then with this identification, $\lambda = (\lambda_1, \lambda_2, \lambda_3)$ where $z^\lambda = z_1^{\lambda_1} z_2^{\lambda_2} z_3^{\lambda_3}$. The weight is dominant if $\lambda_1 \geqslant \lambda_2 \geqslant \lambda_3$. Assume that λ is dominant, and furthermore $\lambda_3 \geqslant 0$ so that λ is a partition. The elements of $\mathcal{B}_\lambda$ may be identified with semistandard Young tableaux with shape λ. So if $\lambda = (3,1,0)$ and $r = 2$, the crystal graph of $\mathcal{B}_\lambda$ is shown in Fig. 1.1. The weight of a tableau T is $\mathrm{wt}(T) = (\mu_1, \mu_2, \mu_3)$ where μ_i is the number of entries in T equal to i.

If $w \in W$, then $\mathcal{B}_\lambda$ has a unique element $v_{w\lambda}$ of weight $w\lambda$. We will call these elements *extremal*. Consider a function f on $\mathcal{B}_\lambda$, which we assume does not vanish on the extremal elements. Then, we may consider

$$E_f = \sum_{v \in \mathcal{B}_\lambda} f(v)\, z^{\mathrm{wt}(v)}.$$

This is a λ-expression. For example, if $f(v) = 1$ for all v, then E_f is the character of the irreducible representation with highest weight λ.

We call a weight *strongly dominant* if it is of the form $\lambda + \rho$ with λ dominant. A dominant weight is strongly dominant if and only if W acts freely on its orbit, or equalently, if it is in the interior of the positive Weyl chamber. Let $\lambda + \rho$ be a strongly dominant weight. We recall that the numerator (30) in the Weyl character formula, as are the p-parts of the Weyl group multiple Dirichlet series, are $(\lambda + \rho)$-expressions. So we may hope to find a function f on $\mathcal{B}_{\lambda+\rho}$ such that E_f equals this numerator or p-part. In some cases, a deformation of the Weyl character formula exists in which the numerator is a sum over $\mathcal{B}_{\lambda+\rho}$. The formula will express χ_λ as a ratio of this sum to a denominator that is a deformation of the Weyl denominator, of the form $\prod_{\alpha\in\Phi^+}(1 - tz^\alpha)$, where t is a deformation parameter. Taking $t = 1$, only the extremal elements of $\mathcal{B}_{\lambda+\rho}$ will make a nonzero contribution, and the numerator will reduce to the numerator in the Weyl character formula. Taking $t = 0$, the only terms that contribute will be those in the image of a map $\mathcal{B}_\lambda \longrightarrow \mathcal{B}_{\lambda+\rho}$, and the sum reduces to a sum over $\mathcal{B}_\lambda$. Thus, when $t = 0$, the formula reduces to (32). Most importantly for us, taking $t = q^{-1}$ and comparing with (29), we see that the sum over $\mathcal{B}_{\lambda+\rho}$ is exactly $\delta^{1/2}(a_\lambda)W^\circ(a_\lambda)$.

Moreover, such formulas exist for metaplectic Whittaker functions. In other words, there is often a way of summing over $\mathcal{B}_{\lambda+\rho}$ and obtaining a Whittaker function on the n-fold cover of some group. There should actually be one such formula for every reduced decomposition of the long Weyl group element into a product of simple reflections, and in some sense, this is true. But, in practice, only certain such decompositions give clean and elegant formulas. Here is a list of cases where nice formulas are known, rigorously or conjecturally:

- Cartan type A_r, any n. See [13, 14]. These produce Whittaker functions on the n-fold covers of GL_{r+1}. In this case, proofs are complete.
- Cartan type B_r, n even. These would produce Whittaker functions on even covers of Sp_{2r}. For general even n, the representation is conjectural, and even in the $n = 2$ case, there is still work to be done. See [10], this volume, for the case $n = 2$.
- Cartan type C_r, n odd. See [2, 3] and Ivanov [40] for a discussion of the Yang–Baxter equation. These will produce Whittaker functions on the n-covers of Sp_{2r+1}. Only the case $n = 1$ is proved.
- Cartan type D_r, any n. See [28], this volume. This case is still largely conjectural.

Let us explain how this works for type A_r for arbitrary r, so $G = \mathrm{GL}_{r+1}$ (or SL_{r+1}). If $n = 1$, the formula in question is *Tokuyama's formula*. Tokuyama [54] expressed his formula as a sum over strict Gelfand-Tsetlin patterns, but it may be reformulated in terms of tableaux or crystals. Using crystals, Tokuyama's formula may be written

$$\left[\prod_{\alpha\in\Phi^+}(1 - tz^\alpha)\right]\chi_\lambda(z) = \sum_{v\in\mathcal{B}_{\lambda+\rho}} G^\flat(v)z^{\mathrm{wt}(v)-w_0\rho}, \tag{33}$$

where w_0 is the long Weyl group element, and the function $G^\flat(v)$ will be described below. Tokuyama's formula is a deformation of the Weyl character formula, which is obtained by specializing $t \longrightarrow 1$; the formula (32) is recovered by specializing $t \longrightarrow 0$. Taking $t = q^{-1}$ to be the cardinality of the residue field for a nonarchimedean local field, and combining Tokuyama's formula with the Casselman–Shalika formula (29), it gives a formula for the p-adic Whittaker function, and similar formulas give the p-part of the multiple Dirichlet series. Since this is the case we are concerned with, we will write q^{-1} instead of t for the deformation parameter, even if occasionally we want to think of it as an indeterminate.

To define $G^\flat(v)$, we first associate with v a *BZL-pattern* (for Berenstein, Zelevinsky [4] and Littelmann [47]). We choose a "long word" by which we mean a decomposition of $w_0 = \sigma_{\omega_1} \cdots \sigma_{\omega_N}$ into a product of simple reflections σ_{ω_i} $(1 \leqslant i \leqslant r)$, where N is the number of positive roots. Let b_1 be the number of times we may apply f_{ω_1} to v, that is, the largest integer such that $f_{\omega_1}^{b_1}(v) \neq 0$. Then, we let b_2 be the largest integer such that $f_{\omega_2}^{b_2} f_{\omega_1}^{b_1}(v) \neq 0$. Continuing this way, we define $b_1, \ldots, b_N$. We may characterize $\mathrm{BZL}(v) = (b_1, b_2, \ldots, b_{\omega_N})$ as the unique sequence of N nonnegative integers such that $f_{\omega_N}^{b_N} \cdots f_{\omega_1}^{b_1}(v)$ is the unique vector $v_{w_0\lambda}$ with lowest weight $w_0\lambda$.

We should be able to give a suitable definition of $G^\flat$ for any Cartan type and any long word, but in practice we only know how to give precise combinatorial definitions in certain cases. We will choose this word:

$$(\omega_1, \ldots, \omega_N) = (1, 2, 1, 3, 2, 1, \ldots)$$

corresponding to the decomposition into simple reflections:

$$w_0 = \sigma_{\omega_1} \cdots \sigma_{\omega_N} = \sigma_1\sigma_2\sigma_1\sigma_3\sigma_2\sigma_1 \cdots \sigma_r\sigma_{r-1} \cdots \sigma_1,$$

where σ_i is the i-th simple reflection in W. We will write $\mathrm{BZL}(v)$ in a tabular array:

$$\mathrm{BZL}(v) = \begin{bmatrix} \ddots & & \vdots \\ b_4 & b_5 & b_6 \\ & b_2 & b_3 \\ & & b_1 \end{bmatrix}. \tag{34}$$

This has the significance that each column corresponds to a single-root operator f_i where $i = 1$ for the rightmost column, $i = 2$ for the next column, and so forth.

Now, we will decorate the pattern by drawing boxes or circles around various b_i according to certain rules that we will now discuss. We describe the circling rule first. It may be proved that b_i satisfies the inequality $b_i \geqslant b_{i+1}$, except in the case that i is a triangular number, so that b_i is the last entry in its row; in the latter case, we only have the inequality $b_i \geqslant 0$. In either case, we circle b_i if its inequality is an equality. In other words, we circle b_i if (in the first case) $b_i = b_{i+1}$ or (in the second case) $b_i = 0$.

For the boxing rule, we box b_i if $e_{\omega_{i+1}} f_{\omega_i}^{b_i} \cdots f_{\omega_1}^{b_1}(v) = 0$.

Let us consider an example. We take $(\omega_1, \omega_2, \omega_1) = (2, 1, 2)$ and v to be the element $\begin{array}{|c|c|c|}\hline 1 & 2 & 2 \\ \hline 3 \\ \cline{1-1}\end{array}$ in the crystal $\mathcal{B}_{(3,1,0)}$ that is highlighted in Fig. 1.1. Then, it is easy to see that

$$\mathrm{BZL}(v) = \begin{bmatrix} 1 & 1 \\ & 1 \end{bmatrix}.$$

We decorate this as follows. Since $b_2 = b_3$, we circle b_2. Moreover, $e_{\omega_1}(v) = 0$ since (referring to the crystal graph) there is no way to move in the e_2 direction. Thus, b_1 is boxed and the decorated BZL pattern looks like this:

$$\mathrm{BZL}(v) = \begin{bmatrix} \textcircled{1} & 1 \\ & \boxed{1} \end{bmatrix}.$$

The boxing and circling rules may seem artificial, but are actually natural, for they have the following interpretation. Kashiwara defined, in addition to the crystals $\mathcal{B}_\lambda$ corresponding to the finite-dimensional irreducible representations, a crystal $\mathcal{B}_\infty$ which is a crystal basis of the quantized universal developing algebra of the lower unipotent part of the Lie algebra of G. There is another crystal $\mathcal{T}_\lambda$ with precisely one element having weight λ. Then, $\mathcal{B}_\infty \otimes \mathcal{T}_{\lambda+\rho}$ is a "universal" crystal with a highest weight vector having weight $\lambda + \rho$. There is then a morphism of crystals $\mathcal{B}_{\lambda+\rho} \longrightarrow \mathcal{B}_\infty \otimes \mathcal{T}_{\lambda+\rho}$. See Kashiwara [41]. This morphism may be made explicit by adopting a viewpoint similar to Littelmann [47]. Indeed, the notion of BZL patterns makes sense for $\mathcal{B}_\infty \otimes \mathcal{T}_{\lambda+\rho}$, and the set of BZL patterns is precisely the cone of patterns (34) such that

$$b_1 \geqslant 0, \qquad b_2 \geqslant b_3 \geqslant 0, \qquad b_4 \geqslant b_5 \geqslant b_6 \geqslant 0, \qquad \cdots.$$

The morphism $\mathcal{B}_{\lambda+\rho} \longrightarrow \mathcal{B}_\infty \otimes \mathcal{T}_{\lambda+\rho}$ is then characterized by the condition that corresponding elements of the two crystals have the same BZL pattern. See Fig. 1.1 of Bump and Nakasuji [23] for a picture of this embedding.

Now, the circling rule may be explained as follows: embedding $v \in \mathcal{B}_{\lambda+\rho}$ into $\mathcal{B}_\infty \otimes \mathcal{T}_{\lambda+\rho}$, an entry in the BZL pattern is circled if and only if BZL(v) lies on the boundary of the cone. Similarly, there is a crystal $\mathcal{B}_{-\infty} \otimes \mathcal{T}_{w_0(\lambda+\rho)}$ with a unique lowest weight vector having weight $w_0(\lambda + \rho)$, and we may embed $B_{\lambda+\rho}$ into this crystal by matching up the lowest weight vectors, and an entry is boxed if and only if BZL(v) lies on the boundary of this opposite cone.

Returning to Tokuyama's formula, define

$$G^\flat(v) = \prod_{i=1}^{N} \left\{ \begin{array}{ll} 1 & \text{if } b_i \text{ is circled but not boxed;} \\ -q^{-1} & \text{if } b_i \text{ is boxed but not circled;} \\ 1 - q^{-1} & \text{if } b_i \text{ is neither circled nor boxed;} \\ 0 & \text{if } b_i \text{ is both circled and boxed.} \end{array} \right\}$$

This was generalized to the $n \geqslant 1$ case as follows. There is, in this generality, a Whittaker function on the metaplectic group, and as in the case $n = 1$, we have

$$\delta^{-1/2}(a_\lambda)W^\circ(a_\lambda) = \sum_{v \in \mathcal{B}_{\lambda+\rho}} G^\flat(v) z^{\mathrm{wt}(v) - w_0\rho}. \tag{35}$$

Now the definition of $G^\flat$ must be slightly changed. Let a be a positive integer. Define $g(a) = q^{-a} g(p^{a-1}, p^a)$ and $h(a) = q^{-a} g(p^a, p^a)$ in terms of the Gauss sum discussed in the last section. These depend only on a modulo n. Then we have

$$G^\flat(v) = \prod_{i=1}^{N} \left\{ \begin{array}{ll} 1 & \text{if } b_i \text{ is circled but not boxed;} \\ g(b_i) & \text{if } b_i \text{ is boxed but not circled;} \\ h(g_i) & \text{if } b_i \text{ is neither circled nor boxed;} \\ 0 & \text{if } b_i \text{ is both circled and boxed.} \end{array} \right\}$$

If $n = 1$, this reduces to our previous definition.

Tokuyama's formula may be given another interpretation, as evaluating the partition function of a statistical mechanical system of the free-fermionic six-vertex model. For this, see [15], Chap. 19 of [14], and the paper [8] in this volume. Since more details may be found in these references, we will be brief. In a statistical mechanical system, there is given a collection of (many) *states*, and each state is assigned a *Boltzmann weight* which is a measure of how energetic the state is; more highly energetic states are less probable.

For example, the two-dimensional Ising model consists of a collection of sites, each of which may be assigned a spin $+$ or $-$. These sites might represent atoms in a ferromagnetic substance, and in the two-dimensional model, they lie in a plane. A state of the system consists of an assignment of spins to every site. At each site, there is a local Boltzmann weight, depending on spin at the site and at its nearest neighbors. This system was analyzed by Onsager, who found, surprisingly, that the partition function could be evaluated explicitly.

Later investigators considered models in which the spins are assigned not to the sites themselves, but to edges in a grid connecting the sites. The Boltzmann weight at the vertex depends on the configuration of spins on the four edges adjacent to the vertex. Thus, if the site is x, we label the four adjacent edges with $\varepsilon_i = +$ or $-$ by the following scheme:

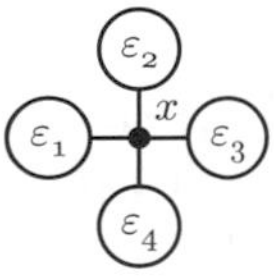

(36)

Of particular interest to us is the *six-vertex* or *two-dimensional ice model* which was solved by Lieb and Sutherland in the 1960s, though it is the treatment of Baxter [1] that is most important to us. There are six admissible configurations. These are given by the following table.

	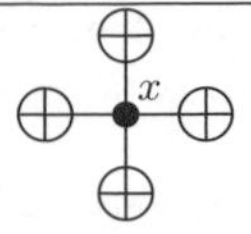	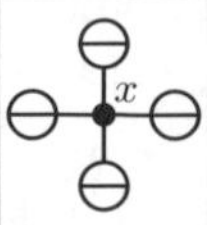	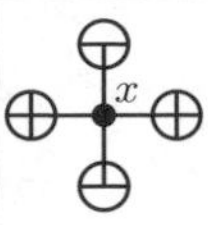	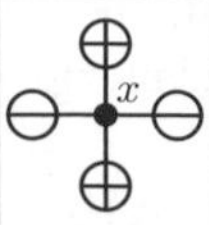	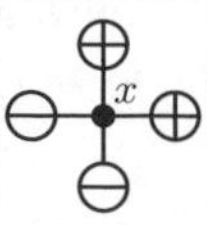	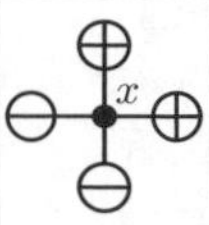
Boltzmann weight	$a_1 = a_1(x)$	a_2	b_1	b_2	c_1	c_2

The set of Boltzmann weights used may vary from site to site, so if (as in the table) the site is x, we may write $a_1(x)$ to indicate this dependence.

If $a_1 = a_2$, $b_1 = b_2$, and $c_1 = c_2$, then the site is called *field-free*. If $a_1a_2 + b_1b_2 = c_1c_2$, the site is called *free-fermionic*. The term comes from physics: in the free-fermionic case, the row transfer matrices (see [8]) are differentiated versions of the Hamiltonians of a quantum mechanical system, the XXZ model, and the quanta for this model are particles of spin 1/2, called fermions.

Hamel and King [38] and Brubaker, Bump, and Friedberg [15] showed that one may exhibit a free-fermionic six-vertex model whose partition function is exactly the Tokuyama expression (33). In [8] in this volume, this is generalized to a system whose partition function is the metaplectic spherical Whittaker function (35). The explanation for this is as follows: there is a map from the set of states of the model to the $\mathcal{B}_{\lambda+\rho}$ crystal. The map is not surjective, but its image is precisely the set of $v \in \mathcal{B}_{\lambda+\rho}$ such that $G^\flat(v) \neq 0$.

Thus, using this bijection of the set of states with those $v \in \mathcal{B}_{\lambda+\rho}$ such that $G^\flat(v) \neq 0$ means that Tokuyama's theorem may be formulated either as the evaluation of a sum over a crystal or as the partition function of a statistical-mechanical system. But there is a subtle and important difference between these two setups. For example, different tools are available. There is a an automorphism of the crystal graph, the *Schützenberger involution*, that takes a vertex of weight μ to one of weight $w_0\mu$, where w_0 is the long Weyl group element; this is sometimes useful in proofs. The set of states of the statistical-mechanical system has no such involution, yet another, more powerful tool becomes available: the *Yang–Baxter equation*.

To describe it, let us associate a matrix with the Boltzmann weights at a site as follows. Let V be a two-dimensional vector space with basis v_+ and v_-. (In the metaplectic case, the scheme proposed in [8] gives its dimension as $2n$.) Associate with each site an endomorphism of $V \otimes V$ as follows. With the vertices labeled as in (36), let R be the linear transformation such that the coefficient of $v_{\varepsilon_3} \otimes v_{\varepsilon_4}$ in $R(v_{\varepsilon_1} \otimes v_{\varepsilon_2})$ is the Boltzmann weight corresponding to the four spins $\varepsilon_1, \varepsilon_2, \varepsilon_3$, and ε_4. Thus, the only nonzero entries in the matrix of R with respect to the basis $v_\pm \otimes v_\pm$ are a_1, a_2, b_1, b_2, c_1, and c_2. We will call an endomorphism R of $V \otimes V$ (or its matrix) an *R-matrix*. We then denote by R_{12}, R_{13}, and R_{23} endomorphisms of $V \otimes V \otimes V$ in which R_{ij} acts on the i-th and j-th components, and the identity 1_V acts on the remaining one. For example, $R_{12} = R \otimes 1_V$.

We are interested in endomorphisms R, S, and T of $V \otimes V$ such that

$$R_{12}S_{13}T_{23} = T_{23}S_{13}R_{12}.$$

It is not hard to check that this is equivalent to (25) in [8]. This was called the *star-triangle relation* by Baxter, and the *Yang–Baxter equation* by others, particularly in the case where R, S, and T are either all equal or drawn from the same parametrized family. In particular, let Γ be a group, and $g \longmapsto R(g)$ a map from Γ into the set of R-matrices such that

$$R_{12}(g)R_{13}(gh)R_{23}(h) = R_{23}(h)R_{13}(gh)R_{12}(g). \tag{37}$$

Then, (37) is called a *parametrized Yang–Baxter equation.*

In Sect. 9.6 of [1], Baxter essentially found parametrized Yang–Baxter equations in the *field-free* case, where $a_1 = a_2 = a$, $b_1 = b_2 = b$, and $c_1 = c_2 = c$. Fix a complex number Δ. Then, his construction gives a parametrized Yang–Baxter equation, with parameter group $\mathbb{C}^\times$, such that the image of R consists of endomorphisms of $V \otimes V$ with corresponding to such field-free R-matrices with $(a^2 + b^2 - c^2)/2ab = \Delta$. This construction led to the development of quantum groups. In the formulation of Drinfeld [33], this instance of the Yang–Baxter equation is related to Hopf algebra $U_q(\widehat{\mathfrak{sl}_2})$. The parameter group indexes modules of this Hopf algebra with $\Delta = \frac{1}{2}(q + q^{-1})$, and Yang–Baxter equation is a consequence of a property (quasitriangularity) of $U_q(\widehat{\mathfrak{sl}_2})$, or a suitable completion.

A parametrized Yang–Baxter equation with parameter group $\mathrm{SL}(2, \mathbb{C})$ was given in Korepin, Boguliubov, and Izergin [44], p. 126. The parametrized R-matrices are contained within the free-fermionic six-vertex model. Scalar R-matrices may be added trivially, so the actual group is $\mathrm{SL}(2, \mathbb{C}) \times \mathrm{GL}_1(\mathbb{C})$. This nonabelian parametrized Yang–Baxter equation was rediscovered in slightly greater generality by Brubaker, Bump, and Friedberg [15], who found a parametrized Yang–Baxter equation for the entire set of R-matrices in the free-fermionic six-vertex model, with parameter group $\mathrm{GL}_2(\mathbb{C}) \times \mathrm{GL}_1(\mathbb{C})$. It is an interesting question how to formulate this in terms of a Hopf algebra, analogous to the field-free case.

Brubaker, Bump, and Friedberg [15] showed that a system may be found, with free-fermionic Boltzmann weights, whose partition function is precisely (33). This fact was generalized Bump, McNamara, and Nakasuji [22], who showed that one may replace the character on the left-hand side by a factorial Schur function. Then, the parametrized free-fermionic Yang–Baxter equation can be used to prove Tokuyama's formula (or its generalization to factorial Schur functions). Moreover, in [8], a different generalization is given in which the partition function represents the metaplectic Whittaker function. In the latter case, however, no Yang–Baxter equation is known if $n > 1$.

When $n = 1$, various facts about Whittaker functions may be proved using the free-fermionic Yang–Baxter equation. One fact that may be checked is that the partition function representing (33), divided by the product on the left-hand side of the equation, is symmetric, in other words invariant under permuting the eigenvalues of z. This is a step in a proof of Tokuyama's theorem. As explained in [8], this fact has a generalization to partition functions representing metaplectic Whittaker functions and seems amenable to the Yang–Baxter equation, but no Yang–Baxter equation is known in this case.

In (35), the definition of $G^\flat$ depends on the choice of a reduced word representing the long Weyl group element. Two particular long words are considered, and the Yang–Baxter equation is used to show that both representations give the same result. If $n > 1$, this remains true, but again the Yang–Baxter equation is unavailable. Consequently, different proofs, based on the Schützenberger involution of the crystal $\mathcal{B}_{\lambda+\rho}$, are given. However, these arguments require extremely difficult combinatorial arguments, and it would be good to have an alternative approach based on the Yang–Baxter equation.

See [10] for another application of the free-fermionic Yang–Baxter equation to metaplectic Whittaker functions, this time on the double cover of $\mathrm{Sp}(2r)$.

1.8 Demazure Operators

Let (π, V) be a principal series representation of $G(F)$, where G is a split semisimple Lie group. The theory described above, including the Casselman–Shalika formula (and its metaplectic generalizations), is for the spherical Whittaker function, that is, the K-fixed vector in the Whittaker model, where $K = G(\mathfrak{o})$.

Let J be the *Iwahori subgroup*, which is the inverse image of $B(\mathbb{F}_q)$ under the map $G(\mathfrak{o}) \longrightarrow G(\mathbb{F}_q)$ that is reduction mod p. We may consider more generally the space V^J of J-fixed vectors in the Whittaker model. These play an important role in the proof of the Casselman–Shalika formula which, we have seen, is a key result in the above discussion.

Until 2011, the investigations that we have been discussing in the above pages concentrated on the unique (up to scalar) K-fixed vector, rather than elements of V^J, though the Iwahori-fixed vectors appeared in the work of Chinta and Offen [18] and McNamara [49] generalizing Casselman and Shalika. Still, the essence of the Casselman–Shalika proof is to finesse as much as possible in order to avoid getting involved with direct calculations of Iwahori Whittaker functions. But it turns out that there is an elegant calculus of Iwahori Whittaker functions, and this is likely to be a key to the relationship between the theory of Whittaker functions and combinatorics.

If $w \in W$, the Demazure operator ∂_w acts on the ring $\mathcal{O}(\hat{T})$ of rational functions on $\hat{T}$. To define it, first consider the case where $w = \sigma_i$ is a simple reflection. Then, if f is a rational function on $\hat{T}(\mathbb{C})$,

$$\partial_{\sigma_i} f(z) = \frac{f(z) - z^{-\alpha_i} f(\sigma_i z)}{1 - z^{-\alpha_i}}.$$

The numerator is divisible by the denominator, so this is again a rational function. The definition of ∂_w is completed by the requirement that if $l(ww') = l(w) + l(w')$, where l is the length function on W, then $\partial_{ww'} = \partial_w \partial_{w'}$.

If λ is a dominant weight, then $\partial_{w_0} z^\lambda$ is the character $\chi_\lambda(z)$, and for general w we will call $\partial_w z^\lambda$ a *Demazure character*. These first arose in the cohomology of line bundles over Schubert varieties, and they have proved to be quite important in combinatorics. As Littelmann and Kashiwara showed, they may be interpreted as operators on functions on crystals. As we will explain, Demazure operators, and the related Demazure-Lusztig operators arise naturally in the theory of Whittaker functions.

Iwahori and Matsumoto observed that V^J is naturally a module for the convolution ring of compactly supported J-bi-invariant functions, and they determined the structure of this ring. Later, Bernstein, Zelevinsky, and Lusztig gave a different presentation of this ring. It is the (extended) affine Hecke algebra $\widetilde{\mathcal{H}}_q$, and it has also turned out to be a key object in combinatorics independent of its origin in the representation theory of p-adic groups. Restricting ourselves to the semisimple case for simplicity, this algebra may be defined as follows. It contains a $|W|$-dimensional subalgebra $\mathcal{H}_q$ with generators $T_1, \dots, T_r$ subject to the quadratic relations

$$T_i^2 = (q-1)T_i + q$$

together with the braid relations: when $i \neq j$,

$$T_i T_j T_i \cdots = T_j T_i T_j \dots,$$

where the number of terms on either side is the order of $\sigma_i \sigma_j$ and as before σ_i is the i-th simple reflection.

The algebra $\widetilde{\mathcal{H}}_q$ is the amalgam of $\mathcal{H}_q$ with an abelian subalgebra ζ^Λ isomorphic to the weight lattice Λ. If $\lambda \in \Lambda$, let ζ^λ be the corresponding element of ζ^Λ. To complete the presentation of $\tilde{\mathcal{H}}$ we have the relation

$$T_i \zeta^\lambda - \zeta^{\sigma_i \lambda} T_i = \zeta^\lambda T_i - T_i \zeta^{\sigma_i \lambda} = \left(\frac{v-1}{1-\zeta^{-\alpha_i}} \right) (\zeta^\lambda - \zeta^{\sigma_i \lambda}), \tag{38}$$

sometimes known as the *Bernstein relation*.

Though historically it first appeared in the representation theory of p-adic groups, the affine Hecke algebra appears in other contexts. For example, the investigation of Kazhdan and Lusztig [42], motivated by Springer's work on the representation theory of Weyl groups, used $\widetilde{\mathcal{H}}_q$ in a fundamental way, and led to applications in different areas of mathematics, such as the topology of flag varieties and the structure of Verma modules. Significantly for the present discussion, Lusztig [48] showed that $\widetilde{\mathcal{H}}_v$ (with v an indeterminate) may be realized as a ring acting on the equivariant K-theory of the flag manifold of $\hat{G}$, and Kazhdan and Lusztig [43, 48] then applied this back to the local Langlands correspondence by constructing the irreducible representations of $G(F)$ having an Iwahori-fixed vector.

The equivariant K-theory of $\hat{G}$ may be described as follows. Let $\mathcal{O}(\hat{T})$ be the ring of rational functions on $\hat{T}(\mathbb{C})$. In our previous notation, it is simply the group algebra of the weight lattice Λ. If X is the flag variety of $\hat{G}$ then $K_{\hat{G}}(X) \cong \mathcal{O}(\hat{T})$. Better still, let $M = \hat{G} \times \mathrm{GL}_1$, where the GL_1 acts trivially on X. Then, $K_M(X) \cong \mathbb{C}[v, v^{-1}] \otimes \mathcal{O}(\hat{T})$, where v is a parameter.

The starting point of the investigations of Kazhdan and Lusztig is a representation of $\widetilde{\mathcal{H}}_v$ on this ring. In this representation on $\mathbb{C}[v, v^{-1}] \otimes \mathcal{O}(\hat{T})$, the generators of $\mathcal{H}_v$ act by certain operators called *Demazure-Lusztig operators*, while the commutative subalgebra ζ^Λ acts by multiplication. We will call this representation of $\widetilde{\mathcal{H}}_v$ on $\mathbb{C}[v, v^{-1}] \otimes \mathcal{O}(\hat{T})$ the *Lusztig representation.*

The same representation of $\widetilde{\mathcal{H}}$ appears in another way, independent of Lusztig's cohomological interpretation. There are two versions of this:

- Ion [39] observed such a representation in the space of Iwahori-fixed vectors of the spherical model of an unramified principal series representation. He concluded that these matrix coefficients are expressed in terms of the nonsymmetric Macdonald polynomials. His methods are based on the double affine Hecke algebra.
- Brubaker, Bump, and Licata [16] found a representation equivalent to the Lusztig representation acting on Whittaker functions. Their method could also be used in the setting of [39].

After Brubaker, Bump, and Licata mentioned the connection between Whittaker functions and Demazure characters, Chinta and Gunnells began looking at the metaplectic case. They found "metaplectic Demazure operators" involving Gauss sums that are related to the Chinta-Gunnells representation. Also, with A. Schilling, Brubaker, Bump, and Licata looked at the possibility that the results of [16] could be reinterpreted in terms of the crystal graph, similarly to the crystal interpretation of Tokuyama's formula. This seems to be a promising line of investigation.

Let us briefly recall the results of [16]. Let $V = V_z$ be as in (27). Let Ω be one of the following two linear functionals on V: it is either the Whittaker functional

$$\Omega(f) = \int_{N(F)} f(wn)\,\psi(n)\,\mathrm{d}n,$$

where ψ is as in (28), or the spherical functional $\Omega(f) = \int_K f(k)\,\mathrm{d}k$. If $w \in W$, let Φ_w be the element of V^J defined as follows. Every element of $G(F)$ may be written as $bw'k$ with $b \in B(F)$, $w' \in W$, and $k \in J$. Then, with τ_z as in (26),

$$\Phi_w(bw'k) = \begin{cases} \delta^{1/2}\tau_z(b) & \text{if } w = w', \\ 0 & \text{otherwise.} \end{cases}$$

The $|W|$ functions Φ_w are a basis of the space of J-fixed vectors in V. (The action $\pi : G(F) \to \mathrm{End}(V)$ is by right translation.) We also define

$$W_w(g) = \Omega(\pi(g)\Phi_w).$$

$$\widetilde{\Phi}_w = \sum_{u \geqslant w} \Phi_u, \qquad \widetilde{W}_w = \sum_{u \geqslant w} W_u,$$

where $u \geqslant w$ is with respect to the Bruhat order.

Let λ be a weight; if Ω is the Whittaker functional, we require λ to be dominant. We may regard $W_w(a_\lambda)$ as an element of $\mathbb{C}[q,q^{-1}] \otimes \mathcal{O}(\hat{T})$. Then, there exist operators $\mathcal{T}_i$ on $\mathbb{C}[q,q^{-1}] \otimes \mathcal{O}(\hat{T})$ such that

$$\mathcal{T}_i^2 = (q^{-1}-1)\mathcal{T}_i + q^{-1}$$

and which also satisfy the braid relations. Therefore, we obtain a representation of $\mathcal{H}_{q^{-1}}$ on $\mathbb{C}[q,q^{-1}] \otimes \mathcal{O}(\hat{T})$. It may be extended to an action of $\widetilde{\mathcal{H}}_{q^{-1}}$. Now if the simple reflection σ_i is a left descent of $w \in W$, that is, $l(\sigma_i w) < l(w)$, then

$$W_{\sigma_i w}(a_\lambda) = \mathcal{T}_i W_w(a_\lambda).$$

(see [16]). The operators $\mathfrak{T}_i$ are slightly different in the two cases (Ω the Whittaker or Spherical functional). In both cases, they are essentially the Demazure-Lusztig operators. For definiteness, we will describe them when Ω is the Whittaker functional. If f is a function on $\hat{T}(\mathbb{C})$, define

$$\partial_i' f(z) = \frac{f(z) - z^{\alpha_i} f(\sigma_i z)}{1 - z^{\alpha_i}} = \frac{f(\sigma_i z) - z^{-\alpha_i} f(z)}{1 - z^{-\alpha_i}}.$$

This is the usual Demazure operator conjugated by the map $z \mapsto -z$. Then, the operators $\mathfrak{T}_i$ are given by

$$\mathfrak{D}_i' = (1 - q^{-1} z^{\alpha}) \partial_\alpha', \qquad \mathfrak{T}_i' = \mathfrak{D}_i' - 1.$$

The Whittaker functions W_w can thus be obtained from W_{w_0} by applying the $\mathfrak{T}_i$. Moreover, W_{w_0} has a particularly simple form:

$$W_{w_0}(a_\lambda) = \begin{cases} \delta^{1/2}(a_\lambda) z^{w_0 \lambda} & \text{if } \lambda \text{ is dominant,} \\ 0 & \text{otherwise.} \end{cases}$$

In conclusion, the Lusztig representation arises naturally in the theory of Whittaker functions or, in Ion's setup, K, J-bi-invariant matrix coefficients. It gives a calculus, whereby the Whittaker functions may be computed recursively from the simplest one W_{w_0}.

It is also important to consider $\widetilde{W}_w$. For example, $\widetilde{W}_1$ is the spherical Whittaker function which we have discussed at length in the previous sections. In the theory of multiple Dirichlet series, it might be useful to substitute $\widetilde{W}_w$ for the p-part at a finite number of places. In the study of the $\widetilde{W}_w$, the remarkable combinatorics of the Bruhat order begins to play an important role. See [16] for further information. An important issue is to extend the theory of the previous sections to the theory of the $\widetilde{W}_w$ and to carry out this unified theory in the metaplectic context.

References

1. R. Baxter. *Exactly solved models in statistical mechanics*. Academic Press Inc. [Harcourt Brace Jovanovich Publishers], London, 1982.
2. J. Beineke, B. Brubaker, and S. Frechette. Weyl group multiple Dirichlet series of type C. *Pacific J. Math.*, To Appear.
3. J. Beineke, B. Brubaker, and S. Frechette. A crystal description for symplectic multiple Dirichlet series, in this volume.
4. A. Berenstein and A. Zelevinsky. Canonical bases for the quantum group of type A_r and piecewise-linear combinatorics. *Duke Math. J.*, 82(3):473–502, 1996.
5. S. Bochner. A theorem on analytic continuation of functions in several variables. *Ann. of Math. (2)*, 39(1):14–19, 1938.
6. B. Brubaker, A. Bucur, G. Chinta, S. Frechette and J. Hoffstein. Nonvanishing twists of GL(2) automorphic L-functions. *Int. Math. Res. Not.*, 78:4211–4239, 2004.
7. B. Brubaker and D. Bump. Residues of Weyl group multiple Dirichlet series associated to $\widetilde{\mathrm{GL}}_{n+1}$. In *Multiple Dirichlet series, automorphic forms, and analytic number theory*, volume 75 of *Proc. Sympos. Pure Math.*, pages 115–134. Amer. Math. Soc., Providence, RI, 2006.
8. B. Brubaker, D. Bump, G. Chinta, S. Friedberg, and G. Gunnells. Metaplectic ice, in this volume.
9. B. Brubaker, D. Bump, G. Chinta, S. Friedberg, and J. Hoffstein. Weyl group multiple Dirichlet series. I. In *Multiple Dirichlet series, automorphic forms, and analytic number theory*, volume 75 of *Proc. Sympos. Pure Math.*, pages 91–114. Amer. Math. Soc., Providence, RI, 2006.
10. B. Brubaker, D. Bump, G. Chinta, and G. Gunnells. Metaplectic Whittaker functions and crystals of type B, in this volume.
11. B. Brubaker, D. Bump, and S. Friedberg. Weyl group multiple dirichlet series. ii. the stable case. *Invent. Math.*, 165(2):325–355, 2006.
12. B. Brubaker, D. Bump, and S. Friedberg. Twisted Weyl group multiple Dirichlet series: the stable case. In *Eisenstein series and applications*, volume 258 of *Progr. Math.*, pages 1–26. Birkhäuser Boston, Boston, MA, 2008.
13. B. Brubaker, D. Bump, and S. Friedberg. Weyl group multiple Dirichlet series, Eisenstein series and crystal bases. *Ann. of Math. (2)*, 173(2):1081–1120, 2011.
14. B. Brubaker, D. Bump, and S. Friedberg. *Weyl group multiple Dirichlet series: type A combinatorial theory*, volume 175 of *Annals of Mathematics Studies*. Princeton University Press, Princeton, NJ, 2011.
15. B. Brubaker, D. Bump, and S. Friedberg. Schur polynomials and the Yang-Baxter equation. *Comm. Math. Phys.*, 308(2):281–301, 2011.
16. B. Brubaker, D. Bump, and A. Licata. Whittaker functions and Demazure operators. *Preprint*, 2011. http://arxiv.org/abs/1111.4230
17. B. Brubaker, S. Friedberg, and J. Hoffstein. Cubic twists of GL(2) automorphic L-functions. *Invent. Math.*, 160(1):31–58, 2005.
18. Ben Brubaker and Daniel Bump. On Kubota's Dirichlet series. *J. Reine Angew. Math.*, 598:159–184, 2006.
19. Alina Bucur and Adrian Diaconu. Moments of quadratic Dirichlet L-functions over rational function fields. *Mosc. Math. J.*, 10(3):485–517, 661, 2010.
20. D. Bump, S. Friedberg, and J. Hoffstein. On some applications of automorphic forms to number theory. *Bull. Amer. Math. Soc. (N.S.)*, 33(2):157–175, 1996.
21. D. Bump, S. Friedberg, and J. Hoffstein. Sums of twisted GL(3) automorphic L-functions. In *Contributions to automorphic forms, geometry, and number theory*, pages 131–162. Johns Hopkins Univ. Press, Baltimore, MD, 2004.
22. D. Bump, P. McNamara and M. Nakasuji. Factorial Schur Functions and the Yang-Baxter Equation, Preprint, 2011. http://arxiv.org/abs/1108.3087

23. D. Bump and M. Nakasuji. Integration on p-adic groups and crystal bases. *Proc. Amer. Math. Soc.*, 138(5):1595–1605, 2010.
24. W. Casselman and J. Shalika. The unramified principal series of p-adic groups. II. The Whittaker function. *Compositio Math.*, 41(2):207–231, 1980.
25. G. Chinta. Mean values of biquadratic zeta functions. *Invent. Math.*, 160(1):145–163, 2005.
26. G. Chinta, S. Friedberg, and J. Hoffstein. Multiple Dirichlet series and automorphic forms. In *Multiple Dirichlet series, automorphic forms, and analytic number theory*, volume 75 of *Proc. Sympos. Pure Math.*, pages 3–41. Amer. Math. Soc., Providence, RI, 2006.
27. G. Chinta and P. Gunnells. Constructing Weyl group multiple Dirichlet series. *J. Amer. Math. Soc.*, 23:189–215, 2010.
28. G. Chinta and P. Gunnells. Littelmann patterns and Weyl group multiple Dirichlet series of type D, in this volume.
29. G. Chinta and O. Offen. A metaplectic Casselman-Shalika formula for GL_r, *Amer. J. Math.*, to appear.
30. G. Chinta and O. Offen. Orthgonal period of a GL(3,Z) Eisenstein series. In Offen Krötz and Sayan, editors, *Representation Theory, Complex Analysis and Integral Geometry I.* Birkhäuser, to appear.
31. A. Diaconu, P. Garrett, and D. Goldfeld. Natural boundaries and integral moments of L-functions, in this volume.
32. C. A. Diaconu, D. Goldfeld, and J. Hoffstein. Multiple Dirichlet series and moments of zeta and L-functions. *Compositio Math.*, 139(3):297–360, 2003.
33. V. G. Drinfeld. Quantum groups. In *Proceedings of the International Congress of Mathematicians, Vol. 1, 2 (Berkeley, Calif., 1986)*, pages 798–820, Providence, RI, 1987. Amer. Math. Soc.
34. C. Eckhardt and S. J. Patterson. On the Fourier coefficients of biquadratic theta series. *Proc. London Math. Soc. (3)*, 64(2):225–264, 1992.
35. B. Fisher and S. Friedberg. Sums of twisted GL(2) L-functions over function fields. *Duke Math. J.*, 117(3):543–570, 2003.
36. S. Friedberg, J. Hoffstein, and D. Lieman. Double Dirichlet series and the n-th order twists of Hecke L-series. *Math. Ann.*, 327(2):315–338, 2003.
37. D. Goldfeld and J. Hoffstein. Eisenstein series of 1/2-integral weight and the mean value of real Dirichlet L-series. *Invent. Math.*, 80(2):185–208, 1985.
38. A. M. Hamel and R. C. King. Bijective proofs of shifted tableau and alternating sign matrix identities. *J. Algebraic Combin.*, 25(4):417–458, 2007.
39. B. Ion. Nonsymmetric Macdonald polynomials and matrix coefficients for unramified principal series. *Adv. Math*, 201(1):36–62, 2006.
40. D. Ivanov. Symplectic ice, in this volume, 2011.
41. M. Kashiwara. On crystal bases. In *Representations of groups (Banff, AB, 1994)*, volume 16 of *CMS Conf. Proc.*, pages 155–197. Amer. Math. Soc., Providence, RI, 1995.
42. D. Kazhdan and G. Lusztig. Representations of Coxeter groups and Hecke algebras. *Invent. Math.*, 53(2):165–184, 1979.
43. D. Kazhdan and G. Lusztig. Proof of the Deligne-Langlands conjecture for Hecke algebras. *Invent. Math.*, 87(2):153–215, 1987.
44. V. Korepin, N. Boguliubov and A. Izergin *Quantum Inverse Scattering Method and Correlation Functions*, Cambridge University Press, 1993.
45. Y. Komori, K. Matsumoto, and H. Tsumura. On Witten multiple zeta-functions associated with semisimple lie algebras iii. in this volume.
46. T. Kubota. *On automorphic functions and the reciprocity law in a number field.* Lectures in Mathematics, Department of Mathematics, Kyoto University, No. 2. Kinokuniya Book-Store Co. Ltd., Tokyo, 1969.
47. P. Littelmann. Cones, crystals, and patterns. *Transform. Groups*, 3(2):145–179, 1998.
48. G. Lusztig. Equivariant K-theory and representations of Hecke algebras. *Proc. Amer. Math. Soc.*, 94(2):337–342, 1985.

49. P. J. McNamara. The metaplectic Casselman-Shalika formula. *Preprint*, 2011. http://arxiv.org/abs/1103.4653.
50. P. J. McNamara. Metaplectic Whittaker functions and crystal bases. *Duke Math. J.*, 156(1):29–31, 2011.
51. J. Neukirch. *Class field theory*, volume 280 of *Grundlehren der Mathematischen Wissenschaften [Fundamental Principles of Mathematical Sciences]*. Springer-Verlag, Berlin, 1986.
52. S. J. Patterson Excerpt from an unwritten letter, in this volume.
53. A. Reznikov. A double Dirichlet series for Hecke L-functions of a CM field. *Preprint*, 2010.
54. T. Tokuyama. A generating function of strict Gelfand patterns and some formulas on characters of general linear groups. *J. Math. Soc. Japan*, 40(4):671–685, 1988.

Chapter 2
A Crystal Definition for Symplectic Multiple Dirichlet Series

Jennifer Beineke, Ben Brubaker, and Sharon Frechette

Abstract We present a definition for Weyl group multiple Dirichlet series (MDS) of Cartan type C, where the coefficients of the series are given by statistics on crystal graphs for certain highest-weight representations of $Sp(2r, \mathbb{C})$. In earlier work (Beineke et al., *Pacific J. Math.*, 2011), we presented a definition based on Gelfand–Tsetlin patterns, and the equivalence of the two definitions is explained here. Finally, we demonstrate how to prove analytic continuation and functional equations for any multiple Dirichlet series with fixed data by reduction to rank one information. This method is amenable to MDS of all types.

Keywords Weyl group multiple Dirichlet series • Crystal graph • Gelfand-Tsetlin pattern • Littelmann polytope • Whittaker function

2.1 Introduction

This paper presents a definition for a family of Weyl group multiple Dirichlet series (henceforth "MDS") of Cartan type C using a combinatorial model for crystal bases due to Berenstein–Zelevinsky [2] and Littelmann [16]. Recall that a Weyl group

J. Beineke
Department of Mathematics, Western New England University,
1215 Wilbraham Road, Springfield, MA 01119, USA,
e-mail: jbeineke@wne.edu

B. Brubaker (✉)
Department of Mathematics, MIT, Cambridge, MA 02139-4307, USA
e-mail: brubaker@math.mit.edu

S. Frechette
Department of Mathematics and Computer Science, College of the Holy Cross,
1 College Street, Worcester, MA 01610, USA
e-mail: sfrechet@mathcs.holycross.edu

D. Bump et al. (eds.), *Multiple Dirichlet Series, L-functions and Automorphic Forms*,
Progress in Mathematics 300, DOI 10.1007/978-0-8176-8334-4_2,

MDS is a Dirichlet series in several complex variables which (at least conjecturally) possesses analytic continuation to a meromorphic function and satisfies functional equations whose action on the complex space is isomorphic to the given Weyl group. In [1], we presented a definition for such a series in terms of a basis for highest-weight representations of $Sp(2r, \mathbb{C})$—type C Gelfand–Tsetlin patterns—and proved that the series satisfied the conjectured analytic properties in a number of special cases. Here we recast that definition in the language of crystal bases and find that the resulting MDS, whose form appears as an unmotivated miracle in the language of Gelfand–Tsetlin patterns, is more naturally defined in this new language.

The family of MDS is indexed by a positive integer r, an odd positive integer n, and an r-tuple of nonzero algebraic integers $\mathbf{m} = (m_1, \ldots, m_r)$ from a ring described precisely in Sect. 2.3. In [1], we further conjectured (and proved for $n = 1$) that this series matches the $(m_1, \ldots, m_r)$th Whittaker coefficient of a minimal parabolic, metaplectic Eisenstein series on an n-fold cover of $SO(2r + 1)$ over a suitable choice of global field. It is known that the definition of MDS we present fails to have the conjectured analytic properties if n is even, reflecting the essential interplay between n and root lengths in our definition (see, for example, Sect. 2.3.6). An alternate definition for Weyl group MDS attached to any root system (with completely general choice of r, n, and $\mathbf{m}$) was given by Chinta and Gunnells [10], who proved they possess analytic continuation and functional equations. Our definition of MDS for type C is conjecturally equal to theirs, and this has been verified in a large number of special cases.

The remainder of the paper is outlined as follows. In Sects. 2.2 and 2.3, we recall the model for the crystal basis from [16] and basic facts about Weyl group MDS for any root system Φ. In Sect. 2.4, we define the MDS coefficients in terms of crystal bases and explain their relation to our earlier definition in [1] (It is instructive to compare this definition with that of [9].). Section 2.5 demonstrates how, for any fixed choice of data determining a single MDS of type C, one may prove that the resulting series satisfies the conjectured functional equations. Similar techniques would be applicable to Weyl group MDS for any root system. As demonstrated in [6, 7], by Bochner's theorem in several complex variables, the existence of such functional equations then leads to a proof of the desired meromorphic continuation to the entire complex space $\mathbb{C}^r$. Thus, we provide a method for proving Part I of Conjecture 1 given in [1] for any fixed choice of initial data specifying a multiple Dirichlet series.

The proof of functional equations for a given Dirichlet series relies on reduction to the rank one case, whose analytic properties were demonstrated by Kubota [15]. Similar techniques were employed in [6–8], where the definition of the Dirichlet series was much simpler having assumed that the defining datum n is sufficiently large. Our methods indicate that the same should be true for arbitrary choice of n and arbitrary root system, leading to several potential applications. First, if one is interested in mean-value estimates for coefficients appearing in a given Weyl group MDS, this method provides the necessary analytic information to apply standard Tauberian techniques. More generally, one may take residues of the Weyl

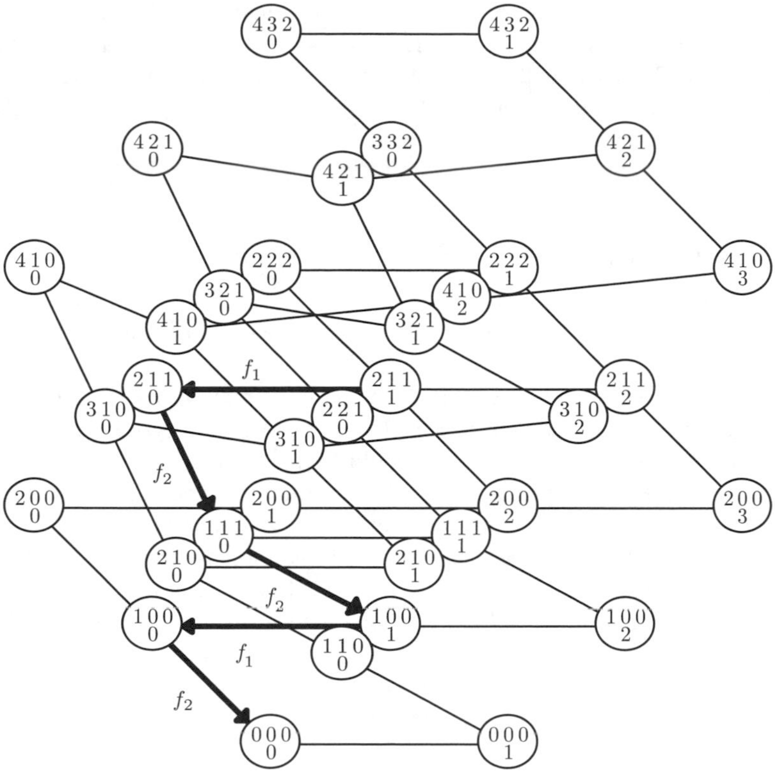

Fig. 2.1 The crystal graph for a highest-weight representation V_λ of $Sp(4)$ with $\lambda = \epsilon_1 + 2\epsilon_2$ (ϵ_i : fundamental dominant weights). The vertices of the graph have been labeled with their corresponding sequence (t_1, t_2, t_3, t_4) obtained by traversing the graph by maximal paths in the Kashiwara lowering operators in the respective order (f_1, f_2, f_1, f_2). This order is determined by the decomposition of $w_0 = \sigma_1\sigma_2\sigma_1\sigma_2$. For each vertex, t_1 is centered in the bottom row, and the top row is (t_2, t_3, t_4) read left to right. The highlighted path demonstrates this for the vertex labeled $(1, 2, 1, 1)$. The picture has been drawn so that vertices that touch represent basis vectors in the same weight space

group MDS to obtain a further class of Dirichlet series with analytic continuation. In computing these residues, it is often useful to first express them in terms of rank one Kubota Dirichlet series given by our method. (For a similar example in type A, see [4].)

Since the initial writing of this paper, a new approach to multiple Dirichlet series using statistical lattice models has emerged (see in particular [5]). In type C, Ivanov [13] has found a two-dimensional lattice model giving rise to the definition of prime-powered coefficients of the MDS presented in this paper (or equivalently in [1]). This leads to an alternate proof of functional equations for the series in the special case $n = 1$ based on the Yang–Baxter equation.

We thank Daniel Bump for helpful conversations and for assistance in preparing Fig. 2.1, which was made using SAGE [19]. We also thank Gautam Chinta, Sol Friedberg, and Paul Gunnells for their shared insights. This work was partially supported by NSF grants DMS-0502730 (Beineke), DMS-0702438, and DMS-0844185 (Brubaker).

2.2 Littelmann's Polytope Basis for Crystals

Given a semisimple algebraic group G of rank r and a simple G-module V_μ of highest weight μ, we may associate a crystal graph X_μ to V_μ. That is, there exists a corresponding simple module for the quantum group $U_q(\mathrm{Lie}(G))$ having the associated crystal graph structure. Roughly speaking, the crystal graph encodes data from the representation V_μ, and should be regarded as a kind of "enhanced character" for the representation; for an introduction to crystal graphs, see [14] or [11]. For now, we merely recall that the vertices of X_μ are in bijection with a basis of weight vectors for the highest-weight representation, and the r "colored" edges of X_μ correspond to simple roots $\alpha_1, \ldots, \alpha_r$ of G. Two vertices b_1, b_2 are connected by a (directed) edge from b_1 to b_2 of color i if the Kashiwara lowering operator f_{α_i} takes b_1 to b_2. If the vertex b has no outgoing edge of color i, we set $f_{\alpha_i}(b) = 0$.

Littelmann gives a combinatorial model for the crystal graph as follows. Fix a reduced decomposition of the long element w_0 of the Weyl group of G into simple reflections σ_i:

$$w_0 = \sigma_{i_1}\sigma_{i_2}\cdots\sigma_{i_N}.$$

Given an element b (i.e., vertex) of the crystal X_μ, let t_1 be the maximal integer such that $b_1 := f_{\alpha_{i_1}}^{t_1}(b) \neq 0$. Similarly, let t_2 be the maximal integer such that $b_2 := f_{\alpha_{i_2}}^{t_2}(b_1) \neq 0$. Continuing in this fashion in order of the simple reflections appearing in w_0, we obtain a string of nonnegative integers $(t_1(b), \ldots, t_N(b))$. We often suppress this dependence on b, and simply write $(t_1, \ldots, t_N)$. Note that by well-known properties of the crystal graph, we are guaranteed that for any reduced decomposition of the long element and an arbitrary element b of the crystal, the path $f_{\alpha_{i_N}}^{t_N} \cdots f_{\alpha_{i_1}}^{t_1}(b)$ through the crystal always terminates at b_{low}, corresponding to the lowest weight vector of the crystal graph X_μ.

Littelmann proves that, for any fixed reduced decomposition, the set of all sequences $(t_1, \ldots, t_N)$ as we vary over all vertices of all highest-weight crystals V_μ associated to G fills out the integer lattice points of a cone in $\mathbb{R}^N$. The inequalities describing the boundary of this cone depend on the choice of reduced decomposition. For a particular "nice" subset of the set of all reduced decompositions, Littelmann shows that the cone is defined by a rather simple set of inequalities. (A precise definition of "nice" and numerous examples may be found in [16], and we will only make use of one such example.) For any fixed highest weight μ, the set of all sequences $(t_1, \ldots, t_N)$ for the crystal X_μ are the integer lattice points of a polytope in $\mathbb{R}^N$. The boundary of the polytope consists of the hyperplanes defined by the cone inequalities independent of μ, together with additional hyperplanes dictated by the choice of μ.

We now describe this geometry in the special case of $Sp_{2r}(\mathbb{C})$. We fix an enumeration of simple roots chosen so that α_1 is the unique long simple root and α_i and α_{i+1} correspond to adjacent nodes in the Dynkin diagram. This example is dealt with explicitly in Sect. 6 of [16] with the following "nice decomposition" of the long element of the associated Weyl group:

$$w_0 = \sigma_1(\sigma_2\sigma_1\sigma_2)(\ldots)(\sigma_{r-1}\ldots\sigma_1\ldots\sigma_{r-1})(\sigma_r\sigma_{r-1}\ldots\sigma_1\ldots\sigma_{r-1}\sigma_r). \tag{1}$$

With $N = r^2$, let $\mathbf{t} = (t_1, t_2, \ldots, t_N)$ be the string generated by traversing the crystal graph from a given weight b to the highest weight μ as described above. An example in rank 2 is given in Fig. 2.1.

In order to describe the cone inequalities for $Sp(2r, \mathbb{C})$ with w_0 as in (1), it is convenient to place the sequence $\mathbf{t} = (t_1, t_2, \ldots, t_N)$ in a triangular array. Following Littelmann [16], construct a triangle Δ consisting of r centered rows of boxes, with $2(r+1-i)-1$ entries in the row i, starting from the top. To any vector $\mathbf{t} \in \mathbb{R}^{r^2}$, let $\Delta(\mathbf{t})$ denote the filled triangle whose entries are the coordinates of $\mathbf{t}$, with the boxes filled from the bottom row to the top row, and from left to right. For notational ease, we reindex the entries of Δ using standard matrix notation; let $c_{i,j}$ denote the jth entry in the ith row of Δ, with $i \leq j \leq 2r - i$. Also, for convenience in the discussion below, we will write $\overline{c}_{i,j} := c_{i,2r-j}$ for $i \leq j \leq r$. Thus, when $r = 3$, we are considering triangles of the form

$c_{1,1}$	$c_{1,2}$	$c_{1,3}$	$\overline{c}_{1,2}$	$\overline{c}_{1,1}$
	$c_{2,2}$	$c_{2,3}$	$\overline{c}_{2,2}$	
		$c_{3,3}$		

so that, for example,

$\mathbf{t} = (2,2,1,1,5,3,2,2,1) \quad \mapsto \quad \Delta(\mathbf{t}) =$

5	3	2	2	1
	2	1	1	
		2		

.

Given this notation, we may now state the cone inequalities.

Proposition 1 (Littelmann, [16], Theorem 6.1). *For $G = Sp(2r, \mathbb{C})$ and w_0 as in (1), the corresponding cone of all sequences $\mathbf{t}$ is given by the set of all triangles $\Delta(\mathbf{t})$ with nonnegative entries $\{c_{i,j}\}$ that are weakly decreasing in rows.*

Recall that for any fundamental dominant weight μ, the set of all paths $\mathbf{t}$ ranging over all vertices of the crystal X_μ are the integer lattice points of a polytope $\mathcal{C}_\mu$ in $\mathbb{R}^N$. We now describe the remaining hyperplane inequalities which define this polytope.

Proposition 2 (Littelmann, [16] Corollary 6.1). *Let $G = Sp_{2r}(\mathbb{C})$ and let w_0 be as in (1). Write $\mu = \mu_1\epsilon_1 + \cdots + \mu_r\epsilon_r$, with ϵ_i the fundamental weights. Then $\mathcal{C}_\mu$ is the convex polytope of all triangles $\Delta(c_{i,j})$ such that the entries in the rows are nonnegative and weakly increasing and satisfy the following upper-bound inequalities for all $1 \le i \le r$ and $1 \le j \le r-1$:*

$$\overline{c}_{i,j} \le \mu_{r-j+1} + s(\overline{c}_{i,j-1}) - 2s(c_{i-1,j}) + s(c_{i-1,j+1}), \tag{2}$$

$$c_{i,j} \le \mu_{r-j+1} + s(\overline{c}_{i,j-1}) - 2s(\overline{c}_{i,j}) + s(c_{i,j+1}), \tag{3}$$

$$\text{and} \quad c_{i,r} \le \mu_1 + s(\overline{c}_{i,r-1}) - s(c_{i-1,r}). \tag{4}$$

In the above, we have set

$$s(\overline{c}_{i,j}) := \overline{c}_{i,j} + \sum_{k=1}^{i-1}(c_{k,j} + \overline{c}_{k,j}), \quad s(c_{i,j}) := \sum_{k=1}^{i}(c_{k,j} + \overline{c}_{k,j}), \quad s(c_{i,r}) := \sum_{k=1}^{i} 2c_{k,r}.$$

We will call these triangular arrays "Berenstein–Zelevinsky–Littelmann patterns" (or "BZL-patterns" for short). The set of patterns corresponding to all vertices in a highest-weight crystal X_μ will be referred to as $BZL(\mu)$.

2.3 Definition of the Multiple Dirichlet Series

In this section, we give the general shape of a Weyl group MDS, beginning with the rank one case. In particular, we reduce the determination of the higher-rank Dirichlet series to its prime-power-supported coefficients, which will be given in the next section as a generating function over BZL-patterns.

2.3.1 Algebraic Preliminaries

Given a fixed positive integer n, let F be a number field containing the $2n$th roots of unity, and let S be a finite set of places containing all ramified places over $\mathbb{Q}$, all archimedean places, and enough additional places so that the ring of S-integers $\mathcal{O}_S$ is a principal ideal domain. Recall that

$$\mathcal{O}_S = \{a \in F \mid a \in \mathcal{O}_v \ \forall v \notin S\},$$

and can be embedded diagonally in $F_S = \prod_{v \in S} F_v$. There exists a pairing

$$(\cdot,\cdot)_S \,:\, F_S^\times \times F_S^\times \longrightarrow \mu_n \text{ defined by } (a,b)_S = \prod_{v \in S} (a,b)_v,$$

where the $(a,b)_v$ are local Hilbert symbols associated to n and v.

To any $a \in \mathcal{O}_S$ and any ideal $\mathfrak{b} \in \mathcal{O}_S$, we may associate the nth-power residue symbol $\left(\frac{a}{\mathfrak{b}}\right)_n$ as follows. For prime ideals $\mathfrak{p}$, the expression $\left(\frac{a}{\mathfrak{p}}\right)_n$ is the unique nth root of unity satisfying the congruence

$$\left(\frac{a}{\mathfrak{p}}\right)_n \equiv a^{(N(\mathfrak{p})-1)/n} \pmod{\mathfrak{p}}.$$

We then extend the symbol to arbitrary ideals $\mathfrak{b}$ by multiplicativity, with the convention that the symbol is 0 whenever a and $\mathfrak{b}$ are not relatively prime. Since $\mathcal{O}_S$ is a principal ideal domain by assumption, we will write

$$\left(\frac{a}{b}\right)_n = \left(\frac{a}{\mathfrak{b}}\right)_n \qquad \text{for } \mathfrak{b} = b\mathcal{O}_S$$

and often drop the subscript n on the symbol when the power is clear from context.

Then if a, b are coprime integers in $\mathcal{O}_S$, we have the nth-power reciprocity law (cf. [17], Theorem 6.8.3)

$$\left(\frac{a}{b}\right) = (b,a)_S \left(\frac{b}{a}\right). \tag{5}$$

Lastly, for a positive integer t and $a, c \in \mathcal{O}_S$ with $c \neq 0$, we define the Gauss sum $g_t(a,c)$ as follows. First, choose a nontrivial additive character ψ of F_S trivial on the $\mathcal{O}_S$ integers (cf. [3] for details). Then the nth-power Gauss sum is given by

$$g_t(a,c) = \sum_{d \bmod c} \left(\frac{d}{c}\right)_n^t \psi\left(\frac{ad}{c}\right), \tag{6}$$

where we have suppressed the dependence on n in the notation on the left.

2.3.2 *Kubota's Rank One Dirichlet Series*

We now present Kubota's Dirichlet series arising from the Fourier coefficient of an Eisenstein series on an n-fold cover of $SL(2, F_S)$. It is the prototypical Weyl group MDS, and many of the general definitions of Sect. 2.3.4 can be understood as natural extensions of those in the rank one case. Moreover, we will make repeated use of the functional equation for the Kubota Dirichlet series when we demonstrate the functional equations for higher-rank MDS by reduction to rank one in Sect. 2.6.

A subgroup $\Omega \subset F_S^\times$ is said to be *isotropic* if $(a,b)_S = 1$ for all $a, b \in \Omega$. In particular, $\Omega = \mathcal{O}_S (F_S^\times)^n$ is isotropic (where $(F_S^\times)^n$ denotes the nth powers in $F_S^\times$). Let $\mathcal{M}_t(\Omega)$ be the space of functions $\Psi : F_S^\times \longrightarrow \mathbb{C}$ that satisfy the transformation property

$$\Psi(\epsilon c) = (c,\epsilon)_S^{-t}\Psi(c) \quad \text{for any } \epsilon \in \Omega, c \in F_S^\times. \tag{7}$$

For $\Psi \in \mathcal{M}_t(\Omega)$, consider the "Kubota Dirichlet series"

$$\mathcal{D}_t(s,\Psi,a) = \sum_{0 \neq c \in \mathcal{O}_s/\mathcal{O}_S^\times} \frac{g_t(a,c)\Psi(c)}{|c|^{2s}}. \tag{8}$$

Here, $|c|$ is the order of $\mathcal{O}_S/c\mathcal{O}_S$, $g_t(a,c)$ is as in (6), and the term $g_t(a,c)\Psi(c)|c|^{-2s}$ is independent of the choice of representative c, modulo S-units. Standard estimates for Gauss sums show that the series is convergent if $\Re(s) > \frac{3}{4}$. To state a precise functional equation, we require the Gamma factor

$$\mathbf{G}_n(s) = (2\pi)^{-2(n-1)s} n^{2ns} \prod_{j=1}^{n-2} \Gamma\left(2s - 1 + \frac{j}{n}\right). \tag{9}$$

In view of the multiplication formula for the Gamma function, we may also write

$$\mathbf{G}_n(s) = (2\pi)^{-(n-1)(2s-1)} \frac{\Gamma(n(2s-1))}{\Gamma(2s-1)}.$$

Let

$$\mathcal{D}_t^*(s,\Psi,a) = \mathbf{G}_m(s)^{[F:\mathbb{Q}]/2} \zeta_F(2ms - m + 1)\mathcal{D}_t(s,\Psi,a), \tag{10}$$

where $m = n/\gcd(n,t)$, $\frac{1}{2}[F:\mathbb{Q}]$ is the number of archimedean places of the totally complex field F, and ζ_F is the Dedekind zeta function of F.

If $v \in S_{\text{fin}}$, the non-archimedean places of S, let q_v denote the cardinality of the residue class field $\mathcal{O}_v/\mathcal{P}_v$, where $\mathcal{O}_v$ is the local ring in F_v and $\mathcal{P}_v$ is its prime ideal. By an *S-Dirichlet polynomial*, we mean a polynomial in q_v^{-s} as v runs through the finitely many places of S_{fin}. If $\Psi \in \mathcal{M}_t(\Omega)$ and $\eta \in F_S^\times$, denote

$$\widetilde{\Psi}_\eta(c) = (\eta, c)_S \, \Psi\left(c^{-1}\eta^{-1}\right). \tag{11}$$

Then, we have the following result (Theorem 1 in [8]), which follows from the work of Brubaker and Bump [3].

Theorem 1. *Let $\Psi \in \mathcal{M}_t(\Omega)$ and $a \in \mathcal{O}_S$. Let $m = n/\gcd(n,t)$. Then $\mathcal{D}_t^*(s,\Psi,a)$ has meromorphic continuation to all s, analytic except possibly at $s = \frac{1}{2} \pm \frac{1}{2m}$, where it may have simple poles. There exist S-Dirichlet polynomials $P_\eta^t(s)$ depending only on the image of η in $F_S^\times/(F_S^\times)^n$ such that*

$$\mathcal{D}_t^*(s,\Psi,a) = |a|^{1-2s} \sum_{\eta \in F_S^\times/(F_S^\times)^n} P_{a\eta}^t(s)\mathcal{D}_t^*(1-s,\widetilde{\Psi}_\eta,a). \tag{12}$$

This result, based on ideas of Kubota [15], relies on the theory of Eisenstein series. The case $t = 1$ is handled in [3]; the general case follows as discussed in the proof of Proposition 5.2 of [7]. Notably, the factor $|a|^{1-2s}$ is independent of the value of t.

2.3.3 Root Systems

Before proceeding to the definition of higher-rank MDS, which uses the language of the associated root system, we first fix notation and recall a few basic results.

Let Φ be a reduced root system contained in a real vector space V of dimension r. The dual vector space $V^\vee$ contains a root system $\Phi^\vee$ in bijection with Φ, where the bijection switches long and short roots. If we write the dual pairing

$$V \times V^\vee \longrightarrow \mathbb{R}: \quad (x,y) \mapsto B(x,y), \tag{13}$$

then $B(\alpha,\alpha^\vee) = 2$. Moreover, the simple reflection $\sigma_\alpha : V \to V$ corresponding to α is given by

$$\sigma_\alpha(x) = x - B(x,\alpha^\vee)\alpha.$$

In particular, σ_α preserves Φ. Similarly, we define $\sigma_{\alpha^\vee} : V^\vee \to V^\vee$ by $\sigma_{\alpha^\vee}(x) = x - B(\alpha,x)\alpha^\vee$ with $\sigma_{\alpha^\vee}(\Phi^\vee) = \Phi^\vee$.

Without loss of generality, we may take Φ to be irreducible (i.e., there do not exist orthogonal subspaces Φ_1, Φ_2 with $\Phi_1 \cup \Phi_2 = \Phi$). Then set $\langle\cdot,\cdot\rangle$ to be the Euclidean inner product on V and $||\alpha|| = \sqrt{\langle\alpha,\alpha\rangle}$ the Euclidean norm, normalized so that $2\langle\alpha,\beta\rangle$ and $||\alpha||^2$ are integral for all $\alpha,\beta \in \Phi$. With this notation, we may alternately write

$$\sigma_\alpha(\beta) = \beta - \frac{2\langle\beta,\alpha\rangle}{\langle\alpha,\alpha\rangle}\alpha \quad \text{for any } \alpha,\beta \in \Phi. \tag{14}$$

We partition Φ into positive roots Φ^+ and negative roots Φ^- and let $\Delta = \{\alpha_1,\ldots,\alpha_r\} \subset \Phi^+$ denote the subset of simple positive roots. Let ϵ_i for $i = 1,\ldots,r$ denote the fundamental dominant weights satisfying

$$\frac{2\langle\epsilon_i,\alpha_j\rangle}{\langle\alpha_j,\alpha_j\rangle} = \delta_{ij}, \quad \delta_{ij} : \text{Kronecker delta.} \tag{15}$$

Any dominant weight λ is expressible as a nonnegative linear combination of the ϵ_i, and a distinguished role in the theory is played by the Weyl vector ρ, defined by

$$\rho = \frac{1}{2}\sum_{\alpha\in\Phi^+}\alpha = \sum_{i=1}^r \epsilon_i. \tag{16}$$

2.3.4 The Form of Higher-Rank Multiple Dirichlet Series

We now begin explicitly defining the MDS, retaining our previous notation. By analogy with the rank 1 definition in (7), given an isotropic subgroup Ω, let $\mathcal{M}(\Omega^r)$ be the space of functions $\Psi : (F_S^\times)^r \longrightarrow \mathbb{C}$ that satisfy the transformation property

$$\Psi(\boldsymbol{\epsilon}\mathbf{c}) = \left(\prod_{i=1}^{r} (\epsilon_i, c_i)_S^{\|\alpha_i\|^2} \prod_{i<j} (\epsilon_i, c_j)_S^{2\langle \alpha_i, \alpha_j \rangle} \right) \Psi(\mathbf{c}) \tag{17}$$

for all $\boldsymbol{\epsilon} = (\epsilon_1, \ldots, \epsilon_r) \in \Omega^r$ and all $\mathbf{c} = (c_1, \ldots, c_r) \in (F_S^\times)^r$.

Given a reduced root system Φ of fixed rank r, an integer $n \geq 1$, $\mathbf{m} \in \mathcal{O}_S^r$, and $\Psi \in \mathcal{M}(\Omega^r)$, then we define a MDS as follows. It is a function of r complex variables $\mathbf{s} = (s_1, \ldots, s_r) \in \mathbb{C}^r$ of the form

$$Z_\Psi(\mathbf{s}; \mathbf{m}) := Z_\Psi(s_1, \ldots, s_r; m_1, \ldots, m_r) = \sum_{\mathbf{c}=(c_1,\ldots,c_r) \in (\mathcal{O}_S/\mathcal{O}_S^\times)^r} \frac{H^{(n)}(\mathbf{c}; \mathbf{m}) \Psi(\mathbf{c})}{|c_1|^{2s_1} \cdots |c_r|^{2s_r}}. \tag{18}$$

The function $H^{(n)}(\mathbf{c}; \mathbf{m})$ carries the main arithmetic content. In general, it is not a multiplicative function but rather a "twisted multiplicative" function. That is, for S-integer vectors $\mathbf{c}, \mathbf{c}' \in (\mathcal{O}_S/\mathcal{O}_S^\times)^r$ with $\gcd(c_1 \cdots c_r, c_1' \cdots c_r') = 1$,

$$H^{(n)}(c_1 c_1', \ldots, c_r c_r'; \mathbf{m}) = \mu(\mathbf{c}, \mathbf{c}') H^{(n)}(\mathbf{c}; \mathbf{m}) H^{(n)}(\mathbf{c}'; \mathbf{m}), \tag{19}$$

where $\mu(\mathbf{c}, \mathbf{c}')$ is an nth root of unity depending on $\mathbf{c}, \mathbf{c}'$. It is given precisely by

$$\mu(\mathbf{c}, \mathbf{c}') = \prod_{i=1}^{r} \left(\frac{c_i}{c_i'} \right)_n^{\|\alpha_i\|^2} \left(\frac{c_i'}{c_i} \right)_n^{\|\alpha_i\|^2} \prod_{i<j} \left(\frac{c_i}{c_j'} \right)_n^{2\langle \alpha_i, \alpha_j \rangle} \left(\frac{c_i'}{c_j} \right)_n^{2\langle \alpha_i, \alpha_j \rangle} \tag{20}$$

where $\left(\frac{\cdot}{\cdot}\right)_n$ is the nth-power residue symbol defined in Sect. 2.3.1. Note that in the special case $\Phi = A_1$, the twisted multiplicativity in (19) and (20) agrees with the usual identity for Gauss sums appearing in the numerator for the rank one case given in (8).

The transformation property of functions in $\mathcal{M}(\Omega^r)$ in (17) above is, in part, motivated by the identity

$$H^{(n)}(\boldsymbol{\epsilon}\mathbf{c}; \mathbf{m}) \Psi(\boldsymbol{\epsilon}\mathbf{c}) = H^{(n)}(\mathbf{c}; \mathbf{m}) \Psi(\mathbf{c}) \quad \text{for all } \boldsymbol{\epsilon} \in \mathcal{O}_S^r, \mathbf{c}, \mathbf{m} \in (F_S^\times)^r.$$

This can be verified using the nth-power reciprocity law from Sect. 2.3.1.

The function $H^{(n)}(\mathbf{c}; \mathbf{m})$ also exhibits a twisted multiplicativity in $\mathbf{m}$. Given any $\mathbf{m}, \mathbf{m}', \mathbf{c} \in \mathcal{O}_S^r$ with $\gcd(m_1' \cdots m_r', c_1 \cdots c_r) = 1$, we let

$$H^{(n)}(\mathbf{c}; m_1m_1', \ldots, m_rm_r') = \prod_{i=1}^{r} \left(\frac{m_i'}{c_i}\right)_n^{-||\alpha_i||^2} H^{(n)}(\mathbf{c}; \mathbf{m}). \tag{21}$$

The definitions in (19) and (21) imply that it is enough to specify the coefficients $H^{(n)}(p^{k_1}, \ldots, p^{k_r}; p^{l_1}, \cdots, p^{l_r})$ for any fixed prime p with $l_i = \mathrm{ord}_p(m_i)$ in order to completely determine $H^{(n)}(\mathbf{c}; \mathbf{m})$ for any pair of S-integer vectors $\mathbf{m}$ and $\mathbf{c}$. These prime-power coefficients are described in terms of data from highest-weight representations associated to $(l_1, \cdots, l_r)$ and will be given precisely in Sect. 2.4.

2.3.5 Weyl Group Actions

In order to precisely state a functional equation for the Weyl group MDS defined in (18), we require an action of the Weyl group W of Φ on the complex parameters $(s_1, \ldots, s_r)$. This arises from the linear action of W, realized as the group generated by the simple reflections $\sigma_{\alpha^\vee}$, on $V^\vee$. From the perspective of Dirichlet series, it is more natural to consider this action shifted by $\rho^\vee$, half the sum of the positive co-roots. Then, each $w \in W$ induces a transformation $V_{\mathbb{C}}^\vee = V^\vee \otimes \mathbb{C} \to V_{\mathbb{C}}^\vee$ (still denoted by w) if we require that

$$B\left(w\alpha, w(\mathbf{s}) - \frac{1}{2}\rho^\vee\right) = B\left(\alpha, \mathbf{s} - \frac{1}{2}\rho^\vee\right).$$

We introduce coordinates on $V_{\mathbb{C}}^\vee$ using simple roots $\Delta = \{\alpha_1, \ldots, \alpha_r\}$ as follows. Define an isomorphism $V_{\mathbb{C}}^\vee \to \mathbb{C}^r$ by

$$\mathbf{s} \mapsto (s_1, s_2, \ldots, s_r), \quad s_i = B(\alpha_i, \mathbf{s}). \tag{22}$$

This action allows us to identify $V_{\mathbb{C}}^\vee$ with $\mathbb{C}^r$, and so the complex variables s_i that appear in the definition of the MDS may be regarded as coordinates in either space. It is convenient to describe this action more explicitly in terms of the s_i, and it suffices to consider simple reflections which generate W. Using the action of the simple reflection σ_{α_i} on the root system Φ given in (14) in conjunction with (22) above gives the following:

Proposition 3. *The action of σ_{α_i} on $\mathbf{s} = (s_1, \ldots, s_r)$ defined implicitly in (22) is given by*

$$s_j \mapsto s_j - \frac{2\langle\alpha_j, \alpha_i\rangle}{\langle\alpha_i, \alpha_i\rangle}\left(s_i - \frac{1}{2}\right), \quad j = 1, \ldots, r. \tag{23}$$

In particular, $\sigma_{\alpha_i} : s_i \mapsto 1 - s_i$. For convenience, we will write σ_i for σ_{α_i}.

2.3.6 Normalizing Factors and Functional Equations

The multiple Dirichlet series must also be normalized using Gamma and zeta factors in order to state precise functional equations. Let

$$n(\alpha) = \frac{n}{\gcd(n, ||\alpha||^2)}, \quad \alpha \in \Phi^+.$$

For example, if $\Phi = C_r$ and we normalize short roots to have length 1, this implies that $n(\alpha) = n$ unless α is a long root and n even (in which case $n(\alpha) = n/2$). By analogy with the zeta factor appearing in (10), for any $\alpha \in \Phi^+$, let

$$\zeta_\alpha(\mathbf{s}) = \zeta\left(1 + 2n(\alpha)B\left(\alpha, \mathbf{s} - \frac{1}{2}\rho^\vee\right)\right),$$

where ζ is the Dedekind zeta function attached to the number field F. Further, for $\mathbf{G}_n(s)$ as in (9), we may define

$$\mathbf{G}_\alpha(\mathbf{s}) = \mathbf{G}_{n(\alpha)}\left(\frac{1}{2} + B\left(\alpha, \mathbf{s} - \frac{1}{2}\rho^\vee\right)\right). \tag{24}$$

Then for any $\mathbf{m} \in \mathcal{O}_S^r$, the normalized multiple Dirichlet series is given by

$$Z_\Psi^*(\mathbf{s}; \mathbf{m}) = \left[\prod_{\alpha \in \Phi^+} \mathbf{G}_\alpha(\mathbf{s})\zeta_\alpha(\mathbf{s})\right] Z_\Psi(\mathbf{s}, \mathbf{m}). \tag{25}$$

For any fixed n, $\mathbf{m}$, and root system Φ, we seek to exhibit a definition for $H^{(n)}(\mathbf{c}; \mathbf{m})$ (or equivalently, given twisted multiplicativity, a definition of H at prime-power coefficients) such that $Z_\Psi^*(\mathbf{s}; \mathbf{m})$ is initially convergent for $\Re(s_i)$ sufficiently large ($i = 1, \ldots, r$), has meromorphic continuation to all of $\mathbb{C}^r$ and satisfies functional equations of the form:

$$Z_\Psi^*(\mathbf{s}; \mathbf{m}) = |m_i|^{1-2s_i} Z_{\sigma_i \Psi}^*(\sigma_i \mathbf{s}; \mathbf{m}) \tag{26}$$

for all simple reflections $\sigma_i \in W$. Here, $\sigma_i \mathbf{s}$ is as in (23) and the function $\sigma_i \Psi$, which essentially keeps track of the rather complicated scattering matrix in this functional equation, is defined as in (37) of [8]. As noted in Sect. 7 of [8], given functional equations of this type together with several assumptions about the form of $H^{(n)}(\mathbf{c}; \mathbf{m})$, one can obtain analytic continuation to a meromorphic function of $\mathbb{C}^r$ with an explicit description of polar hyperplanes via Bochner's convexity principle.

2.4 Definition of the Prime-Power Coefficients

In this section, we use crystal graphs to give a definition for the p-power coefficients $H^{(n)}(p^{\mathbf{k}}; p^{\mathbf{l}})$ in a multiple Dirichlet series for the root system C_r with n odd. More precisely, the p-power coefficients will be given as weighted sums over the BZL-patterns defined in Sect. 2.2 that have weight corresponding to $\mathbf{k}$. Given a fixed r-tuple of integers $\mathbf{l} = (l_1, \ldots, l_r)$, let

$$\lambda = \sum_{i=1}^{r} l_i \epsilon_i, \tag{27}$$

where ϵ_i are fundamental dominant weights. The contributions to $H^{(n)}(p^{\mathbf{k}}; p^{\mathbf{l}})$ are parametrized by basis vectors of the highest-weight representation with highest weight $\lambda + \rho$, where ρ is the Weyl vector for C_r defined in (16). We use the set of BZL-patterns $BZL(\lambda + \rho)$ as our combinatorial model for these basis vectors.

The contributions to each $H^{(n)}(p^{\mathbf{k}}; p^{\mathbf{l}})$ come from a single weight space corresponding to $\mathbf{k} = (k_1, \ldots, k_r)$ in the highest-weight representation $\lambda + \rho$ corresponding to $\mathbf{l}$. Given a BZL-pattern $\Delta = \Delta(c_{i,j})$, define the vector

$$k(\Delta) = (k_1(\Delta), k_2(\Delta), \ldots, k_r(\Delta))$$

with

$$k_1(\Delta) = \sum_{i=1}^{r} c_{i,r}, \quad \text{and } k_j(\Delta) = \sum_{i=1}^{r+1-j} \left(c_{i,r+1-j} + \overline{c}_{i,r+1-j}\right), \quad \text{for } 1 < j \leq r. \tag{28}$$

We define

$$H^{(n)}(p^{\mathbf{k}}; p^{\mathbf{l}}) = H^{(n)}\left(p^{k_1}, \ldots, p^{k_r}; p^{l_1}, \ldots, p^{l_r}\right) = \sum_{\substack{\Delta \in BZL(\lambda+\rho) \\ k(\Delta)=(k_1,\ldots,k_r)}} G(\Delta), \tag{29}$$

where $G(\Delta)$ is a weighting function to be defined presently.

To this end, we will apply certain *decoration rules* to the BZL-patterns. These decorations will consist of *boxes* and *circles* around the individual entries of the pattern, applied according to the following rules:

1. The entry $c_{i,j}$ is circled if $c_{i,j} = c_{i,j+1}$. We understand the entries outside the triangular array to be zeroes, so the right-most entry in a row will be circled if it equals 0.
2. The entry $c_{i,j}$ is boxed if equality holds in the upper-bound inequality of Proposition 2 having $c_{i,j}$ as the lone term on the left-hand side.

We illustrate these rules in the following rank 3 example. Let $(l_1, l_2, l_3) = (0, 1, 1)$. Then there are nine upper-bound inequalities for the polytope $\mathcal{C}_{\lambda+\rho}$. We state them for the five top row elements $c_{1,j}$, leaving the rest to the reader:

$$\overline{c}_{1,1} \leq 2, \quad \overline{c}_{1,2} \leq 2 + \overline{c}_{1,1} \quad c_{1,3} \leq 1 + \overline{c}_{1,2}$$

$$c_{1,2} \leq 2 + \overline{c}_{1,1} - 2\overline{c}_{1,2} + 2c_{1,3}, \quad c_{1,1} \leq 2 - 2\overline{c}_{1,1} + c_{1,2} + \overline{c}_{1,2}.$$

We may now decorate any pattern occurring in $BZL(\lambda + \rho)$. For example, the following BZL-pattern (with decorations) occurs in this set:

$$\begin{array}{|c|c|c|c|c|} \hline \boxed{5} & \boxed{3} & \textcircled{2} & 2 & 1 \\ \hline \multicolumn{1}{c|}{} & 2 & \textcircled{1} & 1 & \multicolumn{1}{c}{} \\ \cline{2-4} \multicolumn{2}{c|}{} & 2 & \multicolumn{2}{c}{} \\ \cline{3-3} \end{array} \quad . \tag{30}$$

To each entry $c_{i,j}$ in a *decorated* $\Delta(\mathbf{c})$, we associate the complex-valued function

$$\gamma(c_{i,j}) = \begin{cases} q^{c_{i,j}} & \text{if } c_{i,j} \text{ is circled (but not boxed),} \\ g_1(p^{c_{i,j}-1}, p^{c_{i,j}}) & \text{if } c_{i,j} \text{ is boxed (but not circled), and } j \neq r, \\ g_2(p^{c_{i,j}-1}, p^{c_{i,j}}) & \text{if } c_{i,j} \text{ is boxed (but not circled), and } j = r, \\ \phi(p^{c_{i,j}}) & \text{if } c_{i,j} \text{ is neither boxed nor circled,} \\ 0 & \text{if } c_{i,j} \text{ is both boxed and circled,} \end{cases} \tag{31}$$

where $g_t(p^\alpha, p^\beta)$ is an nth-power Gauss sum as in (6), $\phi(p^a)$ denotes Euler's totient function for $\mathcal{O}_S/p^a\mathcal{O}_S$, and $q = |\mathcal{O}_S/p\mathcal{O}_S|$. Then at last, we may define the weighting function appearing in (29) by

$$G(\Delta) = \prod_{\substack{1 \leq i \leq r, \\ i \leq j \leq 2r-1}} \gamma(c_{i,j}). \tag{32}$$

For instance, in (30), we find that

$$\gamma(c_{1,1}) = g_1(p^4, p^5), \quad \gamma(c_{1,3}) = q^2, \quad \text{and } \gamma(\overline{c}_{1,2}) = \phi(p^2).$$

Computing the remaining $\gamma(c_{i,j})$'s for Δ in (30), we have

$$G(\Delta) = \{g_1(p^4, p^5)g_1(p^2, p^3)q^2\phi(p^2)\phi(p)\} \cdot \{\phi(p^2)q\phi(p)\} \cdot \phi(p^2).$$

Note that the definition implies that some BZL-patterns Δ will have $G(\Delta) = 0$. For instance, in rank 2 with $l_1 = 3$ and $l_2 = 4$, the decorated pattern

$$\Delta = \begin{array}{|c|c|c|}\hline \boxed{\textcircled{5}} & \boxed{5} & \textcircled{5} \\ \hline & 3 & \\ \cline{2-2}\end{array}$$

occurs, and has $G(\Delta) = 0$.

The definition of $G(\Delta)$ in (32) completes the definition of the prime-power coefficients $H^{(n)}(p^{\mathbf{k}}; p^{\mathbf{l}})$ in (29). According to the twisted multiplicativity given in Sect. 2.3.4, this completely determines the coefficients of the multiple Dirichlet series $Z_\Psi(\mathbf{s}; \mathbf{m})$ defined in (18).

2.5 Equality of the GT and BZL Descriptions

In Sect. 3 of [1], we gave an alternate definition for the p-power coefficients using Gelfand–Tsetlin patterns (henceforth "GT-patterns") as our combinatorial model for the highest-weight representation. In this section, we will demonstrate that the two definitions for p-power coefficients $H^{(n)}(p^{\mathbf{k}}; p^{\mathbf{l}})$ in terms of GT-patterns and BZL-patterns are indeed the same.

A GT-pattern P associated to $Sp(2r, \mathbb{C})$ has the form

$$P = \begin{array}{ccccccccc} a_{0,1} & & a_{0,2} & & \cdots & & a_{0,r} & & \\ & b_{1,1} & & b_{1,2} & \cdots & b_{1,r-1} & & b_{1,r} & \\ & & a_{1,2} & & \cdots & & a_{1,r} & & \\ & & & \ddots & & \ddots & & \vdots & \\ & & & & & & a_{r-1,r} & & \\ & & & & & & & b_{r,r} & \end{array} \tag{33}$$

where the $a_{i,j}, b_{i,j}$ are nonnegative integers and the rows of the pattern interleave. That is, for all $a_{i,j}, b_{i,j}$ in the pattern P above,

$$\min(a_{i-1,j}, a_{i,j}) \geq b_{i,j} \geq \max(a_{i-1,j+1}, a_{i,j+1})$$

and

$$\min(b_{i+1,j-1}, b_{i,j-1}) \geq a_{i,j} \geq \max(b_{i+1,j}, b_{i,j}).$$

A careful summary of patterns of this type arising from branching rules for classical groups can be found in [18] building on the work of [20]:

Let $\lambda + \rho = (l_1 + 1)\epsilon_1 + \cdots + (l_r + 1)\epsilon_r$ and set

$$(L_r, \cdots, L_1) := (l_1 + l_2 + \cdots + l_r + r, \ldots, l_1 + l_2 + 2, l_1 + 1). \tag{34}$$

Then the set of all GT-patterns with top row $(a_{0,1}, \ldots, a_{0,r}) = (L_r, \ldots, L_1)$ forms a basis for the highest-weight representation with highest weight $\lambda + \rho$. We refer to this set of patterns as $GT(\lambda + \rho)$.

Proposition 4 (Littelmann, [16] Corollary 6.2). *The following equations induce a bijection of sets* φ *between* $GT(\lambda + \rho)$ *and* $BZL(\lambda + \rho)$*:*

$$\overline{c}_{i,j} = \sum_{m=1}^{j} (a_{i-1,m} - b_{i,m}), \quad \textit{for } i \le j \le r,$$

and

$$c_{i,j} = \overline{c}_{i,r} + \sum_{m=1}^{r-j} (a_{i,r+1-m} - b_{i,r+1-m}), \quad \textit{for } i < j \le r-1. \tag{35}$$

Remark 1. The map given in Corollary 2 of Sect. 6 in [16] is actually the inverse of the map defined by (35). (Note that there are several typographical errors in the presentation of the map in [16].) In that same section, Littelmann gives an example illustrating this correspondence, in the case of rank 3. This example is given below, with the corrected first entry in the second row:

$$\begin{array}{cccccc} 9 & 5 & 1 & & & \\ & 6 & 5 & 0 & & \\ & & 5 & 3 & & \\ & & & 5 & 2 & \\ & & & 3 & & \\ & & & & 1 & \end{array} \quad \longleftrightarrow \quad \begin{array}{|c|c|c|c|c|} \hline 7 & 7 & 4 & 3 & 3 \\ \hline \multicolumn{1}{c|}{} & 2 & 1 & 0 & \multicolumn{1}{c}{} \\ \cline{2-4} \multicolumn{1}{c}{} & \multicolumn{1}{c|}{} & 2 & \multicolumn{1}{c}{} & \multicolumn{1}{c}{} \\ \cline{3-3} \end{array}$$

Using the bijection of the previous proposition, we may now compare the two definitions for prime-power coefficients of the multiple Dirichlet series.

Proposition 5. *Given a fundamental dominant weight* λ*, let* G_{GT} *be the function defined on Gelfand–Tsetlin patterns* P *in* $GT(\lambda + \rho)$ *in Definition 3 of [1]. Let* $G(\Delta)$ *be the function defined on* BZL*-patterns in (32). Then, with* φ *the bijection of Proposition 4,*

$$G_{GT}((P)) = G(\varphi(P)).$$

Proof. It suffices to check that the cases defining the function on GT-patterns match those for BZL-patterns. Indeed, one must check that "maximal" and "minimal" entries in GT-patterns correspond to boxing and circling, respectively, in BZL-patterns. This is a simple consequence of the bijection in Proposition 4, and we leave the case analysis to the reader.

2.6 Functional Equations by Reduction to Rank One

In this section, we provide evidence toward global functional equations for the multiple Dirichlet series $Z_\Psi(\mathbf{s}; \mathbf{m})$ through a series of computations in a particular rank 2 example. We will demonstrate that these multiple Dirichlet series are, in some sense, built from combinations of rank 1 Kubota Dirichlet series and thus inherit their functional equations. Similar techniques to those presented here would apply for arbitrary rank.

Recall from (23) that, in rank 2, we expect functional equations corresponding to the simple reflections

$$\sigma_1 : (s_1, s_2) \mapsto (1-s_1, s_1+s_2-1/2) \quad \text{and} \quad \sigma_2 : (s_1, s_2) \mapsto (s_1+2s_2-1, 1-s_2), \tag{36}$$

which generate a group acting on $(s_1, s_2) \in \mathbb{C}^2$ isomorphic to the Weyl group of C_2, the dihedral group of order 8.

With notations as before, let $n = 3$, and $\mathbf{m} = (p^2, p^1)$ for some fixed $\mathcal{O}_S$ prime p. Then we will illustrate how our definition of the coefficients $H^{(3)}(\mathbf{c}; p^2, p)$ leads to a multiple Dirichlet series $Z_\Psi(\mathbf{s}; p^2, p)$ satisfying the functional equations

$$Z_\Psi\left(s_1, s_2; p^2, p\right) \to |p^2|^{1-2s_1} Z_{\sigma_1\Psi}\left(1 - s_1, s_1 + s_2 - 1/2; p^2, p\right) \tag{37}$$

and

$$Z_\Psi\left(s_1, s_2; p^2, p\right) \to |p|^{1-2s_2} Z_{\sigma_2\Psi}\left(s_1 + 2s_2 - 1, 1 - s_2; p^2, p\right) \tag{38}$$

corresponding to the above simple reflections according to (26).

Note that Z_Ψ is only initially defined for $\Re(s_i) > c$, $i = 1$ or 2, for some constant c. This is clear from the fact that the prime-power-supported coefficients $H(\ell^{k_1}, \ell^{k_2})$ at a given prime ℓ are nonzero for only finitely many pairs (k_1, k_2). However, (37) and (38) both require meromorphic continuation to larger region which is also a consequence of our expression in terms of rank 1 Dirichlet series. (See [6] for a careful explanation of this for type A_2 MDS.) Together with Bochner's tube domain theorem (cf. Theorem 2.5.10 of [12]), these functional equations imply the meromorphic continuation of the series Z_Ψ to all of $\mathbb{C}^2$, verifying a special case of Conjecture 1 in [1]. The emphasis in this section will be on the functional equations (37) and (38) which rely critically on the combinatorics of the type C root system, so we leave the remaining details of the meromorphic continuation to the reader. It follows from methods quite similar to those of [6–8].

Our strategy is quite simple. To demonstrate the functional equation corresponding to σ_1, write

$$Z_\Psi(s_1, s_2; p^2, p) = \sum_{c_2 \in \mathcal{O}_S/\mathcal{O}_S^\times} |c_2|^{-2s_2} \sum_{c_1 \in \mathcal{O}_S/\mathcal{O}_S^\times} \frac{H^{(3)}(c_1, c_2; p^2, p)\Psi(\mathbf{c})}{|c_1|^{2s_1}} \tag{39}$$

and attempt to realize the inner sum, for any fixed c_2, in terms of rank 1 Kubota Dirichlet series whose one-variable functional equations are all compatible with the global functional equation in (37). Similar methods apply for the other simple reflection. One difficulty with this approach is that our definitions for $H^{(n)}(\mathbf{c}; \mathbf{m})$ up to this point have been "local"—that is, we have only provided explicit definitions for the prime-power-supported coefficients. Of course, our requirement that the

$H^{(n)}(\mathbf{c};\mathbf{m})$ satisfy twisted multiplicativity then uniquely defines the coefficients for any r-tuple of integers $\mathbf{c}$, but there are many complications in attempting to patch together the prime-power-supported pieces to reconstruct a global series.

This strategy was precisely carried out in [7] and [8] for any root system Φ provided n is sufficiently large. Such values of n are referred to as *stable* (see [8] for the precise statement). Indeed, global objects were reconstructed from the prime-power-supported contributions by meticulously checking that all Hilbert symbols and nth-power residue symbols combine neatly into Kubota Dirichlet series with the required twisted multiplicativity. Our purpose here is not to get bogged down in these complications, but rather to show how global functional equations can be anticipated simply by considering the prime-power-supported coefficients. According to [8], the example with $\mathbf{m} = (p^2, p)$ will have a simple description only if $n \geq 7$; hence when $n = 3$, the results of [8] do not apply. Nevertheless, as we will explain, our method of reduction to the rank 1 case is still viable.

2.6.1 Analysis of $H^{(3)}(c_1, c_2; p^2, p)$ with Prime-Power Support

The nature of $H^{(3)}(c_1, c_2; p^2, p)$ with c_1, c_2 powers of a fixed prime depends critically on whether that prime is p, the fixed prime occurring in $\mathbf{m} = (p^2, p)$, or a distinct prime $\ell \neq p$. The prime-power-supported coefficients $H^{(3)}(\ell^{k_1}, \ell^{k_2}; p^2, p)$ at primes $\ell \neq p$ have identical support (k_1, k_2) for any such prime ℓ (as the support depends only on $\mathrm{ord}_\ell(m_1)$ and $\mathrm{ord}_\ell(m_2)$) and a uniform description as products of Gauss sums in terms of ℓ. The (k_1, k_2) coordinates of this support are depicted in Fig. 2.2—the result of the affine linear transformation of the weights in the corresponding highest-weight representation ρ. The vertex in the bottom left corner is placed at $(k_1, k_2) = (0, 0)$. At each of the vertices in the interior, the number shown indicates the number of BZL-patterns associated with that vertex, that is, the multiplicity in the associated weight space. These counts include both singular and nonsingular patterns, though singular patterns give no contribution to the multiple Dirichlet series for any n. Support on the boundary is indicated by black dots, each with a unique corresponding BZL-pattern.

For $n = 3$, each of the eight patterns Δ (four singular, four nonsingular) in the interior of the polygon of support has $G(\Delta) = 0$, so the only nonzero contributions come from the eight boundary vertices. Note that these are just the "stable" vertices, which have $G(\Delta)$ nonzero for all n.

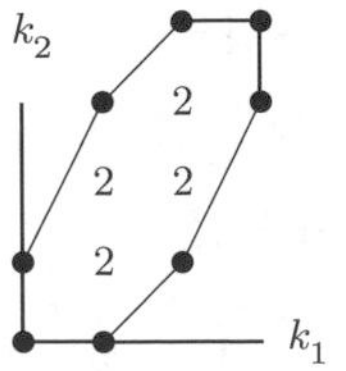

Fig. 2.2 Support (k_1, k_2) for $H^{(3)}(\ell^{k_1}, \ell^{k_2}; p^2, p)$ (with indicated multiplicities of contributing BZL-patterns Δ having $k(\Delta) = (k_1, k_2)$)

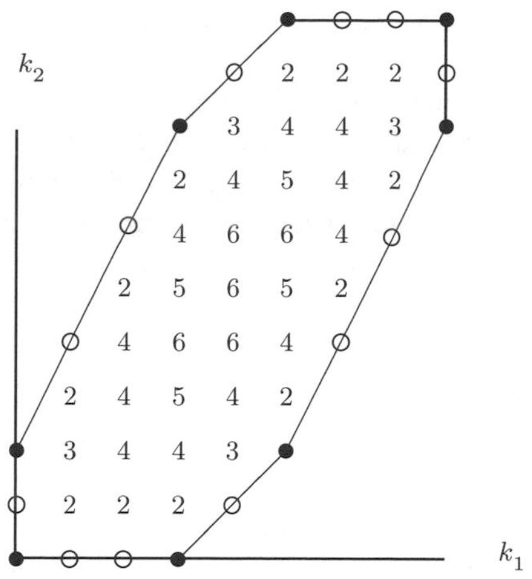

Fig. 2.3 Support (k_1, k_2) for $H^{(3)}(p^{k_1}, p^{k_2}; p^2, p)$ (with indicated multiplicities of contributing BZL-patterns Δ)

The coefficients $H^{(3)}(p^{k_1}, p^{k_2}; p^2, p^1)$ are much more interesting. Recall that these coefficients are parametrized by BZL-patterns with the coordinates of $\lambda + \rho$ given by $(L_2, L_1) = (5, 3)$, as in (34). The supporting vertices (k_1, k_2) for the p-part are shown below in Fig. 2.3. On the support's boundary, stable vertices are indicated by filled circles, and unstable vertices are indicated by open circles, all with multiplicity one.

Again, the choice of $n = 3$ will make $G(\Delta) = 0$ for many of the patterns Δ occurring at these support vertices. Roughly speaking, the nonzero support for any fixed n forms an $n \times n$ regular lattice beginning at the origin. However, this lattice becomes somewhat distorted by the boundary of the polygon, particularly the location of the stable vertices. In fact, our choice of $(L_2, L_1) = (5, 3)$ in this example is so small that this phenomenon is essentially obscured.

2.6.2 *Three Specific Examples*

Returning to the discussion of functional equations, we will first demonstrate a functional equation corresponding to the simple reflection σ_1 taking $s_1 \mapsto 1 - s_1$. Recall our strategy is to show that for any choice of c_2, we may write the inner sum in (39) in terms of Kubota Dirichlet series. For example, let $c_2 = p^8$. By twisted multiplicativity, we see that $H^{(3)}(c_1, p^8; p^2, p)$ will be 0 unless $\mathrm{ord}_\ell(c_1) \leq 1$ for all primes $\ell \neq p$ (as evident from Fig. 2.2, since we seek ℓ-power terms with support $k_2 = 0$). More interestingly, using Fig. 2.3, we see that p-power terms with $k_2 = 8$ must have $3 \leq \mathrm{ord}_p(c_1) \leq 8$. Let us examine the p-power coefficients more closely.

2.6.2.1 The Functional Equation σ_1 with $k_2 = 8$

As seen in Fig. 2.3, $H^{(3)}(p^{k_1}, p^{k_2}; p^2, p)$ with $k_2 = 8$ has support at 6 lattice points $(k_1, 8)$ with a total of 16 BZL-patterns. Having chosen $n = 3$ (so that all Gauss sums appearing are formed with a cubic residue symbol), one checks that only five of these 16 BZL-patterns have nonzero Gauss sum products associated to them. These are listed in the table below.

Δ	$k(\Delta)$	$G(\Delta)$	$G(\Delta)$ for $n = 3$
$\boxed{8}\ \boxed{3}\ \textcircled{0}$ $\textcircled{0}$	$(3, 8)$	$g_2(p^2, p^3)\, g_1(p^7, p^8)$	$-\lvert p\rvert^2 g_1(p^7, p^8)$
$6\ \boxed{5}\ \boxed{2}$ $\textcircled{0}$	$(5, 8)$	$g_1(p^1, p^2)\, g_2(p^4, p^5)\, g_1(p^7, p^6)$	$\lvert p\rvert^6 \phi(p^6)$
$\boxed{8}\ \boxed{3}\ \textcircled{0}$ 3	$(6, 8)$	$g_2(p^2, p^3)\, g_1(p^7, p^8)\, g_2(p^4, p^3)$	$-\lvert p\rvert^2 g_1(p^7, p^8)\phi(p^3)$
$6\ \boxed{5}\ \boxed{2}$ $\boxed{1}$	$(6, 8)$	$g_1(p^1, p^2)\, g_2(p^4, p^5)\, g_1(p^7, p^6)\, g_2(1, p)$	$\lvert p\rvert^6 \phi(p^6) g_2(1, p)$
$\boxed{8}\ \boxed{3}\ \textcircled{0}$ $\boxed{5}$	$(8, 8)$	$g_2(p^2, p^3)\, g_1(p^7, p^8)\, g_2(p^4, p^5)$	$-\lvert p\rvert^2 g_1(p^7, p^8) g_2(p^4, p^5)$

We have computed the final column in the table from the third column, using the following three elementary properties of nth-order Gauss sums at prime powers, which can be proved easily from the definition in (6):

1. If $a \geq b$, then $g_t(p^a, p^b) = \begin{cases} \phi(p^b) & n \mid tb, \\ 0 & n \nmid tb. \end{cases}$
2. For any integers a and t, $g_t(p^{a-1}, p^a) = \lvert p\rvert^{a-1} g_{at}(1, p)$.
3. For any integer t, $g_t(1, p)\, g_{n-t}(1, p) = \lvert p\rvert$.

For notational convenience, let the inner sum in (39) be denoted

$$F(s_1; c_2) = \sum_{c_1 \in \mathcal{O}_S/\mathcal{O}_S^\times} \frac{H^{(3)}(c_1, c_2; p^2, p)\Psi(\mathbf{c})}{|c_1|^{2s_1}}. \tag{40}$$

Fix $c_2 = p^8$ and let

$$F^{(p)}(s_1; p^8) = \sum_{k_1} \frac{H^{(3)}(p^{k_1}, p^8; p^2, p)\Psi(p^{k_1}, p^8)}{|p|^{2k_1 s_1}}. \tag{41}$$

From the table above, this sum is supported at $k_1 = 3, 5, 6$, and 8, so that $F^{(p)}(s_1; p^8)$ equals

$$\frac{-|p|^2 g_1(p^7, p^8)\Psi(p^3, p^8)}{|p|^{6s_1}} \left[1 + \frac{g_2(p^4, p^3)}{|p|^{6s_1}} \frac{\Psi(p^6, p^8)}{\Psi(p^3, p^8)} + \frac{g_2(p^4, p^5)}{|p|^{10s_1}} \frac{\Psi(p^8, p^8)}{\Psi(p^3, p^8)}\right]$$
$$+ \frac{|p|^6 \phi(p^6)\Psi(p^5, p^8)}{p^{10s_1}} \left[1 + \frac{g_2(1, p)}{|p|^{2s_1}} \cdot \frac{\Psi(p^6, p^8)}{\Psi(p^5, p^8)}\right] \tag{42}$$

Ignoring complications from the Ψ function, both bracketed sums may be expressed as the p-part of a Kubota Dirichlet series in s_1. Indeed, letting $\mathcal{D}_2^{(p)}$ denote the prime-power-supported coefficients of the Kubota Dirichlet series $\mathcal{D}_2$ in (8), then

$$\mathcal{D}_2^{(p)}(s_1, \Psi', p^4) = \left[1 + \frac{g_2(p^4, p^3)}{|p|^{6s_1}} \frac{\Psi(p^6, p^8)}{\Psi(p^3, p^8)} + \frac{g_2(p^4, p^5)}{|p|^{10s_1}} \frac{\Psi(p^8, p^8)}{\Psi(p^3, p^8)}\right]$$

for some appropriately defined $\Psi' \in \mathcal{M}_2(\Omega)$, as $\mathcal{D}_2^{(p)}(s_1, \Psi', p^4)$ contains $g_2(p^4, p^{k_1})$ in the numerator, which is nonzero only if $k_1 = 0, 3$, or 5 when $n = 3$. Similarly,

$$\mathcal{D}_2^{(p)}(s_1, \Psi'', 1) = \left[1 + \frac{g_2(1, p)}{|p|^{2s_1}} \cdot \frac{\Psi(p^6, p^8)}{\Psi(p^5, p^8)}\right]$$

for an appropriately defined $\Psi'' \in \mathcal{M}_2(\Omega)$. Thus, according to (42), we may express $F^{(p)}(s_1)$ as the sum of p-parts of Kubota Dirichlet series multiplied by Dirichlet monomials. The reader interested in checking all details regarding the Ψ function should refer to Sect. 5 of [7]; our notation for the one-variable Ψ' or Ψ'' in $M_2(\Omega)$ derived from $\Psi(c_1, c_2)$ is called Ψ^{c_1, c_2} in Lemma 5.3 of [7].

In order to reconstruct the global object $F(s_1; c_2)$ with $c_2 = p^8$, we now turn to the analysis at primes $\ell \neq p$. Since $\text{ord}_\ell(c_2) = 0$, then we can reconstruct $F(s_1; p^8)$ from the twisted multiplicativity in (19) and (21) together with knowledge of terms of the form $H^{(3)}(\ell^{k_1}, 1; p^2, p)$. Then define

$$F^{(\ell)}(s_1; 1) = \sum_{k_1} \frac{H^{(3)}(\ell^{k_1}, 1; p^2, p)\Psi(\ell^{k_1}, p^8)}{|\ell|^{2k_1 s_1}}$$

for all primes $\ell \neq p$. Using twisted multiplicativity in (21),

$$F^{(\ell)}(s_1;1) = \sum_{k_1} \left(\frac{p^2}{\ell^{k_1}}\right)_3^{-2} \left(\frac{p}{1}\right)_3^{-1} \frac{H^{(3)}(\ell^{k_1},1;1,1)\Psi(\ell^{k_1},p^8)}{|\ell|^{2k_1 s_1}}$$

$$= \Psi(1,p^8) + \left(\frac{p^2}{\ell}\right)_3^{-2} H^{(3)}(\ell^1,1;1,1)\Psi(\ell^1,p^8)|\ell|^{-2s_1}$$

$$= \Psi(1,p^8)\left[1 + \left(\frac{p^2}{\ell}\right)_3^{-2} \frac{g_2(1,\ell)}{|\ell|^{2s_1}} \frac{\Psi(\ell^1,p^8)}{\Psi(1,p^8)}\right].$$

To summarize, we have found that

$$F^{(p)}(s_1;p^8) = \frac{-|p|^2 g_1(p^7,p^8)\Psi(p^3,p^8)}{|p|^{6s_1}} \mathcal{D}_2^{(p)}(s_1,\Psi',p^4) + \frac{|p|^6 \phi(p^6)\Psi(p^5,p^8)}{|p|^{10s_1}} \mathcal{D}_2^{(p)}(s_1,\Psi'',1)$$

and

$$F^{(\ell)}(s_1;1) = \Psi(1,p^8)\left[1 + \left(\frac{p^2}{\ell}\right)_3^{-2} \frac{g_2(1,\ell)}{|\ell|^{2s_1}} \frac{\Psi(\ell^1,p^8)}{\Psi(1,p^8)}\right], \text{ for all primes } \ell \neq p.$$

Now using twisted multiplicativity, we can reconstruct $F(s_1;p^8)$. We claim that

$$F(s_1;p^8) = \frac{-|p|^2 g_1(p^7,p^8)\Psi(p^3,p^8)}{|p|^{6s_1}} \mathcal{D}_2(s_1,\Psi',p^4) + \frac{|p|^6 \phi(p^6)\Psi(p^5,p^8)}{|p|^{10s_1}} \mathcal{D}_2(s_1,\Psi'',1).$$

This may be directly verified up to Hilbert symbols (i.e., ignoring Hilbert symbols in the power reciprocity law in (5)) by using twisted multiplicativity to reconstruct $H(c_1,p^8;p^2,p)$ from $F^{(p)}(s_1;p^8)$ and $F^{(\ell)}(s_1;1)$. But to give a full accounting with Hilbert symbols, one needs to verify that the "leftover" Hilbert symbols from repeated applications of reciprocity are precisely those required for the definitions of Ψ' and Ψ'' (again referring to Lemma 5.3 of [7]).

We now return to our general strategy of demonstrating the functional equation σ_1 as in (36). The function $Z_\Psi(s_1,s_2;p^2,p)$ as in (39) with fixed $c_2 = p^8$ yields $F(s_1;p^8)$ as above. We must verify that this portion of $Z_\Psi(s_1,s_2;p^2,p)$ is consistent with the desired global functional equation

$$Z_\Psi\left(s_1,s_2;p^2,p\right) \to |p^2|^{1-2s_1} Z_{\sigma_1\Psi}\left(1-s_1,s_1+s_2-1/2;p^2,p\right)$$

presented at the outset of this section. By Theorem 1,

$$\mathcal{D}_2(s_1, \Psi', p^4) \to |p^4|^{1-2s_1} \mathcal{D}_2(1-s_1, \Psi', p^2)$$

and $|p|^{-6s_1-16s_2} \to |p|^{2-10s_1-16s_2}$ under σ_1. Similarly, $\mathcal{D}_2(s_1, \Psi'', 1) \to \mathcal{D}_2(1-s_1, \Psi'', p^2)$ and $|p|^{-10s_1-16s_2} \to |p|^{-2-6s_1-16s_2}$ under σ_1. Taken together, these calculations imply that

$$\frac{F(s_1; p^8)}{|p^8|^{2s_2}} \to |p^2|^{1-2s_1} \frac{F(1-s_1; p^8)}{|p^8|^{2(s_1+s_2-1/2)}},$$

which is consistent with the global functional equation for Z_Ψ above.

Throughout the above analysis, we chose to restrict to the case where $c_2 = p^8$ to limit the complexity of the calculation. However, identical methods could be used to determine the global object for arbitrary choice of c_2 depending on the order of p dividing c_2, and hence verify the global functional equation for σ_1 in full generality.

Remark 2. With respect to the s_1 functional equation, it turns out to be quite simple to figure out which BZL-patterns contribute to a particular Kubota Dirichlet series appearing in $F(s_1; p^{k_2})$. All such BZL-patterns have identical top rows but differ in the bottom row entry. This entry increases as we increase k_1, as can be verified in our earlier table with $k_2 = 8$. However, as we will see in the next section, functional equations in s_2 and the respective Kubota Dirichlet series used in asserting them obey no such simple pattern.

2.6.2.2 The Functional Equation σ_2 with $k_1 = 3$

We now repeat the methods of the previous section to demonstrate a functional equation under σ_2. As we will show, it is significantly more difficult to organize the local contributions into linear combinations of Kubota Dirichlet series in terms of s_2. Once this is accomplished, the analysis proceeds along the lines of the previous section, so we omit further details.

Let $c_1 = p^3$ be fixed. Mimicking our notation from the previous section, we now set

$$F(s_2; p^3) = \sum_{k_2} \frac{H^{(3)}(p^3, c_2; p^2, p)\Psi(p^3, c_2)}{|c_2|^{2s_2}}. \tag{43}$$

As in the previous section, the bulk of the difficulty lies in analyzing

$$F^{(p)}(s_2; p^3) = \sum_{k_2} \frac{H^{(3)}(p^3, p^{k_2}; p^2, p)\Psi(p^3, p^{k_2})}{|p|^{2k_2 s_2}}.$$

Again referring to Fig. 2.3, coefficients $H^{(3)}(p^{k_1}, p^{k_2}; p^2, p)$ with $k_1 = 3$ involve nine different vertices and a total of 30 BZL-patterns, only six of which give nonzero contributions in the case when $n = 3$. In the table below, we list only those BZL-patterns yielding nonzero Gauss sums. The final column has again been computed from the third column, using the elementary properties of nth-order Gauss sums mentioned in the previous subsection.

Δ	$(k_1, k_2) = k(\Delta)$	$G(\Delta)$	$G(\Delta)$ for $n = 3$
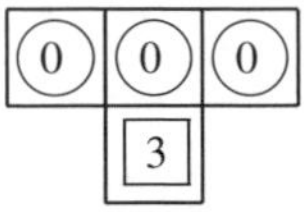	$(3, 0)$	$g_2(p^2, p^3)$	$-\lvert p\rvert^2$
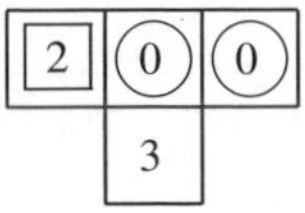	$(3, 2)$	$g_1(p, p^2)\, g_2(p^4, p^3)$	$\lvert p\rvert g_2(1, p)\phi(p^3)$
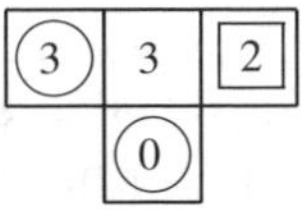	$(3, 5)$	$p^3\, g_1(p, p^2)\, g_1(p^4, p^3)$	$\lvert p\rvert^4 g_2(1, p)\phi(p^3)$
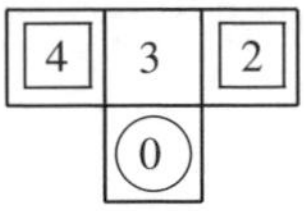	$(3, 6)$	$g_1(p, p^2)\, g_1(p^3, p^4)\, g_2(p^4, p^3)$	$\lvert p\rvert^5\phi(p^3)$
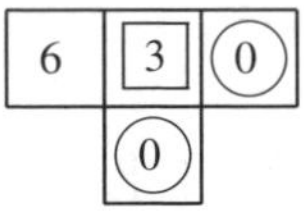	$(3, 6)$	$g_1(p^7, p^6)\, g_2(p^2, p^3)$	$-\lvert p\rvert^2\phi(p^6)$
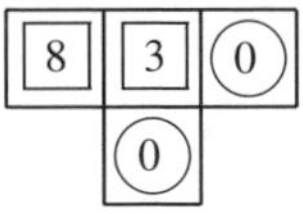	$(3, 8)$	$g_1(p^7, p^8)\, g_2(p^2, p^3)$	$-g_2(1, p)\lvert p\rvert^9$

According to the above table,

$$
\begin{aligned}
&F^{(p)}(s_2; p^3)\\
&\quad = -|p|^2\Psi(p^3, 1) + \frac{|p| g_2(1, p)\phi(p^3)\Psi(p^3, p^2)}{|p|^{4s_2}} + \frac{|p|^4 g_2(1, p)\phi(p^3)\Psi(p^3, p^5)}{|p|^{10s_2}}\\
&\qquad + \frac{|p|^5\phi(p^3)\Psi(p^3, p^6)}{|p|^{12s_2}} - \frac{|p|^2\phi(p^6)\Psi(p^3, p^6)}{|p|^{12s_2}} - \frac{g_2(1, p)|p|^9\Psi(p^3, p^8)}{|p|^{16s_2}}.
\end{aligned}
\tag{44}
$$

By adding and subtracting certain necessary terms at vertices $(3,3)$ and $(3,5)$, and using the fact that $g_1(1,p)g_2(1,p) = |p|$ when $n = 3$, we find that $F^{(p)}(s_2; p^3)$ equals

$$
\begin{aligned}
&-|p|^2\Psi(p^3,1)\left[1+\frac{\phi(p^3)}{|p|^{6s_2}}\frac{\Psi(p^3,p^3)}{\Psi(p^3,1)}+\frac{\phi(p^6)}{|p|^{12s_2}}\frac{\Psi(p^3,p^6)}{\Psi(p^3,1)}+\frac{g_2(1,p)|p|^7}{|p|^{16s_2}}\frac{\Psi(p^3,p^8)}{\Psi(p^3,1)}\right]\\
&+\frac{g_2(1,p)|p|\,\phi(p^3)\Psi(p^3,p^2)}{|p|^{4s_2}}\left[1+\frac{\phi(p^3)}{|p|^{6s_2}}\frac{\Psi(p^3,p^5)}{\Psi(p^3,p^2)}+\frac{g_1(1,p)p^3}{|p|^{8s_2}}\frac{\Psi(p^3,p^6)}{\Psi(p^3,p^2)}\right]\\
&+\frac{|p|^2\phi(p^3)\Psi(p^3,p^3)}{|p|^{6s_2}}\left[1+\frac{|p|g_2(1,p)}{|p|^{4s_2}}\frac{\Psi(p^3,p^5)}{\Psi(p^3,p^3)}\right].
\end{aligned}
\tag{45}
$$

After analyzing the terms in the bracketed sums, ignoring complications from the function Ψ as before, we have

$$
\begin{aligned}
&F^{(p)}(s_2; p^3)\\
&\quad= -|p|^2\Psi(p^3,1)\mathcal{D}_1^{(p)}(s_2,\Psi',p^7)+\frac{g_2(1,p)|p|\phi(p^3)\Psi(p^3,p^2)}{|p|^{4s_2}}\mathcal{D}_1^{(p)}(s_2,\Psi'',p^3)\\
&\quad+\frac{|p|^2\phi(p^3)\Psi(p^3,p^3)}{|p|^{6s_2}}\mathcal{D}_1^{(p)}(s_2,\Psi''',p).
\end{aligned}
\tag{46}
$$

Arguing similarly to the previous section, one can use these local contributions to reconstruct the global Dirichlet series via twisted multiplicativity. The resulting objects satisfy the global functional equation for σ_2 as in (36).

2.6.2.3 The Functional Equation σ_2 with $k_1 = 6$

As a final example, the set of all $H^{(3)}(p^{k_1}, p^{k_2}; p^2, p)$ with $k_1 = 6$ involves 7 support vertices and 18 BZL-patterns. In the case $n = 3$, however, only four of the BZL-patterns have nonzero Gauss sum products associated to them. These are listed in the table below.

Δ	$k(\Delta)$	$G(\Delta)$	$G(\Delta)$ for $n = 3$
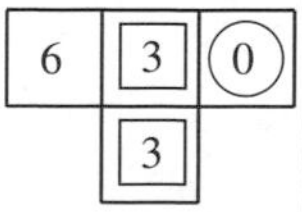	$(6,6)$	$g_1(p^7,p^6)\,g_2(p^2,p^3)\,g_2(p^2,p^3)$	$\lvert p\rvert^4\,\phi(p^6)$
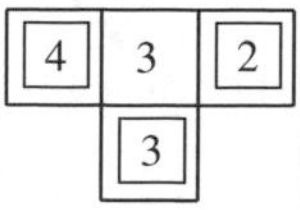	$(6,6)$	$g_1(p^1,p^2)\,g_1(p^3,p^4)\,g_2(p^2,p^3)\,g_2(p^4,p^3)$	$-\lvert p\rvert^7\,\phi(p^3)$

8 3 0 3	$(6, 8)$	$g_1(p^7, p^8)\, g_2(p^2, p^3)\, g_2(p^4, p^3)$	$-\lvert p\rvert^9 g_2(1, p)\phi(p^3)$
6 5 2 1	$(6, 8)$	$g_1(p^1, p^2)\, g_1(p^7, p^6)\, g_2(1, p)\, g_2(p^4, p^5)$	$\lvert p\rvert^{11} g_2(1, p)\phi(p^6)$

Upon first inspection, it is unclear how to package the Gauss sum products neatly into p-parts of Kubota Dirichlet series, as in the previous examples. However, the two nonzero terms at $(6, 6)$ cancel each other out when $n = 3$, as do the two nonzero terms at $(6, 8)$. This seems like a very complicated way to write 0, but we remind the reader that the definition in terms of Gauss sums is "uniform" in n, in the sense that only the order of the multiplicative character in the Gauss sum changes. For other n, the p-part $H^{(n)}(p^{k_1}, p^{k_2}; p^2, p)$ with $k_1 = 6$ will have the same 18 products of Gauss sums, four of which are as shown in the third column of the table above. However, the evaluations as in the last column of the table depend on the choice of n and result in a different organization as Kubota Dirichlet series.

References

1. J. Beineke, B. Brubaker, and S. Frechette, "Weyl Group Multiple Dirichlet Series of Type C," To appear in *Pacific J. Math.*
2. A. Berenstein and A. Zelevinsky, "Tensor product multiplicities, canonical bases and totally positive varieties," *Invent. Math.*, **143**, 77–128 (2001)
3. B. Brubaker and D. Bump, "On Kubota's Dirichlet series," *J. Reine Angew. Math.*, **598** 159–184 (2006).
4. B. Brubaker and D. Bump, "Residues of Weyl group multiple Dirichlet series associated to $\widetilde{\mathrm{GL}}_{n+1}$," Multiple Dirichlet Series, Automorphic Forms, and Analytic Number Theory (S. Friedberg, D. Bump, D. Goldfeld, and J. Hoffstein, ed.), *Proc. Symp. Pure Math.*, vol. 75, 115–134 (2006).
5. B. Brubaker, D. Bump, G. Chinta, S. Friedberg, and P. Gunnells, "Metaplectic Ice," in this volume.
6. B. Brubaker, D. Bump, G. Chinta, S. Friedberg, and J. Hoffstein, "Weyl group multiple Dirichlet series I," Multiple Dirichlet Series, Automorphic Forms, and Analytic Number Theory (S. Friedberg, D. Bump, D. Goldfeld, and J. Hoffstein, ed.), *Proc. Symp. Pure Math.*, vol. 75, 91–114 (2006).
7. B. Brubaker, D. Bump, and S. Friedberg, "Weyl group multiple Dirichlet series II: the stable case," *Invent. Math.*, **165** 325–355 (2006).
8. B. Brubaker, D. Bump, and S. Friedberg, "Twisted Weyl group multiple Dirichlet series: the stable case," *Eisenstein Series and Applications* (Gan, Kudla, Tschinkel eds.), Progress in Math vol. 258, 2008, 1–26.
9. B. Brubaker, D. Bump, and S. Friedberg. "Weyl Group Multiple Dirichlet Series: Type A Combinatorial Theory," Annals of Mathematics Studies, vol. 175. Princeton Univ. Press, Princeton, NJ, 2011.

10. G. Chinta and P. Gunnells, "Constructing Weyl group multiple Dirichlet series," *J. Amer. Math. Soc.*, **23** 189–215, (2010).
11. J. Hong and S.-J. Kang, *Introduction to quantum groups and crystal bases*, Graduate Studies in Mathematics, Vol. 42, American Mathematical Society, 2002.
12. L. Hörmander. *An introduction to complex analysis in several variables*, North-Holland Mathematical Library, Vol. 7, North-Holland Publishing Co., Amsterdam, third edition, 1990.
13. D. Ivanov, Symplectic Ice, in this volume.
14. M. Kashiwara, "On crystal bases," *Representations of groups (Banff, AB, 1994)*, CMS Conf. Proc., Vol. 16, 1995, 155–197.
15. T. Kubota, "On automorphic functions and the reciprocity law in a number field," Lectures in Mathematics, Department of Mathematics, Kyoto University, No. 2. Kinokuniya Book-Store Co. Ltd., Tokyo, 1969.
16. P. Littelmann, "Cones, crystals, and patterns," *Transf. Groups*, **3** 145–179 (1998).
17. J. Neukirch, *Algebraic Number Theory*, Grundlehren der mathematischen Wissenschaften, Vol. 322, Springer-Verlag, 1999.
18. R.A. Proctor, "Young tableaux, Gelfand patterns, and branching rules for classical groups," *J. Algebra*, **164** 299–360 (1994).
19. W. Stein et. al. SAGE Mathematical Software, Version 4.1. http://www.sagemath.org,2009.
20. D.P. Zhelobenko, "Classical Groups. Spectral analysis of finite-dimensional representations," *Russian Math. Surveys*, **17** 1–94 (1962).

Chapter 3
Metaplectic Ice

Ben Brubaker, Daniel Bump, Gautam Chinta, Solomon Friedberg, and Paul E. Gunnells

Abstract We study spherical Whittaker functions on a metaplectic cover of $\mathrm{GL}(r+1)$ over a nonarchimedean local field using lattice models from statistical mechanics. An explicit description of this Whittaker function was given in terms of Gelfand–Tsetlin patterns in (Brubaker et al., Ann. of Math. 173(2):1081–1120, 2011; McNamara, Duke Math. J. 156:29–31, 2011), and we translate this description into an expression of the values of the Whittaker function as partition functions of a six-vertex model. Properties of the Whittaker function may then be expressed in terms of the commutativity of row transfer matrices potentially amenable to proof using the Yang–Baxter equation. We give two examples of this: first, the equivalence of two different Gelfand–Tsetlin definitions, and second, the effect of the Weyl group action on the Langlands parameters. The second example is closely connected

B. Brubaker
Department of Mathematics, MIT, Cambridge, MA 02139-4307, USA
e-mail: brubaker@math.mit.edu

D. Bump (✉)
Department of Mathematics, Stanford University, Stanford, CA 94305-2125, USA
e-mail: bump@math.stanford.edu

G. Chinta
Department of Mathematics, The City College of New York, New York, NY 10031, USA
e-mail: chinta@sci.ccny.cuny.edu

S. Friedberg
Department of Mathematics, Boston College, Chestnut Hill, MA 02467-3806, USA
e-mail: solomon.friedberg@bc.edu

P.E. Gunnells
Department of Mathematics and Statistics, University of Massachusetts,
Amherst, MA 01003, USA
e-mail: gunnells@math.umass.edu

D. Bump et al. (eds.), *Multiple Dirichlet Series, L-functions and Automorphic Forms*,
Progress in Mathematics 300, DOI 10.1007/978-0-8176-8334-4_3,

with another construction of the metaplectic Whittaker function by averaging over a Weyl group action (Chinta and Gunnells, J. Amer. Math. Soc. 23:189–215, 2010; Chinta and Offen, Amer. J. Math., 2011).

Keywords Weyl group multiple Dirichlet series • Crystal graph • Solvable lattice model • Whittaker function • Metaplectic group • Yang–Baxter equation

3.1 Introduction

The study of spherical Whittaker functions of reductive groups over local fields is of fundamental importance in number theory and representation theory. Recently, in two separate series of papers, the authors and their collaborators have studied Whittaker functions on metaplectic covers of such groups. The goal of this paper is to introduce a new method for describing such p-adic metaplectic Whittaker functions: two-dimensional lattice models of statistical mechanics. In such a model, one defines the partition function to be a weighted sum over states of the model. We show that there exists a choice of weights for which the partition functions are metaplectic Whittaker functions. Baxter [2] developed important techniques for evaluating the partition functions of lattice models including the so-called "commutativity of transfer matrices" and the use of the Yang–Baxter equation. We discuss how these methods relate to our descriptions of Whittaker functions and to prior work.

Two different explicit formulas have been given in [10, 15] for the spherical Whittaker function on a metaplectic cover of $\mathrm{GL}(r + 1)$ over a non-archimedean local field. The first of these is expressed in terms of a Weyl group action described in [9], the second in terms of a function on Gelfand–Tsetlin patterns initially introduced in [8]. In fact, this latter representation belongs to a family of explicit formulas, one for each reduced expression of the long element of the Weyl group as a product of simple reflections. Two such reduced expressions in type A are particularly nice and lead to representations of the Whittaker function as sums over Gelfand–Tsetlin patterns. In keeping with earlier works, we refer to the two different descriptions as "Gamma" and "Delta" rules. The main result of [6] is a combinatorial proof that these two definitions are in fact equal. This equality allows one to prove the analytic properties of an associated global object (a multiple Dirichlet series) by applying Bochner's convexity principle.

In the following section, we demonstrate that the Gelfand–Tsetlin patterns we are concerned with are in bijection with admissible states of the six-vertex model having certain fixed boundary conditions. After recalling the description of the metaplectic Whittaker function as a function on Gelfand–Tsetlin patterns in Sect. 3.3, we use the bijection with the lattice model in Sect. 3.4 to express both the Gamma and Delta descriptions of the Whittaker function as partition functions for certain respective choices of Boltzmann weights.

In Sect. 3.5, we take the connection with statistical models further. We show that the necessary result for demonstrating the equivalence of the Gamma and Delta descriptions may be reformulated in terms of the commutativity of transfer matrices. Baxter [2] advocated the use of the Yang–Baxter equation for demonstrating this commutativity. In Sect. 3.6, we explain how this is carried out in the context of the six-vertex model, and we speculate about the possibility of such an equation in the metaplectic case.

Finally, we discuss the Weyl group action on metaplectic Whittaker functions, initially established by Kazhdan and Patterson [12], which plays a critical role in the explicit formulas of [10]. When the degree of the cover is 1, i.e., the linear case, the p-adic spherical Whittaker function is essentially a Schur polynomial by results going back to Shintani [18]. The Weyl group action is thus closely related to the standard permutation action on polynomials in $r+1$ variables. In [7], this Whittaker function (or equivalently, the Schur polynomial multiplied by a q-deformation of the Weyl denominator) is realized as a partition function on a six-vertex model, and its properties are studied via instances of the Yang–Baxter equation. On the other hand, as soon as the degree of the cover is greater than 1, the action looks rather different (cf. (33)–(35)). Nevertheless, we may ask whether these functional equations may also be phrased in terms of transfer matrices and a Yang–Baxter equation, and in this final section, we present evidence toward an affirmative answer.

This work was partially supported by the following grants: NSF grants DMS-0844185 (Brubaker), DMS-0652817 and DMS-1001079 (Bump), DMS-0847586 (Chinta), DMS-0652609 and DMS-1001326 (Friedberg), DMS-0801214 (Gunnells), and NSA grant H98230-10-1-0183 (Friedberg).

3.2 Six-Vertex Model and Gelfand–Tsetlin Patterns

In this section, we demonstrate a bijection between strict Gelfand–Tsetlin patterns and admissible states of the six-vertex model (or "square ice") on a finite square lattice with certain fixed boundary conditions. The boundary conditions on ice were known to Hamel and King, who presented bijections between ice and patterns related to the symplectic group in [11]. A treatment tailored to the aims of the present paper was given in [7], whose terminology we now recall.

A *Gelfand–Tsetlin pattern* of rank r is a triangular array of integers

$$\mathfrak{T} = \left\{ \begin{array}{ccccccc} a_{0,0} & & a_{0,1} & \cdots & a_{0,r-1} & & a_{0,r} \\ & a_{1,1} & & \cdots & & a_{1,r} & \\ & & \ddots & & \iddots & & \\ & & & a_{r,r} & & & \end{array} \right\} \tag{1}$$

in which the rows interleave: $a_{i-1,j-1} \geq a_{i,j} \geq a_{i-1,j}$. The set of all Gelfand–Tsetlin patterns with fixed top row is in bijection with basis vectors of a corresponding highest weight representation of $\mathrm{GL}(r+1,\mathbb{C})$. Indeed, any given top row $(a_{0,0}, a_{0,1}, \ldots, a_{0,r})$ is a partition which may be regarded as a dominant weight of the $\mathrm{GL}(r+1,\mathbb{C})$ weight lattice. Each successive row of a pattern then records a branching rule down to a highest weight representation on a subgroup of rank one less. We will focus mainly on the set of *strict* Gelfand–Tsetlin patterns, whose entries in horizontal rows are strictly decreasing. In terms of representation theory, these patterns result from branching through strictly dominant highest weights. Top rows of strict Gelfand–Tsetlin patterns are then indexed by strictly dominant weights $\lambda+\rho$, where λ is a dominant weight and ρ is the Weyl vector $(r, r-1, \ldots, 0)$.

Now we come to lattice models. The six-vertex model consists of labelings of edges in a square grid where each vertex has adjacent edges in one of six admissible configurations. This model is sometimes referred to as "square ice" where each vertex of the grid represents an oxygen atom and the six admissible ways of labeling adjacent edges correspond to the number of ways in which two of the four edges include a nearby hydrogen atom. If we represent adjacent hydrogen atoms by incoming arrows, and locations where there is no adjacent hydrogen atom by outgoing arrows, the six admissible states are as follows:

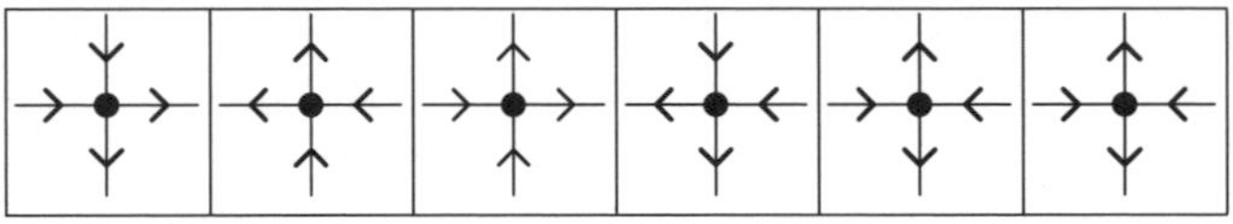

We will use a representation consisting of a lattice whose edges are labeled with signs $+$ or $-$ called *spins*. To relate this to the previous description, interpret a right-pointing or down-pointing arrow as a $+$ and a left-pointing or up-pointing arrow as $-$. We then find the following six configurations. (The index i in the table indicates the row to which the vertex belongs and will be used in later sections.)

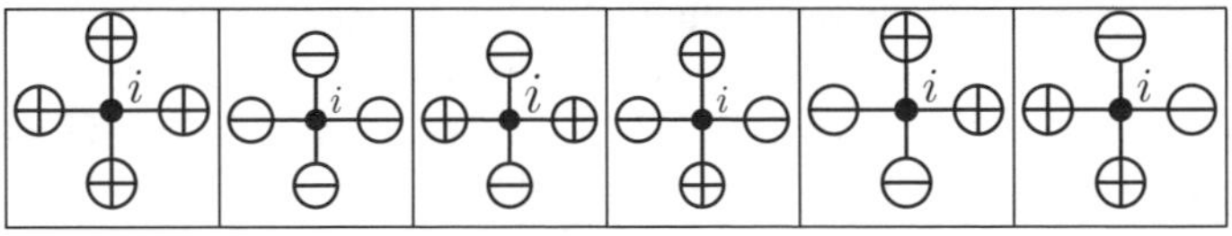

The rectangular lattices we consider will be finite, with boundary conditions chosen so that the admissible configurations are in bijection with strict Gelfand–Tsetlin patterns with fixed rank r and top row $\lambda+\rho$ as above. Here $\lambda = (\lambda_r, \ldots, \lambda_1, \lambda_0)$ with $\lambda_j \geq \lambda_{j-1}$ for all j, and we suppose that $\lambda_0 = 0$.

Boundary Conditions. *The rectangular grid is to have $\lambda_r + r + 1$ columns (labeled 0 through $\lambda_r + r$ increasing from right to left) and $r+1$ rows. Then with $\lambda+\rho = (\lambda_r + r, \lambda_{r-1} + r - 1, \ldots, 0)$, we place a $-$ spin at the top of each column whose label* *is one of the distinct parts of $\lambda + \rho$, i.e., at columns labeled $\lambda_j + j$ for $0 \leq j \leq r$. We place a $+$ spin at the top of each of the remaining columns. Furthermore, we place a $+$ spin at the bottom of every column and on the left-hand side of each row and a $-$ spin on the right-hand side of each row.*

For example, put $r = 2$ and take $\lambda = (3, 2, 0)$ so that $\lambda + \rho = (5, 3, 0)$. Then we have the following boundary conditions for the ice:

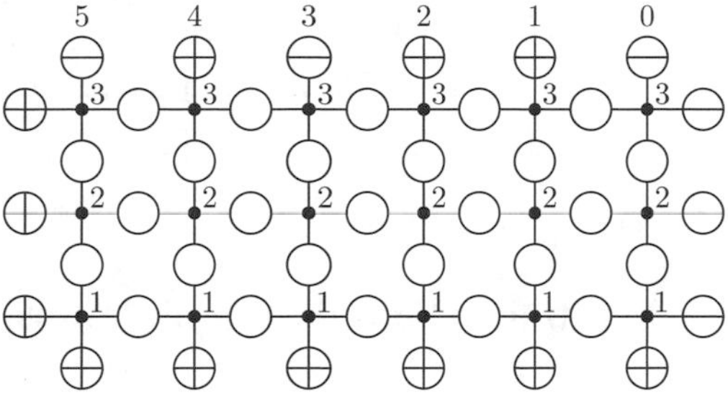

(2)

The column labels are written above each column, and row labels have been placed next to each vertex. These row labels will be used in Sect. 3.4, but need not concern us now. The edge spins have been placed inside circles located along the boundary. The remaining open circles indicate interior spins not determined by our boundary conditions, though any filling of the grid must use only the six admissible configurations in the above table. Such an admissible filling of the finite lattice having above boundary conditions will be referred to as a *state* of ice.

Proposition 1. *Given a fixed rank r and a dominant weight $\lambda = (\lambda_r, \ldots, \lambda_1, 0)$, there is a bijection between strict Gelfand–Tsetlin patterns with top row $\lambda + \rho$ and admissible states of ice having boundary conditions determined by λ as above.*

Proof. We begin with a strict Gelfand–Tsetlin pattern. Each row of the Gelfand–Tsetlin pattern will correspond to the set of spins located between numbered rows of ice, the so-called "vertical spins" since they lie on vertical edges of the grid, as follows. To each entry $a_{i,j}$ in the Gelfand–Tsetlin pattern, we assign a $-$ to the vertical spin between rows labeled $r + 2 - i$ and $r + 1 - i$ in the column labeled $a_{i,j}$. (Recall that we are using decreasing row labels from top to bottom as in the example (2).) The remaining vertical spins are assigned $+$.

It remains to assign horizontal spins, but these are already uniquely determined since the left and right edge horizontal spins have been assigned and each admissible vertex configuration has an even number of adjacent $+$ spins. We must only verify that the resulting configuration uses only the six admissible configurations (from the eight having an even number of $+$ signs) for the corresponding ice. This is easily implied by the interleaving condition on entries in the Gelfand–Tsetlin pattern, which is violated if one of the two inadmissible configurations appears. See Lemma 2 of [7] for more details. □

A simple example illustrates the bijection:

$$\left\{\begin{matrix} 5 & & 3 & & 0 \\ & 3 & & 1 & \\ & & 3 & & \end{matrix}\right\} \longleftrightarrow \tag{3}$$

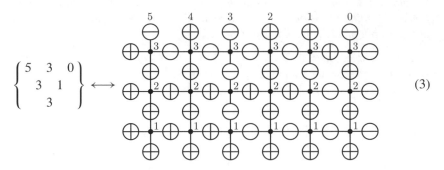

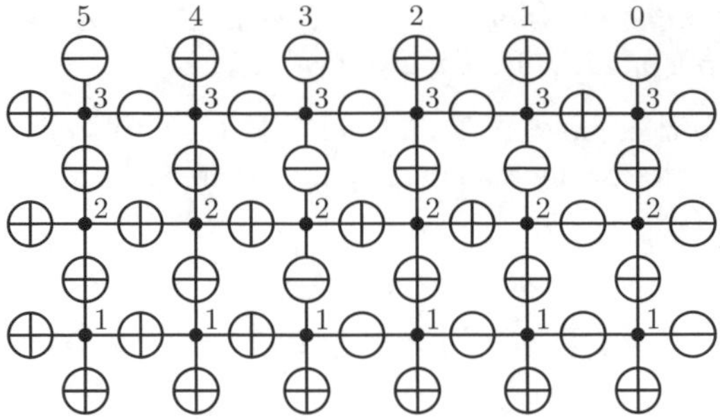

3.3 Metaplectic Whittaker Functions and Patterns

We now discuss the relation between the spherical metaplectic Whittaker function on the n-fold cover of $\mathrm{GL}(r+1)$ over a nonarchimedean local field and Gelfand–Tsetlin patterns. Such a relationship, described globally, was conjectured in [8] and was established in [5]. Though it is possible to pass from the global result to its local analogue, a direct local proof was given by McNamara [15], expressing a metaplectic spherical Whittaker function as a generating function supported on strict Gelfand–Tsetlin patterns. In this section, we recall two formulations of this explicit description, following [6]. In Sect. 3.4, we will explain their translations to square ice via the bijection of Proposition 1.

Let $\widetilde{G(F)}$ denote the n-fold metaplectic cover of $G(F) = \mathrm{GL}(r+1, F)$, where F denotes a nonarchimedean local field having ring of integers $\mathfrak{o}_F$ and residue field of order q. (There are several related such extensions, but we will use the one in [12] where their parameter $c = 0$.) We assume $2n$ divides $q-1$, which guarantees that F contains the group μ_{2n} of $2n$th roots of unity. The group $\widetilde{G(F)}$ is a central extension of $G(F)$ by μ_n:

$$1 \longrightarrow \mu_n \to \widetilde{G(F)} \xrightarrow{\pi} G(F) \longrightarrow 1.$$

We will identify $\mu_n \subset F$ with the group $\mu_n \subset \mathbb{C}$ of complex nth roots of unity by some fixed isomorphism. For convenience, we will sometimes denote $\widetilde{G(F)}$ as just $\widetilde{G}$, and if H is an algebraic subgroup of G, we may denote by $\widetilde{H}$ the preimage of $H(F)$ in $\widetilde{G}$.

For details of the construction of the metaplectic group and results about its representations, see [16] in this volume. Let $B(F)$ be the standard Borel subgroup of upper triangular matrices in $G(F)$ and let $T(F)$ be the diagonal maximal torus. Then $B(F) = T(F)U(F)$, where $U(F)$ is the unipotent radical of $B(F)$. The metaplectic cover splits over various subgroups of $G(F)$; for us, it is relevant that it splits over $U(F)$ and over $K := G(\mathfrak{o}_F)$, the standard maximal compact subgroup. By abuse of notation, we will sometimes denote by K the homomorphic image of K in $\widetilde{G}$ under this splitting.

Let $\mathbf{s} : G(F) \to \widetilde{G}$ be any map such that $\pi \circ \mathbf{s}$ is the identity map on $G(F)$. Then the map $\sigma : G(F) \times G(F) \to \mu_n$ such that

$$\mathbf{s}(g_1)\mathbf{s}(g_2) = \sigma(g_1, g_2)\mathbf{s}(g_1 g_2)$$

is a 2-cocycle defining a class in $H^2(G(F), \mu_n)$. A particular such cocycle was considered by Matsumoto [17], Kazhdan and Patterson [12], and Banks, Levy, and Sepanski [1]. By these references, such as [12] Sect. 0.1, the map $\mathbf{s}$ may be chosen so that the restriction of σ to $T(F)$ is given by the formula

$$\sigma\left(\begin{pmatrix} t_1 & & \\ & \ddots & \\ & & t_{r+1} \end{pmatrix}, \begin{pmatrix} u_1 & & \\ & \ddots & \\ & & u_{r+1} \end{pmatrix}\right) = \prod_{i<j} (t_i, u_j)_n. \tag{4}$$

The cocycle σ also has the property that $\sigma(u, g) = \sigma(g, u) = 1$ if $u \in U(F)$, and so the restriction of $\mathbf{s}$ to $U(F)$ is a homomorphism to $\widetilde{G}$.

We will call a representation π of $\widetilde{G}$ or any subgroup *genuine* if $\pi(\zeta g) = \zeta\pi(g)$ when $\zeta \in \mu_n$. Recall that $\widetilde{T}$ denotes the inverse image under π of the maximal torus $T(F)$.

Let A be a maximal abelian subgroup of $\widetilde{T}$ containing $\widetilde{T} \cap K$. Given a genuine character χ on $A/(\widetilde{T} \cap K)$, we may choose complex numbers $s = (s_1, \ldots, s_r)$ such that

$$\chi\begin{pmatrix} \varpi^{m_1} & & & \\ & \varpi^{m_2} & & \\ & & \ddots & \\ & & & \varpi^{m_{r+1}} \end{pmatrix} = \prod_{i+j \leq r+1} q^{-2m_i s_j}, \quad \varpi : \text{uniformizer in } \mathfrak{o}_F \tag{5}$$

whenever the matrix $\mathrm{diag}(\varpi^{m_1}, \ldots, \varpi^{m_{r+1}})$ is in A. The condition on the integers m_i is that they are in a sublattice of $n\mathbb{Z}^{r+1}$. See [16] for an explicit characterization of this sublattice and for further details on principal series representations of $\widetilde{G}$.

Let $i(\chi)$ be the resulting induced representation from A to $\widetilde{T}$. We extend $i(\chi)$ to the inverse image $\widetilde{B}$ of $B(F)$ in such a way that $\mathbf{s}(U(F))$ acts trivially. We then consider the representation of $\widetilde{G}$ obtained by normalized induction. We call the vector space of the resulting representation $I(\chi)$. It has a one-dimensional space of K-fixed, i.e., spherical, vectors. Let $\phi_K : \widetilde{G} \to i(\chi)$ be a nonzero element of $I(\chi)^K$.

Let w_0 denote the representative in K of the long element of the Weyl group. Then we may construct the spherical metaplectic Whittaker function via the integral

$$\int_{U(F)} \phi_K(w_0 \mathbf{s}(u) g) \psi(u) \mathrm{d}u, \tag{6}$$

where ψ is the character of $U(F)$ given by

$$\psi\left(\begin{pmatrix} 1 & x_{1,2} & \cdots & x_{1,n} \\ & 1 & x_{2,3} \cdots & x_{2,n} \\ & & \ddots & \vdots \\ & & & 1 \end{pmatrix}\right) = \psi_0\left(\sum_i x_{i,i+1}\right)$$

and $\psi_0 : F \to \mathbb{C}$ is an additive character that is trivial on $\mathfrak{o}_F$ but on no larger fractional ideal. Strictly speaking, the integral (6) as we have defined it is an $i(\chi)$-valued function and should be composed with a natural choice of linear functional on $i(\chi)$ to obtain a complex-valued function. The functional maps ϕ_K to $f : \widetilde{G} \to \mathbb{C}$ defined by

$$f(\zeta \mathbf{s}(u)\,\mathrm{diag}(\varpi^{m_1}, \ldots, \varpi^{m_{r+1}})k) = \zeta \prod_{i+j \leq r+1} q^{-(2m_i-1)s_j},$$

where $\zeta \in \mu_n, u \in U(F), k \in K$, and $m_i \in \mathbb{Z}$, according to the Iwasawa decomposition in $\widetilde{G}$. Composing the $i(\chi)$-valued function in (6) with this functional, we thus obtain

$$W(g) = \int_{U(F)} f(w_0 \mathbf{s}(u) g)\psi(u)\,\mathrm{d}u, \tag{7}$$

which we refer to, for brevity, as the *metaplectic Whittaker function*.

The transformation property $W(\mathbf{s}(u)gk) = \psi(u)W(g)$ for all $u \in U(F), k \in K$, implies that it suffices to determine W on the inverse image of the torus $T(F)$. Moreover, since W is genuine, it is sufficient to specify W on $\mathbf{s}(T(F))$. Given $\lambda = \sum_i \lambda_i \omega_i$, where ω_i are fundamental weights, let $t_\lambda = \mathrm{diag}(\varpi^{\lambda_1+\lambda_2+\cdots+\lambda_r}, \varpi^{\lambda_2+\cdots+\lambda_r}, \ldots, 1)$ and set $\mathbf{t}_\lambda = \mathbf{s}(t_\lambda)$. Due to our assumption that F contains the $2n$th roots of unity, $(\varpi, \varpi)_n = 1$, and by (4), it follows that $\mathbf{t}_{\lambda+\mu} = \mathbf{t}_\lambda \mathbf{t}_\mu$.

It is not hard to show that $W(\mathbf{t}_\lambda) = 0$ unless λ is a dominant weight.

Given any dominant weight λ, the metaplectic Whittaker function $W(\mathbf{t}_\lambda)$, may viewed as a Dirichlet series in the r complex variables $\boldsymbol{s} = (s_1, \ldots, s_r)$ used to define the character χ in (5). Then $W(\mathbf{t}_\lambda)$ is equal to the series

$$Z(\boldsymbol{s}; \lambda) = \sum_{\boldsymbol{k}=(k_1,\ldots,k_r)\in\mathbb{N}^r} H(\varpi^{k_1}, \ldots, \varpi^{k_r}; \lambda)\, q^{k_1(1-2s_1)+\cdots+k_r(1-2s_r)}, \tag{8}$$

where the complex-valued function $H(\varpi^{k_1}, \ldots, \varpi^{k_r}; \lambda)$ will be defined presently. The definition of H was first given in the context of metaplectic Eisenstein series in [8] and later [5], and a different definition is given in [9]. The series (8) is a p-part of a *Weyl group multiple Dirichlet series* as defined in those papers. It is a Dirichlet polynomial since, as we will see, only finitely many values of H are nonzero. The equality of $Z(\boldsymbol{s}; \lambda)$ and $W(\mathbf{t}_\lambda)$ is the main result of [15].

The positive integer n will continue to denote the degree of the metaplectic cover. We define the Gauss sum

$$g(a,b) = \int_{\mathfrak{o}_F^\times} (u,\varpi)_n^b \, \psi_0\left(\varpi^{a-b}u\right) \mathrm{d}u,$$

where $(\cdot,\cdot)_n$ denotes the nth power Hilbert symbol, and we normalize the Haar measure so that $\mu(\mathfrak{o}_F) = 1$. As a further shorthand, for any positive integer b, we set

$$g(b) = g(b-1,b), \qquad h(b) = g(a,b) \quad \text{for any } a \geq b. \tag{9}$$

Note that for a fixed base field F, these values depend only on b mod n. If n divides b, in particular if $n = 1$, we have

$$g(b) = -\frac{1}{q}, \quad h(b) = 1 - \frac{1}{q}. \tag{10}$$

We caution the reader that the q-powers that appear in the $g(a,b)$ are normalized differently than in the previous works [5, 6, 8]; these are the functions denoted $g^\flat$ and $h^\flat$ in [6]. The function h is a degenerate Gauss sum whose values may be made explicit, while (if $n \nmid b$) the function $g(b)$ is a nontrivial nth order Gauss sum.

Any strict Gelfand–Tsetlin pattern $\mathfrak{T}$ with entries indexed as in (1), we associate a weighting function γ to each entry $a_{i,j}$ with $i \geq 1$ as follows:

$$\gamma(a_{i,j}) = \begin{cases} g(b_{i,j}) \text{ if } a_{i,j} = a_{i-1,j-1}, \\ h(b_{i,j}) \text{ if } a_{i-1,j} \neq a_{i,j} \neq a_{i-1,j-1}, \\ 1 \qquad\quad \text{if } a_{i,j} = a_{i-1,j}, \end{cases} \quad \text{where } b_{i,j} = \sum_{l=j}^{r} (a_{i,l} - a_{i-1,l}). \tag{11}$$

Then we define

$$G^\Gamma(\mathfrak{T}) = \prod_{i=1}^{r} \prod_{j=i}^{r} \gamma(a_{i,j}). \tag{12}$$

If $\mathfrak{T}$ is a Gelfand–Tsetlin pattern that is not strict, we define $G^\Gamma(\mathfrak{T}) = 0$. We also define

$$\boldsymbol{k}^\Gamma(\mathfrak{T}) = (k_1^\Gamma(\mathfrak{T}), \ldots, k_r^\Gamma(\mathfrak{T})) \quad \text{where} \quad k_i^\Gamma(\mathfrak{T}) = \sum_{l=i}^{r} a_{i,l} - a_{0,l}. \tag{13}$$

In particular, note that both G^Γ and $\boldsymbol{k}^\Gamma$ are defined using differences of elements above and to the right of $a_{i,j}$. The superscript Γ may be regarded as indicator that these quantities are defined using such "right-hand" differences.

We present these definitions in this *ad hoc* fashion in order to give a brief and self-contained treatment, but in fact, they have very natural descriptions when reinterpreted as functions on a Kashiwara crystal graph. See [6] for an extensive discussion.

As an example, consider the Gelfand–Tsetlin pattern $\mathfrak{T}$ in (3). Then

$$(b_{1,1}, b_{1,2}, b_{2,2}) = (1,1,2) \text{ so that } G^\Gamma(\mathfrak{T}) = h(1)g(2) \text{ and } (k_1,k_2) = (1,3). \tag{14}$$

Theorem 1. (Brubaker, Bump, and Friedberg [5]; McNamara [15]) *Given a dominant weight λ and a fixed r-tuple of nonnegative integers $\boldsymbol{k} = (k_1, \ldots, k_r)$, the function $H(\varpi^{k_1}, \ldots, \varpi^{k_r}; \lambda)$ appearing in the $\mathfrak{p}$-adic Whittaker function $W(\mathbf{t}_\lambda)$ is given by*

$$H(\varpi^{\boldsymbol{k}}; \lambda) := H(\varpi^{k_1}, \ldots, \varpi^{k_r}; \lambda) = \sum_{\boldsymbol{k}^\Gamma(\mathfrak{T}) = \boldsymbol{k}} G^\Gamma(\mathfrak{T}),$$

where the sum is over all Gelfand–Tsetlin patterns with top row corresponding to $\lambda + \rho$ satisfying the subscripted condition.

There is a second explicit description of $H(\varpi^{\boldsymbol{k}}; \lambda)$ in terms of "left-hand" differences using functions G^Δ and $\boldsymbol{k}^\Delta$ that are analogous to those defined in (12) and (13), respectively. Assuming that $\mathfrak{T}$ is strict, set

$$\delta(a_{i,j}) = \begin{cases} g(c_{i,j}) & \text{if } a_{i,j} = a_{i-1,j}, \\ h(c_{i,j}) & \text{if } a_{i-1,j} \neq a_{i,j} \neq a_{i-1,j-1}, \\ 1 & \text{if } a_{i,j} = a_{i-1,j-1}, \end{cases} \quad \text{where } c_{i,j} = \sum_{l=1}^{j} (a_{i-1,l-1} - a_{i,l}) \tag{15}$$

and define

$$G^\Delta(\mathfrak{T}) = \prod_{i=1}^{r} \prod_{j=i}^{r} \delta(a_{i,j}). \tag{16}$$

If $\mathfrak{T}$ is not strict, define $G^\Delta(\mathfrak{T}) = 0$. We also set

$$\boldsymbol{k}^\Delta(\mathfrak{T}) = (k_1^\Delta(\mathfrak{T}), \ldots, k_r^\Delta(\mathfrak{T})), \quad \text{where} \quad k_i^\Delta(\mathfrak{T}) = \sum_{l=1}^{i} a_{0,l-1} - a_{r+1-i,r+1-l}. \tag{17}$$

The main theorem of [6] is as follows:

Theorem 2. (Statement A of Brubaker, Bump, and Friedberg [6]) *Given a dominant weight λ and a fixed r-tuple of nonnegative integers $\boldsymbol{k} = (k_1, \ldots, k_r)$,*

$$\sum_{\boldsymbol{k}^\Gamma(\mathfrak{T}) = \boldsymbol{k}} G^\Gamma(\mathfrak{T}) = \sum_{\boldsymbol{k}^\Delta(\mathfrak{T}) = \boldsymbol{k}} G^\Delta(\mathfrak{T}), \tag{18}$$

where the sums each run over all Gelfand–Tsetlin patterns with top row corresponding to $\lambda + \rho$ satisfying the subscripted condition.

As an immediate corollary, we have a second description of the p-adic Whittaker function in terms of G^Δ and $\boldsymbol{k}^\Delta$. We refer to these two recipes on the left- and right-hand sides of (18) as the Γ- and Δ-rules, respectively.

In fact, there are many other descriptions for the Whittaker function, though these are generally much more difficult to write down as explicitly. Indeed, as explained in [3, 14], there exist bases for highest weight representations corresponding to any reduced expression for the long element w_0 of the Weyl group of $\mathrm{GL}(r+1)$—S_r, the symmetric group on r letters—as a product of simple reflections σ_i. These make use of the Kashiwara crystal graph and are commonly called *string bases*. Using these bases, one may make a correspondence between long words and recipes for the Whittaker function (cf. [6, Chap. 2]). From this perspective, the Γ-rule corresponds to the word

$$w_0 = \sigma_1(\sigma_2\sigma_1)\cdots(\sigma_r\sigma_{r-1}\cdots\sigma_1),$$

whereas the Δ-rule corresponds to the word

$$w_0 = \sigma_r(\sigma_{r-1}\sigma_r)\cdots(\sigma_1\sigma_2\cdots\sigma_r).$$

These two words are as far apart as possible in the lexicographic ordering of all reduced decompositions. The proof of Theorem 2 as given in [6] uses a blend of combinatorial arguments to give various equivalent forms of the identity (18) as we move through the space of long words. We highlight various aspects of the proof in more detail now.

The proof is by induction on the rank r. The inductive hypothesis allows us to equate any two recipes for the Whittaker function whose associated long words differ by a sequence of relations obtained from a lower rank case. For example, assuming the rank 2 case allows us to perform a braid relation $\sigma_1\sigma_2\sigma_1 = \sigma_2\sigma_1\sigma_2$, which could be applied to the word corresponding to the Γ-rule above. After a series of such identities, we arrive at two descriptions for the Whittaker function as a weighted sum over Gelfand–Tsetlin patterns that agree on the bottom $r-2$ rows of the pattern. Thus, we may restrict our attention to the top three rows of a rank r pattern. We refer to such three-row arrays of interleaving integers, where we fix both the top and bottom of the three rows, as "short Gelfand–Tsetlin patterns" and re-index the three rows as follows:

$$\mathfrak{t} = \left\{ \begin{array}{ccccccccc} \ell_0 & & \ell_1 & & \cdots & & \ell_{r-1} & & \ell_r \\ & a_1 & & a_2 & & a_{r-1} & & a_r & \\ & & m_1 & & m_2 & & m_{r-1} & & \end{array} \right\}. \tag{19}$$

These two recipes for the Whittaker function will be called $G^{\Gamma\Delta}$ (as this recipe uses a right-hand rule for the entries a_i and a left-hand rule for the entries m_j) and $G^{\Delta\Gamma}$ (where the use of rules is reversed). To be exact, using the definitions in (11) and (15), we have

$$G^{\Gamma\Delta}(\mathfrak{t}) = \prod_{i=1}^{r}\gamma(a_i)\prod_{j=1}^{r-1}\delta(m_j) \quad \text{and} \quad G^{\Delta\Gamma}(\mathfrak{t}) = \prod_{i=1}^{r}\delta(a_i)\prod_{j=1}^{r-1}\gamma(m_j).$$

Rather than define functions $k^{\Gamma\Delta}$ and $k^{\Delta\Gamma}$ on short patterns in analogy to the recipes above, it is enough to specify the middle row sum as the other rows are fixed.

Before stating the reduction, we require one final ingredient. There is a natural involution q_r on short Gelfand–Tsetlin patterns of rank r, given by acting on middle row entries according to

$$q_r : \quad a_i \longmapsto \max(\ell_{i-1}, m_{i-1}) + \min(\ell_i, m_i) - a_i =: a_i',$$

where if $i = 0$, we understand that $\max(\ell_0, m_0) = \ell_0$, and if $i = r$, $\min(\ell_r, m_r) = \ell_r$. This involution q_r is used by Berenstein and Kirillov (cf. [13]) to define a Schützenberger involution on Gelfand–Tsetlin patterns. Brubaker, Bump, and Friedberg use the involution q_r to give the following reduction of Statement A in [6].

Theorem 3. (Brubaker, Bump, and Friedberg; Statement B of [6]) *Fix an $(r+1)$-tuple of positive integers $\boldsymbol{\ell} = (\ell_0, \ldots, \ell_r)$, an $(r-1)$-tuple of positive integers $\boldsymbol{m} = (m_1, \ldots, m_{r-1})$, and a positive integer k. Then*

$$\sum_{\sum a_i = k} G^{\Gamma\Delta}(\mathfrak{t}) = \sum_{\sum a_i' = k'} G^{\Delta\Gamma}(q_r(\mathfrak{t})),$$

where a_i' are the entries of $q_r(\mathfrak{t})$, $k' = \sum_i \ell_i + \sum_j m_j - k$, and the sums range over all short patterns with top row $\boldsymbol{\ell}$ and bottom row $\boldsymbol{m}$, satisfying the indicated condition.

See [4] and Chap. 6 of [6] for a full proof of the reduction from Statement A to Statement B. As noted above, the proof of Statement B proceeds through a series of additional reductions which occupy 13 chapters of [6]. In brief, for "generic" short patterns $\mathfrak{t}$, the Schützenberger involution q_r gives a finer equality $G^{\Gamma\Delta}(\mathfrak{t}) = G^{\Delta\Gamma}(q_r(\mathfrak{t}))$, which implies the equality of Statement B summand by summand. By "generic," we mean that the entries of the short pattern are in general position—in particular, for all i, $\ell_i \neq m_i$ using the notation of (19). Note that the Schützenberger involution does not necessarily preserve strictness for all patterns in the remaining nongeneric cases, and one needs much more subtle arguments to handle these short patterns. For such patterns, Statement B is not in fact true summand by summand, and one does need to sum over all short patterns with fixed row sum to obtain equality.

As an alternative to establishing Statement B, we mention that one could also prove Theorem 2 by computing the Whittaker integral in two ways, mimicking the techniques of [5], thus obtaining a proof via decomposition theorems in algebraic groups which respect the metaplectic cover. In subsequent sections of this paper, we propose a third way of viewing these theorems using ice-type models, which portends new connections between number theory/representation theory and statistical physics.

3.4 Ice and Metaplectic Whittaker Functions

In statistical mechanics, one attempts to infer global behavior from local interactions. In the context of lattice models, this means that we attach a *Boltzmann weight* to each vertex in the grid, and for each admissible state of the model, we consider the product of all Boltzmann weights ranging over all vertices of the grid. Then one can attempt to determine the *partition function* of the lattice model, which is simply the sum over all admissible states of the associated weights. In this section, we explain how to obtain the metaplectic spherical Whittaker function of Sect. 3.3 as the partition function of a lattice model with boundary conditions as defined in Sect. 3.2.

We make use of the two sets of Boltzmann weights B^{Γ} and B^{Δ}. When these weights are applied to an admissible state of ice, we refer to the resulting configuration as *Gamma ice* or *Delta ice*, respectively. In order to indicate which set of weights is being used at a particular vertex, we use $\bullet$ for Gamma ice and $\circ$ for Delta ice.

Gamma ice	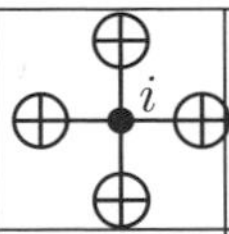	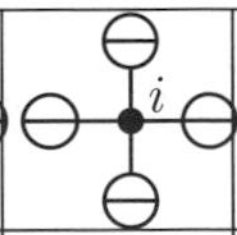	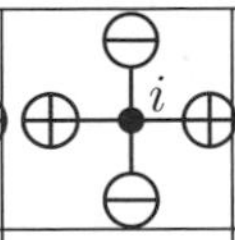	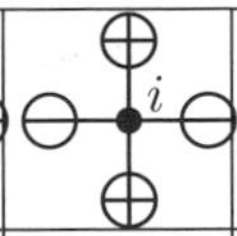	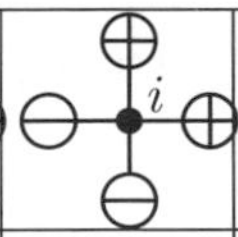	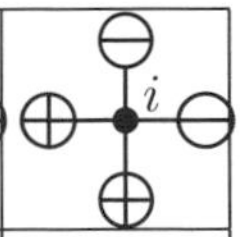
Boltzmann weight B^{Γ}	1	z_i	$g(a)$	z_i	$h(a)z_i$	1
Delta ice	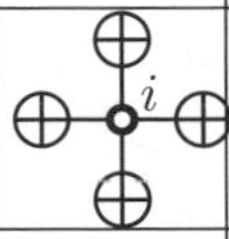	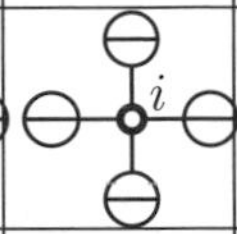	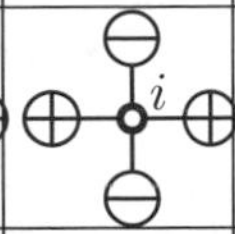	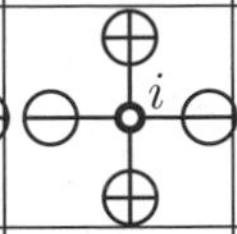	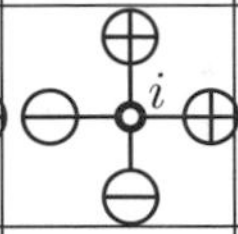	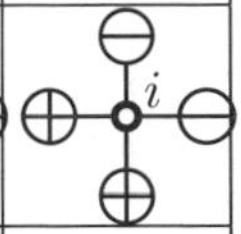
Boltzmann weight B^{Δ}	1	$g(a)z_i$	1	z_i	$h(a)z_i$	1

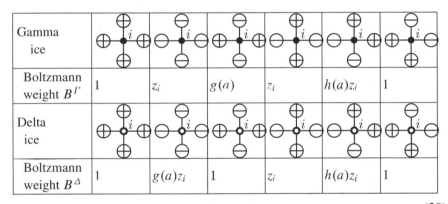

(20)

In giving these Boltzmann weights, we have made use of the notation in (9). For Gamma ice, the constant a equals the number of $+$ signs in the ith row to the right of the vertex $\bullet$. For Delta ice, the constant a equals the number of $-$ signs in the ith row to the left of the vertex $\circ$. In either case, we refer to this constant as the "charge" at the vertex. Note by our definitions in (9), the Boltzmann weights only depend on the charge mod n. The weights B^{Γ} and B^{Δ} also depend on parameters z_i, where i indicates the row in which the vertex is found. For Gamma ice, the row numbers decrease from $r+1$ to 1 as we move from top to bottom as in the example (2), while for Delta ice, the row numbers increase from 1 to $r+1$. These z_i are referred to as "spectral parameters." We often suppress the dependence of B^{Γ} and B^{Δ} on the spectral parameters z_i, $1 \le i \le r+1$. Let $\mathbf{z} = (z_1, \dots, z_{r+1})$.

Given an admissible state of Gamma ice (or Delta ice, respectively) $\mathcal{S}$, we define

$$\mathcal{G}^{\Gamma}(\mathcal{S},z) = \prod_{v\in\mathcal{S}} B^{\Gamma}(v), \quad \mathcal{G}^{\Delta}(\mathcal{S},z) = \prod_{v\in\mathcal{S}} B^{\Delta}(v), \tag{21}$$

where the product (in either case) is taken over all vertices in the state of ice $\mathcal{S}$.

Proposition 2. *Under the bijection of Proposition 1, with strict pattern $\mathfrak{T}$ corresponding to an admissible state of Gamma ice $\mathcal{S}$, then $G^{\Gamma}(\mathfrak{T})$ as defined in (12) is related to $\mathcal{G}^{\Gamma}(\mathcal{S},z)$ in (21) as follows:*

$$\mathcal{G}^{\Gamma}(\mathcal{S},z) = G^{\Gamma}(\mathfrak{T}) z_{r+1}^{d_0(\mathfrak{T})-d_1(\mathfrak{T})} z_r^{d_1(\mathfrak{T})-d_2(\mathfrak{T})} \cdots z_2^{d_{r-1}(\mathfrak{T})-d_r(\mathfrak{T})} z_1^{d_r(\mathfrak{T})},$$

where $d_i(\mathfrak{T})$ is the sum of the entries in the ith row of the pattern $\mathfrak{T}$.

Similarly, for an admissible state of Delta ice $\mathcal{S}$, $G^{\Delta}(\mathfrak{T})$ as defined in (16) is related to $\mathcal{G}^{\Delta}(\mathcal{S},z)$ in (21) by

$$\mathcal{G}^{\Delta}(\mathcal{S},z) = G^{\Delta}(\mathfrak{T}) z_1^{d_0(\mathfrak{T})-d_1(\mathfrak{T})} z_2^{d_1(\mathfrak{T})-d_2(\mathfrak{T})} \cdots z_r^{d_{r-1}(\mathfrak{T})-d_r(\mathfrak{T})} z_{r+1}^{d_r(\mathfrak{T})}.$$

We first illustrate this for Gamma ice with our working example from (3) in rank 2. The admissible Gamma ice $\mathcal{S}$ and its associated Boltzmann weights are pictured below.

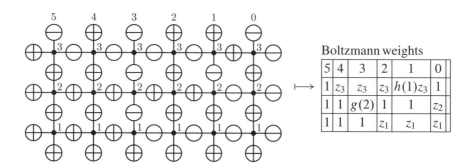

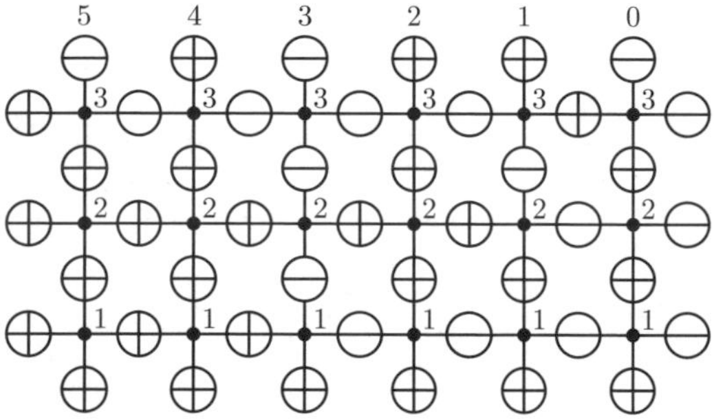

$\longmapsto$

Boltzmann weights

5	4	3	2	1	0
1	z_3	z_3	z_3	$h(1)z_3$	1
1	1	$g(2)$	1	1	z_2
1	1	1	z_1	z_1	z_1

Taking the product over all these weights, we obtain $\mathcal{G}^{\Gamma}(\mathcal{S},z) = h(1)g(2)z_3^4 z_2 z_1^3$, which indeed matches $G^{\Gamma}(\mathfrak{T}) z_3^{d_0(\mathfrak{T})-d_1(\mathfrak{T})} z_2^{d_1(\mathfrak{T})-d_2(\mathfrak{T})} z_1^{d_2(\mathfrak{T})}$ with $\mathfrak{T}$ as in (3) and $G^{\Gamma}(\mathfrak{T})$ as in (14).

Remark 1. The relevant terms in the metaplectic Whittaker function take the form

$$G^{\Gamma}(\mathfrak{T}) q^{k_1^{\Gamma}(\mathfrak{T})(1-2s_1)+\cdots+k_r^{\Gamma}(\mathfrak{T})(1-2s_r)} \tag{22}$$

as given in Theorem 1. However, $\boldsymbol{k}^{\Gamma}(\mathfrak{T}) = (k_1^{\Gamma},\ldots,k_r^{\Gamma})$ may be easily recovered from our fixed choice of highest weight $\lambda+\rho = (\ell_1,\ldots,\ell_r,0)$ and the row sums $d_i := d_i(\mathfrak{T})$ used in the monomial above. Indeed,

$$k_1^{\Gamma} = d_1 - (\ell_2+\cdots+\ell_r), \quad k_2^{\Gamma} = d_2 - (\ell_3+\cdots+\ell_r), \ldots, \quad k_r^{\Gamma} = d_r.$$

Hence, upon performing this simple transformation, we may recover the monomials in q^{1-2s_i} in (22) from those in z_j appearing in $\mathcal{G}^\Gamma(\mathcal{S},z)$ of the above proposition. A similar set of transformations holds for the Delta rules.

Proof. Proposition 2 is a consequence of the bijection given in Proposition 1. We sketch the proof for Gamma ice, as the proof for Delta ice is similar. Recall that $-$ vertical spins correspond to the entries of the pattern, so the values $\gamma(a_{i,j})$ given in (11) should appear in the Boltzmann weights for vertices sitting above a $-$ vertical spin. The particular cases of (11) to be used are determined by the vertical spin *above* the vertex in question. We now show that each $b_{i,j}$ in (11) matches the charge, the number of $+$ signs to the right of the vertex in row i. Equivalently, we must show that every spin between column $a_{i,j}$ and column $a_{i-1,j}$ in row i is assigned a $+$. So suppose that $a_{i,j} > a_{i-1,j}$ and let v be the vertex in row i, column $a_{i,j}$, and let w be the vertex in row i, column $a_{i-1,j}$. Then the north and south spins for v are $(+,-)$ which, by the six admissible configurations in Gamma ice, forces the east spin to be $+$. All the vertices between v and w have north and south spins $(+,+)$ according to our bijection. The east spin $+$ for v becomes the west spin for the neighboring vertex v' to the right of v, forcing the east spin of v' to be $+$ as well. This effect propagates down the row, forcing all row spins between v and w to be $+$. Finally, we must show that the spectral parameters for $\mathcal{G}$ are given by differences of consecutive row sums. This is Lemma 3 of [7]. □

Given a fixed set of boundary conditions for the vertex model and an assignment B of Boltzmann weights associated to each admissible vertex, we refer to the set of all admissible states $\mathcal{S}$ as a "system." Given a system $\mathfrak{S}$, its partition function $Z(\mathfrak{S})$ is defined as

$$Z(\mathfrak{S}) := Z(\mathfrak{S},z) = \sum_{\mathcal{S}\in\mathfrak{S}} B(\mathcal{S},z), \quad \text{with} \quad B(\mathcal{S},z) := \prod_{v\in\mathcal{S}} B(v), \tag{23}$$

where this latter product is taken over all vertices v in the state $\mathcal{S}$. In particular, let $\mathfrak{S}^\Gamma$ denote the system with boundary conditions as in Sect. 3.2, Boltzmann weights B^Γ, and rows labeled in descending order from top to bottom. Similarly, let $\mathfrak{S}^\Delta$ denote the system with the same boundary conditions, but with Boltzmann weights B^Δ, and rows labeled in ascending order from top to bottom. Using this language, we may now summarize the results of the past two sections in a single theorem.

Theorem 4. *Given a dominant weight λ for GL_{r+1}, the metaplectic Whittaker function $W(\mathbf{t}_\lambda)$ is expressible as either of the two partition functions $Z(\mathfrak{S}^\Gamma)$ or $Z(\mathfrak{S}^\Delta)$.*

This is merely the combination of Theorems 1 and 2 together with Proposition 2.

3.5 Transfer Matrices

Baxter considered the problem of computing partition functions for solvable lattice models (cf. [2]). His approach is based on the idea of using the Yang–Baxter equation (called the "star-triangle identity" by Baxter) to prove the commutativity of *row transfer matrices*. We will show that basic properties of metaplectic Whittaker functions can be interpreted as commutativity of such transfer matrices, and at least when the metaplectic degree $n = 1$, the Yang–Baxter equation can be used to give proofs of these.

The row transfer matrices shall now be described. Let us consider a row of vertices that all have the same Boltzmann weights. If $B = (a_1, a_2, b_1, b_2, c_1, c_2)$, then we use the assignment of Boltzmann weights in the following table:

	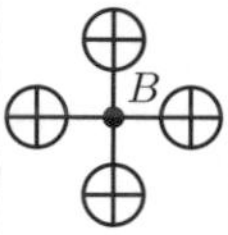	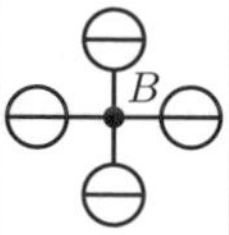	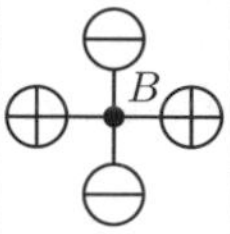	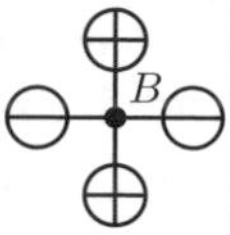	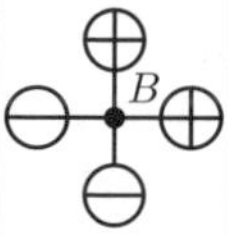	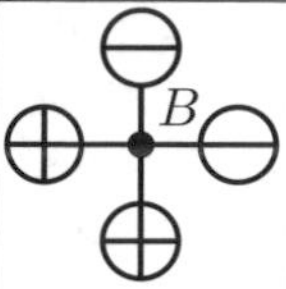
Boltzmann weight	a_1	a_2	b_1	b_2	c_1	c_2

The vertical edge spins in the top and bottom boundaries will be collected into vectors $\alpha = (\alpha_N, \ldots, \alpha_0)$ and $\beta = (\beta_N, \ldots, \beta_0)$. The subscripts correspond to the columns which, we recall, are numbered in ascending order from right to left. For example, if $\alpha = (-, +, -, +, +, -)$ and $\beta = (+, -, +, +, +, -)$, we would consider the partition function of the following one-layer system of ice:

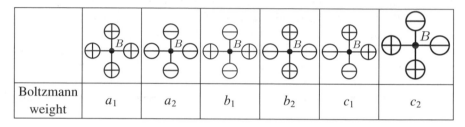

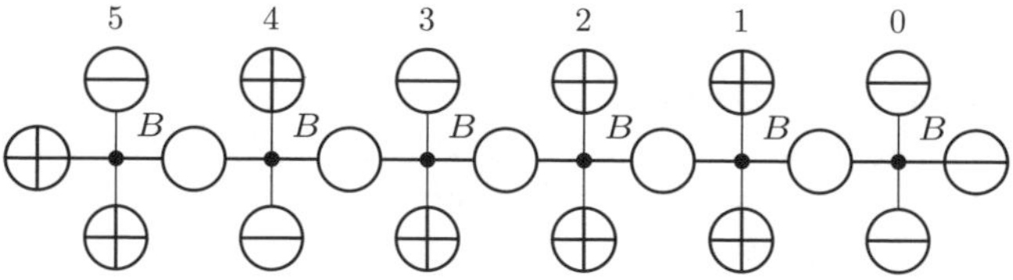

(24)

Let $V_B(\alpha, \beta)$ denote the partition function. Recall that we compute this as follows. We complete the state by assigning values to the interior edges (unlabeled in this figure) and sum over all such completions. Let V_B be the $2^{N+1} \times 2^{N+1}$ matrix whose entries are all possible partition functions $V_B(\alpha, \beta)$, where the choices of α and β index the rows and columns of the matrix, respectively. This is referred to as the *transfer matrix* for the one-layer system of size N with Boltzmann weights B at every vertex.

Now let us consider a two-layer system:

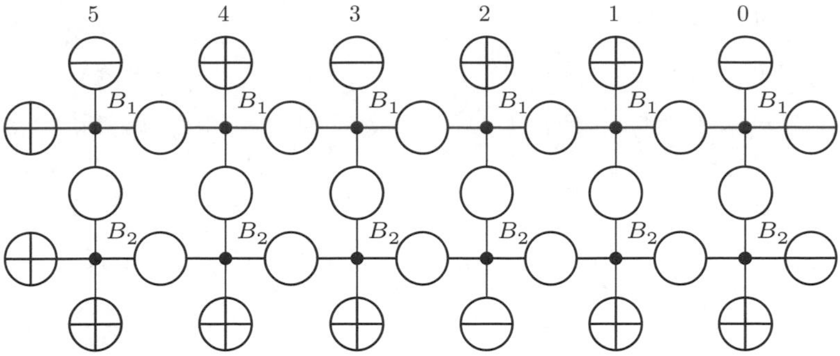

(25)

Note that we are using two sets of Boltzmann weights B_1 and B_2 for the top and bottom layers, respectively. We may try to express the partition function $V(\alpha, \gamma)$ for the two-layer system pictured above having top row α as in (24) and bottom row $\gamma = (+, +, +, -, +, +)$ in terms of one-layer partition functions. However, each one-layer system is only determined upon a choice of vertical spins lying between the rows. One such choice of edge spins is β as in the one-layer system in (24), but we must sum over all possible choices to get the partition function of the two-layer system. Therefore,

$$V(\alpha, \gamma) = \sum_{\beta} V_{B_1}(\alpha, \beta) V_{B_2}(\beta, \gamma),$$

which is precisely the entry $V(\alpha, \gamma)$ in the product of the two transfer matrices V_{B_1} and V_{B_2}.

Cases where the transfer matrices commute are of special interest. Indeed, this commutativity means that one can interchange the roles of Boltzmann weights B_1 and B_2 in (25), and the value of the product of the transfer matrices is unchanged. Baxter considers the case where $B_1 = (a, a, b, b, c, c)$ and $B_2 = (a', a', b', b', c', c')$ for arbitrary choices of a, a', b, b', c, c'. His boundary conditions are different from ours. Baxter's boundary conditions are toroidal; that is, the boundary spins at the left and right edges of the each row are equal, so those edges may be identified, treated as interior edges, and hence, summed over. With this modification, Baxter proves that if $\Delta = \Delta'$, where $\Delta = (a^2 + b^2 - c^2)/2ab$ and Δ' is similarly defined with a', b' and c', then the transfer matrices commute. Obtaining a sufficiently large family of commuting transfer matrices is a step towards evaluating the partition function, since by doing so, one can make the eigenspaces one-dimensional. Thus the problem of simultaneously diagonalizing them has a unique solution and therefore becomes tractable.

Let us now show how Statement B may be formulated in terms of commuting transfer matrices. We consider a two-layer system having a layer of Gamma ice and a layer of Delta ice, thus:

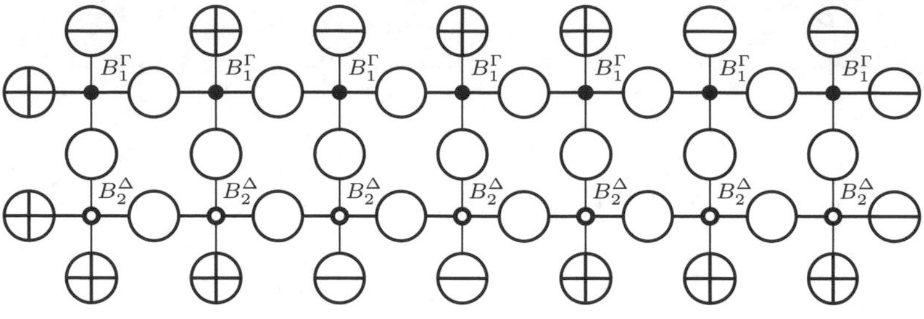

(26)

We use the values in (20); in the top row, the spectral parameter is z_1, and in the bottom row, it is z_2. Regarding the boundary conditions, as always, the rows of ice must have $+$ at the left edge and $-$ at the right edge. Furthermore, we fix a choice of spins for the top edge and the bottom edge of this two-layered ice such that the top edge has two more $-$ than the bottom row. In this example, the locations of the $-$ along the top edge are (reading from right to left) $0, 1, 4, 6$ and along the bottom edge, they are at $3, 4$. We have labeled each vertex with $\bullet\, \Gamma_1$ and $\circ \Delta_2$ to remind the reader of the Boltzmann weights that we are using. We will call this system $\mathfrak{S}^{\Gamma\Delta}$ and its partition function $Z(\mathfrak{S}^{\Gamma\Delta})$.

On the other hand, we may consider the same configuration with the roles of the Boltzmann weights for Gamma and Delta ice switched, as in the figure below. Note that the boundary conditions remain the same as in (26). We will refer to this system as $\mathfrak{S}^{\Delta\Gamma}$.

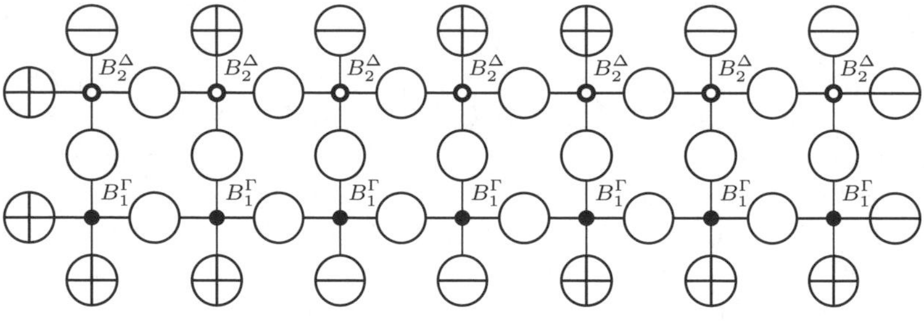

Theorem 5. *Given top and bottom boundary values as vectors of spins α and γ and Boltzmann weights B_1^Γ and B_2^Δ as in (20), let $\mathfrak{S}^{\Gamma\Delta}$ and $\mathfrak{S}^{\Delta\Gamma}$ be the systems described above. Then $Z(\mathfrak{S}^{\Gamma\Delta}) = Z(\mathfrak{S}^{\Delta\Gamma})$.*

We prove this by showing that the claim is equivalent to Statement B, stated as Theorem 3 here and proved by combinatorial means in [6]. Note in particular that we have reformulated Statement B as the commutativity of two transfer matrices.

Proof. We associate two strictly decreasing vectors of integers with α and γ, which we call $\boldsymbol{l}$ and $\boldsymbol{m}$. Namely, let $\boldsymbol{l} = (l_0, l_1, l_2, \dots)$, where the l_is are the integers such that $\alpha_{l_i} = -$, arranged in descending order; $\boldsymbol{m}$ is defined similarly with regard to γ. Thus, in the example (26) above, there are $-$ spins in the $6, 4, 1, 0$ columns of the top row and so $\boldsymbol{l} = (6, 4, 1, 0)$, while $\boldsymbol{m} = (4, 3)$. Similarly, given any admissible state of the system, let β be the middle row of edge spins and associate in similar fashion a sequence $\boldsymbol{a} = (a_1, a_2, \dots)$ according to the location of $-$ signs in β.

We observe that the sequences $\boldsymbol{l}, \boldsymbol{a}, \boldsymbol{m}$ interleave. This holds for the same reason that the rows of the pattern interleave in Proposition 1; it is a consequence of Lemma 2 of [7]. Therefore, the legal states of either system $\mathfrak{S}$ are in bijection with the (strict) short Gelfand–Tsetlin patterns

$$\mathfrak{t} = \left\{ \begin{array}{ccccccccc} \ell_0 & & \ell_1 & & \cdots & & \ell_{r-1} & & \ell_r \\ & a_1 & & a_2 & & a_{r-1} & & a_r & \\ & & m_1 & & m_2 & & m_{r-1} & & \end{array} \right\}.$$

These are *not* in bijection with the terms of the sum $G^{\Gamma\Delta}(\mathfrak{t})$ appearing in Theorem 3 because there is no condition on the middle row sum. Rather, the states of ice give all possible middle row sums. However, letting $\mathcal{G}^{\Gamma\Delta}(\mathcal{S}, z)$ denote the Boltzmann weight for a state of $\Gamma\Delta$ ice, this may be regarded as a homogeneous polynomial in the two spectral parameters z_1 and z_2 of our two-row system. In the notation of Proposition 2, this monomial is $z_1^{d_0(\mathfrak{t})-d_1(\mathfrak{t})} z_2^{d_1(\mathfrak{t})-d_2(\mathfrak{t})}$, where $d_i(\mathfrak{t})$ denotes the ith row sum in the short Gelfand–Tsetlin pattern above. Clearly, the middle row sum can be recovered from knowledge of this monomial for fixed choice of boundary conditions α and γ, which dictate the top and bottom row of the short pattern. A similar correspondence may be obtained for the $\Delta\Gamma$ system whose short patterns $\mathfrak{t}'$ are associated to the monomial $z_2^{d_0(\mathfrak{t}')-d_1(\mathfrak{t}')} z_1^{d_1(\mathfrak{t}')-d_2(\mathfrak{t}')}$. Of course, the boundary conditions remain constant whether we are using the $\Gamma\Delta$ or $\Delta\Gamma$ system, so $d_0(\mathfrak{t}) = d_0(\mathfrak{t}')$ and $d_2(\mathfrak{t}) = d_2(\mathfrak{t}')$. Thus, the monomials

$$z_1^{d_0(\mathfrak{t})-d_1(\mathfrak{t})} z_2^{d_1(\mathfrak{t})-d_2(\mathfrak{t})} \quad \text{and} \quad z_2^{d_0(\mathfrak{t}')-d_1(\mathfrak{t}')} z_1^{d_1(\mathfrak{t}')-d_2(\mathfrak{t}')}$$

agree precisely when

$$d_1(\mathfrak{t}) = d_0(\mathfrak{t}) + d_2(\mathfrak{t}) - d_1(\mathfrak{t}'),$$

which is exactly the condition on the sum in Theorem 3. Hence we see that the commutativity of transfer matrices—the statement that $Z(\mathfrak{S}^{\Gamma\Delta}) = Z(\mathfrak{S}^{\Delta\Gamma})$—is an equality of two homogeneous polynomials, and the matching of each monomial corresponds to the identity of Statement B for each possible middle row sum.

3.6 The Yang–Baxter Equation

The proof of Theorem 5, the commutativity of transfer matrices, uses the equivalence with Theorem 3 and hence implicitly relies on all of the combinatorial methods of [6] in order to obtain this result. In this section, we want to explore the extent to which Baxter's methods for solving statistical lattice models, most notably the Yang–Baxter equation, may be used to prove the commutativity of transfer matrices.

In our context of two-dimensional square lattice models, the Yang–Baxter equation may be viewed as a fundamental identity between partition functions on two very small pieces of ice—each having six boundary edges to be fixed, three internal edges, and three vertices each with an assigned set of Boltzmann weights.

Definition 1 (Yang–Baxter Equation). Let R, S, and T be three collections of Boltzmann weights associated to each admissible vertex. Then for every fixed combination of boundary conditions $\sigma, \tau, \alpha, \beta, \rho, \theta$, we have the following equality of partition functions:

Z (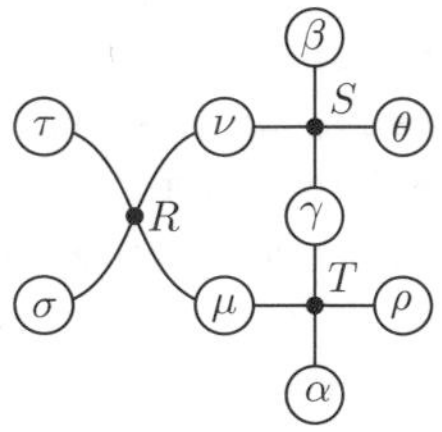

) $= Z$ (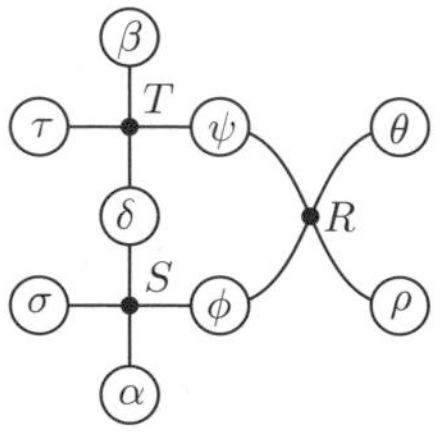

) . (27)

Recall that these partition functions are sums of Boltzmann weights over all admissible states. Hence, the left-hand side is a sum over all choices of internal edge labels μ, ν, γ, while the right-hand side is a sum over internal edge labels ϕ, ψ, δ. Note that the roles of S and T are interchanged on the two sides of the equality.

In the diagram above one vertex, labeled R has been rotated by 45° for ease of drawing the systems. It should be understood in the same way as S and T—it has a Boltzmann weight associated to a set of admissible adjacent edge labels. However, vertices of this type have a distinguished role to play in the arguments that follow, so we use the term R-*vertex* to refer to any vertex rotated by 45° like R in (27).

Once equipped with the Yang–Baxter equation, the commutativity of transfer matrices, i.e., invariance of the partition function under interchange of rows, may be proved under certain assumptions. We illustrate the method with a three-layer system of ice $\mathfrak{S}$ having boundary conditions and admissible vertices like those of the system $\mathfrak{S}^{\Gamma}$, to give the basic idea. Suppose we wanted to analyze the effect of swapping the second and third rows in the following configuration:

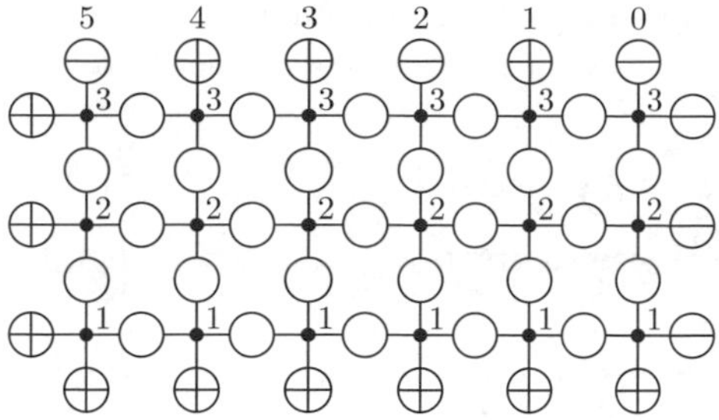

(28)

Suppose there exists only one admissible R-vertex having positive spins on the right, without loss of generality, we take it to have all positive spins. Then the partition function $Z(\mathfrak{S})$ for (28) multiplied by the Boltzmann weight for the R-vertex with all $+$ spins is equal to the partition function for the following configuration of ice:

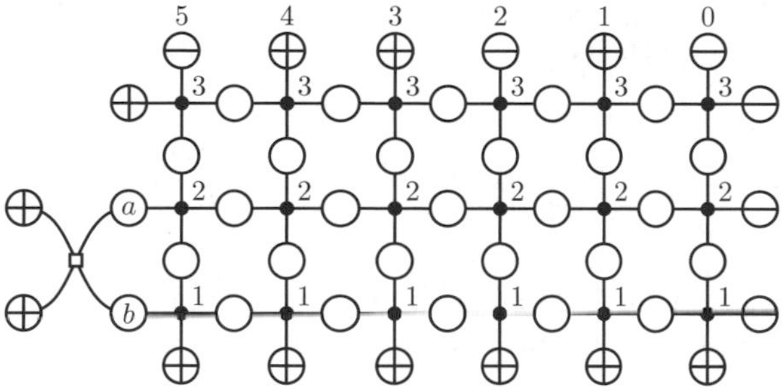

(29)

(By assumption, the only legal values for a and b are $+$, so every state of this problem determines a unique state of the original problem.) Now we apply the Yang–Baxter equation to move this R-vertex rightward, to obtain equality with the following configuration:

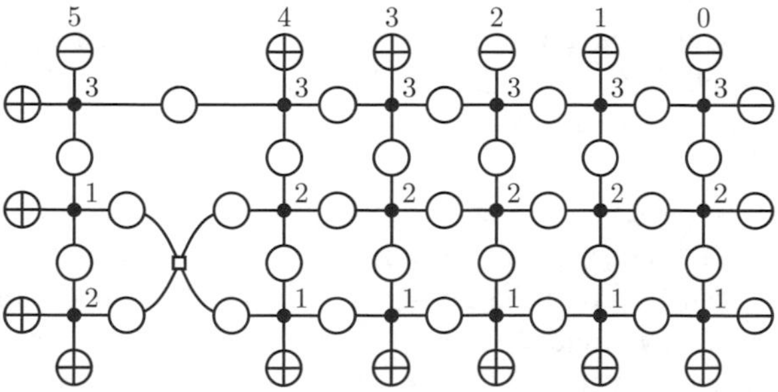

Repeatedly applying the Yang–Baxter equation, we eventually obtain the configuration in which the R-vertex is moved entirely to the right.

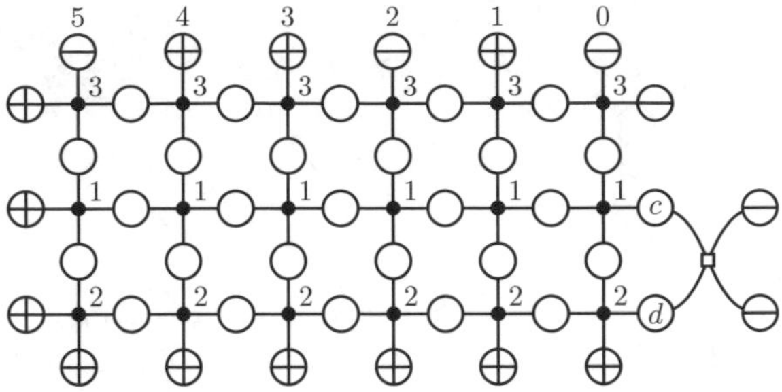

(30)

In drawing the above picture, we have again assumed that there is just one legal configuration for the R-vertex having two − spins on the left, and assumed the spins of this R-vertex were all −. If we let $\mathfrak{S}'$ denote the system with the same boundary conditions as $\mathfrak{S}$ shown in (28) but with the second and third row Boltzmann weights interchanged, we have shown

B_R 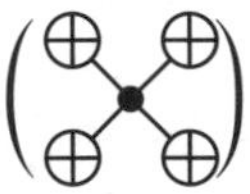 $Z(\mathfrak{S}) = B_R$ 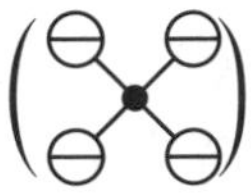 $Z(\mathfrak{S}')$, (31)

where B_R denotes the assignment of Boltzmann weight to each configuration. In particular, if the two admissible R-vertices coming from the left- and right-hand sides of (31) have equal Boltzmann weights, we obtain the exact equality of the two configurations, i.e., the commutativity of transfer matrices.

We now explore the possibility of obtaining a Yang–Baxter equation with S and T in (27) corresponding to the Boltzmann weights B^{Γ} and B^{Δ}, respectively, from (20). In light of our previous argument, this would give an alternate proof of Theorem 3. However, the Boltzmann weights in (20) depend not only on spins $+$ or $-$ on adjacent edges but also on a "charge" a mod n. Recall from Sect. 3.4 that using B^{Γ} weights, charge records the number of $+$ signs in a row between the given vertex and the $-$ boundary spin at the right-hand edge the row. Using B^{Δ} weights charge counts the number of $-$ signs between the vertex and the $+$ boundary at the left.

In order to demonstrate a Yang–Baxter equation, we need Boltzmann weights that are purely local—i.e., depend only on properties of adjacent edges—so we need a different way of interpreting charge. We do this by labeling horizontal edges with both a spin and a number mod n. We declare the Boltzmann weight of these vertices to be 0 unless the edge labels a, b mod n to the immediate left and right of the vertex reflect the way charge is counted for the given spins. For example, with B^{Γ} weights, $a = b + 1$ if the spin below a is $+$ and $a = b$ if the spin below a is $-$. Using this interpretation, we record the nonzero vertices for both sets of Boltzmann weights:

Gamma ice	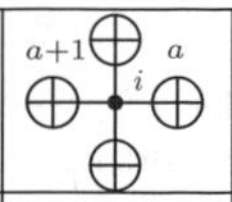	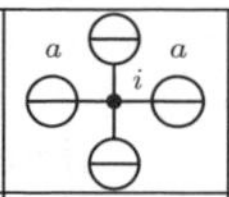	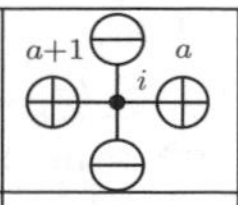	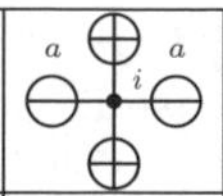	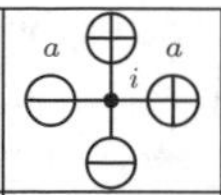	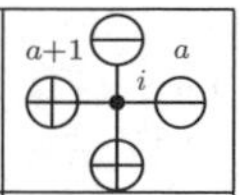
B^Γ weight	1	z_i	$g(a)$	z_i	$h(a)z_i$	1
Delta ice	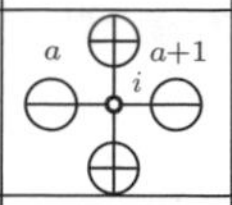	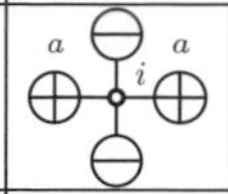	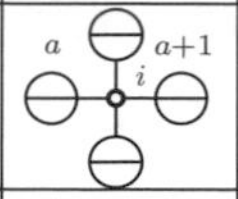	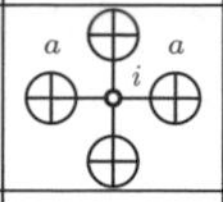	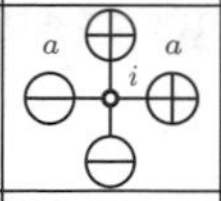	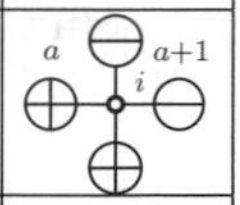
B^Δ weight	1	$g(a)z_i$	1	z_i	1	$h(a)z_i$

(32)

The above vertices are admissible for *any* choice of a mod n (and the integers $a+1$ are, of course, understood to be mod n as well). This means that we are generalizing the six-vertex model, since due to the dependence on a, each vertex has more than six admissible states.

For $n = 1$, the charge labels on horizontal edges are trivial as the Gauss sums $g(a)$ and $h(a)$ are independent of a as evaluated in (10). For this special case, it was shown in [7] that a Yang–Baxter equation exists with weights S and T as in (27) taken to be B^Γ and B^Δ from the table above. We refer the reader to [7] for the corresponding Boltzmann weights R for which the Yang–Baxter equation is satisfied. Thus, we obtain an alternate proof of Theorem 4, or equivalently Theorem 3, using methods from lattice models.

In general, we know from [6] that Theorem 4 is true for any positive integer n. It would be extremely interesting to find a local relation like (27) similarly proving that the transfer matrices commute, and this is currently under investigation by the authors.

3.7 Weyl Group Invariance and the Yang–Baxter Equation

Kazhdan and Patterson [12, Lemma 1.3.3] describe how the metaplectic Whittaker functions transform under the action of the Weyl group. This invariance—which does not follow directly from the description of the coefficients H given in Theorem 1—plays a key role in the proof of the metaplectic Casselman-Shalika formula for GL_{r+1} by Chinta and Offen [10], and was the main inspiration for the Weyl group action in [9].

In this section, we restate this Weyl group invariance in terms of the partition functions defined in the previous sections. We content ourselves to describe how a simple reflection acts on the partition function. Let σ_i denote the simple reflection in the Weyl group corresponding to the ith simple root. We let σ_i act on the spectral parameter $z = (z_1, z_2, \ldots, z_{r+1})$ by $\sigma_i(z) = (z_1, \ldots, z_{i-1}, z_{i+1}, z_i, z_{i+2}, \ldots, z_{r+1})$, i.e., the ith and $(i + 1)$st coordinates are transposed. Here, the notation $Z(\mathfrak{S}, z)$ refers to the partition function associated to the system $\mathfrak{S}$, where $\mathfrak{S}$ is either of the two systems $\mathfrak{S}^\Gamma$ or $\mathfrak{S}^\Delta$ introduced in Sect. 3.4.

Further define, for $j = 0, \ldots, n - 1$,

$$P^{(j)}(x, y) = x^j y^{n-j} \frac{1 - q^{-1}}{x^n - q^{-1} y^n} \quad \text{and} \quad Q^{(j)}(x, y) = g(j) \frac{x^n - y^n}{x^n - q^{-1} y^n}, \tag{33}$$

where we again use the shorthand notation of (9) and interpret $g(0) := g(n) = -q^{-1}$. The functions P, Q are closely related to the functions τ_s^1, τ_s^2 of [12, Lemma 1.3.3].

For each $1 \le i \le r$, we may decompose the partition function

$$Z(\mathfrak{S}, z) = \sum_{0 \le j < n} Z_i^{(j)}(\mathfrak{S}, z), \tag{34}$$

where $Z_i^{(j)}(\mathfrak{S}, z)$ is the sum over all states $\mathcal{S} \in \mathfrak{S}$ such that $B(\mathcal{S}, z)$ is equal to a constant times $z_1^{a_1} \cdots z_{r+1}^{a_{r+1}}$, where $a_i - a_{i+1} \equiv j \pmod{n}$. Then the Whittaker function satisfies

$$Z_i^{(j)}(\mathfrak{S}, \sigma_i(z)) = P^{(j)}(z_{i+1}, z_i) \cdot Z_i^{(j)}(\mathfrak{S}, z) + Q^{(j)}(z_{i+1}, z_i) \cdot Z_i^{(n-j)}(\mathfrak{S}, z). \tag{35}$$

We now consider the extent to which the functional equations (35) can be interpreted in the language of transfer matrices. First, we consider the case $n = 1$. The decomposition on the right-hand side of (34) has only one term, namely Z itself, since the congruence condition is automatically satisfied by all monomials for any i. The ith functional equation (35) becomes

$$Z(\mathfrak{S}, \sigma_i(z)) = (P^{(0)}(z_{i+1}, z_i) + Q^{(0)}(z_{i+1}, z_i)) Z(\mathfrak{S}, z)$$

or

$$(z_i - z_{i+1}/q) Z(\mathfrak{S}, z) = (-z_i/q + z_{i+1}) Z(\mathfrak{S}, \sigma_i(z)). \tag{36}$$

Recalling the effect of σ_i on z defined above, the partition function on the right-hand side is the result of swapping the spectral parameters associated to rows i and $i + 1$ in the system $\mathfrak{S}$. Note that (36) is not exactly the same as "commutation of two transfer matrices" because we do not have the identity $Z(\mathfrak{S}, z) = Z(\mathfrak{S}, \sigma_i(z))$. Indeed, the partition function Z is not a symmetric function, but it is very close to one: it is a Schur polynomial times a q-deformation of the Weyl denominator (cf. [7]).

Nevertheless, with assumptions as in Sect. 3.6, we may ask for a Yang–Baxter equation leading to a proof of (36). That is, we seek sets of Boltzmann weights

R, S, and T satisfying (27) where $S = T = B^{\Gamma}$ or $S = T = B^{\Delta}$. Comparing (36) with (31), we further require Boltzmann weights B_R for the R-vertices such that

B_R 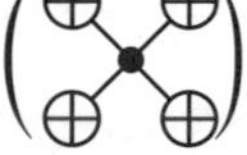 $= z_i - z_{i+1}/q,$

B_R 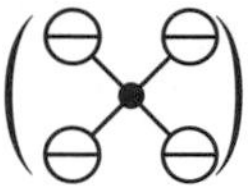 $= -z_i/q + z_{i+1}.$

It follows from results in [7] that we may use the following coefficients in the R-vertex for Gamma ice:

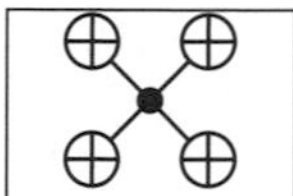	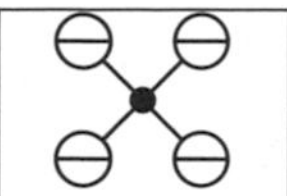	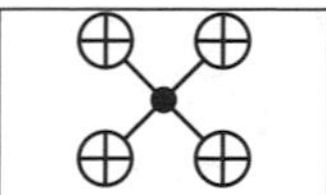	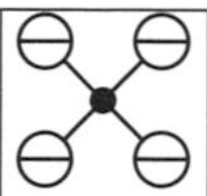	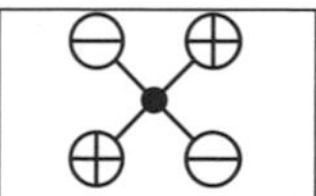	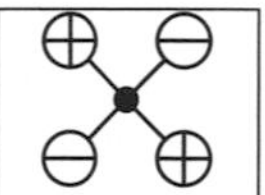
$z_i - q^{-1}z_{i+1}$	$z_{i+1} - q^{-1}z_i$	$q^{-1}(z_{i+1} - z_i)$	$z_{i+1} - z_i$	$(1 - q^{-1})z_{i+1}$	$(1 - q^{-1})z_i$

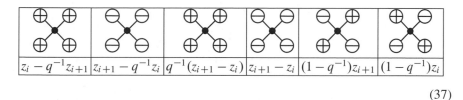

(37)

We are taking all $t_i = -q^{-1}$ in Table 1 in [7] and observe that the order of the rows in this paper is opposite those in that paper. Our convention here is the same as in [6].

For $n > 1$, the situation is more complicated, but rather suggestive. In general, $Z_i^{(j)}(\mathfrak{S}) \neq Z_i^{(n-j)}(\mathfrak{S})$, so we cannot rewrite (35) to look like (31) and (36). However, according to (33), the denominators of P and Q appearing in the i-th functional equation (35) are equal and independent of j. For any j, they are $z_i^n - z_{i+1}^n/q$. Thus, clearing denominators, we may rewrite (35) as follows:

$$(z_{i+1}^n - q^{-1}z_i^n)Z_i^{(j)}(\mathfrak{S}, \sigma_i(z)) \\ = p^{(j)}(z_{i+1}, z_i) \cdot Z_i^{(j)}(\mathfrak{S}, z) + q^{(j)}(z_{i+1}, z_i) \cdot Z_i^{(n-j)}(\mathfrak{S}, z), \qquad (38)$$

where

$$p^{(j)}(z_{i+1}, z_i) = (1 - q^{-1})z_{i+1}^j z_i^{n-j}, \qquad q^{(j)}(z_{i+1}, z_i) = g(j)(z_{i+1}^n - z_i^n).$$

Let $\mathcal{S}$ be a state of the system, and as before, let $a_1, \dots, a_{r+1}$ be the exponents of $z_1, \dots, z_{r+1}$ in $B(\mathcal{S}, z)$. We make the following observation: In the weights (32), there is a contribution of z_i if and only if the charge is not augmented as we move across the vertex. Since (in Gamma ice) the charges at the right edge will have value 0, it follows that the charges at the left edge will have value c_i, where $a_i + c_i$ is the number of vertices in the row. Therefore,

$$a_i - a_{i+1} = c_{i+1} - c_i, \qquad (39)$$

and we may therefore write

$$Z_i^{(j)}(\mathfrak{S}, z) = \sum_{c_{i+1} - c_i \equiv j \bmod n} B(\mathcal{S}, z).$$

We will now explain how, with a suitable R-vertex, (38) could also be interpreted as an identity similar to (31), but now with sets of Boltzmann weights involving charges. We will describe the characteristics that such an R-vertex might have. For simplicity, we will assume that n is odd.

The value will depend on the spins and charges of the adjacent edges. Let us assume first that the spins on these four edges are all $+$, with charges d_{i+1}, d_i, d'_{i+1}, d'_i as follows:

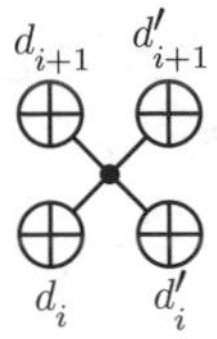

(40)

If $j = d_{i+1} - d_i$ and $j' = d'_{i+1} - d'_i$, then we require that the Boltzmann weight of this vertex v is zero unless $j' \equiv j$ or $n - j \bmod n$. Moreover, in these cases, we require that the Boltzmann weight of (40) is

$$\begin{cases} p^{(j)}(z_{i+1}, z_i) \text{ if } j \equiv j' \bmod n \\ q^{(j)}(z_{i+1}, z_i) \text{ if } j \equiv n - j' \bmod n \end{cases}$$

except when $j \equiv 0$. In this case, the weight will be

$$p^{(0)}(z_{i+1}, z_i) + q^{(0)}(z_{i+1}, z_i) = z_i^n - q^{-1} z_{i+1}^n,$$

since $g(0) = -q^{-1}$.

Regarding the case where the vertex has spin $-$ on all four adjoining edges, we require that the Boltzmann weight of

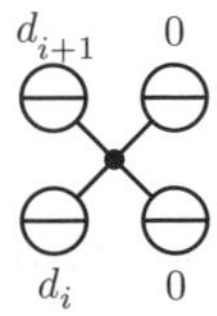

is zero unless $d_i = d_{i+1} = 0$, in which case, it is $z_{i+1}^n - q^{-1} z_i^n$.

Assuming that the R-vertex has the above properties, we may now express the functional equation in a form similar to (31). Let us fix the vertical edge spins above the z_{i+1} row and below the z_i row and work with just the two relevant rows; let $\mathfrak{S}'$

denote the two-layer system consisting of just rows $i+1$ and i with these boundary spins fixed. In order to establish (35), or equivalently (38), it suffices to show

$$(z_{i+1}^n - q^{-1} z_i^n) Z_i^{(j)}(\mathfrak{S}', \sigma_i(\mathbf{z}))$$
$$= p^{(j)}(z_{i+1}, z_i) \cdot Z_i^{(j)}(\mathfrak{S}', \mathbf{z}) + q^{(j)}(z_{i+1}, z_i) \cdot Z_i^{(n-j)}(\mathfrak{S}', \mathbf{z}).$$

Since $Z(\mathfrak{S}', \mathbf{z})$ is a homogeneous polynomial in the z_i, and since only a_i and a_{i+1} are allowed to vary, we have $a_i + a_{i+1}$ equal to a constant. Since we are assuming that n is odd, there will be a unique pair of charges c_i and c_{i+1} mod n such that (39) is satisfied, and such that $c_{i+1} - c_i \equiv j$ modulo n.

Now let us consider the partition function of the system

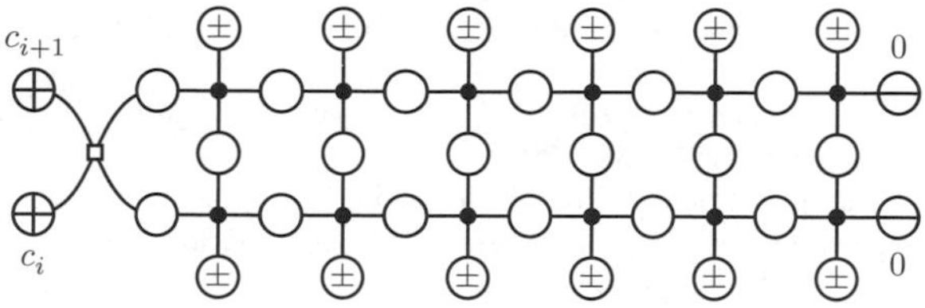

obtained by attaching the R-vertex to the left of $\mathfrak{S}'$. From the above discussion, this equals

$$p^{(j)}(z_{i+1}, z_i) \cdot Z_i^{(j)}(\mathfrak{S}', \mathbf{z}) + q^{(j)}(z_{i+1}, z_i) \cdot Z_i^{(n-j)}(\mathfrak{S}', \mathbf{z}).$$

Similarly, the partition function of the system

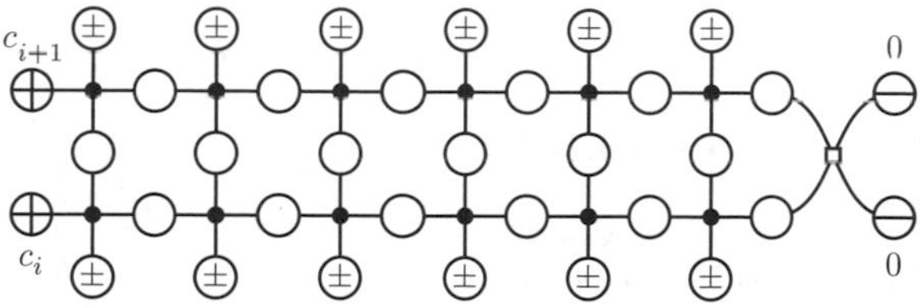

is

$$(z_{i+1}^n - q^{-1} z_i^n) Z_i^{(j)}(\mathfrak{S}', \sigma_i(\mathbf{z})).$$

The equality of these partition functions implies (38).

At this writing, we do not know if the values of the R-vertex that we have described can be completed to a full R-matrix such that the appropriate Yang–Baxter equation is satisfied. We know that this can be done when $n = 1$, and since (38) is true, it seems very plausible that this can be done in general. Thus, we may conjecture that within this scheme, or some similar one, it is possible to formulate a Yang–Baxter equation adapted to these weights that gives a proof of (38). Such a "metaplectic" Yang–Baxter equation might well have importance beyond the problems that we have discussed in this paper.

References

1. W. D. Banks, J. Levy, and M. R. Sepanski, *Block-compatible metaplectic cocycles*, J. Reine Angew. Math. **507** (1999), 131–163.
2. R. J. Baxter, *Exactly solved models in statistical mechanics*, Academic Press Inc. [Harcourt Brace Jovanovich Publishers], London, 1989, Reprint of the 1982 original.
3. A. Berenstein and A. Zelevinsky, *Tensor product multiplicities, canonical bases and totally positive varieties*, Invent. Math. **143** (2001), no. 1, 77–128.
4. B. Brubaker, D. Bump, and S. Friedberg, *Gauss sum combinatorics and metaplectic Eisenstein series*, Automorphic forms and L-functions I. Global aspects, Contemp. Math., vol. 488, Amer. Math. Soc., Providence, RI, 2009, pp. 61–81.
5. B. Brubaker, D. Bump, and S. Friedberg, *Weyl group multiple Dirichlet series, Eisenstein series and crystal bases*, Ann. of Math. (2) **173** (2011), no. 2, 1081–1120.
6. B. Brubaker, D. Bump, and S. Friedberg, *Weyl Group Multiple Dirichlet Series: Type A Combinatorial Theory*. Annals of Mathematics Studies, vol. 225, Princeton University Press, 2011.
7. B. Brubaker, D. Bump, and S. Friedberg, *Schur polynomials and the Yang-Baxter equation*, Comm. Math. Phys., 308(2):281–301, 2011.
8. B. Brubaker, D. Bump, S. Friedberg, and J. Hoffstein, *Weyl group multiple Dirichlet series. III. Eisenstein series and twisted unstable A_r*, Ann. of Math. (2) **166** (2007), no. 1, 293–316.
9. G. Chinta and P. E. Gunnells, *Constructing Weyl group multiple Dirichlet series*, J. Amer. Math. Soc. **23** (2010), no. 1, 189–215.
10. G. Chinta and O. Offen, *A metaplectic Casselmann–Shalika formula for GL_r*, Amer. J. Math., to appear.
11. A. M. Hamel and R. C. King, *U-turn alternating sign matrices, symplectic shifted tableaux and their weighted enumeration*, J. Algebraic Combin. **21** (2005), no. 4, 395–421.
12. D. A. Kazhdan and S. J. Patterson, *Metaplectic forms*, Inst. Hautes Études Sci. Publ. Math. (1984), no. 59, 35–142.
13. A. N. Kirillov and A. D. Berenstein, *Groups generated by involutions, Gelfand-Tsetlin patterns, and combinatorics of Young tableaux*, Algebra i Analiz **7** (1995), no. 1, 92–152.
14. P. Littelmann, *Cones, crystals, and patterns*, Transform. Groups **3** (1998), no. 2, 145–179.
15. P. J. McNamara, *Metaplectic Whittaker functions and crystal bases*, Duke Math. J. **156** (2011), no. 1, 29–31.
16. P. McNamara, *Principal series representations of metaplectic groups over local fields*, in this volume.
17. H. Matsumoto, *Sur les sous-groupes arithmétiques des groupes semi-simples déployés*, Ann. Sci. École Norm. Sup. (4) **2** (1969), 1–62.
18. T. Shintani, *On an explicit formula for class-*1 *"Whittaker functions" on GL_n over P-adic fields*, Proc. Japan Acad. **52** (1976), no. 4, 180–182.

Chapter 4
Metaplectic Whittaker Functions and Crystals of Type B

Ben Brubaker, Daniel Bump, Gautam Chinta, and Paul E. Gunnells

Abstract The spherical metaplectic Whittaker function on the double cover of $\mathrm{Sp}(2r, F)$, where F is a nonarchimedean local field, is considered from several different points of view. Previously, an expression, similar to the Casselman–Shalika formula, had been given by Bump, Friedberg, and Hoffstein as a sum is over the Weyl group. It is shown that this coincides with the expression for the p-parts of Weyl group multiple Dirichlet series of type B_r as defined by the averaging method of Chinta and Gunnells. Two conjectural expressions as sums over crystals of type B are given and another as the partition function of a free-fermionic six-vertex model system.

Keywords Weyl group multiple Dirichlet series • Crystal graph • Solvable lattice model • Metaplectic Whittaker function • Chinta-Gunnells averaging method

Let n be an integer and let F be a nonarchimedean local field whose characteristic is not a prime dividing n. Let μ_k be the group of kth roots of unity in the algebraic closure of F; we assume that $\mu_{2n} \subset F$. Let G be a split, simply connected

B. Brubaker
Department of Mathematics, MIT, Cambridge, MA 02139–4307, USA
e-mail: brubaker@math.mit.edu

D. Bump (✉)
Department of Mathematics, Stanford University, Stanford, CA 94305–2125, USA
e-mail: bump@math.stanford.edu

G. Chinta
Department of Mathematics, The City College of New York, New York, NY 10031, USA
e-mail: chinta@sci.ccny.cuny.edu

P.E. Gunnells
Department of Mathematics and Statistics, University of Massachusetts,
Amherst, MA 01003, USA
e-mail: gunnells@math.umass.edu

D. Bump et al. (eds.), *Multiple Dirichlet Series, L-functions and Automorphic Forms*,
Progress in Mathematics 300, DOI 10.1007/978-0-8176-8334-4_4,

semisimple algebraic group over F. We assume that G is actually defined over the ring $\mathfrak{o}$ of integers in F in such a way that $K = G(\mathfrak{o})$ is a special maximal compact subgroup of $G(F)$.

Matsumoto [26] constructed an n-fold metaplectic cover $\tilde{G}(F)$ of $G(F)$. For this, we only need $\mu_n \subset F$, but the hypothesis $\mu_{2n} \subset F$ simplifies the metaplectic cocycle and the resulting formulas. We are interested in values of a spherical Whittaker function W on $\tilde{G}(F)$.

Let $G = \mathrm{Sp}_{2r}$ and let the cover degree $n = 2$. In this case, we connect a known description of the Whittaker function to the theory of multiple Dirichlet series:

- Bump, Friedberg, and Hoffstein [12] gave a description of the Whittaker function, essentially as a sum of at most 2^r irreducible characters of $\mathrm{Sp}(2r)$, that is, of Cartan type C_r.
- Chinta and Gunnells [17] gave a recipe for the p-parts of Weyl group multiple Dirichlet series for any root system Φ and any positive integer n. In the special case $\Phi = B_r$ and $n = 2$, we show that this agrees with the description of the Whittaker function in [12].

In addition to these descriptions, we have three other conjectural formulas for the metaplectic Whittaker function on Sp_{2r} using the root system of type B_r. Let λ denote a dominant weight for this root system and let t_λ be an element of the split maximal torus parametrized by λ. Then:

- The value of the Whittaker function W at t_λ may be expressed as a sum over the Kashiwara crystal $\mathcal{B}_\lambda$ (Conjecture 1).
- The value $W(t_\lambda)$ may be expressed as a sum over the Kashiwara crystal $\mathcal{B}_{\lambda+\rho}$, where ρ is the Weyl vector (Conjecture 2).
- The value $W(t_\lambda)$ may be expressed as the partition function for a statistical lattice model–square ice with U-turn boundary (Conjecture 3).

The second and third conjectural descriptions are easily seen to be equivalent but give rise to very different considerations. We offer partial proofs of these conjectures in the following sections. The conjectures are further convincingly supported by extensive calculations using SAGE. An interesting feature of this situation is the interplay between type B descriptions and type C descriptions.

We thank the anonymous referee for a careful reading of this chapter. This work was supported by NSF grants DMS-0801214, DMS-0844185, DMS-0847586, and DMS-1001079.

4.1 The Classical Case: The Casselman–Shalika Formula

Before considering the metaplectic case, let us review the situation when $n = 1$, so that $G(F)$ and $\tilde{G}(F)$ are the same. Let Λ be the weight lattice of the connected L-group ${}^LG^\circ$. It is the group $X({}^LT)$ of rational characters of a maximal torus ${}^LT$

of ${}^L G^\circ$. If $\lambda \in \Lambda$ and $z \in {}^L T$, we will denote by z^λ the value of λ at z. Let Φ be the root system of ${}^L G^\circ$, so that the root system of G is the dual root system $\hat{\Phi}$.

If T is an F-split torus of G, then $\Lambda \cong T(F)/T(\mathfrak{o})$. If $\lambda \in \Lambda$, let t_λ be a representative of its coset in $T(F)$. Unramified quasicharacters of $T(F)$ correspond to elements of ${}^L T$. Indeed, an unramified quasicharacter ξ of $T(F)$ is a quasicharacter that is trivial on $T(\mathfrak{o})$, that is, a character of Λ, and so there is an element $z \in {}^L T$ such that $\xi(t_\lambda) = z^\lambda$. In this case, we write $\xi = \xi_z$.

If α is a positive root, then the coroot $\alpha^\vee$ is a positive root of G with respect to T. Let $X_{\alpha^\vee}$ be the corresponding root eigenspace in Lie(G), and let N be the maximal unipotent subgroup with Lie algebra $\bigoplus_{\alpha\in\Phi^+} X_{\alpha^\vee}$. Then $B = TN$ is a Borel subgroup.

Let ψ_N be a nondegenerate character of N. Then ψ_N is trivial on $\exp(X_{\alpha^\vee})$ if α is a positive root that is not simple. If α is a simple positive root, then we may arrange that ψ_N is trivial on $\exp(X_{\alpha^\vee}) \cap K$ but no larger subgroup of $\exp(X_{\alpha^\vee})$.

Let $\xi = \xi_z$ be a character of $T(F)$, which we extend to a character of $B(F)$ by taking $N(F)$ to be in the kernel. Let δ be the modular quasicharacter of $B(F)$. The normalized induced representation $\pi(\xi)$ consists of all locally constant functions $f : G(F) \longrightarrow \mathbb{C}$ such that $f(bg) = (\xi\delta^{1/2})(b)f(g)$, with $G(F)$ acting by right translation. The standard spherical vector f° is the unique function such that $f^\circ(k) = 1$ for $k \in K$. Let w_0 be a representative of the long Weyl group element. We may assume that $w_0 \in K$. Then the spherical Whittaker function is

$$W(g) = \int_{N(F)} f^\circ(w_0 n g)\psi_N(n)\,\mathrm{d}n. \tag{1}$$

If $\xi = \xi_z$ then the integral is convergent provided $|z^\alpha| < 1$ for $\alpha \in \Phi^+$. For other z, it may be defined by analytic continuation from this domain.

According to the formula of Casselman and Shalika [13], we have $W(t_\lambda) = 0$ unless the weight λ is dominant, and if λ is dominant, then

$$W(t_\lambda) = \prod_{\alpha\in\Phi+} (1 - q^{-1}z^\alpha)\chi_\lambda(z), \tag{2}$$

where χ_λ is the irreducible character of ${}^L G^\circ$ with highest weight λ and q is the cardinality of the residue field.

Let $\mathcal{B}_\lambda$ be the Kashiwara crystal with highest weight λ, so that

$$\chi_\lambda(z) = \sum_{v\in\mathcal{B}_\lambda} z^{\mathrm{wt}(v)}, \tag{3}$$

where wt denotes the weight function on the crystal. Ignoring the normalizing constant $\prod_{\alpha\in\Phi+}(1 - q^{-1}z^\alpha)$ in (2), this could be regarded as a formula for the Whittaker function.

We note that by the Weyl character formula

$$\prod_{\alpha\in\Phi^+}(1-z^\alpha)\chi_\lambda(z)=\sum_{w\in W}(-1)^{l(w)}z^{w(\rho+\lambda)+\rho}, \qquad \rho=\frac{1}{2}\sum_{\alpha\in\Phi^+}\alpha.$$

The factor $\prod_{\alpha\in\Phi^+}(1-z^\alpha)$ is the *Weyl denominator*, and the factor $\prod_{\alpha\in\Phi^+}(1-q^{-1}z^\alpha)$ which appears in (2) is a deformation of this factor.

We are therefore interested in deformations of the Weyl character formula in which the deformed denominator appears. A typical such formula will have the form

$$\prod_{\alpha\in\Phi^+}(1-q^{-1}z^\alpha)\chi_\lambda(z)=\sum_{v\in\mathcal{B}_{\lambda+\rho}}G(v)z^{\mathrm{wt}(v)}, \tag{4}$$

where $\mathcal{B}_{\lambda+\rho}$ is the Kashiwara crystal with highest weight $\lambda+\rho$ and wt is the standard weight function on the crystal. We will call a function G on $\mathcal{B}_{\lambda+\rho}$ which satisfies this identity a *Tokuyama function*. The archetype is the formula of Tokuyama [28], where it was stated in the language of Gelfand–Tsetlin patterns and translated into the crystal language in [9]. This is for Cartan type A. For Cartan types C and D, see [3, 15] in this volume.

For general n, we may define the metaplectic Whittaker function by an integral generalizing (1), and then ask for a formula of the form

$$W(t_\lambda)=\delta^{1/2}(t_\lambda)\sum_{v\in\mathcal{B}_{\lambda+\rho}}G(v)z^{\mathrm{wt}(v)}. \tag{5}$$

We will give analogs of both (3) and (4) for the metaplectic Whittaker function on the double cover of $\mathrm{Sp}_{2r}(F)$. However, this is the *only* metaplectic example where we have an analog of (3), whereas analogs of (4) may be found in many cases of group and degree of metaplectic cover:

- $G=\mathrm{SL}_n$ and any n: [7, 9, 10]
- $G=\mathrm{Spin}(2r+1)$ and n odd: [3] (rigorously for $n=1$ or n sufficiently large)
- $G=\mathrm{Spin}(2r)$ and n even: [15]
- $G=\mathrm{Sp}(2r)$ and n even: this chapter (rigorously for $n=2$)

4.2 The Metaplectic Whittaker Function

We review the formula for the metaplectic Whittaker function on the double cover of $\mathrm{Sp}_{2r}(F)$ which was found by Bump, Friedberg, and Hoffstein. We are assuming that $\mu_4\subset F$, which simplifies the formula slightly, since the quadratic Hilbert symbol $(-1,a)_2=(a,a)_2=1$ because -1 is a square.

Let $\mathrm{Sp}_{2r} = \{g \in \mathrm{GL}_{2r} \mid gJ^t g = J\}$, with $J = \begin{pmatrix} & -J_r \\ J_r & \end{pmatrix}$, $J_r = \begin{pmatrix} & & 1 \\ & \cdot^{\cdot^{\cdot}} & \\ 1 & & \end{pmatrix}$.

The metaplectic cocycle defining the double cover satisfies

$$\sigma\left(\begin{pmatrix} x_1 & & & & & \\ & \ddots & & & & \\ & & x_r & & & \\ & & & x_r^{-1} & & \\ & & & & \ddots & \\ & & & & & x_1^{-1} \end{pmatrix}, \begin{pmatrix} y_1 & & & & & \\ & \ddots & & & & \\ & & y_r & & & \\ & & & y_r^{-1} & & \\ & & & & \ddots & \\ & & & & & y_1^{-1} \end{pmatrix}\right) = \prod (x_i, y_i)_2.$$

The double cover $\widetilde{\mathrm{Sp}}_{2r}(F)$ consists of pairs (g, ε) with $g \in \mathrm{Sp}_{2r}(F)$ and $\varepsilon = \pm 1$. The multiplication is $(g, \varepsilon)(g', \varepsilon') = (gg', \varepsilon\varepsilon'\sigma(g, g'))$. Let $\Lambda_C = \mathbb{Z}^r$; in the next section, we will interpret this as the weight lattice of Cartan type C_r. An element $\lambda = (\lambda_1, \ldots, \lambda_r) \in \Lambda_C$ is *dominant* if $\lambda_1 \geqslant \cdots \geqslant \lambda_r \geqslant 0$. We define the "alternator"

$$\mathcal{A} = \sum_{w \in W} (-1)^{l(w)} w \tag{6}$$

as a member of the group algebra of the Weyl group W. As a group acting on the spectral parameters $z = (z_1, \ldots, z_r)$, W is the group generated by the $r!$ permutations and the 2^r transformations $z_i \to z_i^{\pm 1}$. The r simple reflections $s_1, \ldots, s_r \in W$ correspond to $s_i : z_i \leftrightarrow z_{i+1}$ for $i = 1, \ldots, r-1$ and $s_r : z_r \mapsto 1/z_r$. We will denote $z^\lambda = \prod z_i^{\lambda_i}$ for $\lambda \in \Lambda_C$. Let $\rho_C = (r, r-1, \ldots, 1)$ denote the Weyl vector. By the Weyl denominator formula,

$$\Delta_C(z) := \sum_{w \in W} (-1)^{\ell(w)} w(z^{\rho_C}) = z^{-\rho_C} \prod_{i=1}^{r} (1 - z_i^2) \prod_{i<j} (1 - z_i z_j)(1 - z_i z_j^{-1}).$$

We sometimes simply write the denominator as Δ_C, when clear from the context.

If $\lambda \in \Lambda_C$, let

$$t_\lambda = \begin{pmatrix} p^{\lambda_1} & & & & & \\ & \ddots & & & & \\ & & p^{\lambda_r} & & & \\ & & & p^{-\lambda_r} & & \\ & & & & \ddots & \\ & & & & & p^{-\lambda_1} \end{pmatrix}.$$

We fix an additive character ψ on F. This gives rise to a nondegenerate character ψ_N on the subgroup $N(F)$ of upper triangular unipotent matrices n of $\mathrm{Sp}_{2r}(F)$ by $\psi_N(n) = \psi(n_{12} + n_{23} + \cdots n_{r,r+1})$. The cocycle $\sigma(n, g) = \sigma(g, n) = 1$ for

$n \in N(F)$ and g arbitrary, so the map $N(F) \longrightarrow \widetilde{\mathrm{Sp}}_{2r}(F)$ given by $n \mapsto (n,1)$ is a homomorphism, and we may identify $N(F)$ with its image.

If $a \in F^\times$, let $\gamma(a) = \sqrt{|a|}\int \psi(ax^2)\,\mathrm{d}x / \int \psi(x^2)\,\mathrm{d}x$ where the integral is taken over any sufficiently large fractional ideal. Let $\mathbf{s} : T(F) \longrightarrow \widetilde{\mathrm{Sp}}_{2r}(F)$ be the map $t \mapsto \mathbf{s}(t) = (t,1)$. Then $\gamma(ab)/\gamma(a)\gamma(b) = (a,b)_2$, the local quadratic Hilbert symbol.

Theorem 1. (Bump, Friedberg, Hoffstein) *If $\lambda \in \Lambda_C$ is dominant, we have*

$$W(t_\lambda) = \delta^{1/2}(t_\lambda)\frac{1}{\Delta_C}\mathcal{A}\left(z^{\lambda+\rho_C}\prod_{i=1}^{r}\left(1-q^{-1/2}z_i^{-1}\right)\right)W(1).$$

Moreover,

$$W(1) = \left(\prod_{i=1}^{r}\gamma(p^{\lambda_i})^{-1}\right)\prod_i\left(1+q^{-\frac{1}{2}}z_i\right)\prod_{i<j}\left(1-q^{-1}z_iz_j^{-1}\right)\left(1-q^{-1}z_iz_j\right).$$

If λ is not dominant, then $W(t_\lambda) = 0$.

Let us combine the two most important parts of this formula and write

$$W(\lambda) = \prod_i\left(1+q^{-\frac{1}{2}}z_i\right)\prod_{i<j}\left(1-q^{-1}z_iz_j^{-1}\right)\left(1-q^{-1}z_iz_j\right)$$
$$\times\frac{1}{\Delta_C}\mathcal{A}\left(z^{\lambda+\rho_C}\prod_{i=1}^{r}\left(1-q^{-1/2}z_i^{-1}\right)\right). \tag{7}$$

We note that in this context, λ is integral but (7) makes sense if λ is half-integral. Furthermore, the Whittaker function can be extended to the larger group GSp_{2r}. It is natural to expect that our results can be extended to GSp_{2r} and that the values of (7) when λ is half-integral are to be interpreted as values of the Whittaker function on GSp_{2r}. Although we cannot confirm this when λ is half-integral, we will make some observations about the values of (7) in this case.

4.3 An Embarrassment of L-Groups

Although Langlands only defined an L-group for algebraic groups, there is a natural candidate for an L-group of $\tilde{G}(F)$ when G is split. For $G = \mathrm{Sp}_{2r}$, it is natural to assume that the L-group should be

$$\begin{cases} \mathrm{Sp}_{2r}(\mathbb{C}) & \text{if } n \text{ is even,} \\ \mathrm{Spin}_{2r+1}(\mathbb{C}) & \text{if } n \text{ is odd.} \end{cases}$$

For example, the alternation of the Cartan type of the L-group is suggested by Savin [27], who found that the Cartan type of the genuine part of the Iwahori Hecke algebra was isomorphic to that of $\mathrm{Sp}_{2r}(F)$ if n is odd and of $\mathrm{Spin}_{2r+1}(F)$ if n is even, suggesting that the L-group of the metaplectic n-fold cover of Sp_{2r} should be isomorphic to the L-groups of these groups. Thus, we may provisionally expect that in generalizing the Casselman–Shalika formula to the double cover of Sp_{2r}, the role of ${}^L G^\circ$ should be played by $\mathrm{Sp}_{2r}(\mathbb{C})$, and indeed, such a generalization was found by Bump, Friedberg, and Hoffstein [12].

It is therefore a little surprising that in generalizing (5), the relevant crystal $\mathcal{B}_\lambda$ is not of type C_r but rather of type B_r! In explaining this, both the representations of $\mathrm{Sp}_{2r}(\mathbb{C})$ (type C_r) and $\mathrm{Spin}_{2r+1}(\mathbb{C})$ (type B_r) will play a role.

We will compare these representation theories by the *ad hoc* method of identifying the ambient spaces of their weight lattices. The weight lattice Λ_C of type C_r is $\mathbb{Z}^r$. The lattice Λ_C has index two in the weight lattice Λ_B of type B_r. The lattice Λ_B consists of $\lambda = (\lambda_1, \dots, \lambda_r) \in \frac{1}{2}\mathbb{Z}^r$ such that all $\lambda_i - \lambda_j \in \mathbb{Z}$. The Weyl group W of type B_r is the same as the Weyl group of type C_r; acting on Λ_B or Λ_C, it is generated by simple reflections $s_1, \dots, s_r$ where s_i acting on $\Lambda = \mathbb{Z}^r$ interchanges λ_i and λ_{i+1} in $\lambda = (\lambda_1, \dots, \lambda_r)$ when $i < r$, and s_r sends $\lambda_r \to -\lambda_r$. The Weyl vector ρ of any root system is half the sum of the positive roots, and so for B_r and C_r, respectively, we have

$$\rho_B = \left(r - \frac{1}{2}, r - \frac{3}{2}, \dots, \frac{1}{2}\right), \qquad \rho_C = (r, r-1, \dots, 1).$$

If $\lambda \in \Lambda_C$ is a dominant weight, then the irreducible character of $\mathrm{Sp}_{2r}(\mathbb{C})$ with highest weight λ will be denoted χ_λ^C, and similarly, if $\lambda \in \Lambda_B$ is a dominant weight, the irreducible character of $\mathrm{Spin}_{2r+1}(\mathbb{C})$ with highest weight λ will be denoted χ_λ^B. In either case, let g be an element of the relevant group. Let $z = (z_1, \dots, z_r)$ be such that the eigenvalues of g are $z_i^{\pm 1}$ in the symplectic case or such that the eigenvalues of the image of g in $\mathrm{SO}_{2r+1}(\mathbb{C})$ are $z_i^{\pm 1}$ and 1 in the spin case. Then the Weyl character formula asserts that

$$\chi_\lambda^C(g) = \frac{1}{\Delta_C}\mathcal{A}(z^{\rho_C+\lambda}) \qquad \text{or} \qquad \chi_\lambda^B(g) = \frac{1}{\Delta_B}\mathcal{A}(z^{\rho_B+\lambda})$$

depending on which case we are in, where the Weyl denominators are

$$\begin{aligned}\Delta_C &= \mathcal{A}(z^{\rho_C}) \\ &= \prod_{i<j}\left[\left(z_i^{1/2}z_j^{-1/2} - z_i^{-1/2}z_j^{1/2}\right)\left(z_i^{1/2}z_j^{1/2} - z_i^{-1/2}z_j^{-1/2}\right)\right]\prod_i\left(z_i - z_i^{-1}\right),\end{aligned}$$

$$\begin{aligned}\Delta_B =& \mathcal{A}(z^{\rho_B})\\ =& \prod_{i<j}\left[\left(z_i^{1/2}z_j^{-1/2}-z_i^{-1/2}z_j^{1/2}\right)\left(z_i^{1/2}z_j^{1/2}-z_i^{-1/2}z_j^{-1/2}\right)\right]\prod_i\left(z_i^{1/2}-z_i^{-1/2}\right).\end{aligned}$$

In particular,

$$\frac{\Delta_C}{\Delta_B}=\prod_{i=1}^{r}\left(z_i^{1/2}+z_i^{-1/2}\right)=\frac{z^{\rho_C}}{z^{\rho_B}}\prod_{i=1}^{r}\left(1+z_i^{-1}\right). \tag{8}$$

On the face of it, the last formula has little meaning, since the Weyl denominators live on different groups. We will use it in the next section.

4.4 Ambivalence of the L-Group

Let G be a reductive group over a nonarchimedean local field F. Let us consider the role of the L-group in the Casselman–Shalika formula. The semisimple conjugacy classes of ${}^L G^\circ$ parametrize the spherical representations of $G(F)$. Let π be a spherical representation and $z = z_\pi$ the parametrizing conjugacy class. Then the values of the irreducible characters of $G(F)$ on z equal the values of the spherical Whittaker function of π.

So we should seek a similar interpretation in the metaplectic case. Let $G = \mathrm{Sp}_{2r}(F)$ and let $\tilde{G}(F)$ be the double cover. Either $\mathrm{Sp}_{2r}(\mathbb{C})$ or $\mathrm{SO}_{2r+1}(\mathbb{C})$ will serve to parametrize the principal series representations of G.

We first seek an interpretation of the factor

$$\prod_i\left(1+q^{-\frac{1}{2}}z_i\right)\prod_{i<j}\left(1-q^{-1}z_iz_j^{-1}\right)\left(1-q^{-1}z_iz_j\right) \tag{9}$$

appearing in (7) as a deformation of a Weyl denominator. The Weyl denominators of types B and C are, respectively, $z^{-\rho_B}$ and $z^{-\rho_C}$ times

$$\prod_i(1-z_i)\prod_{i<j}(1-z_iz_j^{-1})(1-z_iz_j),\quad\text{and}\quad\prod_i(1-z_i^2)\prod_{i<j}(1-z_iz_j^{-1})(1-z_iz_j).$$

Now, there are two ways of looking at (9). We may write it as

$$\prod_i\left(1-q^{-\frac{1}{2}}z_i\right)^{-1}\times\prod_i\left(1-q^{-1}z_i^2\right)\prod_{i<j}\left(1-q^{-1}z_iz_j^{-1}\right)\left(1-q^{-1}z_iz_j\right),$$

and the factor in front may be interpreted as the p-part of a quadratic L-function. The remaining terms in the product give the typical deformation of the Weyl denominator of type C, and taking the classical limit $q^{-1}\to 1$ recovers the familiar denominator formula in type C. On the other hand, we may let $q^{-\frac{1}{2}} \to -1$, in which case (9) becomes the Weyl denominator of type B.

A similar dual interpretation pertains with the factor

$$\frac{1}{\Delta_C}\mathcal{A}\left(z^{\lambda+\rho_C}\prod_{k=1}^{r}\left(1-q^{-1/2}z_i^{-1}\right)\right). \tag{10}$$

On the one hand, if we expand the product, we get a sum

$$\sum_{S\subset\{1,2,3,\ldots,r\}}(-q^{1/2})^{|S|}\frac{1}{\Delta_C}\mathcal{A}\left(z^{\lambda+\rho_C}\prod_{i\in S}z_i^{-1}\right). \tag{11}$$

Each term is either zero or an irreducible character of $\mathrm{Sp}_{2r}(\mathbb{C})$ by the Weyl character formula. Hence, (10) may be regarded as a sum of $\leqslant 2^r$ irreducible characters of $\mathrm{Sp}_{2r}(\mathbb{C})$ and thus has a type C flavor. But, on the other hand, let us again specialize $q^{\frac{1}{2}} \to -1$. Then using (8), the factor (10) becomes

$$\frac{1}{\Delta_B}\mathcal{A}\left(z^{\lambda+\rho_B}\right) = \chi_\lambda^B(z). \tag{12}$$

This formula generalizes to a formula like (3) for the metaplectic Whittaker function in the form (10). We will discuss this point in a subsequent section.

4.5 The Weyl Group Averaging Method

Chinta and Gunnells [16,17] gave a construction of the p parts for multiple Dirichlet series which applies to any root system and choice of fixed positive integer n. In this section, we show that their construction gives the metaplectic Whittaker function of $\widetilde{\mathrm{Sp}}_{2r}(F)$ when the root system is of type B_r and the cover degree $n = 2$. (For additional articles on relations between multiple Dirichlet series and Whittaker functions, see also Chinta and Offen [18] and Chinta, Friedberg, and Gunnells [14].)

This method begins by defining an action on rational functions of the spectral parameters, which we review in the case at hand.

As before, let T be the maximal torus of diagonal elements in SO_{2r+1}, whose eigenvalues are $z_1,\ldots,z_r,1,z_r^{-1},\ldots,z_1^{-1}$. Let T' be the preimage of T in Spin_{2r+1}. The coordinate ring $\mathbb{C}[T']$ of T' is then generated by $z_i^{\pm 1}$ and by $\sqrt{z_1\cdots z_r}$. We remark that $\mathbb{C}[T']$ can be identified with the group ring $\mathbb{C}[\Lambda_B]$, with the z_i, $1\leq i<r$ corresponding to the first $r-1$ fundamental weights and the product $\sqrt{z_1\cdots z_r}$ corresponding to the spin representation (as before, we think of Λ_C sitting as a sublattice of Λ_B of index 2).

Let $\mathbb{C}(T')$ be the fraction field of $\mathbb{C}[T']$. Consider the rational map $\mathbb{C}(T') \to \mathbb{C}(T')$ that takes $z_i \to z_i$ for $i = 1, \ldots, r-1$ and $\sqrt{z_1 \cdots z_r} \to -\sqrt{z_1 \cdots z_r}$. We write this map as $f(z) \mapsto f(\varepsilon z)$, slightly abusing notation from before. Then we define an action of W by

$$(f|s_i)(z) = f(s_i z), \quad 1 \leq i < r,$$

and

$$(f|s_r)(z) = \frac{1 - q^{-1/2} z_r^{-1}}{1 - q^{-1/2} z_r} f^+(s_r z) + \frac{1}{z_r} f^-(s_r z),$$

where

$$f^+(z) = \frac{f(z) + f(\varepsilon z)}{2}, \quad f^-(z) = \frac{f(z) - f(\varepsilon z)}{2}.$$

The braid relations are satisfied, and so this definition extends to a right action $f \mapsto f|w$ for all $w \in W$. Now, the description of [17] for the p-part of the multiple Dirichlet series may be written as

$$\mathcal{H}(\lambda, z) = z^{\lambda+\rho_C} \sum_{w \in W} \frac{z^{-\lambda-\rho_C}|w}{\Delta_C(wz)}. \tag{13}$$

We further define a q-deformation of the Weyl denominator Δ_C by

$$D(z;q) = \prod_{i=1}^{r} \left(1 - q^{-1} z_i^2\right) \prod_{i<j} (1 - q^{-1} z_i z_j)(1 - q^{-1} z_i z_j^{-1}).$$

Let

$$P = \prod_{i=1}^{r} \left(1 - q^{-1/2} z_i\right).$$

Lemma 1. *If $f = f^+$, then*

$$\frac{(f|w^{-1})(z)}{f(wz)} = \frac{P(wz)}{P(z)}.$$

Proof. If $p(w,z) = P(wz)/P(z)$, then p satisfies the cocycle condition $p(ww', z) = p(w, w'z)\, p(w', z)$. The left-hand side also satisfies the same cocycle relation, so we are reduced to the case where w is a simple reflection, in which case, it follows easily from the definition.

Theorem 2. *We have*

$$D(z;q)\, \mathcal{H}(\lambda, z) = (-1)^r z^{\lambda+\rho_C}\, W(\lambda),$$

where $W(\lambda)$ is the Whittaker value defined in (7).

Proof. The function $\mathcal{H}(\lambda, z)$ in (13) may be rewritten in the form

$$\frac{z^{\lambda+\rho_C}}{\Delta_C}\sum_{w\in W}(-1)^{l(w)}\left(z^{-\lambda-\rho_C}|w\right) = \frac{z^{\lambda+\rho_C}}{\Delta_C\,P(z)}\mathcal{A}(P(z)z^{-\lambda-\rho_C}), \tag{14}$$

where the latter equality follows by replacing w by w^{-1} and using the previous lemma. Again, we have employed the notation for the alternator as in (6). Note in particular that for any rational function f, $\mathcal{A}(f(z)) = (-1)^r\mathcal{A}(f(w_0 z))$. Since $w_0 z = (z_1^{-1}, \dots, z_r^{-1})$, the expression on the right-hand side of (14) equals

$$\frac{(-1)^r z^{\lambda+\rho_C}}{\prod_{i=1}^r(1-q^{-1/2}z_i)}\frac{1}{\Delta_C}\mathcal{A}\left(z^{\lambda+\rho_C}\prod_{i=1}^r(1-q^{-1/2}z_i^{-1})\right).$$

Multiplying by $D(z)$ and simplifying, the statement follows.

4.6 BZL Patterns

Let w_0 be the long Weyl group element. Choose a decomposition reduced decomposition $w_0 = s_{\omega_1}\cdots s_{\omega_N}$ into a product of simple reflections where $1 \leqslant \omega_i \leqslant r$ (the rank). Let

$$\omega = (\omega_1, \dots, \omega_N)$$

be the corresponding *reduced word* for w_0.

Let $\mathcal{B}_\lambda$ be the crystal of an irreducible finite-dimensional representation of highest weight λ for any Cartan type and let W be the corresponding Weyl group. We will denote the Kashiwara (root) operators by e_i and f_i. There are maps $\mathcal{B} \longrightarrow \mathcal{B} \sqcup \{0\}$. There is a unique element $v_\lambda \in \mathcal{B}$ corresponding to the highest weight λ.

To each vertex $v \in \mathcal{B}_\lambda$ and each reduced word ω, we associate an integer sequence as follows. Let k_1 be the largest integer such that $e_{\omega_1}^{k_1}(v) \neq 0$. Then let k_2 be the largest integer such that $e_{\omega_2}^{k_2}e_{\omega_1}^{k_1}(v) \neq 0$ and so forth. Upon using all root operators in the order specified by the long word decomposition, $e_{\omega_N}^{k_N}\cdots e_{\omega_1}^{k_1}(v) = v_\lambda$, the vertex corresponding to the highest weight vector. The sequence $(k_1, \dots, k_N)$ determines v and can be arrayed in a pattern to give a convenient way of parametrizing elements of the crystal. These patterns were studied by Littelmann [24] and by Berenstein and Zelevinsky [4]. We will refer to the sequence $(k_1, \dots, k_N)$ as a *BZL string* or *BZL pattern* and write $(k_1, \dots, k_N) = \mathrm{BZL}_\omega(v)$. This construction applies equally well to any symmetrizable Kac–Moody group, but we focus entirely on type B root systems and their crystal graphs in this section.

Theorem 3. *Let $\mathcal{B}$ be a crystal graph of type B. There exists a unique function σ on $\mathcal{B}$ taking values in the nonnegative integers with the following properties. If v_λ is the highest weight vector, then $\sigma(v_\lambda) = 0$. If $x, y \in \mathcal{B}$ and $f_i(x) = y$ with $i < r$, then $\sigma(x) = \sigma(y)$. If $e_r(x) = 0$ and $y = f_r^k(x)$, then*

$$\sigma(y) = \begin{cases} \sigma(x) & \text{if } k \text{ is even,} \\ \sigma(x) + 1 & \text{if } k \text{ is odd.} \end{cases}$$

Let us illustrate this with an example:

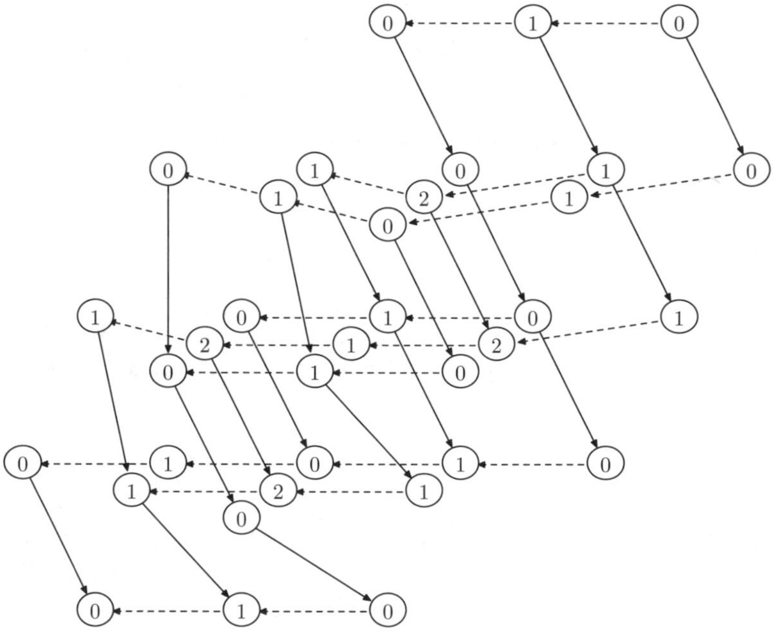

This illustrates the crystal with the highest weight $\lambda = (2, 1)$ for B_2. We draw $x \longrightarrow y$ with a solid arrow if $y = f_1(x)$ and with a dashed arrow if $y = f_2(x)$. The vertex in the upper right-hand corner is v_λ. The values of σ are shown for every element.

Proof. We will give one definition of σ for each reduced decomposition

$$w_0 = s_{\omega_1} s_{\omega_2} \cdots s_{\omega_{r^2}}$$

of the long element. We will show that these definitions are all equivalent, then deduce the statement of the theorem. We start with the BZL string of $v \in \mathcal{B}$ corresponding to this word. Thus, corresponding to the word

$$\omega = (\omega_1, \omega_2, \ldots, \omega_{r^2}),$$

we produce the sequence $k_1, k_2, \ldots, k_{r^2}$ with each k_n defined by

$$e_{\omega_n}^{k_n} \cdots e_{\omega_1}^{k_1} v \neq 0, \qquad e_{\omega_n}^{k_n+1} \cdots e_{\omega_1}^{k_1} v = 0,$$

so that $e_{\omega_{r^2}}^{k_{r^2}} \cdots e_{\omega_1}^{k_1} v = v_\lambda$ is the highest weight element of the crystal base. Define

$$\sigma_\omega(v) = \sum_{\omega_j = r} \begin{cases} 1 \text{ if } k_j \text{ is odd,} \\ 0 \text{ if } k_j \text{ is even.} \end{cases} \tag{15}$$

We wish to assert that if ω and $\omega' = (\omega'_1, \ldots, \omega'_{r^2})$ are two reduced decompositions, then $\sigma_\omega = \sigma_{\omega'}$.

The proof will involve a reduction to the rank two case, so let us first prove that the statement is true for crystals of type B_2. In this case, there are only two reduced words, and we may assume that $\omega = \{1, 2, 1, 2\}$ and $\omega' = \{2, 1, 2, 1\}$. In this case, Littelmann (cf. Sect. 2 of [24]) proved that

$$\begin{aligned} k'_1 &= \max(k_4, k_3 - k_2, k_2 - k_1), \\ k'_2 &= \max(k_3, k_1 - 2k_2 + 2k_3, k_1 + 2k_4), \\ k'_3 &= \min(k_2, 2k_1 - k_3 + k_4, k_4 + k_1), \\ k'_4 &= \min(k_1, 2k_2 - k_3, k_3 - 2k_4). \end{aligned}$$

From this, it follows easily that the number of odd elements of the set $\{k'_2, k'_4\}$ is the same as the number of odd elements of the set $\{k_1, k_3\}$, that is, $\sigma_\omega = \sigma_{\omega'}$.

We turn now to the proof that $\sigma_\omega = \sigma_{\omega'}$ for arbitrary rank r. Consider the equivalence relation on all reduced words representing w_0 generated by $\omega \sim \omega'$ if ω' is obtained from ω by replacing a string $\{l, m, l, m, \ldots\}$ of length equal to the order N of $s_l s_m$ in the Weyl group by the string $\{m, l, m, l, \ldots\}$ of the same length. By a theorem of Tits, any two reduced decompositions are equivalent under this relation. As a consequence, it is sufficient to show that $\sigma_\omega = \sigma_{\omega'}$ when ω' is obtained by replacing an occurrence of l, m, l by m, l, m (if $m = l + 1 < r$), or an occurrence of l, m, l, m by m, l, m, l (if $l = r - 1, m = r$), or an occurrence of l, m by m, l when $|l - m| > 1$.

Suppose that $i_t = l$, $i_{t+1} = m$, etc. are the elements of ω that are changed in ω'. The elements i_{t-1} and i'_{t-1} of ω and ω' preceding this string (if it is not initial) are not l nor m and similarly for the elements following it. Let

$$v_h = e_{i_h}^{k_h} \cdots e_{i_1}^{k_1} v, \qquad e_{i_h}^{k_h+1} \cdots e_{i_1}^{k_1} v = 0,$$

$$v'_h = e_{i'_h}^{k'_h} \cdots e_{i'_1}^{k'_1} v, \qquad e_{i'_h}^{k'_h+1} \cdots e_{i'_1}^{k'_1} v = 0,$$

so $v_0 = v'_0 = v$ and $v_{r^2} = v'_{r^2} = v_\lambda$. We will argue that the sequences $(v_0, v_2, \ldots, v_{r^2})$ and $(v'_0, v'_2, \ldots, v'_{r^2})$ and the sequences $(k_1, k_2, \ldots, k_{r^2})$ and $(k'_1, k'_2, \ldots, k'_{r^2})$ are identical, except at indices t through $t + N - 2$ where ω and ω' differ.

To see this, remove all edges of the crystal graph except those labeled l and m, which produces a crystal graph $\mathcal{B}'$ of type A_2, B_2, $A_1 \times A_1$, or $A_1 \times B_1$. Clearly, $v_{t-1} = v'_{t-1}$ since ω and ω' agree up to this point. Let $\mathcal{B}''$ be the connected component of $\mathcal{B}'$ containing this. Then v_{t+N-1} is the highest weight vector in $\mathcal{B}''$ and so is v'_{t+N-1}. It is now clear that the portion of the BZL pattern which lies within this crystal is the only part of $k_1, \ldots, k_{r^2}$ which is different from $k'_1, \ldots, k'_{r^2}$, and we have only to show that the number of k_i within this subpattern with $\omega_i = r$ such that k_i is odd is the same as for the k'_i. That is, we have reduced to the rank two case. If $\mathcal{B}'$ is of type B_2, we have proven this, and the other three cases are trivial, since an A_2 or $A_1 \times A_1$ crystal has no edges of type r, while an $A_1 \times B_1$ crystal is just a Cartesian product.

Now, let $1 \leqslant i \leqslant r$. To verify the assertion that $\sigma(x) = \sigma(y)$ if $f_i(x) = y$, choose a word ω whose first element $\omega_1 = i$. If $(k_1, \ldots, k_{r^2}) = \mathrm{BLZ}_\omega(x)$, then $(k_1 + 1, k_2, \ldots, k_{r^2}) = \mathrm{BZL}_\omega(y)$. Since $\sigma(x)$ is the number of odd k_i with $\omega_i = r$, it is obvious that $\sigma(x) = y$. On the other hand, suppose that $e_r(x) = 0$. Choosing ω such that $\omega_1 = r$, we have $\mathrm{BZL}(x) = (k_1, \ldots, k_{r^2})$ with $k_1 = 0$, while $\mathrm{BZL}(f_r^k(x)) = (k, k_2, \ldots, k_{r^2})$, and so obviously, $\sigma(f_r^k(x)) = \sigma(x)$ if k is odd and $\sigma(x) + 1$ if k is even.

We recall that the Weyl group acts on the crystal: each simple reflection s_i acts by reversing the i-root strings. It is shown that this action gives rise to a well-defined action of W on the crystal in Littelmann [25].

Proposition 1. *If λ is integral, then the function σ is constant on W orbits of the crystal.*

Proof. It is clear from the definition that reversing the i-root string through $v \in \mathcal{B}_\lambda$ does not change $\sigma(v)$ if $i < r$ since σ is constant on the root string in that case. If $i = r$, then the fact that λ is integral means that each root string has odd length, and therefore $\sigma(s_i(v)) = \sigma(v)$ in this case also since $v - s_r(v) = k\alpha_r$ with k even. (Note that if λ is half-integral, the Weyl group action does not preserve σ.)

Conjecture 1. Assume that λ is integral. Then

$$\frac{1}{\Delta_C}\mathcal{A}\left(z^{\lambda+\rho_C}\prod_{k=1}^{r}\left(1-q^{-1/2}z_i^{-1}\right)\right) = \sum_{v\in\mathcal{B}_\lambda}\left(-q^{1/2}\right)^{\sigma(v)} z^{\mathrm{wt}(v)}.$$

This expresses the metaplectic Whittaker function (except for its normalizing constant) as a sum over the crystal. As noted in a previous section, the left-hand side may be expanded as a polynomial in q whose coefficients are composed of irreducible characters. Hence, the above proposition may be viewed as partial evidence for the conjecture. It has also been verified numerically for many choices of λ.

4.7 Decorated BZL Patterns

Let us disregard the normalizing constant (9) for the time being, and consider (10) to be the value of the p-adic Whittaker function at t_λ in $\widetilde{\mathrm{Sp}}_{2r}(F)$, where λ is a dominant weight for $\mathrm{Spin}_{2r+1}(\mathbb{C})$. Strictly speaking, this only makes sense if λ is integral. However, if λ is half-integral, it is probable that this scenario can be extended, taking t_λ in $\widetilde{\mathrm{GSp}}_{2r}(F)$. In any case, (10) is defined whether λ is integral or half-integral.

We saw in (12) that when $q^{1/2}$ is specialized to -1, the value of (10) becomes the character χ_λ^B of an irreducible representation of $\mathrm{Spin}_{2r+1}(\mathbb{C})$. We will reinterpret this fact in terms of crystals, showing that for any q, the expression (10) may be interpreted as a deformation of χ_λ^B. Indeed, we give a conjectural expression for the metaplectic Whittaker function evaluated at t_λ as a sum over vertices in the crystal $\mathcal{B}_\lambda$ by making use of certain decorated BZL patterns.

That is, we decorate the BZL string $k_1, \ldots, k_N$ by drawing boxes or circles around some of the entries according to the following rules. For the boxing rule, if

$$f_{\omega_i} e_{\omega_{i-1}}^{k_{i-1}} \cdots e_{\omega_1}^{k_1}(v) = 0,$$

then we box k_i. Concretely, this means that the path from v to v_λ that goes through

$$v, e_{\omega_1}^{k_1}(v), e_{\omega_2}^{k_2} e_{\omega_1}^{k_1}, \ldots$$

includes the entire ω_i-string passing through the vertex $e_{\omega_i}^{k_i} \cdots e_{\omega_1}^{k_1}(v)$. In this sense, we roughly think of the value k_i as being as large as possible and cannot be increased.

The circling rule may be regarded also very roughly as signifying that the value k_i is as small as possible and cannot be decreased. To make this precise for type B_r and for one particularly nice reduced word ω, we take a closer look at the BZL patterns as treated by Littelmann [24].

We will use the Bourbaki ordering of the weights, so that the fundamental dominant weights are $\omega_1, \ldots, \omega_r$ with $\omega_1 = (1, 0, \ldots, 0)$ the highest weight of the standard representation and $\omega_r = \left(\frac{1}{2}, \ldots, \frac{1}{2}\right)$ the highest weight of the spin representation. Then the reduced decomposition that we will use is

$$w_0 = s_r(s_{r-1}s_r s_{r-1})(s_{r-2}s_{r-1}s_r s_{r-1}s_{r-1}) \cdots (s_1 \cdots s_r \cdots s_1).$$

Thus, $\omega = (r, r-1, r, r-1, r-2, r-1, r, r-1, r-2, \ldots)$ and $N = r^2$. An alternative indexing will sometimes be convenient, so we will write alternatively

$$\mathrm{BZL}(v) = (k_1, \ldots, k_{r^2}) = (k_{r,r}, k_{r-1,r-1}, k_{r-1,r}, k_{r-1,r+1}, \ldots).$$

Following Littelmann, we put the entries into a triangular array, from bottom to top and left to right, thus

$$\left\{\begin{matrix} k_{1,1} & \cdots & & k_{1,r} & k_{1,r+1} & \cdots & k_{1,2r-1} \\ & \ddots & & \vdots & & \iddots & \\ & & k_{r-1,r-1} & k_{r-1,r} & k_{r-1,r+1} & & \\ & & & k_{r,r} & & & \end{matrix}\right\} = \left\{\begin{matrix} \vdots \\ k_5\ k_6\ k_7\ k_8\ k_9 \\ k_2\ k_3\ k_4 \\ k_1 \end{matrix}\right\}$$

Littelmann proved that the entries in each row satisfy the following inequalities (independent of the choice of highest weight λ):

$$2k_{i,i} \geqslant 2k_{i,i+1} \geqslant \cdots \geqslant 2k_{i,r-1} \geqslant k_{i,r} \geqslant 2k_{i,r+1} \geqslant \cdots \geqslant 2k_{i,2r-i} \geqslant 0.$$

Note that every value is doubled except the middle one.

We circle the BZL string entry k_i if the corresponding lower bound inequality is an equality. Let us make this explicit in the case $r = 3$. In this case,

$$\mathrm{BZL}(v) = (k_{3,3}, k_{2,2}, k_{2,3}, k_{2,4}, k_{1,1}, k_{1,2}, k_{1,3}, k_{1,4}, k_{1,5}) = (k_1, k_2, \ldots, k_9),$$

and the array is:

$$\left\{\begin{matrix} k_{1,1}\ k_{1,2}\ k_{1,3}\ k_{1,4}\ k_{1,5} \\ k_{2,2}\ k_{2,3}\ k_{2,4} \\ k_{3,3} \end{matrix}\right\} = \left\{\begin{matrix} k_5\ k_6\ k_7\ k_8\ k_9 \\ k_2\ k_3\ k_4 \\ k_1 \end{matrix}\right\}. \tag{16}$$

We have

$$k_{3,3} \geqslant 0,$$

and if $k_{3,3} = 0$, we circle it. We have $2k_{2,2} \geqslant k_{2,3}$, and if this is an equality, we circle $k_{2,2}$. Similarly, $k_{2,3} \geqslant 2k_{2,4}$, and if this is equality, we circle $k_{2,3}$.

We attach a simple root of the B_r root system to each column of the array, in this order:

$$\alpha_1, \ldots, \alpha_{r-1}, \alpha_r, \alpha_{r-1}, \ldots, \alpha_1.$$

The assignment is chosen so that a BZL string entry is in the column labeled by α_i if the corresponding element of the long word ω is i. Thus, letting c_i be the sum of the ith column, we have

$$\mathrm{wt}(v) = \lambda - (c_1 + c_{2r-1})\alpha_1 - (c_2 + c_{2r-1})\alpha_2 - \cdots - c_r\alpha_r.$$

Only α_r is a short root.

4.8 A Tokuyama Function on BZL Patterns

Let p be a prime element in the nonarchimedean local field F. Let $(\, , \,)_n$ be the local nth power Hilbert symbol. For any m and nonzero $c \in \mathfrak{o}$, we define the nth order Gauss sum

$$g_t(m,c) = \sum_{\substack{x \bmod c \\ \gcd(x,c)=1}} \psi\left(\frac{mx}{c}\right) (x,c)_n^t.$$

We will only need these for $t = 1, 2$. For a nonnegative integer a, we also use the shorthand notations

$$g_t(a) = g_t(p^{a-1}, p^a), \qquad h_t(a) = g_t(p^a, p^a).$$

In the special case $n = 2$, then all these Gauss sums may be made explicit. The Gauss sum $g_1(1, p)$ is a square root of q, which we will denote $q^{1/2}$; by choosing the additive character ψ correctly, we may arrange that it is the positive square root. Assuming $n = 2$, we then have

$$g_1(a) = q^{a-\frac{1}{2}}, \qquad h_1(a) = \begin{cases} q^{a-1}(q-1) & \text{if } a \text{ is even,} \\ 0 & \text{otherwise,} \end{cases}$$

and

$$g_2(a) = -q^{a-1}, \qquad h_2(a) = q^{a-1}(q-1).$$

We now assume that n is even, and that $\mathcal{B}$ is crystal of type B_r. If $v \in \mathcal{B}$, we define

$$G(v) = \prod_{k \in \mathrm{BZL}(v)} \begin{cases} q^{-k} h_t(k) & \text{if } k \text{ is unboxed and uncircled,} \\ q^{-k} g_t(k) & \text{if } k \text{ is boxed but not circled,} \\ 1 & \text{if } k \text{ is circled but not boxed,} \\ 0 & \text{if } k \text{ is both boxed and circled,} \end{cases}$$

where the subscript $t = t(k)$ in the cases above is 1 if the root corresponding to k is α_r, and $t = 2$ otherwise. This means that $t = 1$ if k is in the middle column of the BZL array (16) and $t = 2$ otherwise. Note that these differ from the weighting functions used in [9] in three ways:

- Due to the presence of both long and short roots, we must use two kinds of Gauss sums, indexed by t (as done in [6]).
- The function $G(v)$ has been normalized by q^{-k} which ultimately simplifies the formulas (and was called $G^\flat(v)$ in Chap. 1 of [9]).
- We have made our BZL patterns using the e_i instead of the f_i, which has the effect of permuting the contributions $G(v)$ assigned to any given weight vector v.

Now, let λ be a dominant weight. Then we claim that $G(v)$ is a Tokuyama function for the metaplectic Whittaker function. More precisely:

Conjecture 2. Assume that λ is integral. Then with $W(\lambda)$ as in (7), we have

$$W(\lambda) = \sum_{v \in \mathcal{B}_{\lambda+\rho}} G(v) z^{-\operatorname{wt}(v)}.$$

In order to make progress on this conjecture, we translate the problem to the realm of statistical lattice models. For more on the relationships between bases of highest weight representations and statistical lattice models, see [5] in this volume.

4.9 Ice Models

We will now give an alternative description of the Whittaker function as the partition function of a statistical system in the six-vertex model.

The use of the Yang–Baxter equation to evaluate the partition functions for the six-vertex model was initiated by Baxter [1]. The so-called "domain wall" boundary conditions, which are different from those used by Baxter, were introduced by Korepin, who in the field-free case obtained recursion relations for the $N \times N$ partition function; Izergin used these relations to evaluate the partition function as a determinant. See [21], Sect. VII.10, and Kuperberg [22], who explains how to use the Yang–Baxter equation to prove the Korepin–Izergin determinant formula and to use it to enumerate alternating sign matrices.

Meanwhile, Tokuyama [28] obtained a generalization of the Weyl character formula that expresses the character of the irreducible representation of $\mathrm{GL}(n, \mathbb{C})$ with highest weight λ (a fixed dominant weight) times a deformation of the Weyl denominator as a sum over Gelfand–Tsetlin patterns. Hamel and King [19] reformulated Tokyuma's result as evaluating the partition function of a six-vertex model. Their proofs were combinatorial, based on *jeu de taquin*. It was shown by Brubaker, Bump, and Friedberg [8] that the results of Tokuyama and Hamel and King may be proved using the Yang–Baxter equation.

The Yang–Baxter equation used by Brubaker, Bump, and Friedberg is a different case from that of the Korepin–Izergin determinant formula since it requires Boltzmann weights that are not "field-free," but which are "free-fermionic." (These terms are defined in Sect. 7 of [11] in this volume.) The boundary conditions are a generalization of the domain wall boundary conditions considered by Korepin and Izergin, that is, the special case $\lambda = 0$.

Regarding the free-fermionic Yang–Baxter equation, Korepin and Izergin had found a parametrized Yang–Baxter equation for the free-fermionic six-vertex model with parameter group $SL(2)$. See [21], page 126. A slightly more general parametrized Yang–Baxter equation for the free-fermionic six-vertex model, with parameter group $GL(2) \times GL(1)$, was used by Brubaker, Bump, and Friedberg [8] to prove the results of Tokuyama and Hamel and King.

Beyond [22], Kuperberg [23] considered other lattice models that he used to enumerate other classes of alternating sign matrices. One that is particularly relevant for us is the "U-turn model." Hamel and King [19] found deformations of the Weyl character formula for the group $\mathrm{Sp}(2r, \mathbb{C})$ representing a character times a deformation of the Weyl denominator, the partition function of a U-turn model. Ivanov [20] then proved a variant of this formula using the free-fermionic Yang–Baxter equation.

In view of the Casselman–Shalika formula, the partition function computed by Ivanov may be regarded as the value of a spherical Whittaker function on an odd orthogonal group. This interpretation of the results of Hamel and King and Ivanov is explained in Brubaker, Beineke, and Frechette [2]. See also [3] where an interpretation of this Whittaker function as a sum over crystals is given.

In this chapter, we will use a variant of the model considered by Hamel and King, Ivanov and Brubaker, and Beineke and Frechette in order to express the Whittaker functions on the metaplectic double covers of $\mathrm{Sp}(2r)$. The model that we give is very similar to that in Ivanov, the only difference between the systems being the Boltzmann weights at the U-turn or "cap" vertices.

We consider a rectangular grid having $2r$ rows and the number of columns to be determined. The intersections of the rows and columns of the grid will be called *vertices*. The vertices in the odd-numbered rows will be designated "Gamma ice" (labeled $\bullet$, and those in even numbered rows (labeled $\circ$) will be designated "Delta ice." Each pair of rows will be closed at the right edge by a "cap" containing a single vertex. Thus, if $r = 2$, the array looks like:

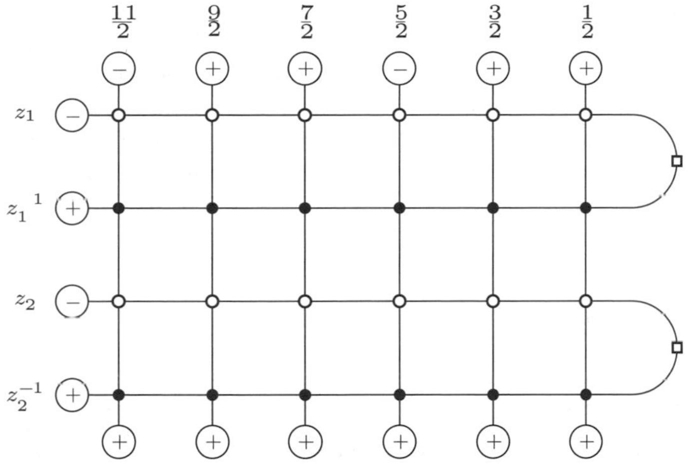

We have labeled the boundary edges by certain signs $\pm$. The interior edges will also be labeled with signs, but these signs will be variable, whereas the boundary edge signs are fixed and are part of the data describing the system.

The boundary edge signs are to be assigned as follows. We put alternating signs $-, +, -, +, \ldots$ on the left edge, so that the rows of Delta ice begin with $-$ and the rows of Gamma ice begin with $+$. We put $+$ signs at the bottom of each column. For the top, we label the columns with half integers beginning with $\frac{1}{2}$ at the right and increasing by 1 from right to left. Given a highest weight λ of type B, we put $-$ in the columns labeled from values in $\lambda + \rho_B$. Thus, if $r = 2$ and $\lambda = (4, 2)$, then $\lambda + \rho_B = \left(\frac{11}{2}, \frac{5}{2}\right)$, and so we put $-$ in those columns, as indicated in the figure above. The remaining top edges are labeled $+$.

A *state* of the system is an assignment of spins $\pm$ to the remaining interior edges. For the Gamma and Delta vertices, the assignments will be taken from Table 1 in Ivanov [20] in this volume.

For the cap vertices, which we will label with a $\square$, the two adjacent edges must have the same sign, and here we use different weights from Ivanov, as follows:

$\square$ Cap vertex	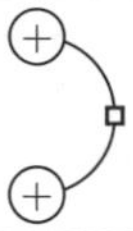	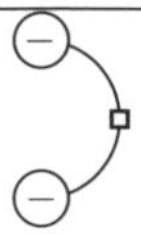
Boltzmann weight	$-\sqrt{-t}z_i^{1/2}$	$z_i^{-1/2}$

For the moment, we may regard $t, z_1, \ldots, z_r$ as arbitrary parameters to be determined later. Given a highest weight λ of type B, we may fix boundary spins as above. Then an admissible state is one in which each vertex in the state has a Boltzmann weight taken from the above table. Let $\mathfrak{S}_\lambda = \mathfrak{S}_\lambda(z_1, \ldots, z_r, t)$ be the set of all states.

Given a state $S \in \mathfrak{S}_\lambda$ of the system, the Boltzmann weight $\mathrm{BW}(S)$ of the state is the product over all vertices of the weights of the vertex. The *partition function* $Z(\mathfrak{S})$ is the sum of the $\mathrm{BW}(S)$ over all states S.

As before, we let the Weyl group W of type B_r act on the parameters $z_1, \ldots, z_r$; it is generated by permutations of the z_i and the 2^r transformations $z_i \to z_i^{\pm 1}$.

Theorem 4. *The product*

$$z^{\rho_B} \prod_i (1 - i\sqrt{t} z_i^{-1}) \left[\prod_{i>j} (1 + t z_i z_j)(1 + t z_i z_j^{-1}) \right] Z(\mathfrak{S}) \tag{17}$$

is invariant under the action of W.

The ideas of this proof are similar to those in [8] and [20], where the "caduceus" braid also appears.

Proof. Structurally, the proof is the same as that of the result in Ivanov [20]. Due to the difference in the Boltzmann weights at the cap vertices, the formulas turn out differently, and some auxiliary results are different between the two proofs.

We first show invariance under the simple reflections which interchange z_i and z_{i+1}.

The parametrized Yang–Baxter equation in [8] implies the following statement, which is the same as Lemma 1 in [20]. Given any pair $X, Y \in \{\Gamma, \Delta\}$, we may make three types of vertices: XY, X, and Y, each of whose Boltzmann weights is given by the above tables. Call these flavors of vertices R, S, and T, respectively. Let $\varepsilon_1, \ldots, \varepsilon_6$ be six choices of sign $\pm$. Then the following two partition functions (each involving respective Boltzmann weights at three vertices) are equal:

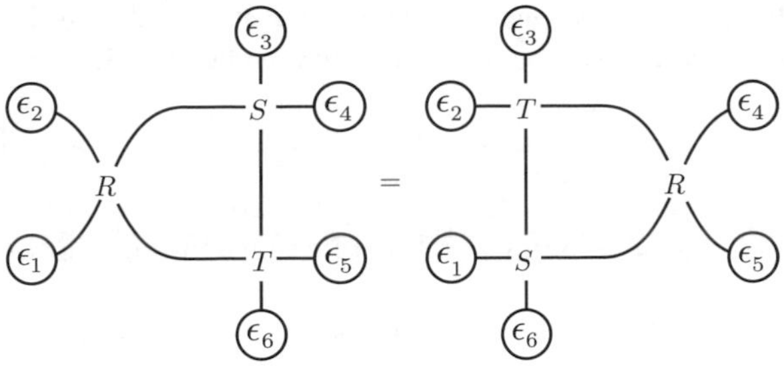

This means that (on each side of the equation) we sum over all assignments of signs to the three interior edges. The reversal of the spectral parameters and of the order of the S and T vertices is indicated.

Now, consider four rows of the system, which have (alternately) Δ, Γ, Δ, Γ vertices, with spectral parameters z_i, z_i^{-1}, z_j, and z_j^{-1}. (So $j = i + 1$.) To the left of these four rows, we attach the following "caduceus" braid, which was employed in this context for type C by Ivanov [20]:

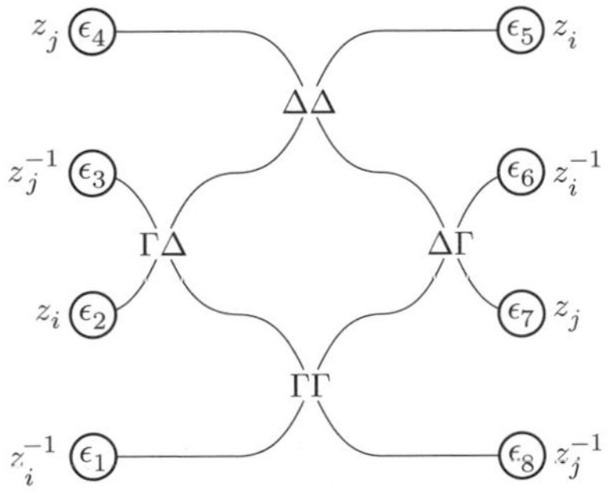

See (6) of [20] for a picture of the resulting configuration, with the caduceus braid attached.

We observe that there is only one legal configuration for this system which has $(\varepsilon_4, \varepsilon_3, \varepsilon_2, \varepsilon_1) = (-, +, -, +)$. This configuration is:

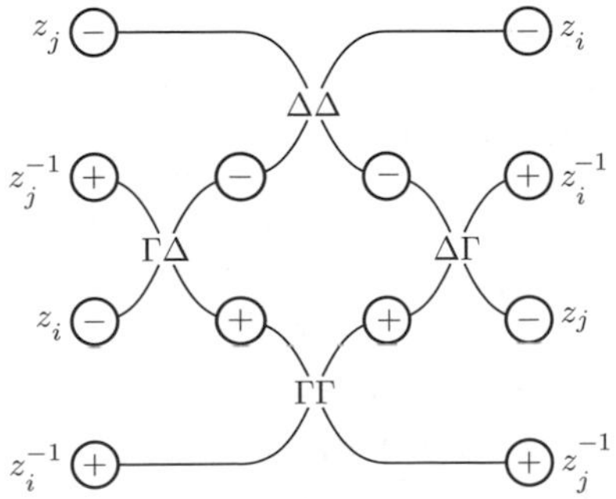

The partition function for this piece of ice is therefore just the product of the values at the four vertices, which can be read off from the above table:

$$(tz_j + z_i^{-1})(z_i + tz_j)(tz_i^{-1} + z_j^{-1})(tz_i + z_j^{-1}). \tag{18}$$

Hence, we may attach the caduceus to the left of four rows in our original U-turn ice configuration, and the resulting partition function multiplies the original partition function by this factor.

Using repeated application of the Yang–Baxter equation as in Ivanov [20], the braid may be moved across resulting in a configuration as in (7) of Ivanov [20]. In the process, the z_i and z_j spectral parameters are interchanged—effectively, the two pairs of rows are switched. To analyze the resulting partition function requires the following lemma, which is the analog of Lemma 3 in Ivanov [20].

Lemma 2. *Let $\varepsilon_1, \varepsilon_2, \varepsilon_3, \varepsilon_4 \in \{+, -\}$. Then the partition function of the system on the left in the following diagram:*

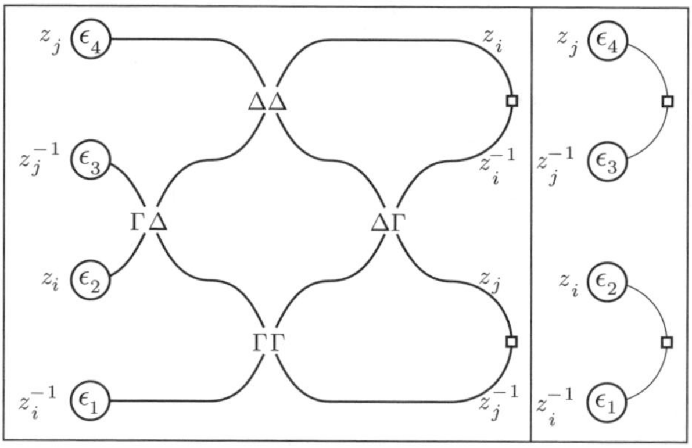

equals

$$(tz_i + z_j^{-1})(tz_i + z_j)(tz_i^{-1} + z_j^{-1})(tz_j + z_i^{-1}). \tag{19}$$

times the partition function of the system on the right.

This lemma allows us to pass the braid all the way through the U-turn ice, resulting in an interchange of parameters z_i and $z_j = z_{i+1}$. Thus, our original partition function is related to that of U-turn ice with z_i and z_{i+1} swapped by the ratio of (18) to (19). This ratio equals

$$\frac{z_j + tz_i}{z_i + tz_j} = \frac{z_j}{z_i} \cdot \frac{1 + tz_i z_j^{-1}}{1 + tz_j z_i^{-1}} = \frac{\mathbf{z}^{s_i \rho_B}}{\mathbf{z}^{\rho_B}} \frac{1 + tz_i z_j^{-1}}{1 + tz_j z_i^{-1}},$$

which means that the product (17) is invariant under this interchange.

Now, we consider the effect of the interchange $z_r \leftrightarrow z_r^{-1}$. For this, we begin by transforming the very bottom row of Γ vertices with spectral parameters z_r into Δ

vertices with the spectral parameter z_r^{-1} by flipping the signs of all the horizontal edges in the row. Thus, we will be using the following weights before and after the described change:

Γ vertex (before)	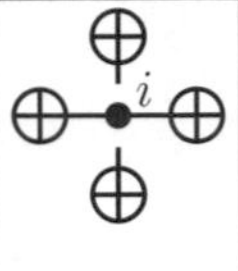	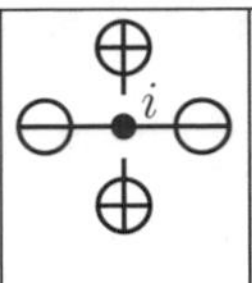	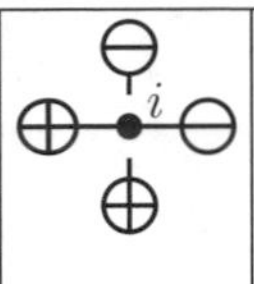	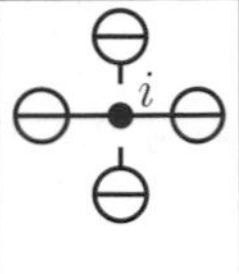	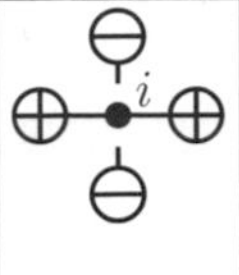	
Boltzmann weight	1	z_r^{-1}	1	z_r^{-1}	t	$z_r^{-1}(t+1)$
Δ vertex (after)	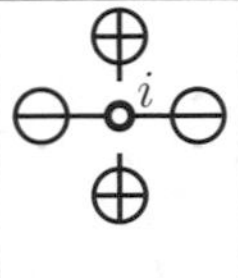	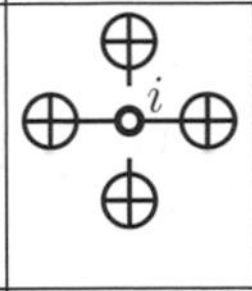	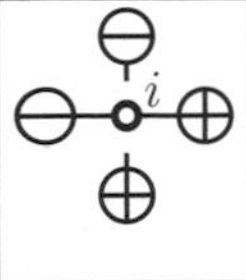	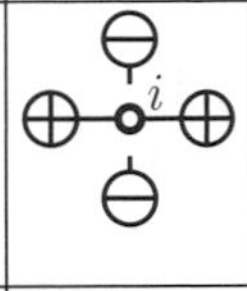	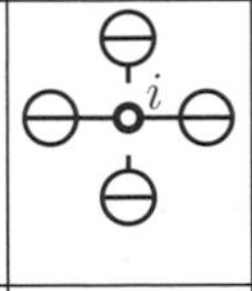	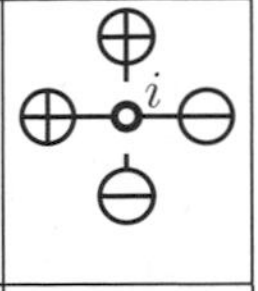
Boltzmann weight	1	z_r^{-1}	1	$z_r^{-1}t$	1	$z_r^{-1}(t+1)$

This change has no effect on the Boltzmann weights because of the boundary conditions. Indeed, only $+$ signs occur in the bottom edge spins, and therefore only the first three types of vertices in the table above occur. In order to compensate for the change, we must replace the cap vertices with the following modified ones, which we label by ■ instead of □:

■ Cap vertex	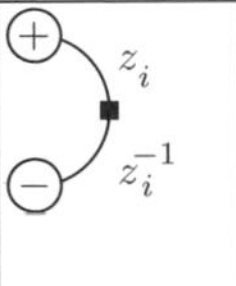	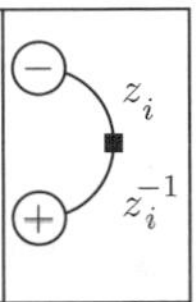
Boltzmann weight	$-\sqrt{-t}z_r^{1/2}$	$z_r^{-1/2}$

Now, we attach a $\Delta\Delta$ vertex to the left, using the following Boltzmann weights:

$\Delta\Delta$	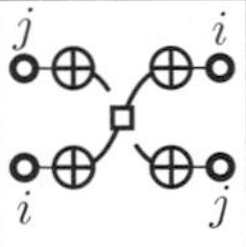	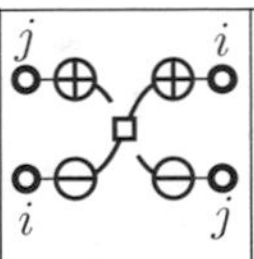	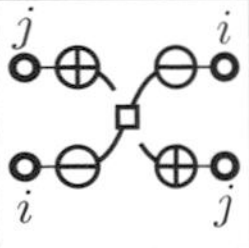	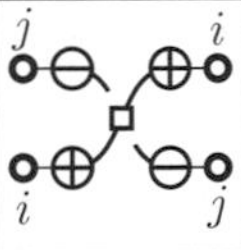	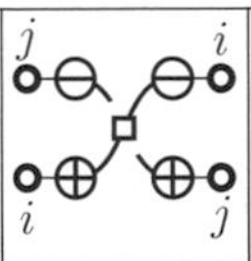	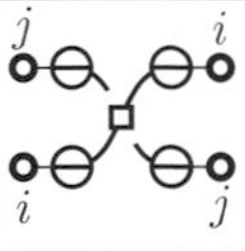
Boltzmann weight	$tz_r + z_r^{-1}$	$z_r^{-1}(t+1)$	$tz_r^{-1} - tz_r$	$z_r - z_r^{-1}$	$(t+1)z_r$	$z_r + tz_r^{-1}$

As the signs in the bottom row have all been reversed, we are attaching the braid on the left to a pair of rows each beginning with $-$. There is only one such admissible $\Delta\Delta$ vertex—the last in the table above. Attaching this braid then multiplies our original partition function by $z_r + tz_r^{-1}$. We use the Yang–Baxter equation repeatedly to push this $\Delta\Delta$ vertex across the bottom two rows until it encounters the cap.

Then we have the following configuration, referred to by Kuperberg [23] as the "fish equation":

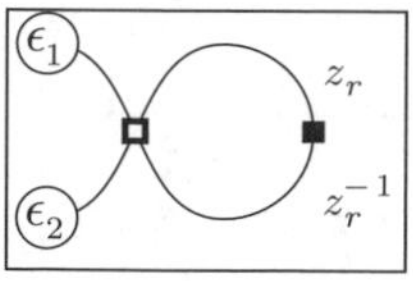

It may be checked that the value of this configuration is

$$\left(1-\sqrt{-t}z_r\right)\left(1+\sqrt{-t}z_r^{-1}\right)$$

times the value of the single ■ vertex. After this is substituted, we may then repeat the flipping of all signs in the bottom row, turning the vertices in this row back into Γ vertices, but now with parameter z_r^{-1} changed to z_r.

Therefore, $z_r + tz_r^{-1}$ times $Z(\mathfrak{S})$ equals $(1-\sqrt{-t}z_r)(1+\sqrt{-t}z_r^{-1})$ times the partition function with z_r replaced by its inverse. This implies that (17) is invariant under $z_r \to z_r^{-1}$.

Conjecture 3. Take $t = -\frac{1}{q}$. Then, $Z(\mathfrak{S})$ equals

$$\frac{\mathbf{z}^{w(\rho_B)}}{\Delta_C}\prod_i\left(1+q^{-1/2}z_i\right)\left[\prod_{i<j}\left(1-q^{-1}z_iz_j\right)\left(1-q^{-1}z_iz_j^{-1}\right)\right]$$
$$\mathcal{A}\left(\mathbf{z}^{\lambda+\rho_C}\prod_{i=1}^{r}\left(1-q^{-1/2}z_i^{-1}\right)\right).$$

It follows from Theorem 4 that $Z(\mathfrak{S})$ is divisible by the product

$$\prod_i\left(1+q^{-1/2}z_i\right)\left[\prod_{i<j}\left(1-q^{-1}z_iz_j\right)\left(1-q^{-1}z_iz_j^{-1}\right)\right],$$

and the quotient is a polynomial in $q^{-1/2}$ and z_i, z_i^{-1} that is invariant under the Weyl group.

Theorem 5. *The conjecture is true if* $r \leq 3$.

We omit the proof, but we note that both sides as polynomials in the arbitrary parameter $q^{-1/2}$, we may confirm the identity in the conjecture for the special values $q^{-1/2} = 0$ or 1. If $r \leqslant 3$, we may prove the conjecture by bounding the size of the possible degree of $q^{1/2}$ in the resulting partition function and using the known pair of special values in $q^{-1/2}$.

Thus, this ice-type model conjecturally represents the Whittaker function. This conjecture implies Conjecture 2 using the bijection between states of U-turn ice with boundary corresponding to λ and vertices in the crystal $\mathcal{B}_\lambda$ of type B having the $G(v) \neq 0$. This bijection is implicit in [5] given the bijection between Gelfand–Tsetlin patterns and BZL patterns described by Littelmann [24].

References

1. R. Baxter. *Exactly solved models in statistical mechanics*. Academic Press Inc. [Harcourt Brace Jovanovich Publishers], London, 1982.
2. J. Beineke, B. Brubaker, and S. Frechette. Weyl group multiple Dirichlet series of type C. *Pacific J. Math.*, To appear.
3. J. Beineke, B. Brubaker, and S. Frechette. A crystal definition for symplectic multiple Dirichlet series, in this volume.
4. A. Berenstein and A. Zelevinsky. String bases for quantum groups of type A_r. In *I. M. Gel′fand Seminar*, volume 16 of *Adv. Soviet Math.*, pages 51–89. Amer. Math. Soc., Providence, RI, 1993.
5. B. Brubaker, D. Bump, G. Chinta, S. Friedberg, and P. Gunnells. Metaplectic Ice, in *this volume*.
6. B. Brubaker, D. Bump, and S. Friedberg. Weyl group multiple Dirichlet series. II. The stable case. *Invent. Math.*, 165(2):325–355, 2006.
7. B. Brubaker, D. Bump, and S. Friedberg. Gauss sum combinatorics and metaplectic Eisenstein series. In *Automorphic forms and L-functions I. Global aspects*, volume 488 of *Contemp. Math.*, pages 61–81. Amer. Math. Soc., Providence, RI, 2009.
8. B. Brubaker, D. Bump, and S. Friedberg. Schur polynomials and the Yang-Baxter equation. *Comm. Math. Phys.*, 308(2):281–301, 2011.
9. B. Brubaker, D. Bump, and S. Friedberg. *Weyl Group Multiple Dirichlet Series: Type A Combinatorial Theory*. Annals of Mathematics Studies v. 175, Princeton University Press, Princeton, NJ, 2011.
10. B. Brubaker, D. Bump, S. Friedberg, and J. Hoffstein. Weyl group multiple Dirichlet series III: Eisenstein series and twisted unstable A_r. *Annals of Math.*, 166:293–316, 2007.
11. D. Bump, Multiple Dirichlet Series, in this volume.
12. D. Bump, S. Friedberg, and J. Hoffstein. p-adic Whittaker functions on the metaplectic group. *Duke Math. J.*, 63(2):379–397, 1991.
13. W. Casselman and J. Shalika. The unramified principal series of p-adic groups. II. The Whittaker function. *Compositio Math.*, 41(2):207–231, 1980.
14. G. Chinta, S. Friedberg, and P. E. Gunnells. On the p-parts of quadratic Weyl group multiple Dirichlet series. *J. Reine Angew. Math.*, 623:1–23, 2008.
15. G. Chinta and P. E. Gunnells. Littelmann patterns and Weyl group multiple Dirichlet series of type D, in this volume.
16. G. Chinta and P. E. Gunnells. Weyl group multiple Dirichlet series constructed from quadratic characters. *Invent. Math.*, 167(2):327–353, 2007.
17. G. Chinta and P. E. Gunnells. Constructing Weyl group multiple Dirichlet series. *J. Amer. Math. Soc.*, 23(1):189–215, 2010.
18. G. Chinta and O. Offen, *A metaplectic Casselmann–Shalika formula for GL_r*, Amer. J. Math., to appear.
19. A. M. Hamel and R. C. King. Bijective proofs of shifted tableau and alternating sign matrix identities. *J. Algebraic Combin.*, 25(4):417–458, 2007.
20. D. Ivanov. Symplectic ice, in this volume, 2011.
21. V. Korepin, N. Boguliubov and A. Izergin *Quantum Inverse Scattering Method and Correlation Functions*, Cambridge University Press, 1993.
22. G. Kuperberg. Another proof of the alternating-sign matrix conjecture, *Internat. Math. Res. Notices*, 1996(3):139–150 (1996).

23. G. Kuperberg. Symmetry classes of alternating-sign matrices under one roof. *Ann. of Math. (2)*, 156(3):835–866, 2002.
24. P. Littelmann. Cones, crystals, and patterns. *Transform. Groups*, 3(2):145–179, 1998.
25. P. Littelmann. Paths and root operators in representation theory. *Ann. of Math. (2)*, 142(3):499–525, 1995.
26. H. Matsumoto. Sur les sous-groupes arithmétiques des groupes semi-simples déployés. *Ann. Sci. École Norm. Sup. (4)*, 2:1–62, 1969.
27. G. Savin. Local Shimura correspondence. *Math. Ann.*, 280(2):185–190, 1988.
28. T. Tokuyama. A generating function of strict Gelfand patterns and some formulas on characters of general linear groups. *J. Math. Soc. Japan*, 40(4):671–685, 1988.

Chapter 5
Littelmann Patterns and Weyl Group Multiple Dirichlet Series of Type D

Gautam Chinta and Paul E. Gunnells

Abstract We formulate a conjecture for the local parts of Weyl group multiple Dirichlet series attached to root systems of type D. Our conjecture is analogous to the description of the local parts of type A series given by Brubaker et al. (Ann. of Math. 166(1):293–316, 2007) in terms of Gelfand–Tsetlin patterns. Our conjecture is given in terms of patterns for irreducible representations of even orthogonal Lie algebras developed by Littelmann (Transform. Groups 3(2):145–179, 1998).

Keywords Weyl group multiple Dirichlet series • Crystal graph • Gelfand-Tsetlin pattern • Littelmann pattern

5.1 Introduction

We begin with some notation. Let Φ be a reduced root system of rank r and n a positive integer. Let F be a number field containing the $2n$-th roots of unity. Let S be a set of places of F containing the Archimedean places and those that ramify over $\mathbb{Q}$, as well as sufficiently many more places to ensure that the ring of S-integers $\mathcal{O}_S$ has class number 1. Let $\mathbf{m} = (m_1, \ldots, m_r)$ be a fixed nonzero r-tuple of elements of $\mathcal{O}_S$. Let $\mathbf{s} = (s_1, \ldots, s_r)$ be an r-tuple of complex variables.

Given the data above, one can form a *Weyl group multiple Dirichlet series*. This is a Dirichlet series in the r variables s_i with a group of functional equations

---

G. Chinta
Department of Mathematics, The City College of New York, New York, NY 10031, USA
e-mail: chinta@sci.ccny.cuny.edu

P.E. Gunnells (✉)
Department of Mathematics and Statistics, University of Massachusetts, Amherst, MA 01003, USA
e-mail: gunnells@math.umass.edu

D. Bump et al. (eds.), *Multiple Dirichlet Series, L-functions and Automorphic Forms*, Progress in Mathematics 300, DOI 10.1007/978-0-8176-8334-4_5,

isomorphic to the Weyl group W of Φ. More precisely, one can define a set of functions of the form

$$Z(\mathbf{s}; \mathbf{m}, \Psi) = Z^n_\Phi(\mathbf{s}; \mathbf{m}, \Psi) = \sum_{\mathbf{c}} \frac{H(\mathbf{c}; \mathbf{m})\Psi(\mathbf{c})}{\prod |c_i|^{s_i}},$$

where each c_i ranges over nonzero elements of $\mathcal{O}_S$ modulo units, Ψ is taken from a certain finite-dimensional complex vector space Ω of functions on $(F_S^\times)^r$, and H is an important function we shall say more about shortly. Then, the collection of all such Z as Ψ ranges over a basis of Ω satisfies a group of functional equations isomorphic to W with an appropriate scattering matrix. For more about why Weyl group multiple Dirichlet series are interesting objects, as well as a discussion about the basic framework for their construction, we refer to [9].

The heart of the construction of Z is the function H. This function must be carefully defined to ensure that Z satisfies the correct group of functional equations. The heuristic of [9] dictates how to define H on the *powerfree* tuples $\mathbf{c}, \mathbf{m}$ (those tuples such that the product $c_1 \cdots c_r m_1 \cdots m_r$ is squarefree). Moreover, it is further specified in [9] how the values of H on the *prime power* tuples $\mathbf{c} = (\varpi^{k_1}, \dots, \varpi^{k_r})$, $\mathbf{m} = (\varpi^{l_1}, \dots, \varpi^{l_r})$, where $\varpi \in \mathcal{O}_S$ is a prime, determine H on all tuples.

Thus, writing ℓ for a tuple of nonnegative integers $(l_1, \dots, l_r)$ and letting ϖ^ℓ denote the tuple $(\varpi^{l_1}, \dots, \varpi^{l_r})$, the construction of Z reduces to defining the multivariate generating function

$$N(x_1, \dots, x_r; \ell) := \sum_{k_i \geq 0} H\left(\varpi^{k_1}, \dots, \varpi^{k_r}; \varpi^\ell\right) x_1^{k_1} \cdots x_r^{k_r}. \tag{1}$$

At present, there are two different approaches to defining the generating function (1) and thus to constructing Weyl group multiple Dirichlet series. Both are related to characters of representations of the semisimple complex Lie algebra attached to Φ. Let $\omega_i, i = 1, \dots, r$ be the fundamental weights of Φ and let θ be the strictly dominant weight $\sum (l_i + 1)\omega_i$.

- The *Gelfand–Tsetlin* approach [4,5,7], which works for $\Phi = A_r$, gives formulas for the coefficients $H(\varpi^{k_1}, \dots, \varpi^{k_r}; \varpi^\ell)$. These formulas are written in terms of Gauss sums and statistics extracted from Gelfand–Tsetlin patterns for the representation of $\mathfrak{sl}_{r+1}(\mathbb{C})$ of lowest weight $-\theta$.
- The *averaging* approach [11–13,15], which works for all Φ, uses a "metaplectic" deformation of the Weyl character formula to construct a rational function with known denominator, whose numerator is then taken to define N.

Both approaches have their advantages and limitations. The Gelfand–Tsetlin construction gives very explicit formulas for H, formulas that (remarkably) are uniform in n and that lead to a direct connection with the global Fourier coefficients of Borel Eisenstein series on the n-fold cover of SL_{r+1} [8], but suffers from the obvious disadvantage that it only works for type A. The averaging approach, on the other hand, works for all Φ, quickly leads to the definition of Z, yet has the drawback that it seems difficult to get similarly explicit formulas for the coefficients of N. By combining recent work of Chinta and Offen [14] and McNamara [17], we know that in type A, the two definitions of N coincide, although it seems difficult to give a direct combinatorial proof.

This note arose from our attempts to understand the Gelfand–Tsetlin approach to (1). In the course of studying [5], it became plain to us that the most suitable language to understand the constructions in [5] is that of Kashiwara's *crystal graphs*, as encoded in the generalization of the Gelfand–Tsetlin basis due to Littelmann [16], which we call *Littelmann patterns*. These patterns, which reformulate and extend earlier work of Berenstein–Zelevinsky [1–3], are well suited for constructing H. Indeed, the definitions in [5] become much more transparent when phrased in terms of these patterns, as one can see in [8].

To test the relevance of this observation, we decided to try to formulate a Littelmann analogue of the Gelfand–Tsetlin construction when Φ is a root system of type D. The main result of this note is thus Conjecture 1, which explicitly describes the generating function $N(x_1, \ldots, x_r; \ell)$ for the ϖ-part of the type D Weyl group multiple Dirichlet series constructed using the averaging method. We remark that for $n = 1$, proving Conjecture 1 would give a type D analogue of a theorem of Tokuyama [18].

To formulate Conjecture 1, we began by translating the definition in [5] into Littelmann patterns. We then compared the putative ϖ-part with that constructed in [11, 13]. This allowed us to adjust the contributions of certain patterns until we reached agreement for all examples.

We have some limited evidence for the truth of Conjecture 1. First, for $D_2 \simeq A_1 \times A_1$, the conjecture is easily seen to be true. Next, we have tested the conjecture for D_3 when $n \leq 4$ and for D_4 when $n \leq 2$ by computing the ϖ-parts for many tuples ℓ using the averaging method and comparing with the predictions of Conjecture 1. In all cases, including many not used in experimental development of Conjecture 1, there was complete agreement. Note that $D_3 \simeq A_3$, so the ϖ-part of the D_3-series has already been described explicitly using the results of [5], and in this guise has already been compared extensively with ϖ-parts constructed by averaging. Nevertheless, agreement in rank 3 between ϖ-parts constructed using Conjecture 1 and using averaging is a nontrivial check since D_3 Littelmann patterns are quite different from A_3 Gelfand–Tsetlin patterns.

Finally, Brubaker and Friedberg have recently computed the global Whittaker coefficients of Eisenstein series on covers of GL_4 by inducing from the parabolic subgroup of type $GL_2 \times GL_2$ [6]. Their computations—which build on earlier work of Bump–Hoffstein [10] and are the first attempts to extend the results of [8] beyond type A and to work with other parabolic subgroups—express the Whittaker coefficients in terms of certain exponential sums. In the course of their work, Brubaker and Friedberg found that the integrals can be broken up in accordance with the decomposition of $H(\varpi^{k_1}, \ldots, \varpi^{k_r}; \omega^{\ell})$ given by Conjecture 1 and that if one does so, the contributions to the global Whittaker coefficient exactly agree with Conjecture 1. We find this connection between Eisenstein series and ϖ-parts to be strongly convincing evidence of the correctness of Conjecture 1.

We thank Ben Brubaker and Sol Friedberg for helpful conversations. We thank K.-H. Lee for a careful reading of an earlier version of this chapter. Both authors thank the NSF for support through grants DMS-0847586 and DMS-0801214.

5.2 Littelmann Patterns

Let $\mathfrak{g}$ be the simple complex Lie algebra of type D_r, in other words the Lie algebra of the group $\mathrm{SO}_{2r}(\mathbb{C})$. Let θ be a dominant weight of $\mathfrak{g}$ and let V_θ be the irreducible $\mathfrak{g}$-module of highest weight θ. In [16, Sect. 7], Littelmann describes a way to index a basis of V_θ using patterns that are analogous to the classical Gelfand–Tsetlin patterns for the Lie algebra of $\mathrm{SL}_r(\mathbb{C})$. A Littlemann pattern is a length ℓ sequence of nonnegative integers, where ℓ is the length of the longest word of the Weyl group. Essentially this sequence records the lengths of certain strings of edges in the crystal graph. We refer to [16] for more details.

First, we label vertices of the Dynkin diagram of $\mathfrak{g}$ with the integers $1, \ldots, r$. We label the upper node of the right prong 1, the lower node of the prong 2, the node at the elbow of the prong 3, and then the remaining nodes increase from 4 to r, reading right to left (Fig. 5.1). We remark that this is not the standard labeling by Bourbaki, which begins with 1 at the left of the diagram.

A Littlemann pattern T for D_r consists of a collection of integers $a_{i,j}$, where $1 \leq i \leq r-1$ and $i \leq j \leq 2r-2$. We picture T by drawing the integers placed in $r-1$ rows of centered boxes. The first row contains $2r-2$ boxes, the second $2r-4$ boxes, and so on down to the $(r-1)$st row, which contains two boxes. The integers are placed in the boxes so that $a_{i,i}$ is placed in the leftmost box of the ith row, and then, the remaining integers $a_{i,j}$ are put in the boxes in order as j increases. We define an involution on each row by $\overline{a}_{i,j} = a_{i,2r-1-j}$.

To index a weight vector in V_θ, there are two sets of inequalities the $a_{i,j}$ must satisfy. The first is independent of θ: in each row we must have

$$a_{i,i} \geq a_{i,i+1} \geq \cdots \geq a_{i,r-2} \geq a_{i,r-1}, a_{i,r} \geq a_{i,r+1} \geq \cdots \geq a_{i,2r-1-i} \geq 0 \qquad (2)$$

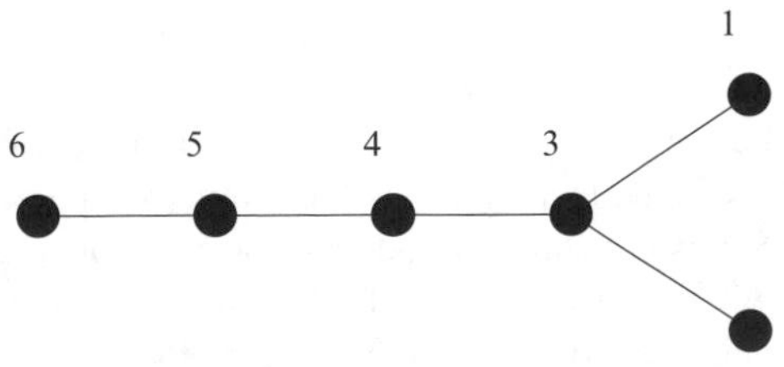

Fig. 5.1 The diagram for D_6

Fig. 5.2 An admissible pattern for D_4

6	6	5	4	4	2
	3	3	2	1	
		1	3		

or, using the bar notation,

$$a_{i,i} \geq a_{i,i+1} \geq \cdots \geq a_{i,r-2} \geq a_{i,r-1}, \overline{a}_{i,r-1} \geq \overline{a}_{i,r-2} \geq \cdots \geq \overline{a}_{i,i} \geq 0.$$

In other words, the $a_{i,j}$ are weakly decreasing in the rows, with the exception that no comparison is made between $a_{i,r-1}$ and $a_{i,r}$. Both of these entries, however, are required to be $\leq a_{i,r-2}$ and $\geq a_{i,r+1}$.

Definition 1. A pattern T is *admissible* if T satisfies (2) for all i.

Figure 5.2 shows an admissible pattern for D_4.

The next set of inequalities involves the highest weight θ. Write

$$\theta = \sum m_k \omega_k,$$

where the ω_k are the fundamental weights. Then, an admissible T will correspond to a weight vector in V_θ if T satisfies

$$\overline{a}_{i,j} \leq m_{r-j+1} + s(\overline{a}_{i,j-1}) - 2s(a_{i-1,j}) + s(a_{i-1,j+1}) \quad \text{for } j \leq r-2, \tag{3}$$

$$a_{i,j} \leq m_{r-j+1} + s(a_{i,j+1}) - 2s(\overline{a}_{i,j}) + s(\overline{a}_{i,j-1}) \quad \text{for } j \leq r-2, \tag{4}$$

$$a_{i,r-1} \leq m_2 + s(\overline{a}_{i,r-2}) - 2t(a_{i-1,r-1}), \text{ and} \tag{5}$$

$$a_{i,r} \leq m_1 + s(\overline{a}_{i,r-2}) - 2t(a_{i-1,r}), \tag{6}$$

where we write for $j < r-1$

$$s(\overline{a}_{i,j}) = \overline{a}_{i,j} + \sum_{k=1}^{i-1}(a_{k,j} + \overline{a}_{k,j}),$$

$$s(a_{i,j}) = \sum_{k=1}^{i}(a_{k,j} + \overline{a}_{k,j}),$$

$$s(a_{i,r-1}) = s(\overline{a}_{i,r-1}) = \sum_{k=1}^{i} a_{k,r-1} + a_{k,r},$$

$$t(a_{i,r-1}) = \sum_{k=1}^{i} a_{k,r-1}, \quad t(a_{i,r}) = \sum_{k=1}^{i} a_{k,r}.$$

Definition 2. A pattern T is *θ-admissible* if T is admissible and its entries satisfy (3)–(6).

Note that the inequalities for the ith row only involve the entries of T on the ith and $(i-1)$st rows. Moreover, when ordered in terms of increasing i, there is a unique inequality in which a given entry $a_{i,j}$ appears on the left.

Definition 3. We say that an entry in a θ-admissible pattern is *critical* if this first inequality is actually an equality.

To complete our discussion of Littelmann patterns, we must assign a weight $\lambda(T)$ to each pattern T. This is a vector $\lambda(T) = (\lambda_1, \ldots, \lambda_r)$ of nonnegative integers, where

$$\lambda_k = \begin{cases} \sum_{i=1}^{r-1}(a_{i,r+1-k} + \overline{a}_{i,r+1-k}) & k = 3, 4, \ldots, r \\ \sum_{i=1}^{r-1} a_{i,r-2+k} & k = 1, 2. \end{cases}$$

We write $|\lambda| = \lambda_1 + \cdots + \lambda_r$. In our conjecture, if a pattern T occurs for the twist $\theta = \sum m_i \omega_i$, it will contribute to the coefficient of $\mathbf{x}^{\lambda(T)} := x_1^{\lambda_1} \cdots x_r^{\lambda_r}$ in the numerator $N(\mathbf{x}, \ell)$, where $\ell = (l_1, \ldots, l_r)$ and $l_i = m_i - 1$. For instance, the pattern in Fig. 5.2 contributes to the coefficient of $x_1^9 x_2^9 x_3^{14} x_4^8$, with x_1 corresponding to left middle column of three entries and x_2 to the right middle column of three entries.

5.3 The Decorated Graph of a Pattern

Let T be a θ-admissible pattern. We want to associate to T a graph $\Gamma(T)$. The graph $\Gamma(T)$ will also potentially be endowed with *decorations*, which will be circled vertices. The vertices of $\Gamma(T)$ correspond to the pairs (i, j) indexing entries of T; the graph will have at least one connected component for each row of T.

We begin by describing how each row determines a subgraph. Consider the ith row of T. Each entry $a_{i,j}$ in this row corresponds to a vertex. We draw the corresponding vertices in a row, with the two vertices in the middle corresponding to the incomparable entries $a_{i,r-1}, a_{i,r}$ entries arranged vertically. For definiteness, we assign $a_{i,r-1}$ to the top vertex and $a_{i,r}$ to the bottom vertex. See Fig. 5.3 for the arrangement for the top row of a pattern for D_6.

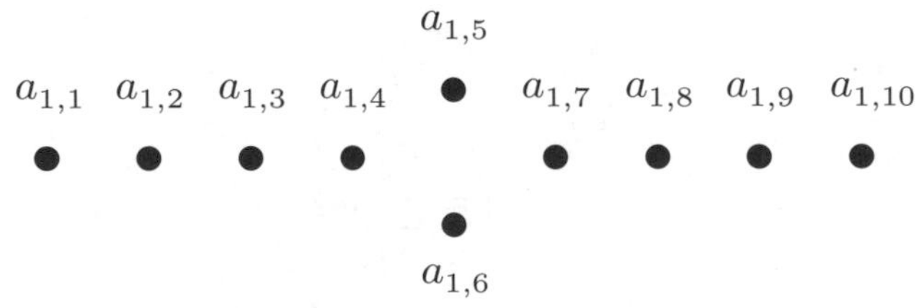

Fig. 5.3 The vertices for the top row of D_6

Fig. 5.4 A symmetric multiple leaner

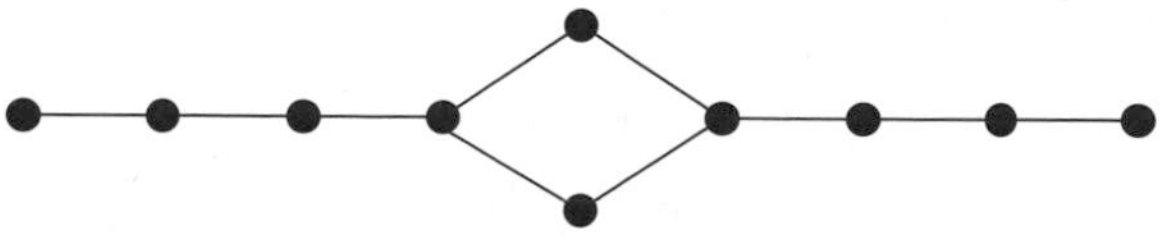

6	6	5	4	4	2
	3	3	2	1	
		1	3		

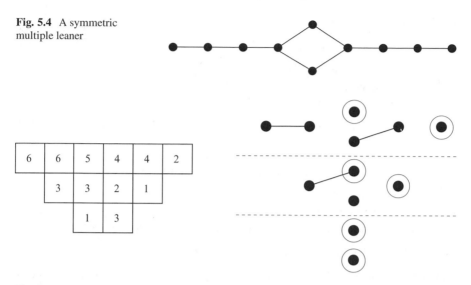

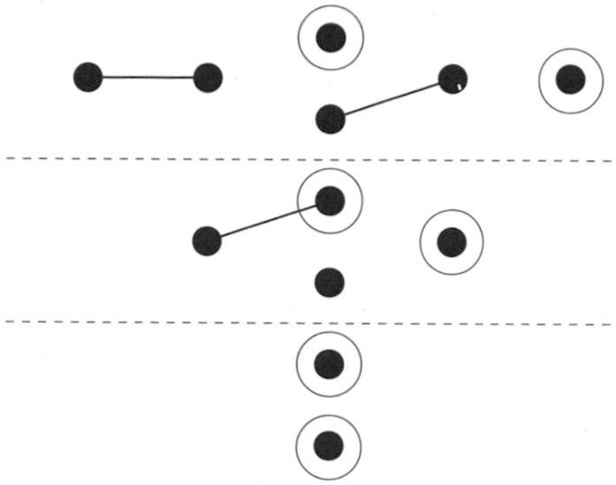

Fig. 5.5 The decorated graph of the pattern in Fig. 5.2

Now, join two vertices by an edge if their corresponding pattern entries appear consecutively in the inequalities (2), are equal, and are comparable in (2). Note that we do not join the vertices corresponding to $a_{i,r-1}, a_{i,r}$ by an edge if they happen to be equal since they are not comparable in (2). This gives a graph for this row. We then do the same for each row of T. The result is $\Gamma(T)$ without decorations.

Certain symmetric connected components that arise in the construction of $\Gamma(T)$ will play a special role in our conjecture.

Definition 4. Let T be an admissible pattern and suppose $a_{i,j} = \overline{a}_{i,j}$ for some i, j with $j \neq r-1, r$. Then, the component of $\Gamma(T)$ containing $a_{i,j}, \overline{a}_{i,j}$ is called a *multiple leaner*. The *legs* of a multiple leaner are the subgraphs joining the vertices $a_{i,j}$ to $a_{i,r-2}$ and $a_{i,r+1}$ to $\overline{a}_{i,j}$. If in addition $a_{i,j-1} \neq a_{i,j}$ and $\overline{a}_{i,j-1} \neq \overline{a}_{i,j}$, then we say the multiple leaner is *symmetric*. We define the *length* $l(C)$ of a symmetric multiple leaner to be half the number of its vertices.

The term *leaning* is inspired by Brubaker, Bump, Friedberg and Hoffstein [5]; see also Sect. 5.5. Figure 5.4 shows an example of a symmetric multiple leaner of length 5, when all the entries in the top row of a pattern for D_6 are equal. Note that the minimal length of a symmetric multiple leaner is 2 and that multiple leaners can appear in patterns for D_3 but not for D_2.

To complete the construction of $\Gamma(T)$, we must describe how to add the decorations. This is very simple: we circle each vertex whose corresponding entry is critical in the sense of Definition 3.

Figure 5.5 shows an example of the decorated graph of the Littelmann pattern in Fig. 5.2. We assume that a highest weight θ has been specified so that the circled vertices in the graph correspond to critical entries.

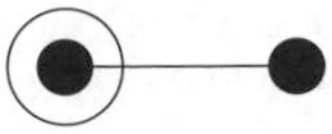

Fig. 5.6 See Definition 5

5.4 Strictness

In [5], certain patterns for a given weight are discarded and do not contribute to the relevant coefficient of N; such patterns are called *nonstrict* in [5]. In type A, strictness corresponds to an easily stated property for Gelfand–Tsetlin patterns. If one interprets the definition of strictness in [5] in terms of type A Littelmann patterns, one sees that a type A pattern is nonstrict exactly when

- An entry is simultaneously 0 and critical. Or
- There are two adjacent entries that are equal, with the left entry critical

We take these to be our definition for type D patterns as well.

Definition 5. A type D Littelmann pattern T is called *strict* if the following conditions hold:

- The graph $\Gamma(T)$ contains no circled vertex whose corresponding entry $a_{i,j}$ equals 0.
- No component of $\Gamma(T)$ that is not a multiple leaner contains a subgraph of the form shown in Fig. 5.6 (in this figure, the rightmost vertex is less than the left vertex in the partial order from (2)).

Note that the subgraph from Fig. 5.6 is allowed to appear in multiple leaners.

5.5 Leaning and Standard Contributions

In the following for $k = 1, \ldots, n-1$, we write g_k for the Gauss sum $g(\varpi^{k-1}, \varpi^k)$ (see, for example, [11] for the definition of the Gauss sums). For convenience, we extend the notation and define g_0 to be -1. It is also convenient to define g_m for $m \geq n$ by $g_m = g_k$, where $m = k \bmod n$ and $k = 0, \ldots, n-1$. We let q be the norm of ϖ.

Let T be a strict pattern, and let $\Gamma(T)$ be the associated decorated graph. For any connected component C of $\Gamma(T)$, let y_C be the *rightmost* vertex, in the sense of the order induced by the inequalities (2). If C has two rightmost vertices, meaning that it is in the ith row and contains entries $a_{i,r-2} = a_{i,r-1} = a_{i,r} \neq a_{i,r+1}$, then we define the rightmost vertex to be the vertex corresponding to $a_{i,r-1}$, that is, the upper vertex in Fig. 5.3.

Definition 6. Let T be a pattern and $\Gamma = \Gamma(T)$ the associated decorated graph. Fix n and let y be an entry of T. We define the *standard contribution* $\sigma(y)$ by the following rule:

- If the vertex corresponding to $y \neq 0$ is uncircled, then we put $\sigma(y) = 1 - 1/q$ if n divides y and $\sigma(y) = 0$ otherwise.
- If the vertex corresponding to $y \neq 0$ is circled, then we put $\sigma(y) = g_k/q$, where $y = k \bmod n$ and $k = 0, \ldots, n-1$.

Note that $\sigma(y)$ depends on n and θ, even though we omit them from the notation.

We are almost ready to state our conjecture. There is one more phenomenon that plays a role, namely, *leaning*. Essentially, leaning means that if entries are consecutive and equal in a Littelmann pattern T, where consecutive means adjacent in (2), then only one should contribute to the corresponding coefficient of $N(\mathbf{x}; \ell)$. This is why we introduce the graph $\Gamma(T)$. Its connected components keep track of these equalities among entries.

Thus, we are led to consider contributions of the connected components of $\Gamma(T)$, not just the entries. There is further slight twist that the contribution of a multiple leaning component is different from that of all other components:

Definition 7. Let C be a connected component of $\Gamma(T)$. The *standard contribution* $\sigma(C)$ *of* C is defined as follows:

- If C is not a multiple leaner, then we put $\sigma(C) = \sigma(y_C)$, where y_C is the rightmost entry of C.
- If C is a multiple leaner that is not symmetric, let y_C be the entry on the endpoint of its shorter leg. Then, we define $\sigma(C) = \sigma(y_C)$.
- If C is a symmetric multiple leaner, then let y_C be its rightmost entry $a_{i,j}$ and υ_C (upsilon = Greek y) to be the entry $a_{i,j-1}$. Then we define

$$\sigma(C) = \begin{cases} \sigma(y_C)(1 - 1/q^{l(C)}) & \text{if } y_C \text{ is uncircled,} \\ \sigma(y_C)\sigma(\upsilon_C)(1/q^{l(C)-1}) & \text{if } y_C \text{ is circled,} \end{cases}$$

where $l(C)$ is defined to the half the number of vertices of C (Definition 4).

We are now ready to state our conjecture:

Conjecture 1. Let $N(\mathbf{x}; \ell) = \sum_\lambda a_\lambda \mathbf{x}^\lambda$ be the ϖ-part constructed by averaging [11, 13] for the Weyl group multiple Dirichlet series $Z_\Phi^n(\mathbf{s}; \mathbf{m}, \Psi)$. Then, we have

$$a_\lambda = q^{|\lambda|} \sum_T \prod_{C \subset \Gamma(T)} \sigma(C), \tag{7}$$

where the sum is taken over all strict patterns T of weight λ and with highest weight $\theta = \sum(l_i + 1)\omega_i$, and the product is taken over the connected components of $\Gamma(T)$.

Example 1. Suppose the pattern in Fig. 5.2 appears for a highest weight θ such that the decorated graph appears in Fig. 5.5. Suppose $n = 2$. Then, the contribution of this pattern to the coefficient of $x_1^9 x_2^9 x_3^{14} x_4^8$ will be

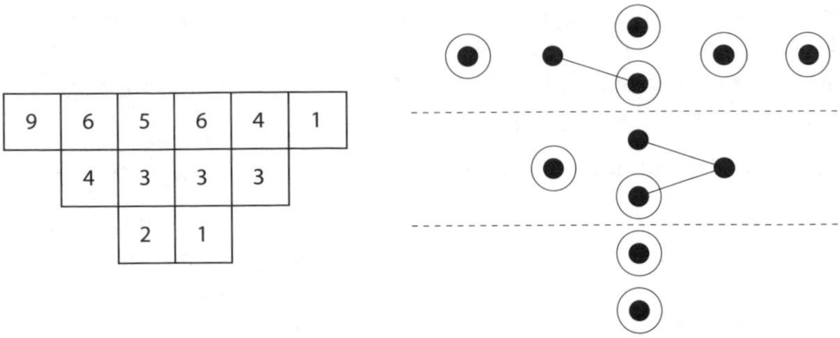

Fig. 5.7 A nonstrict pattern

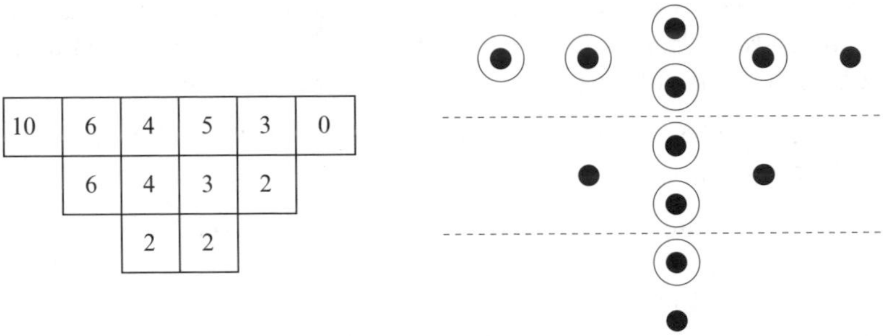

Fig. 5.8 T_1

$$q^{40}\left(1-\frac{1}{q}\right)^3\left(-\frac{1}{q}\right)\left(\frac{g_1}{q}\right)^5.$$

Example 2. We consider another example for $n = 2$. Suppose the twisting parameter is $\ell = (1,0,2,0)$, which corresponds to the highest weight $\theta = 2\omega_1 + \omega_2 + 3\omega_3 + \omega_4$. We will compute the coefficient a_λ of the monomial $\mathbf{x}^\lambda = x_1^{10}x_2^{10}x_3^{17}x_4^{10}$. Note that $|\lambda| = 47$.

There are 27 Littelmann patterns that we must consider. Six of these patterns are nonstrict, for instance, the pattern shown in Fig. 5.7. Of the remaining 21, only two give nonzero contributions; these patterns T_1, T_2 appear in Figs. 5.8–5.9. Note that Fig. 5.9 contains a multiple leaner of length 2. All of the other 19 patterns have an odd entry that is not circled, and thus have a connected component in $\Gamma(T)$ with standard contribution equal to zero.

Each vertex in $\Gamma(T_1)$ is its own connected component. We see three uncircled even nonzero entries, five circled even nonzero entries, and three circled odd entries. Thus, T_1 contributes

$$q^{47}\left(1-\frac{1}{q}\right)^3\left(-\frac{1}{q}\right)^5\left(\frac{g_1}{q}\right)^3$$

to a_λ.

10	6	4	5	3	0
	4	4	4	4	
		2	1		

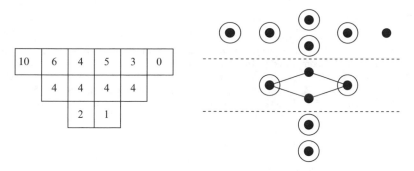

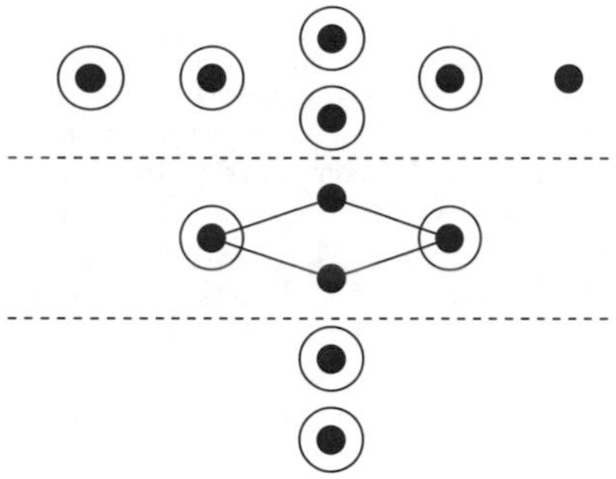

Fig. 5.9 T_2

The pattern T_2 has a multiple leaner C of length $l(C) = 2$. Its rightmost entry y_C is circled, and the entry υ_C is uncircled. We have $\sigma(y_C) = -1/q$, $\sigma(\upsilon_C) = (1 - 1/q)$. These appear in (7) multiplied by the additional factor $1/q$ to account for the length of C. Each of the remaining vertices is its own connected component, and we have no uncircled nonzero even entries, four circled nonzero even entries, and three circled odd entries. Thus, T_2 contributes

$$q^{47}\left(-\frac{1}{q}\right)^4\left(\frac{g_1}{q}\right)^3\left(-\frac{1}{q}\right)\left(1-\frac{1}{q}\right)\left(\frac{1}{q}\right)$$

to a_λ. After simplifying, we find

$$a_\lambda = -q^{36}\left(q^3 - 2q^2 + 2q - 1\right)g_1^3,$$

in agreement with the ϖ-part from [13].

Example 3. We conclude by describing an example for D_4 when $n = 1$. We put $\ell = (0,0,0,0)$ (the "untwisted" case) so that the highest weight is $\omega_1+\omega_2+\omega_3+\omega_4$. The polynomial $N(\mathbf{x};\ell)$ is supported on 601 monomials. There are 4,096 Littelmann patterns to consider, 2,216 of which are nonstrict. The remaining patterns each give a nonzero contribution to N. The resulting polynomial can be written succinctly as

$$N(\mathbf{x};\ell) = \prod_{\alpha>0}\left(1 - q^{d(\alpha)-1}\mathbf{x}^\alpha\right),$$

where the product is taken over the positive roots. Here $d(\alpha) = k_1 + k_2 + k_3 + k_4$ if α is the linear combination of simple roots $k_1\alpha_1 + k_2\alpha_2 + k_3\alpha_3 + k_4\alpha_4$, and $\mathbf{x}^\alpha$ refers to the monomial $x_1^{k_1}x_2^{k_2}x_3^{k_3}x_4^{k_4}$.

References

1. A. Berenstein and A. Zelevinsky, *Tensor product multiplicities and convex polytopes in partition space*, J. Geom. Phys. **5** (1988), no. 3, 453–472.
2. A. Berenstein and A. Zelevinsky, *String bases for quantum groups of type A_r*, I. M. Gelfand Seminar, Adv. Soviet Math., vol. 16, Amer. Math. Soc., Providence, RI, 1993, pp. 51–89.
3. A. Berenstein and A. Zelevinsky, *Canonical bases for the quantum group of type A_r and piecewise-linear combinatorics*, Duke Math. J. **82** (1996), no. 3, 473–502.
4. A. Berenstein and A. Zelevinsky, *Weyl Group Multiple Dirichlet Series: Type A Combinatorial Theory*, Annals of Mathematics Studies, vol. 225, Princeton University Press, 2011.
5. B. Brubaker, D. Bump, S. Friedberg, and J. Hoffstein, *Weyl group multiple Dirichlet series. III. Eisenstein series and twisted unstable A_r*, Ann. of Math. (2) **166** (2007), no. 1, 293–316.
6. B. Brubaker,personal communication.
7. B. Brubaker, D. Bump, and S. Friedberg, *Twisted Weyl group multiple Dirichlet series: the stable case*, Eisenstein series and applications, Progr. Math., vol. 258, Birkhäuser Boston, Boston, MA, 2008, pp. 1–26.
8. B. Brubaker, D. Bump, and S. Friedberg, *Weyl group multiple Dirichlet series, Eisenstein series and crystal bases*, Ann. of Math. (2) **173** (2011), no. 2, 1081–1120.
9. B. Brubaker, D. Bump, G. Chinta, S. Friedberg, and J. Hoffstein, *Weyl group multiple Dirichlet series. I*, Multiple Dirichlet series, automorphic forms, and analytic number theory, Proc. Sympos. Pure Math., vol. 75, Amer. Math. Soc., Providence, RI, 2006, pp. 91–114.
10. D. Bump and J. Hoffstein, *Some conjectured relationships between theta functions and Eisenstein series on the metaplectic group*, Number theory (New York, 1985/1988), Lecture Notes in Math., vol. 1383, Springer, Berlin, 1989, pp. 1–11.
11. G. Chinta and P. E. Gunnells, *Constructing Weyl group multiple Dirichlet series*, J. Amer. Math. Soc. **23** (2010), no. 1, 189–215.
12. G. Chinta and P. E. Gunnells, *Weyl group multiple Dirichlet series of type A_2*. In Number Theory, Analysis and Geometry: In Memory of Serge Lang. Goldfeld, Jorgenson, Jones, Ramakrishnan, Ribet, and Tate, J. (Eds.). Springer (2012).
13. G. Chinta and P. E. Gunnells, *Weyl group multiple Dirichlet series constructed from quadratic characters*, Invent. Math. **167** (2007), no. 2, 327–353.
14. G. Chinta and O. Offen, *A metaplectic Casselmann-Shalika formula for GL_r*, Amer. J. Math., to appear.
15. G. Chinta, S. Friedberg, and P. E. Gunnells, *On the p-parts of quadratic Weyl group multiple Dirichlet series*, J. Reine Angew. Math. **623** (2008), 1–23.
16. P. Littelmann, *Cones, crystals, and patterns*, Transform. Groups **3** (1998), no. 2, 145–179.
17. Peter J. McNamara. Metaplectic Whittaker functions and crystal bases. *Duke Math. J.*, 156(1):29–31, 2011.
18. T. Tokuyama, *A generating function of strict Gelfand patterns and some formulas on characters of general linear groups*, J. Math. Soc. Japan **40** (1988), no. 4, 671–685.

Chapter 6
Toroidal Automorphic Forms, Waldspurger Periods and Double Dirichlet Series

Gunther Cornelissen and Oliver Lorscheid

Abstract The space of toroidal automorphic forms was introduced by Zagier in the 1970s: a GL_2-automorphic form is toroidal if it has vanishing constant Fourier coefficients along all embedded non-split tori. The interest in this space stems (amongst others) from the fact that an Eisenstein series of weight s is toroidal for a given torus precisely if s is a nontrivial zero of the zeta function of the quadratic field corresponding to the torus.

In this chapter, we study the structure of the space of toroidal automorphic forms for an arbitrary number field F. We prove that this space admits a decomposition into a subspace of Eisenstein series (and derivatives) and a subspace of cusp forms. The subspace of Eisenstein series is generated by all derivatives up to order n-1 of an Eisenstein series of weight s and class group character ω for certain n, s, ω, namely, precisely when s is a zero of order n of the L-series $L_F(\omega, s)$. The subspace of cusp forms consists of exactly those cusp forms π whose central L-value is zero: $L(\pi, 1/2) = 0$.

The proofs are based on an identity of Hecke for toroidal integrals of Eisenstein series and a result of Waldspurger about toroidal integrals of cusp forms combined with nonvanishing results for twists of L-series proven by the method of double Dirichlet series.

Keywords Toroidal automorphic forms • Double Dirichlet series • Eisenstein series • Non-vanishing results • Twists of L-series

---

G. Cornelissen (✉)
Mathematisch Instituut, Universiteit Utrecht, Postbus 80010, 3508 TA Utrecht, Nederland
e-mail: g.cornelissen@uu.nl

O. Lorscheid
Instituto Nacional de Matematica Pura e Aplicada, Estrada Dona Castorina 110, CEP: 22460-320, Rio de Janeiro, Brazil
e-mail: oliver@mpim-bonn.mpg.de

D. Bump et al. (eds.), *Multiple Dirichlet Series, L-functions and Automorphic Forms*, Progress in Mathematics 300, DOI 10.1007/978-0-8176-8334-4_6,

6.1 Introduction

A classical theorem of Hecke (cf. Hecke [13] Werke p. 201) shows that on the modular curve $X(1)$, the integral of an Eisenstein series along a closed geodesic of discriminant $d > 0$ is essentially the zeta function of the number field $\mathbf{Q}(\sqrt{d})$. As was observed by Don Zagier in [26], the formula fits into a more general framework, where integrals of automorphic forms for global fields over tori (of any discriminant) evaluate to L-series. The approach in [26] is to define a space of so-called *toroidal* automorphic forms by the vanishing of these integrals for varying tori, and the author calls for an (independent) understanding of the space of toroidal automorphic forms, after which one may hopefully apply the gained knowledge in combination with the generalization of Hecke's formula to deduce something about zeta functions. For example, if the irreducible subrepresentations of the space of toroidal automorphic forms are tempered, the Riemann hypothesis follows. This seems for now an elusive programme, but see [7] for a case study for some function fields.

In this chapter, we study the space of toroidal automorphic forms in its own right. We use our increased knowledge (compared to the 1970s, when [26] was written) about the decomposition of the space of automorphic forms for a general number field (by Franke), toroidal integrals of cusp forms (by Waldspurger in his study of the Shimura correspondence) and nonvanishing of quadratic twists (by the method of multiple Dirichlet series, essentially in the works of Friedberg, Hoffstein and Lieman) to prove the following theorem, which summarizes our main results. Note that in light of the previous paragraph, we avoid using any unproven hypothesis about the zeros of L-series.

Theorem 1.1. *The space of toroidal automorphic forms for an arbitrary number field decomposes into an Eisenstein part and a cuspidal part. More precisely,*

(i) (cf. Theorem 4.1) *The Eisenstein part of the space of toroidal automorphic forms is spanned by all derivatives of Eisenstein series of weight $s_0 \in \mathbf{C}$ and class group character ω, precisely up to the order of vanishing of the L-series corresponding to ω at s_0.*
(ii) (cf. Theorem 5.2) *No nontrivial residues of Eisenstein series are toroidal.*
(iii) (cf. Theorem 6.1) *The cuspidal part of the space of toroidal automorphic forms is spanned by those cusp forms π for which the central value of its L-series vanish: $L(\pi, 1/2) = 0$.*

We will use the next section to set up notation and give precise definitions of the spaces involved.

Remark 1.2. How many derivatives of Eisenstein series will be toroidal? It is reasonable to expect that the only multiplicities in the zeros of L-functions of a number field F arise from multiplicities in the decomposition of the regular representation of the Galois group of the normal closure of $F/\mathbf{Q}$ (following the Artin formalism of factorization of L-series). The Rudnick–Sarnak theory of statistical distribution of zeros of principal primitive L-series (cf. [20, 21], Sect. 5) indicates that the zeros of different such principal primitive L-series should be uncorrelated. For example, for $F = \mathbf{Q}$, all zeros of the Riemann zeta function are expected to be simple.

But of course, as soon as $F/\mathbf{Q}$ is non-abelian, there will be such multiplicities arising from irreducible representations of the Galois group of higher dimension; cf. also the possibility that a Galois extension $N/\mathbf{Q}$ contains two distinct subfields that are arithmetically equivalent (corresponding to two subgroups of the Galois group G from which the trivial representation induces the same representation of G, cf. [14]), hence have the same zeta function, whose zeros will then occur with multiplicity in the zeta function of N.

Remark 1.3. Our way of averaging toroidal integrals is a two-step method: by *first* relating them to twists of L-series and *then* using standard techniques to average those. Is there a direct way to average toroidal integrals, for example, by a Rankin–Selberg unfolding of an Eisenstein series twisted with toroidal integrals? We did not work this out. But, for example, the toroidal integral of a holomorphic weight two cusp form f over a torus corresponding to $\mathbf{Q}(\sqrt{d})$ for some $d > 0$ is the integral of the differential form ω corresponding to f over the closed geodesic Γ_d in the modular curve corresponding to d. In this specific case, one wants to have an asymptotic result for a sum of "modular symbols" $\sum_{d<X} \int_{\Gamma_d} \omega$. Compare with [4], where Eisenstein series twisted by modular symbols are studied.

Remark 1.4. There is an obvious generalization of T-toroidal automorphic forms on $\mathrm{GL}(n)$ for a maximal indecomposable torus $T \subset \mathrm{GL}(n)$. One may wonder whether the methods presented in this chapter can be generalized to the $\mathrm{GL}(n)$-case.

Part of the ingredients of our proof is already in the literature: on the one hand, Wielonsky [25] generalized Hecke's formula (Theorem 3.2 (i)): toroidal integrals of Eisenstein series on $\mathrm{GL}(n)$ that are induced by a parabolic subgroup of type $(n-1, 1)$ equal certain L-series; see Lachaud [17] for a recent treatment. On the other hand, Friedberg, Hoffstein and Lieman [11] introduced double Dirichlet series w.r.t. n-th order twists of Hecke L-series and showed that they admit a meromorphic continuation and satisfy a functional equation.

However, we are not aware of a generalization of Hecke's theorem 3.2 to *all* Eisenstein series on $\mathrm{GL}(n)$ or of Waldspurger's work on toroidal integrals of cusp forms (6.1) to $\mathrm{GL}(n)$. Maybe there is a way to circumvent the relation to L-series; compare the previous remark and the method in [5].

Remark 1.5. In [15] (extended version in [17]) and [16], Lachaud ties up the theory of toroidal automorphic forms with Connes' trace form programme [6] in the study of zeros of zeta functions.

Remark 1.6. For global function fields, methods more akin to the geometric Langlands programme allow one to prove that the space of toroidal automorphic forms is finite dimensional, and one can control the linear relations between Eisenstein series in a very precise way, leading to an actual dimension formula for the Eisenstein part of the space of toroidal automorphic forms, cf. [18].

6.2 Definition of Toroidal Automorphic Forms

Notations 2.1. Let F be a number field, F_v be the completion at v, $\mathbf{A} = \mathbf{A}_F$ the adeles of F. Set $G = \mathrm{GL}(2)$ and let Z be its center. Let $\mathcal{A}$ be the space of automorphic forms for G over F with trivial central character.

Notations 2.2. Let $T \subset G$ be a maximal non-split torus defined over F, χ_T the corresponding character on the idele class group and $E = F_T$ the corresponding quadratic field extension of F. This means that there is a nonsquare $d \in F$ such that E/F is generated by a square root of d and that $T(F)$ is conjugated in $G(F)$ to the standard torus

$$T_d(F) = \left\{ \begin{pmatrix} a & b \\ bd & a \end{pmatrix} \in G(F) \right\}.$$

If $T = T_d$, then write χ_d for χ_T.

Definition 2.3. The *zeroth Fourier coefficient w.r.t.* T (or T*-toroidal integral*) of an automorphic form $f \in \mathcal{A}$ is defined to be the function

$$f_T(g) := \int_{T_F Z(\mathbf{A}) \backslash T(\mathbf{A})} f(tg)\, \mathrm{d}t$$

for $g \in G(\mathbf{A})$.

Definition 2.4.

(i) The *space of* T*- or* E*-toroidal automorphic forms for* F is

$$\mathcal{A}_{\mathrm{tor}}(T) := \mathcal{A}_{\mathrm{tor}}(E) := \{ f \in \mathcal{A} : f_T(g) = 0,\ \forall g \in G(\mathbf{A}) \}.$$

(ii) The *space of toroidal automorphic forms for* F is

$$\mathcal{A}_{\mathrm{tor}} := \bigcap_{E/F} \mathcal{A}_{\mathrm{tor}}(E),$$

where the intersection is taken over all quadratic field extensions E/F.

Remark 2.5. These definitions are independent of the choice of torus corresponding to E/F since they are conjugacy invariant.

Remark 2.6. There is also a definition of T-toroidal automorphic form for a split torus T, but one has to be careful, since the toroidal integral f_T as defined above over a split torus T can diverge. This can be taken care of by subtracting suitable parabolic Fourier coefficients before integrating (cf. [18, Sect. 1.5] for the definition in the function field case). One could thus consider the space of automorphic forms that are T-toroidal for all maximal tori T, split or not. Due to the results of the present text and [18, Sect. 6.2] (which transfer to the number field case), this space coincides with $\mathcal{A}_{\mathrm{tor}}$, and we forgo describing the more involved theory for split tori.

The space $\mathcal{A}$ is an automorphic representation of $G(\mathbf{A})$ for right translation by $G(\mathbf{A})$. By applying [9] to GL(2), the space $\mathcal{A}$ decomposes into a direct sum of automorphic representations as follows:

$$\mathcal{A} = \mathcal{A}_0 \oplus \mathcal{E} \oplus \mathcal{R},$$

where $\mathcal{A}_0$ is the space of cusp forms, $\mathcal{E}$ is the space generated by the derivatives of Eisenstein series and $\mathcal{R}$ is generated by the residues of these Eisenstein series and their "derivatives". We will give the precise definitions of $\mathcal{E}$ and $\mathcal{R}$ in Sects. 6.3 and 6.5, respectively.

Multiplicity one holds for GL(2); hence, if π is any subrepresentation of $\mathcal{A}$, it inherits this decomposition since it is determined by its isomorphism type. In order to investigate the space of toroidal automorphic forms $\mathcal{A}_{\text{tor}}$, which is an automorphic representation by its very definition, it thus suffices to investigate

$$\mathcal{A}_{0,\text{tor}} := \mathcal{A}_{\text{tor}} \cap \mathcal{A}_0, \qquad \mathcal{E}_{\text{tor}} := \mathcal{E} \cap \mathcal{A}_{\text{tor}} \text{ and } \mathcal{R}_{\text{tor}} := \mathcal{R} \cap \mathcal{A}_{\text{tor}}$$

separately, since $\mathcal{A}_{\text{tor}} = \mathcal{A}_{0,\text{tor}} \oplus \mathcal{E}_{\text{tor}} \oplus \mathcal{R}_{\text{tor}}$.

6.3 Toroidal Integrals of Eisenstein Series: A Formula of Hecke

Notations 3.1. Let χ denote a character of the idele class group $I = F^\times \backslash \mathbf{A}^\times$. We can write $\chi = \omega \cdot |\,|^{s_0 - \frac{1}{2}}$ for some finite-order character ω, and we will sometimes regard a function of χ as a function of s_0 (assuming ω to be fixed). We set $\mathrm{Re}(\chi) = \mathrm{Re}(s_0) - \frac{1}{2}$. We also remark that the shift in s_0 by $-\frac{1}{2}$ is in accordance with the usual convention in the adelic theory of Eisenstein series, putting the center of symmetry at 0.

We define the *principal series* $\mathcal{P}(\chi)$ by

$$\left\{ f : G(\mathbf{A}) \overset{\text{smooth}}{\longrightarrow} \mathbf{C} \colon \forall \left(\begin{smallmatrix} a & b \\ & d \end{smallmatrix}\right), g \in G(\mathbf{A}), f\left(\left(\begin{smallmatrix} a & b \\ & d \end{smallmatrix}\right) g\right) = \chi(a/d)\,|a/d|^{1/2} f(g) \right\},$$

where a function $f : G(\mathbf{A}) \to \mathbf{C}$ is smooth if it is smooth in the usual sense at archimedean places and locally constant at finite places. Let $f \in \mathcal{P}(\chi)$ be embedded in a *flat section* f_χ of the principal series, i.e., there exists a function $f_\chi(s)$ of $s \in \mathbf{C}$ such that $f_\chi(s) \in \mathcal{P}(\chi\,|\,|^s)$ with $f = f_\chi(0)$ and $f_\chi(s)(e) = f(e)$ for all $s \in \mathbf{C}$, where $e = \left(\begin{smallmatrix} 1 & 0 \\ 0 & 1 \end{smallmatrix}\right)$. Note that every $f \in \mathcal{P}(\chi)$ is embedded into a unique flat section. In the following, we will write $f = f_\chi(0) \in \mathcal{P}(\chi)$ to refer to this situation. We define the *(completed) Eisenstein series* as

$$E(g, f) = L\left(\chi^2, \frac{1}{2}\right) \cdot \sum_{\gamma \in B(F) \backslash G(F)} f(\gamma g)$$

in terms of meromorphic continuation. Here, $L(\chi^2, 1/2)$ denotes the *completed L-series*, i.e., including the factors at infinity. Note that $E(g, f)$ is defined for all $f = f_\chi(0)$ unless $\chi^2 = |\,|^{\pm 1}$, when the Eisenstein series has a simple pole. At these values of χ, the residues of the Eisenstein series define automorphic forms, which will be investigated in Sect 6.5—these values of χ characterize the cases where the principal series $\mathcal{P}(\chi)$ is *not* irreducible as an automorphic representation of $G(\mathbf{A})$.

Also, note that the symbol "E" is now in use for both a field E/F and a function $E(g, f)$, but this should cause no confusion.

We now compute the toroidal integrals of Eisenstein series. Statement (i) in the theorem below is the adelic formulation of a theorem of Hecke ([13]). In this formulation, it was first stated by Zagier [26].

Theorem 3.2. *Let T be a maximal torus in G corresponding to the quadratic field extension E/F. Let $\chi_E : I \to \mathbf{C}^\times$ be the quadratic character whose kernel equals the norms of $\mathbf{A}_E$. For every $f = f_\chi(0) \in \mathcal{P}(\chi)$, every $g \in G_\mathbf{A}$ and every character $\chi : I \to \mathbf{C}^\times$, there exists a holomorphic function $e_T(g, f_\chi(s))$ of $s \in \mathbf{C}$ with the following properties:*

(i) *For all χ such that $\chi^2 \neq |\,|^{\pm 1}$,*

$$E_T(g, f) = e_T(g, f)\, L\left(\chi, \frac{1}{2}\right)\, L\left(\chi\chi_E, \frac{1}{2}\right).$$

(ii) *For every $g \in G_\mathbf{A}$ and $\chi : I \to \mathbf{C}^\times$, there is a $f \in \mathcal{P}(\chi)$ such that $e_T(g, f) \neq 0$.*

Proof. The strategy of the proof is as follows: we first prove (i) for $\mathrm{Re}(\chi)$ sufficiently large, so we are in the region of absolute convergence. We rewrite the Eisenstein series conveniently as a certain adelic integral. Then, a change of variables identifies its toroidal integral with a Tate integral for an L-series of E.

Rewriting the Eisenstein series. First, we explain how to represent $E(g, f)$ by a certain adelic integral, and then the statement in (i) is a simple application of a change of variables. Let φ be a Schwartz-Bruhat function on $\mathbf{A}^2$. Then,

$$F(g, \varphi, \chi) = \int\limits_{Z(\mathbf{A})} \varphi((0, 1)zg)\chi(\det zg)\, |\det zg|^{1/2}\, \mathrm{d}z \tag{1}$$

is a Tate integral for $L(\chi^2, 1)$, which converges if the real part of χ is larger than $1/2$ (the square of the character χ occurs because in the above integrand $\chi(\det(z)) = \chi(z)^2$). One verifies easily that $F(\cdot, \varphi, \chi)$ is an element of $\mathcal{P}(\chi)$.

In [24, Chap. VII, Sect. 7], Weil defines a particular test function φ_0 (the "standard function") with the property that for $e = \left(\begin{smallmatrix}1 & 0\\ 0 & 1\end{smallmatrix}\right)$,we have $F(e, \varphi_0, \chi) = c_F^{-1} L(\chi^2, 1)$ for a non-zero constant c_F that only depends on the field F. Thus,

$$c_F \cdot F(g, \varphi_0, \chi) = L(\chi^2, 1) \cdot f(g)$$

for an $f \in \mathcal{P}(\chi)$ with $f(e) = 1$; in particular, $F(\cdot, \varphi_0, \chi)$ is a non-trivial element of $\mathcal{P}(\chi)$.

We note that for every $g \in G(\mathbf{A})$, the function $\varphi_0(\cdot g)$ is a Schwartz-Bruhat function, too. Since $\mathcal{P}(\chi)$ is irreducible, the integrals $F(g, \varphi, \chi)$ for varying φ exhaust all products of the form $L(\chi^2, 1) f(g)$, where $f \in \mathcal{P}(\chi)$. Thus, there is a Schwartz-Bruhat function $\varphi = \varphi(f)$ for every $f \in \mathcal{P}(\chi)$ such that

$$E(g, f) = \sum_{\gamma \in B(F)\backslash G(F)} F(\gamma g, \varphi, \chi), \tag{2}$$

where the equality has to be interpreted in terms of meromorphic continuation. Since for all φ the Tate integral (1) is a multiple of $L(\chi^2, 1)$, there exists $f \in \mathcal{P}(\chi)$ for every Schwartz-Bruhat function φ such that equation (2) holds true.

Computing the toroidal integral. With this reformulation at hand, we can prove the theorem precisely as it has been done by Zagier in [26]: by identifying $T(F)$ with $E^\times$ and $T(\mathbf{A}_F)$ with $\mathbf{A}_E^\times$, and using Fubini's theorem, the toroidal integral of $E(g, f)$ is changed into a Tate integral for an L-series of E. Explicitly, $E_T(g, f)$ is given by

$$\int\limits_{T(F)Z(\mathbf{A})\backslash T(\mathbf{A})} \sum_{\gamma \in B(F)\backslash G(F)} \int\limits_{Z(\mathbf{A})} \varphi((0,1) z\gamma t g) \chi(\det z\gamma t g) \, |\det z\gamma t g|^{1/2} \, \mathrm{d}z \, \mathrm{d}t.$$

Via $T(F) = E^\times$ and

$$B(F)\backslash G(F) = F^\times \backslash F^2 - \{(0,0)\} = F^\times \backslash E^\times,$$

we identify the "domain of integration" with

$$T(F)Z(\mathbf{A})\backslash T(\mathbf{A}) \times B(F)\backslash G(F) \times Z(\mathbf{A}) \cong \mathbf{A}_E^\times,$$

and note that this identification is compatible with measures. Hence,

$$E_T(g, f) = \chi(\det g) \, |\det(g)|^{1/2} \cdot \int\limits_{\mathbf{A}_E^\times} \Phi(t) \, (\chi \circ N_{E/F})(t) \, |t|_E^{1/2} \, \mathrm{d}t$$

for a certain Schwartz-Bruhat function Φ on $\mathbf{A}_E^\times$. This is a Tate integral for $L_E(\chi \circ N_{E/F}, 1/2)$ and factors in a product of two L-series as in (i), thus giving (i) for $\mathrm{Re}(\chi)$ sufficiently large. Statement (i) now follows for all characters by meromorphic continuation in s since both sides are meromorphic functions of $s \in \mathbf{C}$.

Nonvanishing of the "constant". For (ii), note that the nonvanishing of $e_T(g, f)$ follows from the fact that we can choose φ such that the test function Φ is again the standard function as described by Weil [24, Chap. VII, Sect. 7], and then

$$e_T(g, f_\chi(s)) = c_E^{-1} \chi(\det(g)) \, |\det(g)|^{s+1/2} \tag{3}$$

is a non-vanishing holomorphic function of s. □

In order to accord for possible multiple zeros of L-series, we need to also take into account higher derivatives of Eisenstein series.

Notations 3.3.

(i) We denote by $E^{(n)}(g, f)$ the n-th derivative of $E(g, f_\chi(s))$ w.r.t. s at $s = 0$.
(ii) We denote by $\mathcal{E}$ the space of automorphic forms that is generated by all derivatives $E^{(n)}(\cdot, f)$ of Eisenstein series, where $n \geq 0$ and $f \in \mathcal{P}(\chi)$ with χ varying through all idele class group characters whose square is not trivial.
(iii) Similarly, we denote by $L^{(n)}(\chi, 1/2)$ the n-th derivative of $L(\chi, 1/2 + s)$ at $s = 0$ and by $e_T^{(n)}(g, f)$ the n-th derivative of the function $e_T(g, f_\chi(s))$ at $s = 0$.

Definition 3.4. We say χ is a zero of $L(\cdot, 1/2)$ of order n if $L^{(i)}(\chi, 1/2) = 0$ for all $i < n$ but $\neq 0$ for $i = n$, and then we write $\mathrm{ord}_\chi L(\cdot, 1/2) = n$.

Proposition 3.5. *Let T be a maximal non-split torus in G and n a non-negative integer.*

(i) *For all $g \in G_{\mathbb{A}}$ and χ such that $\chi^2 \neq |\ |^{\pm 1}$, we have*

$$E_T^{(n)}(g, f) = \sum_{\substack{i+j+k=n \\ i,j,k\geq 0}} \frac{n!}{i!\,j!\,k!}\, e_T^{(i)}(g, f)\, L^{(j)}\left(\chi, \frac{1}{2}\right)\, L^{(k)}\left(\chi\chi_T, \frac{1}{2}\right).$$

(ii) *The n-th derivative $E^{(n)}(g, f)$ of an Eisenstein series is E-toroidal if and only if χ is a zero of $L(\cdot, 1/2) \cdot L(\cdot\chi_E, 1/2)$ of order at least n.*

Proof. The first part follows from the Leibniz rule. For (ii), use (i) and observe the following:

- The derivatives $e_T^{(i)}(g, f)$ are nonzero as function of f, as is easily seen from (3).
- If $E^{(m)}(g, f)$ is E-toroidal, then so is $E^{(n)}(g, f)$ for all $n < m$ since $E^{(m)}$ generates an automorphic subrepresentation of the space of toroidal automorphic forms that contains all derivatives of lower order.

This finishes the proof. □

6.4 Application of Double Dirichlet Series: Toroidal Eisenstein Series

In this section, we will prove the following:

Theorem 4.1. *Let $f \in \mathcal{P}(\chi)$. The n-th derivative $E^{(n)}(g, f)$ of an Eisenstein series is toroidal if and only if χ is a zero of $L(\cdot, \frac{1}{2})$ of order at least n. Hence,*

$$\mathcal{E}_{\mathrm{tor}} = \left\langle E^{(n)}(\cdot, f) \ :\ \exists \chi, f \in \mathcal{P}(\chi) \text{ and } n \leq \mathrm{ord}_\chi L\left(\cdot, \frac{1}{2}\right)\right\rangle.$$

Proof. Recall that we know from the computation of toroidal integrals in Proposition 3.5 when the n-th derivative $E^{(n)}(\cdot, f)$ of an Eisenstein series is E-toroidal. It suffices to prove that for all χ, there exists a quadratic E/F such that $L(\chi\chi_E, 1/2) \neq 0$. This follows from Theorem 4.2. □

As before, we now write $\chi = \omega \cdot |\,|^{s_0 - \frac{1}{2}}$ for a finite character ω, where we consider $s = s_0 \in \mathbf{C}$ as varying parameter. We use the notation $L(\omega, s)$ for $L(\chi, 1/2)$.

Theorem 4.2. *Let F denote a number field, let ω denote a class group character on F. Then, there exists a quadratic field extension E/F such that $L(\omega\chi_E, s) \neq 0$.*

Remark 4.3. Before we start with the proof, we make some incomplete historical remarks. Nonvanishing of quadratic twists can be proven by sieve methods, but only for a restricted set of number fields. A method that works more uniformly is that of multiple Dirichlet series (so Fourier coefficients of metaplectic Eisenstein series); unfortunately, the result we need is not literally in the existing literature on multiple Dirichlet series, but rather arises from a combination of existing methods. We observe that for $\mathrm{Re}(s) \neq \frac{1}{2}$, the result can be proven using "unweighted" double Dirichlet series (see Chinta, Friedberg and Hoffstein [2] Theorem 1.1). To extend to the (for us interesting) range $\mathrm{Re}(s) = \frac{1}{2}$, one needs to analytically continue the double Dirichlet series with weights; for higher order characters and a general number field, this is done by Friedberg, Hoffstein and Lieman in [11], and for quadratic twists of function fields by Fisher and Friedberg [8]. We will combine the methods of these latter two sources. Since proofs of many facts are literally the same, we will not repeat them here, but we will set up all required notations. One can go a (small) step further and combine the method with Tauberian theorems to establish lower bounds on the number of nonvanishing twists of bounded conductor, but we will not need those. Also, we refrain from discussing averaging of toroidal integrals *directly*, without first relating them to L-series, cf. also Remark 1.3.

Proof (of Theorem 4.2). First note that the L-series we consider are "completed" by the correct archimedean factors, so they do not have trivial zeros outside the critical strip.

Because of the functional equation, we can assume that $\frac{1}{2} \leq \mathrm{Re}(s) \leq 1$. Since $L(\omega\chi_E, s)$ does not vanish on $\mathrm{Re}(s) = 1$ ([19] Chap. 1, Sect. 4), we can even assume $\frac{1}{2} \leq \mathrm{Re}(s) < 1$.

The strategy of the proof is now the following: we assume by contradiction that all nontrivial twists $L(\omega\chi_E, s)$ vanish at $s = s_0$. Then, the (analytically continued) double Dirichlet series $Z^0_\omega(s, w)$ (to be defined below), with the trivial twist extracted, vanishes identically in w for $s = s_0$. But it has residue at $w = 1$ a non-zero constant times $L(\omega, 2s_0)$. Hence, we find $L(\omega, 2s_0) = 0$, and this is impossible if $\frac{1}{2} \leq \mathrm{Re}(s_0) < 1$.

To define the double Dirichlet series in a rigorous way, we follow [11], Sect. 1. Since the class number of F is not necessarily one, the most natural double Dirichlet series (and the one that has a natural analytic continuation and set of functional equations) does not only sum over quadratic twists χ_E, but rather over more general characters on a ray class group. We now introduce this series first.

Let $S = S_f \cup S_\infty$ denote a finite set of places of F that contains all infinite places S_∞ of F and a set S_f of finite places such that the ring of S_f-integers has class number one. For v a finite place corresponding to an ideal $\mathfrak{p}_v$, let $q_v = |\mathcal{O}/\mathfrak{p}_v|$. Set $C = \prod_{v \in S_f} \mathfrak{p}_v^{n_v}$, where $n_v = 1$ for v not above 2, and for v above 2, n_v is so large that any $a \in F_v$ with $\mathrm{ord}_v(a-1) \geq n_v$ is a square in F_v. Let H_C denote the ray class group of modulus C, let $h_C := |H_C|$ and set

$$R_C = H_C \otimes \mathbf{Z}/2 = \mathbf{Z}/a_1 \times \cdots \times \mathbf{Z}/a_r.$$

Choose generators b_i for $\mathbf{Z}/a_i$, and choose a set $\mathcal{E}_0$ of ideals prime to S that represent the b_i. For any $E_0 \in \mathcal{E}_0$, let m_{E_0} denote an element of F^* that generates the (principal) ideal $E_0\mathcal{O}_S$. Let $\mathcal{E}$ denote a set of representatives of R_C that are of the form $E = \prod E_0^{n_{E_0}}$, where the E_0 are elements of $\mathcal{E}_0$ and the n_{E_0} are natural numbers, and set $m_E = \prod m_{E_0}^{n_{E_0}}$ with the convention that $\mathcal{O} \in \mathcal{E}$ with $m_{\mathcal{O}} = 1$. Then, $E\mathcal{O}_S = (m_E)$. Let $I(S)$ denote the set of fractional ideals coprime to S_f. For $d, e \in I(S)$ coprime, write $d = (a)EG^2$ with $E \in \mathcal{E}, a \in F^*, a \equiv 1 \bmod C, G \in I(S)$, and define

$$\chi_d(e) = \left(\frac{d}{e}\right) := \left(\frac{am_E}{e}\right).$$

This is well defined (cf. [11], Proposition 1.1); it does not depend on the decomposition of d (but it does depend on the choice of m_E). For d principal, χ_d is the usual quadratic character for $F(\sqrt{d})/F$ (cf. Notation 2.2).

Let $L_S(\omega, s)$ denote the L-series of F for the class group character ω, but with the Euler factors corresponding to the places in S removed. For $d \in I(S)$, let S_d denote the set of primes above d. Let $J(S)$ denote the set of integral ideals in $I(S)$. For $d \in J(S)$, let $|d|$ denote its norm. Write $d = d_0 d_1^2$ with d_0 squarefree.

We define the weight factor to be

$$a(\omega, s, d) := \sum_{\substack{e_i \in J(S) \\ e_1 e_2 | d_1}} \frac{\mu(e_1)\chi_d(e_1)\omega\left(e_1 e_2^2\right)}{|e_1|^s |e_2|^{2s-1}},$$

where μ is the Möbius function. Let ρ denote a character on the idele class group unramified outside S; this will be used later on to filter out principal ideals, which are the ones we are interested in. We define the double Dirichlet series as

$$Z_{\omega,\rho}(s, w) := \sum_{d \in J(S)} \frac{L_{S \cup S_d}(\omega\chi_d, s)\rho(d)}{|d|^w} \cdot a(\omega, s, d).$$

This is convergent for $\mathrm{Re}(s)$ and $\mathrm{Re}(w)$ sufficiently large (say, ≥ 1). The following properties are proven in exactly the same way as in [11] (the only difference to [11], which treats the case of twists by characters of higher order, is the set of functional equations, which here is of order 12 instead of 32, cf. [11], Remark 2.6):

(i) The function $Z_{\omega,\rho}(s,w)$ admits a meromorphic continuation to $\mathbf{C}^2$.
(ii) The poles of $Z_{\omega,\rho}(s,w)$ are located on the union of the lines:

$$w=0, w=1, s=0, s=1, w+s=\frac{1}{2} \text{ and } w+s=\frac{3}{2}.$$

(iii) If $\rho \neq 1$, then $Z_{\omega,\rho}(s,w)$ is holomorphic at $w=1$; if $\rho=1$ and $s\neq 1/2$, $Z_{\omega,\rho}(s,w)$ has a simple pole at $w=1$. If $\rho=1$ and $s=\frac{1}{2}$, $Z_{\omega,\rho}(s,w)$ has a double pole at $w=1$, and $L_S(\omega^2,2s)$ has a simple pole. We have

$$\lim_{w\to 1}(w-1)Z_{\omega,\rho}(s,w) = \begin{cases} 0 & \text{if } \rho\neq 1; \\ (\mathrm{Res}_{w=1}\,\zeta_F(w))\cdot\left(\prod_{v\in S_f}\zeta_{F,v}(1)^{-1}\right)\cdot L_S(\omega^2,2s) & \text{if } \rho=1 \end{cases}$$

(cf. [3], Sect. 5; see also Sect. 5.3 in *loc. cit.* for a computation of the principal part of $Z_{\omega,\rho}(s,w)$ around $w=1$, which we will not need here).

Let

$$Z^0_\omega(s,w) := \sum_{\substack{d\in J(S)\\ [d]-0}} \frac{L_{S\cup S_d}(\omega\chi_d,s)}{|d|^w}\cdot a(\omega,s,d), \tag{4}$$

denote the modified double Dirichlet series, where we only sum over principal ideals d (indicated by the fact that their class $[d]$ is trivial in R_C). Note that by plugging in the decomposition of the characteristic function of the class [0] as $h_C^{-1}\sum_{\rho\in\widehat{R_C}}\rho$, we find

$$Z^0_\omega(s,w)=\frac{1}{h_C}\sum_{\rho\in\widehat{R_C}} Z_{\omega,\rho}(s,w).$$

Hence, $Z^0_\omega(s,w)$ inherits an analytic continuation from $Z_{\omega,\rho}(s,w)$. Using the above computation of residues, we find that

$$\lim_{w\to 1}(w-1)Z^0_\omega(s_0,w)=c\cdot L(\omega^2,2s_0), \tag{5}$$

where c is some nonzero constant.

We can now finish the proof of the theorem. We are assuming that all "principal" twists $L(\omega\chi_d,s_0)$ ($[d]=0$, χ_d nontrivial) vanish at some s_0 with $\frac{1}{2}\leq \mathrm{Re}(s_0)<1$. Note first that this obviously implies the vanishing of all twists for the modified L-series with the $S\cup S_d$-Euler factors removed.

A slight complication arises since the s_0 we consider are outside of the region of absolute convergence of the series $Z^0_\omega(s,w)$, but we can use a convexity estimate to get that for such s_0 and $\mathsf{Re}(w)$ *large enough*, the double Dirichlet series $Z^0_\omega(s_0, w)$ as defined in (4) will also converge. This is because of Phragmén-Lindelöf estimates in the d-aspect of the form

$$|L_{S\cup S_d}(\omega\chi_d, s_0)a(\omega, s_0, d)| \ll |d|$$

(cf. [3], Sect. 3.3), so $\mathsf{Re}(w) > 2$ will do.

Now, recall our hypothesis that $L_{S\cup S_d}(\omega\chi_d, s_0) = 0$ for all principal d with $\chi_d \neq 1$, i.e., d is not a square. Hence, if we subtract the terms for which $d = e^2$ is a square from (4), we find the identically zero function

$$Z^0_\omega(s_0, w) - \sum_{\substack{e\in J(S)\\ [e^2]=0}} \frac{L_{S\cup S_e}(\omega, s_0)}{|e|^{2w}} \cdot a(\omega, s_0, e^2) = 0 \tag{6}$$

(first for $\mathsf{Re}(w)$ sufficiently large, hence after analytic continuation, for all w).

We now prove that the term we have subtracted off does not have a pole at $w = 1$; actually, it converges absolutely at $w = 1$. Write the term as

$$L_S(\omega, s_0) \cdot \sum_{\substack{e\in J(S)\\ [e^2]=0}} \frac{B_e}{|e|^{2w}} \cdot a(\omega, s_0, e^2), \tag{7}$$

where B_e consists of the reciprocals of the (finitely many) $(S_e \setminus S)$-Euler factors of $L(\omega, s_0)$. Essentially, the absolute convergence is due to the fact that the exponent of $|e|$ is ≈ 2 if $w \approx 1$ and the other factors are small in $|e|$. We present some details. First of all, note that $L_S(\omega, s_0)$ does not have a pole for $\frac{1}{2} \leq \mathsf{Re}(s_0) < 1$. We estimate for $\mathsf{Re}(s_0) \geq 1/2$

$$|B_e| = \left| \prod_{\mathfrak{p}\in S_e\setminus S} \left(1 - \frac{\omega(\mathfrak{p})}{|\mathfrak{p}|^{s_0}}\right) \right| \leq 2^{\varpi(|e|)[F:\mathbf{Q}]} \leq d(|e|)^{[F:\mathbf{Q}]} \ll |e|^{1/3}$$

(where $\varpi(n)$ is the number of positive prime divisors of an integer n and $d(n)$ is the number of positive divisors of n, e.g. [12] Sect. 22.13). We estimate the other factor as

$$|a(\omega, s_0, e^2)| = \left| \sum_{\substack{e_1,e_2\in J(S)\\ e_1e_2|e}} \frac{\mu(e_1)\omega(e_1e_2^2)}{|e_1|^{s_0}|e_2|^{2s_0-1}} \right| \leq \sum_{\substack{e_1,e_2\in J(S)\\ e_1e_2|e}} 1$$

$$\leq d(|e|)^{2[F:\mathbf{Q}]} \ll |e|^{1/3}$$

since $\left|\mu(e_1)\omega(e_1 e_2^2)\right| \leq 1$ and $|e_1|^{s_0}|e_2|^{2s_0-1} \geq 1$ for $\mathrm{Re}(s_0) \geq \frac{1}{2}$. We combine this into the estimate

$$\sum_{\substack{e\in J(S)\\ [e^2]=0}} \left| \frac{B_e}{|e|^{2w}} \cdot a(\omega, s_0, e^2) \right| \quad \ll \quad k \sum_{\substack{e\in J(S)\\ [e^2]=0}} \left| \frac{1}{|e|^{2w-2/3}} \right|,$$

for some constant k, the latter sum being an absolutely convergent series for $w = 1$.

We conclude from this and (6) that also $Z_\omega^0(s_0, w)$ does not have a pole at $w = 1$, i.e., $\lim_{w\to 1}(w-1)Z_\omega^0(s_0, w) = 0$. Then, by (5), we find that $L(\omega^2, 2s_0) = 0$ with $1 \leq \mathrm{Re}(2s_0) < 2$, which is impossible. This finishes the proof. □

6.5 Toroidal Residues of Eisenstein Series

Let χ be a character of the idele class group such that $\chi^2 = |\,|^{\pm 1}$ and $f = f_\chi(0) \in \mathcal{P}(\chi)$. Then, the Eisenstein series $E(g, f_\chi(s))$ has a simple pole at $s = 0$, but the residue

$$R(g, f) := \mathrm{Res}_{s=0}\, E(g, f_\chi(s))$$

is an automorphic form. More generally, we consider the "derivatives"

$$R^{(n)}(g, f) := \lim_{s\to 0} \frac{d^n}{ds^n}\left(s \cdot E(g, f_\chi(s))\right)$$

for $n \geq 0$.

Definition 5.1. Define $\mathcal{R}$ as the space of automorphic forms that is generated by the functions $R^{(n)}(g, f)$, where $n \geq 0$, f ranges through $\mathcal{P}(\chi)$ and χ ranges through all characters on the idele group such that $\chi^2 = |\,|^{\pm 1}$.

Theorem 5.2. $\mathcal{R}_{\mathrm{tor}} = \{0\}$.

Proof. We compute the toroidal integral of $R(g, f)$ for a torus T corresponding to a quadratic field E:

$$\begin{aligned} R_T(g, f) &= \lim_{s\to 0}\ s\ E_T(g, f_\chi(s)) \\ &= \lim_{s\to 0}\ s\ e_T(g, f_\chi(s))\ L\left(\chi, s+\frac{1}{2}\right) L\left(\chi\chi_E, s+\frac{1}{2}\right) \\ &= e_T(g, f)\ \mathrm{Res}_{s=0}\ L\left(\chi, s+\frac{1}{2}\right) L\left(\chi\chi_E, s+\frac{1}{2}\right). \end{aligned}$$

Recall that $\chi^2 = |\,|^{\pm 1}$; hence, $\chi\,|\,|^{\mp 1/2}$ is quadratic, so either trivial or by class field theory is equal to χ_E for some quadratic field extension E/F. In the case that E is nontrivial, we have that

$$L\left(\chi\chi_E, s+\frac{1}{2}\right) = \zeta_F\left(s+\frac{1}{2}\pm\frac{1}{2}\right),$$

where ζ_F is as usual the completed zeta function of F with poles at 0 and 1, so

$$\zeta_F\left(s+\frac{1}{2}\pm\frac{1}{2}\right)$$

has a pole at $s = 0$ (for both choices of sign). Hence, by the above formula for $R_T(g, f)$, the toroidal integral of $R_T(g, f)$ for this E cannot vanish since the other factor $L(\chi, 1/2)$ does not vanish.

If $\chi|\,|^{\mp 1/2}$ is trivial, then we choose an arbitrary non-split torus T. We find that

$$L(\chi, s+1/2) = \zeta_F\left(s+\frac{1}{2}\pm\frac{1}{2}\right)$$

has a pole at $s = 0$, but the other factor in the above computation of the toroidal integral,

$$L\left(\chi\chi_E, s+\frac{1}{2}\right) = L\left(\chi_E, s+\frac{1}{2}\pm\frac{1}{2}\right)$$

does not vanish at $s = 0$.

This also implies that $R^{(n)}(g, f)$ cannot be toroidal, since $R(g, f)$ is contained in the automorphic representation generated by this automorphic form. □

6.6 Application of Waldspurger Periods: Toroidal Cusp Forms

In this section, we prove the following:

Theorem 6.1. *A cuspidal representation $\pi \subset \mathcal{A}_0$ is toroidal if and only if*

$$L(\pi, 1/2) = 0.$$

Thus, we have

$$\mathcal{A}_{0,\mathrm{tor}} = \left\langle \pi \in \mathcal{A}_0 \ : \ L\left(\pi, \frac{1}{2}\right) = 0 \right\rangle.$$

Proof. A formula of Waldspurger ([22, Proposition 7]) shows that the T-period of an automorphic form f in an irreducible cuspidal representation π is a nonzero multiple of

$$L\left(\pi, \frac{1}{2}\right) \cdot L\left(\pi \otimes \chi_T, \frac{1}{2}\right).$$

Thus, π is toroidal if $L(\pi, 1/2) = 0$.

We are left to prove the reverse implication. Assume that $L(\pi, 1/2) \neq 0$. Note that all automorphic representation of $\mathrm{GL}(n)$ with trivial central character are self-contragredient. Thus, the functional equation of the L-series of π in $s = 1/2$ is

$$L(\pi, 1/2) = \epsilon(\pi, 1/2) \cdot L(\pi, 1/2),$$

and necessarily $\epsilon(\pi, 1/2) = 1$. This allows us to apply a theorem of Friedberg and Hoffstein: let π_v be representations of $G(F_v)$ such that $\pi \simeq \otimes' \pi_v$, where v ranges over all places; let S be the (finite) set of places v such that π_v is square integrable; let ξ be the trivial Hecke character. Then, [10, Theorem B (1)] states that there are infinitely many different nonconjugate tori T such that $L(\pi \otimes \chi_T, 1/2) \neq 0$, and such that for all $v \in S$, the local character χ_v is trivial, i.e., T_v is split. In particular, there is such a non-split torus T.

We want to apply to apply [22, Theorem 2, p. 221] (in the "situation globale," where the quaternion algebra is chosen to split), which implies that π is not T-toroidal. To do so, we have to verify that condition (i) in loc. cit. is satisfied. By [22, Lemme 8 (iii)], condition (i) is satisfied for all local factors π_v that are not square integrable. If $v \in S$, then T_v is split and [22, Lemme 8 (ii)] implies that condition (i) holds for square integrable π_v.

This shows that π is not toroidal, which concludes the proof of the theorem. □

Remark 6.2. Theorem [23, Theorem 4, p. 288] of Waldspurger says that there exists a χ_d for $d \in F^\times$ such that $L(\pi \otimes \chi_d, 1/2) \neq 0$, but without the claim that χ_d is nontrivial. However, if there exists one such d (square or not), there exist infinitely many, as may be seen from Lemma 7.1 in [1], which shows that the Dirichlet series occurring as Mellin transform of the series $t(\sigma)$ on p. 289 of the proof in [23] cannot have only finitely many nonzero coefficients.

Remark 6.3. There do exist number fields F for which there exist nontrivial toroidal cusp forms. We only need π to be a cusp form with $L(\pi, 1/2) = 0$. This happens when the root number is -1, which is, for example, the case for a cuspidal lift of a classical holomorphic cusp form to an imaginary quadratic field (cf. also Waldspurger [23], Remark on p. 282). The argument also shows that there are no toroidal cusp $F = \mathbf{Q}$.

Acknowledgments We thank Gautam Chinta for help with multiple Dirichlet series and Wee Teck Gan for his remarks on the proof of Theorem 6.1.

References

1. Ben Brubaker, Alina Bucur, Gautam Chinta, Sharon Frechette, and Jeffrey Hoffstein. Nonvanishing twists of GL(2) automorphic L-functions. *Int. Math. Res. Not.*, (78):4211–4239, 2004.
2. Gautam Chinta, Solomon Friedberg, and Jeffrey Hoffstein. Asymptotics for sums of twisted L-functions and applications. In *Automorphic representations, L-functions and applications: progress and prospects*, volume 11 of *Ohio State Univ. Math. Res. Inst. Publ.*, pages 75–94. de Gruyter, Berlin, 2005.
3. Gautam Chinta, Solomon Friedberg, and Jeffrey Hoffstein. Multiple Dirichlet series and automorphic forms. In *Multiple Dirichlet series, automorphic forms, and analytic number theory*, volume 75 of *Proc. Sympos. Pure Math.*, pages 3–41. Amer. Math. Soc., Providence, RI, 2006.
4. Gautam Chinta and Dorian Goldfeld. Grössencharakter L-functions of real quadratic fields twisted by modular symbols. *Invent. Math.*, 144(3):435–449, 2001.

5. Gautam Chinta and Omer Offen. Orthogonal period of a $GL_3(\mathbf{Z})$ Eisenstein series. In *Representation theory, complex analysis and integral geometry* (eds. Bernhard Krötz, Omer Offen and Eitan Sayag), Birkhäuser, 2012.
6. Alain Connes. Trace formula in noncommutative geometry and the zeros of the Riemann zeta function. *Selecta Math. (N.S.)* 5:29–106, 1999.
7. Gunther Cornelissen and Oliver Lorscheid. Toroidal automorphic forms for some function fields. *J. Numb. Th.*, 129:1456–1463, 2009.
8. Benji Fisher and Solomon Friedberg. Double Dirichlet series over function fields. *Compos. Math.*, 140(3):613–630, 2004.
9. Jens Franke. Harmonic analysis in weighted L_2-spaces. *Ann. Sci. École Norm. Sup. (4)*, 31(2):181–279, 1998.
10. Solomon Friedberg and Jeffrey Hoffstein. Nonvanishing theorems for automorphic L-functions on GL(2). *Ann. of Math. (2)*, 142(2):385–423, 1995.
11. Solomon Friedberg, Jeffrey Hoffstein, and Daniel Lieman. Double Dirichlet series and the n-th order twists of Hecke L-series. *Math. Ann.*, 327(2):315–338, 2003.
12. Godfrey H. Hardy and Edward M. Wright. *An Introduction to the Theory of Numbers.* Oxford University Press, 1980 (Fifth edition).
13. Erich Hecke. Über die Kroneckersche Grenzformel für reelle quadratische Körper und die Klassenzahl relativ-abelscher Körper. *Verhandl. Naturforschenden Gesell. Basel*, 28:363–372, 1917.
14. Norbert Klingen. *Arithmetical similarities*. Oxford Mathematical Monographs. The Clarendon Press Oxford University Press, New York, 1998.
15. Gilles Lachaud. Zéros des Fonctions L et formes toriques. *C. R. Acad. Sci. Paris Sér. I Math* 335: 219–222, 2002.
16. Gilles Lachaud. Spectral analysis and the Riemann hypothesis. *J. Comput. Appl. Math.* 160:175–190, 2003.
17. Gilles Lachaud. Zéros de fonctions L and formes toroïdales. `arxiv:0907.0536`, 2009.
18. Oliver Lorscheid. *Toroidal automorphic forms for function fields*. Israel *J. of Math.* (to appear).
19. M. Ram Murty and V. Kumar Murty. *Non-vanishing of L-functions and applications*, volume 157 of *Progress in Mathematics*. Birkhäuser Verlag, Basel, 1997.
20. Michael Rubinstein and Peter Sarnak. Chebyshev's bias. *Experiment. Math.*, 3(3):173–197, 1994.
21. Zeév Rudnick and Peter Sarnak. Zeros of principal L-functions and random matrix theory. *Duke Math. J.*, 81(2):269–322, 1996. A celebration of John F. Nash, Jr.
22. Jean-Loup Waldspurger. Sur les valeurs de certaines fonctions L automorphes en leur centre de symétrie. *Compositio Math.*, 54(2):173–242, 1985.
23. Jean-Loup Waldspurger. Correspondances de Shimura et quaternions. *Forum Math.*, 3(3):219–307, 1991.
24. André Weil. *Basic number theory*. Classics in Mathematics. Springer-Verlag, Berlin, 1995.
25. Frank Wielonsky. Séries d'Eisenstein, intégrales toroïdales et une formule de Hecke. *Enseign. Math. (2)* 31:93–135, 1985.
26. D. Zagier. Eisenstein series and the Riemann zeta function. In *Automorphic forms, representation theory and arithmetic (Bombay, 1979)*, volume 10 of *Tata Inst. Fund. Res. Studies in Math.*, pages 275–301. Tata Inst. Fundamental Res., Bombay, 1981.

Chapter 7
Natural Boundaries and Integral Moments of L-Functions

Adrian Diaconu, Paul Garrett, and Dorian Goldfeld

Abstract It is shown, under some expected technical assumption, that a large class of multiple Dirichlet series which arise in the study of moments of L-functions have natural boundaries. As a remedy, we consider a new class of multiple Dirichlet series whose elements have nice properties: a functional equation and meromorphic continuation. This class suggests a notion of integral moments of L-functions.

Keywords Multiple Dirichlet series • Integral moments of L-functions • Good's method • Eisenstein series • GL(3)

7.1 Introduction

The problem of obtaining asymptotic formulae (as $T \longrightarrow \infty$) for the integral moments

$$\int_0^T \left|\zeta\left(\frac{1}{2}+it\right)\right|^{2r} \mathrm{d}t \qquad (\text{for } r = 1,2,3,\ldots) \tag{1}$$

is approximately 100 years old and very well known. See [4] for a good exposition of this problem and its history. Following [1], it was proved by Carlson that for $\sigma > 1 - \frac{1}{r}$,

$$\int_0^T |\zeta(\sigma+it)|^{2r}\,\mathrm{d}t \sim \left[\sum_{n=1}^{\infty} d_r(n)^2 n^{-2\sigma}\right]\cdot T \qquad (T \longrightarrow \infty).$$

A. Diaconu (✉) • P. Garrett
School of Mathematics, University of Minnesota, Minneapolis, MN 55455, USA
e-mail: cad@umn.edu; garrett@umn.edu

D. Goldfeld
Goldfeld, Department of Mathematics, Columbia University, New York, NY 10027, USA
e-mail: goldfeld@columbia.edu

D. Bump et al. (eds.), *Multiple Dirichlet Series, L-functions and Automorphic Forms*, Progress in Mathematics 300, DOI 10.1007/978-0-8176-8334-4_7,

Furthermore,

$$\sum_{n=1}^{\infty} d_r(n)^2 n^{-s} = \zeta(s)^{r^2} \prod_p P_r\left(p^{-s}\right),$$

where

$$P_r(x) = (1-x)^{2r-1} \sum_{n=0}^{r-1} \binom{r-1}{n}^2 x^r.$$

Now, Estermann [12] showed that the Euler product $\prod_p P_r(s)$ is absolutely convergent for $\Re(s) > \frac{1}{2}$, and that it has meromorphic continuation to $\Re(s) > 0$. He also proved the disconcerting theorem that for $r \geq 3$, the Euler product $\prod_p P_r(s)$ has the line $\Re(s) = 0$ as natural boundary. Estermann's result was generalized by Kurokowa (see [19, 20]) to a much larger class of Euler products. This situation, where an innocuous looking L-function has a natural boundary, is now called the Estermann phenomenon. A very interesting instance of the Estermann phenomenon is for L-functions formed with the arithmetic Fourier coefficients $a(n), n{=}1,2,3,\ldots,$ of an automorphic form on, say, $GL(2)$. The L-functions

$$\sum_{n=1}^{\infty} a(n) n^{-s}, \qquad \sum_{n=1}^{\infty} |a(n)|^2 n^{-s},$$

both have good properties: meromorphic continuation and functional equation, but for $r \geq 3$, the Dirichlet series

$$\sum_{n=1}^{\infty} |a(n)|^r n^{-s} \tag{2}$$

has a natural boundary. Thus, the L-function defined in (2) does not have the correct structure when $r \geq 3$. It is now generally believed that the *correct* notion of (2) is the rth symmetric power L-function as in [24].

Another approach to obtain asymptotics for (1) is to study the meromorphic continuation in the complex variable w of the zeta integral

$$\mathcal{Z}_r(w) = \int_1^{\infty} \left|\zeta\left(\frac{1}{2} + it\right)\right|^{2r} t^{-w}\, \mathrm{d}t, \tag{3}$$

for r, a positive rational integer. This integral is easily shown to be absolutely convergent for $\Re(w)$ sufficiently large. Such an approach was pioneered by Ivić, Jutila, and Motohashi [16–18, 23] and somewhat later in [10].

One aim of this chapter is to give evidence that for $r \geq 3$, the function $\mathcal{Z}_r(w)$ has a natural boundary along $\Re(w) = \frac{1}{2}$. For simplicity of exposition, we shall consider

(3) only in the special case $r = 3$. There is an infinite class of other examples of this phenomenon to which this method should generalize. For instance,

$$\int_1^{\infty} \left|\zeta_{\mathbb{Q}(i)}\left(\frac{1}{2}+it\right)\right|^4 t^{-w}\,\mathrm{d}t = \int_1^{\infty} \left|\zeta\left(\frac{1}{2}+it\right) L\left(\frac{1}{2}+it, \chi_{-4}\right)\right|^4 t^{-w}\,\mathrm{d}t,$$

which is compatible with $\mathcal{Z}_4(w)$, should also have a natural boundary.

The fact that the Estermann phenomenon occurs for the integrals (1) and (3) suggests that for $r \geq 3$, the *classical* $2r$-th integral moment of zeta

$$\int_0^{T} \left|\zeta\left(\frac{1}{2}+it\right)\right|^{2r} \mathrm{d}t \tag{4}$$

does not have the *correct structure*. It is therefore doubtful that *substantial* advances in the theory of the Riemann zeta function will come from further investigations of (4).

The final goal of this chapter is to provide an alternative to (4) in the same spirit that the symmetric power L-function is an alternative to (2). Accordingly, in Sect. 7.3, we introduce a natural notion of integral moments for GL_3 automorphic forms over $\mathbb{Q}$ (see Theorem 1) with good analytic properties, i.e., it has a spectral expansion (obtained by combining Theorems 1 and 2). In the case of GL_2 over $\mathbb{Q}$, our notion coincides with the classical second integral moment.

The third author was partially supported by NSF grant 1001036.

7.2 Multiple Dirichlet Series with Natural Boundaries

For $s_1, \ldots, s_r$ and $w \in \mathbb{C}$ with sufficiently large real parts, let

$$Z(s_1, \ldots, s_r, w) = \int_1^{\infty} \zeta(s_1+it)\zeta(s_1-it)\cdots\zeta(s_r+it)\zeta(s_r-it)\,t^{-w}\,\mathrm{d}t. \tag{5}$$

This multiple Dirichlet series was considered in [10], and is more convenient than $\mathcal{Z}_r(w)$. Specializing $r = 3$, we can write

$$Z(s_1, s_2, s_3, w)$$
$$= \sum_{m,n} \frac{1}{(mn)^{\Re(s_1)}} \int_1^{\infty} \left(\frac{m}{n}\right)^{it} \zeta(s_2+it)\zeta(s_2-it)\zeta(s_3+it)\zeta(s_3-it)\,t^{-w}\,\mathrm{d}t.$$

The reason $\mathcal{Z}_3(w)$ should have a natural boundary is simple. The inner integral admits meromorphic continuation to $\mathbb{C}^3$. For $s_2 = s_3 = \frac{1}{2}$, this function should have infinitely many poles on the line $\Re(w) = \frac{1}{2}$, the positions depending on m, n.

As $m, n \longrightarrow \infty$, the number of poles in any fixed interval will tend to infinity. Summing over m, n, all these poles form a natural boundary. Accordingly, the main difficulty is to meromorphically continue the integral

$$\int_1^\infty \left(\frac{m}{n}\right)^{it} \zeta(s_2+it)\zeta(s_2-it)\zeta(s_3+it)\zeta(s_3-it)\, t^{-w}\, \mathrm{d}t, \tag{6}$$

as a function of s_2, s_3, w to $\mathbb{C}^3$ (see also Motohashi [22, 23], where in the integral (6) t^{-w} is replaced by a Gaussian weight). When $m = n = 1$, the meromorphic continuation of (6) was already established by Motohashi in [21]. Although this integral can certainly be studied by his method, the approach we follow is based on the more general ideas developed in [6–9, 14]. Using our techniques, it is possible to study in a *unified* way very general integrals attached to integral moments.

One can establish the meromorphic continuation of the slightly more general integral

$$\int_1^\infty \left(\frac{m}{n}\right)^{it} L(s_1+it, f)\, L(s_2-it, f)\, t^{-w}\, \mathrm{d}t, \tag{7}$$

where f is an automorphic form on $GL_2(\mathbb{Q})$ and $L(s, f)$ is the L-function attached to f. This implies the meromorphic continuation of an integral of type

$$\int_1^\infty L(s_1+it, f)\, L(s_2-it, f) \left|\sum_{n\le N} a_n n^{it}\right|^2 t^{-w}\, \mathrm{d}t$$

(with $a_n \in \mathbb{C}$ for $1 \le n \le N$). In fact, it is technically easier to study the integral (7) when f is a cuspform on $SL_2(\mathbb{Z})$ than the corresponding analysis of (6). Accordingly, to illustrate our point, for simplicity, we shall discuss the case when f is a holomorphic cuspform of even weight κ for $SL_2(\mathbb{Z})$. Then f has a Fourier expansion

$$f(z) = \sum_{\ell=1}^{\infty} a_\ell e^{2\pi i \ell z}, \quad (z = x+iy,\ y>0).$$

For m, n, two coprime positive integers, consider the congruence subgroup

$$\Gamma_{m,n} = \left\{ \begin{pmatrix} a & b \\ c & d \end{pmatrix} \in SL_2(\mathbb{Z}) \;\middle|\; b \equiv 0 \pmod{m},\ c \equiv 0 \pmod{n} \right\}.$$

Then, the function $F_{\frac{n}{m}}(z) := y^\kappa\, \overline{f\left(\frac{n}{m}z\right) f(z)}$ is $\Gamma_{m,n}$–invariant. For $v \in \mathbb{C}$, let $\varphi(z)$ be a function satisfying

$$\varphi(\rho z) = \rho^v \varphi(z), \quad (\text{for } \rho > 0 \text{ and } z = x+iy,\ y>0),$$

and (formally) define the Poincaré series

$$P(z;\varphi) \;=\; \sum_{\gamma\in Z\backslash\Gamma_{m,n}} \varphi(\gamma z), \tag{8}$$

where Z is the center of $\Gamma_{m,n}$. To ensure convergence, one can choose, for instance,

$$\varphi(z) = y^{v}\left(\frac{y}{\sqrt{x^2+y^2}}\right)^{w}, \tag{9}$$

where $v, w \in \mathbb{C}$ with sufficiently large real parts. These Poincaré series were introduced by Anton Good in [14].

Let $\langle\ ,\ \rangle$ denote the Petersson scalar product for automorphic forms for the group $\Gamma_{m,n}$. As in [9], we have the following.

Proposition 1. *Let m and n be two coprime positive integers, and let $P(z;\varphi)$, $F_{\frac{n}{m}}$, and $\Gamma_{m,n}$ be as defined above. For $\sigma > 0$ sufficiently large and φ defined by (9), we have*

$$\left\langle P(\cdot,\varphi),\ F_{\frac{n}{m}}\right\rangle = \frac{\pi(2\pi)^{-(v+\kappa+1)}\Gamma(w+v+\kappa-1)}{2^{w+v+\kappa-2}}\cdot\left(\frac{m}{n}\right)^{\sigma}$$
$$\cdot\int_{-\infty}^{\infty}\left(\frac{m}{n}\right)^{it} L(\sigma+it,f)L(v+\kappa-\sigma-it,f)$$
$$\cdot\frac{\Gamma(\sigma+it)\Gamma(v+\kappa-\sigma-it)}{\Gamma\left(\frac{w}{2}+\sigma+it\right)\Gamma\left(\frac{w}{2}+v+\kappa-\sigma-it\right)}\,dt.$$

As we already pointed out, the above proposition (with appropriate modifications) remains valid if the cuspform f is replaced by a *truncation* of the usual Eisenstein series $E(z,s)$ (for instance, on the line $\Re(s)=\frac{1}{2}$), or a Maass form. On the other hand, using Stirling's formula, it can be shown that the kernel in the above integral is (essentially) asymptotic to t^{-w}, as $t \longrightarrow \infty$. This fact holds whether f is holomorphic or not. It follows that the meromorphic continuation of (7) can be obtained from the meromorphic continuation (in $w \in \mathbb{C}$) of the Poincaré series (8).

The meromorphic continuation of the Poincaré series (8) can be obtained by spectral theory[1] as in [9]. To describe the contribution from the discrete part of the spectrum, let

$$\eta(z) = y^{\frac{1}{2}}\sum_{\ell\neq 0}\rho(\ell)\,K_{i\mu}(2\pi|\ell|y)\,e^{2\pi i\ell x}$$

[1]The Poincaré series $P(z,\varphi)$ is *not* square integrable. Just after an obvious Eisenstein series is subtracted, the remaining part is not only in L^2 but also has sufficient decay so that its integrals against Eisenstein series converge absolutely (see [7–9]).

($K_\mu(y)$ is the K-Bessel function) be a Maass cuspform (for the group $\Gamma_{m,n}$) which is an eigenfunction of the Laplacian with eigenvalue $\frac{1}{4} + \mu^2$. We shall need the well-known transforms

$$\int_{-\infty}^{\infty} (x^2+1)^{-w} \mathrm{e}^{-2\pi i \ell x y}\, \mathrm{d}x = \frac{2\pi^w}{\Gamma(w)} (|\ell| y)^{w-\frac{1}{2}} K_{\frac{1}{2}-w}(2\pi|\ell| y) \quad \left(\Re(w) > \frac{1}{2}\right),$$

and

$$\int_0^\infty y^v K_{i\mu}(y)\, K_{\frac{1}{2}-w}(y) \frac{\mathrm{d}y}{y}$$
$$= \frac{2^{v-3}\,\Gamma\left(\frac{\frac{1}{2}-i\mu+v-w}{2}\right)\Gamma\left(\frac{\frac{1}{2}+i\mu+v-w}{2}\right)\Gamma\left(\frac{-\frac{1}{2}-i\mu+v+w}{2}\right)\Gamma\left(\frac{-\frac{1}{2}+i\mu+v+w}{2}\right)}{\Gamma(v)},$$

the latter being valid provided $\Re(v+w) > \frac{1}{2}$, $\Re(w-v) < \frac{1}{2}$, and μ is real, i.e., we assume the Selberg $\frac{1}{4}$–conjecture. Unfolding the integral and applying the above transforms, one obtains

$$\frac{\langle P(\cdot,\varphi),\eta\rangle}{\langle \eta,\eta\rangle}$$
$$= \frac{1}{\langle \eta,\eta\rangle} \int_0^\infty \int_{-\infty}^{\infty} y^{v+\frac{1}{2}} \left(\frac{y}{\sqrt{x^2+y^2}}\right)^w \sum_{\ell\neq 0} \overline{\rho(\ell)}\, K_{i\mu}(2\pi|\ell| y)\, e^{-2\pi i \ell x} \frac{\mathrm{d}x\mathrm{d}y}{y^2}$$
$$= \frac{1}{\langle \eta,\eta\rangle} \sum_{\ell\neq 0} \overline{\rho(\ell)} \int_0^\infty \int_{-\infty}^{\infty} y^{v+\frac{1}{2}} (1+x^2)^{-\frac{w}{2}} K_{i\mu}(2\pi|\ell| y)\, \mathrm{e}^{-2\pi i \ell x y} \frac{\mathrm{d}x\mathrm{d}y}{y}$$
$$= \frac{2\pi^{\frac{w}{2}}}{\langle \eta,\eta\rangle\cdot\Gamma(\frac{w}{2})} \sum_{\ell\neq 0} \overline{\rho(\ell)}\, |\ell|^{\frac{w-1}{2}} \int_0^\infty y^{v+\frac{w}{2}} K_{i\mu}(2\pi|\ell| y)\, K_{\frac{1-w}{2}}(2\pi|\ell| y) \frac{\mathrm{d}y}{y}$$
$$= \frac{\pi^{-v}}{2\langle \eta,\eta\rangle} L\left(v+\frac{1}{2},\bar{\eta}\right)$$
$$\cdot \frac{\Gamma\left(\frac{\frac{1}{2}-i\mu+v}{2}\right)\Gamma\left(\frac{\frac{1}{2}+i\mu+v}{2}\right)\Gamma\left(\frac{-\frac{1}{2}-i\mu+v+w}{2}\right)\Gamma\left(\frac{-\frac{1}{2}+i\mu+v+w}{2}\right)}{\Gamma(v+\frac{w}{2})\Gamma(\frac{w}{2})}. \tag{10}$$

Here, $L(s,\eta)$ is the L-function associated to η. Note that the above computation is valid (all integrals and infinite sums converge absolutely) provided v, w have large real parts. The identity (10) then extends by analytic continuation. The ratio of products of gamma functions in the right-hand side of (10) has simple poles at $v + w = \frac{1}{2} \pm i\mu$ with corresponding residues:

$$\frac{\pi^{-v}}{\langle \eta, \eta \rangle} \cdot \frac{\Gamma(\pm i\mu)\Gamma\left(\frac{\frac{1}{2} \mp i\mu + v}{2}\right)}{\Gamma\left(\frac{\frac{1}{2} \pm i\mu - v}{2}\right)} \cdot L\left(v + \frac{1}{2}, \bar{\eta}\right).$$

For $v = 0$ and $\Re(w) \geq \frac{1}{2}$, it is expected that the above residues are almost always nonzero and that $\langle \eta, F_{\frac{n}{m}} \rangle \neq 0$ for almost all η ranging over a basis of Maass cuspforms for $\Gamma_{m,n}$. It also follows from Weyl's law that the number of such poles with imaginary part in the interval $[-T, T]$ is $\approx T^2$ as $T \longrightarrow \infty$. Summing over m, n, we see from the above argument that the function

$$\sum_{m,n} m^{-2\Re(s_1)} \left\langle P(\cdot, \varphi),\ F_{\frac{n}{m}} \right\rangle,$$

with the choices $\sigma = \kappa/2$ and $v = 0$ is expected to have a natural boundary at $\Re(w) = \frac{1}{2}$. In a similar manner, one may show that the function $Z(s_1, 1/2, 1/2, w)$, in particular, should have meromorphic continuation to *at most* $\Re(s_1) \geq \frac{1}{2}$ and $\Re(w) > \frac{1}{2}$.

7.3 A Notion of Integral Moment for $GL_3(\mathbb{Q})$

In [6], we propose a mechanism to obtain asymptotics for integral moments of GL_r $(r \geq 2)$ automorphic L-functions over an arbitrary number field. Our treatment follows the viewpoint of [7], where second integral moments for GL_2 are presented in a form enabling application of the structure of adele groups and their representation theory. We establish relations of the form

$$\text{moment expansion} = \int_{Z_{\mathbb{A}} GL_r(k) GL_r(\mathbb{A})} \text{Pé} \cdot |f|^2 = \text{spectral expansion},$$

where Pé is a Poincaré series on GL_r over number field k, for cuspform f on $GL_r(\mathbb{A})$. Roughly, the *moment expansion* is a sum of weighted moments of convolution L-functions $L(s, f \otimes F)$, where F runs over a basis of cuspforms on GL_{r-1}, as well as further continuous-spectrum terms. Indeed, the moment-expansion side itself does involve a spectral decomposition on GL_{r-1}. The *spectral expansion* side follows immediately from the spectral decomposition of the Poincaré series, and (surprisingly) consists of only three parts: a leading term, a sum arising from cuspforms on GL_2, and a continuous part from GL_2. That is, no cuspforms on GL_ℓ with $2 < \ell \leq r$ contribute.

As mentioned at the end of the introduction, in the case of GL_2 over $\mathbb{Q}$, the above expression gives (for f spherical) the spectral decomposition of the classical integral moment

$$\int_{-\infty}^{\infty} \left| L(\frac{1}{2}+it, f) \right|^2 g(t)\,\mathrm{d}t$$

for suitable smooth weights $g(t)$.

Integral moments for GL_3. In the simplest case beyond GL_2, take f a spherical cuspform on GL_3 over $\mathbb{Q}$. We construct a weight function $\Gamma(s, v, w, f_\infty, F_\infty)$ depending upon complex parameters s, v, and w and upon the *Archimedean* data for both f and cuspforms F on GL_2, such that the second integral moment attached to f arises as an integral:

$$\int_{Z_{\mathbb{A}}GL_3(\mathbb{Q})\;GL_3(\mathbb{A})} \mathrm{Pé}(g)\,|f(g)|^2\,\mathrm{d}g$$

$$= \sum_{F \text{ on } GL_2} \frac{1}{2\pi i} \int_{\Re(s)=\frac{1}{2}} |L(s, f\otimes F)|^2 \cdot \Gamma(s, 0, w, f_\infty, F_\infty)\,\mathrm{d}s$$

$$+\frac{1}{4\pi i}\frac{1}{2\pi i}\sum_{k\in\mathbb{Z}} \int_{\Re(s_1)=\frac{1}{2}} \int_{\Re(s_2)=\frac{1}{2}} |L(s_1, f\otimes E^{(k)}_{1-s_2})|^2$$

$$\cdot\, \Gamma(s_1, 0, w, f_\infty, E^{(k)}_{1-s_2,\infty})\,\mathrm{d}s_2\,\mathrm{d}s_1$$

(see also the statement of Theorem 1). Here, for $\Re(s_2) = 1/2$, write $1 - s_2$ in place of $\bar{s}_2$, to maintain holomorphy in complex-conjugated parameters. In this vein, over $\mathbb{Q}$, it is reasonable to put

$$L(s_1, f\otimes \bar{E}^{(k)}_{s_2}) = L\left(s_1, f\otimes E^{(k)}_{1-s_2}\right)$$

$$= \frac{L(s_1 - s_2 + \frac{1}{2}, f)\cdot L(s_1 + s_2 - \frac{1}{2}, f)}{\zeta(2-2s_2)} \quad \text{(finite-prime part)}$$

since the natural normalization of the Eisenstein series $E^{(k)}_{s_2}$ on GL_2 contributes the denominator $\zeta(2s_2)$. In the above expression, F runs over an orthonormal basis for all level-one cuspforms on GL_2, with *no* restriction on the right K_∞-type. The Eisenstein series $E^{(k)}_s$ run over all level-one Eisenstein series for $GL_2(\mathbb{Q})$ with no restriction on K_∞-type denoted here by k. The weight function $\Gamma(s, v, w, f_\infty, F_\infty)$ can be described as follows. Let $U(\mathbb{R})$ denote the subgroup of $GL_3(\mathbb{R})$ of matrices of the form $\begin{pmatrix} ccI_2 & * \\ & 1 \end{pmatrix}$. For $w \in \mathbb{C}$, define φ on $U(\mathbb{R})$ by

$$\varphi\begin{pmatrix} I_2 & x \\ & 1 \end{pmatrix} = (1 + ||x||^2)^{-\frac{w}{2}},$$

and set

$$\psi\begin{pmatrix}1 & x_1 & x_3\\ & 1 & x_2\\ & & 1\end{pmatrix} = e^{2\pi i(x_1+x_2)}.$$

Then, the weight function is (essentially)

$$\begin{aligned}
&\Gamma(s,v,w,\ f_\infty, F_\infty)\\
&=|\rho_F(1)|^2\cdot\int_0^\infty\int_0^\infty\int_{O_2(\mathbb{R})}\int_0^\infty\int_0^\infty\int_{O_2(\mathbb{R})}\left(t^2y\right)^{v-s+\frac{1}{2}}\cdot\left(t'^2y'\right)^{s-\frac{1}{2}}\mathcal{K}(h,m)\\
&\cdot W_{f,\mathbb{R}}\begin{pmatrix}ty & & \\ & t & \\ & & 1\end{pmatrix} W_{F,\mathbb{R}}\left(\begin{pmatrix}y & \\ & 1\end{pmatrix}\cdot k\right)\\
&\cdot \overline{W}_{f,\mathbb{R}}\begin{pmatrix}t'y' & & \\ & t' & \\ & & 1\end{pmatrix} \overline{W}_{F,\mathbb{R}}\left(\begin{pmatrix}y' & \\ & 1\end{pmatrix}\cdot k'\right)\\
&\cdot dk\,\frac{dy}{y^2}\,\frac{dt}{t}\,dk'\,\frac{dy'}{y'^2}\,\frac{dt'}{t'},
\end{aligned}$$

where: $\rho_F(1)$ is the first Fourier coefficient of F,

$$h=\begin{pmatrix}ty & & \\ & t & \\ & & 1\end{pmatrix}\begin{pmatrix}k & \\ & 1\end{pmatrix},\qquad m=\begin{pmatrix}t'y' & & \\ & t' & \\ & & 1\end{pmatrix}\begin{pmatrix}k' & \\ & 1\end{pmatrix},$$

and

$$\mathcal{K}(h,m) = \int_{U(\mathbb{R})}\varphi(u)\,\psi\left(huh^{-1}\right)\overline{\psi}(mum^{-1})\,du.$$

Here, $W_{f,\mathbb{R}}$ and $W_{F,\mathbb{R}}$ denote the Whittaker functions at ∞ attached to f and F, respectively. We can sum the above weight Γ over (GL_2) right K_∞-types, and denote the resulting weight by M.

To obtain higher moments of automorphic L-functions such as ζ, we replace the cuspform f by a truncated Eisenstein series or wave packet of Eisenstein series. For example, for GL_3, the continuous part of the above moment expansion gives the following natural integral:

$$\int_{\Re(s)=\frac{1}{2}}\int_{-\infty}^{\infty}\left|\frac{\zeta(s+it)^3\cdot\zeta(s-it)^3}{\zeta(1-2it)}\right|^2 M(s,t,w)\,dt\,ds,$$

where M is the smooth weight above obtained by summing over the right K_∞-types the function Γ above with $v = 0$. We stress the fact that this higher moment of the Riemann zeta is just *one term* in a GL_3 moment expansion which involves a complete sum over a GL_2 orthonormal basis (it corresponds to level-one Eisenstein series $E_s^{(k)}$ for $GL_2(\mathbb{Q})$), and unfortunately, there seems to be no way to isolate this term from the rest of the sum.

For applications to Analytic Number Theory, one finds it useful to present, in classical language, the derivation of the *explicit* moment identity, when $r = 3$ over $\mathbb{Q}$. To do so, let $G = GL_3(\mathbb{R})$ and define the standard subgroups

$$P = \left\{ \begin{pmatrix} 2\times 2 & * \\ & 1\times 1 \end{pmatrix} \right\}, \quad U = \left\{ \begin{pmatrix} I_2 & * \\ & 1 \end{pmatrix} \right\}, \quad H = \left\{ \begin{pmatrix} 2\times 2 & \\ & 1 \end{pmatrix} \right\},$$

$Z =$ center of G. Let N be the unipotent radical of standard minimal parabolic in H, i.e., the subgroup of upper-triangular unipotent elements in H, and set $K = O_3(\mathbb{R})$.

For $w \in \mathbb{C}$, define φ on U by

$$\varphi\begin{pmatrix} I_2 & x \\ & 1 \end{pmatrix} = (1 + ||x||^2)^{-\frac{w}{2}}.$$

We extend φ to G by requiring right K-invariance and left equivariance:

$$\varphi(mg) = \left| \frac{\det A}{d^2} \right|^v \cdot \varphi(g), \quad \left(v \in \mathbb{C},\ g \in G,\ m = \begin{pmatrix} A & \\ & d \end{pmatrix} \in ZH \right).$$

More generally, we can take *suitable* functions (see [5, 7]) φ on U and extend them to G by right K-invariance and the same left equivariance.

For $\Re(v)$ and $\Re(w)$ sufficiently large, define the Poincaré series

$$\text{Pé}(g) = \text{Pé}(g; v, w) = \sum_{\gamma \in H(\mathbb{Z})\backslash SL_3(\mathbb{Z})} \varphi(\gamma g), \quad (g \in G) \tag{11}$$

where $H(\mathbb{Z})$ is the subgroup of $SL_3(\mathbb{Z})$ whose elements belong to H. Note that $H(\mathbb{Z}) \approx SL_2(\mathbb{Z})$. To see that the series defining Pé(g) converges absolutely and uniformly on compact subsets of G/ZK, one can use the Iwasawa decomposition to make a simple comparison with the maximal parabolic Eisenstein series.

For a cuspform f of type $\mu = (\mu_1, \mu_2)$ on $SL_3(\mathbb{Z})$ (right ZK-invariant), consider the integral

$$I = I(v, w) = \int_{ZSL_3(\mathbb{Z})\backslash G} \text{Pé}(g)\, |f(g)|^2\, dg. \tag{12}$$

Unwinding the Poincaré series, we write

$$I = \int_{ZH(\mathbb{Z})\backslash G} \varphi(g)\,|f(g)|^2\,\mathrm{d}g.$$

Next, we will use the Fourier expansion (see [13]):

$$f(g) = \sum_{\gamma\in N(\mathbb{Z})H(\mathbb{Z})} \sum_{\ell_1=1}^{\infty} \sum_{\ell_2\neq 0} \frac{a(\ell_1,\ell_2)}{|\ell_1\ell_2|}\cdot W_\mu(L\gamma g) \tag{13}$$

(with $a(\ell_1,\ell_2) = a(\ell_1,-\ell_2)$) where $N(\mathbb{Z})$ is the subgroup of upper-triangular unipotent elements in $H(\mathbb{Z})$, $L = \mathrm{diag}(\ell_1\ell_2,\ell_1,1)$, and W_μ is the Whittaker function. Then, the integral I further unwinds to

$$I = \sum_{\ell_1,\ell_2} \frac{a(\ell_1,\ell_2)}{|\ell_1\ell_2|} \int_{ZN(\mathbb{Z})\backslash G} \varphi(g)\,W_\mu(Lg)\,\bar{f}(g)\,\mathrm{d}g. \tag{14}$$

Now, let P_1 be the (minimal) parabolic subgroup of G of upper-triangular matrices, and let K_1 be the subgroup of K fixing the row vector $(0,0,1)$. Using the Iwasawa decomposition

$$G = P_1\cdot K, \qquad P = (HZ)\cdot U = P_1\cdot K_1,$$

we can write (up to a constant) the right-hand side of (14) as

$$I = \sum_{\ell_1,\ell_2} \frac{a(\ell_1,\ell_2)}{|\ell_1\ell_2|} \int_{(N(\mathbb{Z})\backslash H)\times U} \varphi(hu)\,W_\mu(Lhu)\,\bar{f}(hu)\,\mathrm{d}h\,\mathrm{d}u. \tag{15}$$

The constant involved is $\left(\int_{K_1} 1\,\mathrm{d}k\right)^{-1}$.

One of the key ideas is to decompose the left $H(\mathbb{Z})$-invariant function $\bar{f}(hu)$ along $H(\mathbb{Z})H$. Accordingly, we have the spectral decomposition

$$\bar{f}(hu) = \int_{(\eta)} \eta(h) \int_{H(\mathbb{Z})\backslash H} \bar{\eta}(m)\,\bar{f}(mu)\,\mathrm{d}m\,\mathrm{d}\eta \tag{16}$$

$$= \sum_{\ell_1',\ell_2'} \overline{\frac{a(\ell_1',\ell_2')}{|\ell_1'\ell_2'|}} \int_{(\eta)} \eta(h) \int_{N(\mathbb{Z})\backslash H} \bar{\eta}(m)\,\overline{W}_\mu(L'mu)\,\mathrm{d}m\,\mathrm{d}\eta. \tag{17}$$

Plugging (17) into (15), we can decompose

$$I = \sum_{\ell_1, \ell_2} \sum_{\ell_1', \ell_2'} \frac{a(\ell_1, \ell_2)}{|\ell_1 \ell_2|} \frac{\overline{a(\ell_1', \ell_2')}}{|\ell_1' \ell_2'|} I_{\ell_1, \ell_2, \ell_1', \ell_2'}, \tag{18}$$

where, for fixed $\ell_1, \ell_2, \ell_1', \ell_2'$,

$$I_{\ell_1, \ell_2, \ell_1', \ell_2'} = \int_{(\eta)} \int_{(N(\mathbb{Z})\backslash H)\times U} \int_{N(\mathbb{Z})\backslash H} \varphi(hu)\, W_\mu(Lhu)\, \eta(h)\, \overline{W}_\mu(L'mu)\, \overline{\eta}(m)\, \mathrm{d}h\, \mathrm{d}m\, \mathrm{d}u\, \mathrm{d}\eta. \tag{19}$$

The integral over U in (19) is

$$\begin{aligned} &\int_U \varphi(u)\, W_\mu(Lhu)\, \overline{W}_\mu(L'mu)\, \mathrm{d}u \\ &\quad= W_\mu(Lh)\, \overline{W}_\mu(L'm) \int_U \varphi(u)\, \psi\left(Lhuh^{-1}L^{-1}\right) \overline{\psi}(L'mum^{-1}L'^{-1})\, \mathrm{d}u \\ &\quad= W_\mu(Lh)\, \overline{W}_\mu(L'm) \int_{-\infty}^{\infty} \int_{-\infty}^{\infty} \cdots\ \mathrm{d}x_2\, \mathrm{d}x_3 \\ &\quad= W_\mu(Lh)\, \overline{W}_\mu(L'm)\, \mathcal{K}(Lh, L'm), \end{aligned}$$

where

$$\psi \begin{pmatrix} 1 & x_1 & x_2 \\ & 1 & x_3 \\ & & 1 \end{pmatrix} = \mathrm{e}^{2\pi i (x_1 + x_3)}.$$

Therefore,

$$I_{\ell_1, \ell_2, \ell_1', \ell_2'} = \int_{(\eta)} \int_{N(\mathbb{Z})\backslash H} \int_{N(\mathbb{Z})\backslash H} \varphi(h)\, \mathcal{K}(Lh, L'm)\, W_\mu(Lh)\, \eta(h)\, \overline{W}_\mu(L'm)\, \overline{\eta}(m)\, \mathrm{d}h\, \mathrm{d}m\, \mathrm{d}\eta. \tag{20}$$

For $n \in N$ and $h \in H$, we have

$$\begin{aligned} \varphi(nh) &= \varphi(h), \\ \mathcal{K}(Lnh, L'm) &= \mathcal{K}(Lh, L'm), \\ W_\mu(Lnh) &= \psi\left(LnL^{-1}\right) W_\mu(Lh). \end{aligned}$$

Hence,

$$\int_{N(\mathbb{Z})\backslash H}\int_{N(\mathbb{Z})\backslash H} \varphi(h)\,\mathcal{K}(Lh,\,L'm)\,W_\mu(Lh)\,\eta(h)\,\overline{W}_\mu(L'm)\,\overline{\eta}(m)\,\mathrm{d}h\,\mathrm{d}m$$
$$= \int_{N\backslash H}\int_{N\backslash H} \varphi(h)\,\mathcal{K}(Lh,\,L'm)\,W_\mu(Lh)\,\overline{W}_\mu(L'm)$$
$$\cdot \int_{N(\mathbb{Z})\backslash N} \psi\left(LnL^{-1}\right)\eta(nh)\,\mathrm{d}n \;\cdot \int_{N(\mathbb{Z})\backslash N} \overline{\psi}\left(L'n'L'^{-1}\right)\overline{\eta}(n'm)\,\mathrm{d}n'\,\mathrm{d}h\,\mathrm{d}m. \tag{21}$$

To simplify (21), let

$$h = \begin{pmatrix} ty & & \\ & t & \\ & & 1 \end{pmatrix}\begin{pmatrix} k & \\ & 1 \end{pmatrix}, \quad m = \begin{pmatrix} t'y' & & \\ & t' & \\ & & 1 \end{pmatrix}\begin{pmatrix} k' & \\ & 1 \end{pmatrix}, \quad (k,\,k' \in O_2(\mathbb{R})).$$

The functions η above are of the form $|\det|^{-s} \otimes F$ with $s \in i\mathbb{R}$. In what follows, for convergence purposes, the real part of the parameter s will necessarily be shifted to a fixed (large) $\sigma = \Re(s)$. The shifting occurs in (17) (there is a hidden vertical integral in the integral over η).

Remark. For every K-type κ, we choose F in an orthonormal basis consisting of common eigenfunctions for all Hecke operators T_n. Furthermore, this basis is normalized as in Corollary 4.4 and (4.69) [11] with respect to Maass operators.

Note that

$$\int_{N(\mathbb{Z})\backslash N} \psi\left(LnL^{-1}\right)F(nh)\,\mathrm{d}n = \frac{\rho_F(-\ell_2)}{\sqrt{|\ell_2|}}\,W^{\pm}_{F,\mathbb{R}}\left(\begin{pmatrix} |\ell_2|\,y & \\ & 1 \end{pmatrix}\cdot k\right),$$

$$\int_{N(\mathbb{Z})\backslash N} \overline{\psi}\left(L'n'L'^{-1}\right)\bar{F}(n'm)\,\mathrm{d}n' = \frac{\overline{\rho_F(-\ell_2')}}{\sqrt{|\ell_2'|}}\,\overline{W}^{\pm}_{F,\mathbb{R}}\left(\begin{pmatrix} |\ell_2'|\,y' & \\ & 1 \end{pmatrix}\cdot k'\right), \tag{22}$$

where $W^{\pm}_{F,\mathbb{R}}$ are the GL_2 Whittaker functions attached to F. These functions can be expressed in terms of the *classical* Whittaker function

$$W_{\alpha,\beta}(y) = \frac{y^\alpha\,\mathrm{e}^{-\frac{y}{2}}}{2\pi i}\int_{-i\infty}^{i\infty} \frac{\Gamma(u)\,\Gamma(-u-\alpha-\beta+\frac{1}{2})\,\Gamma(-u-\alpha+\beta+\frac{1}{2})}{\Gamma(-\alpha-\beta+\frac{1}{2})\,\Gamma(-\alpha+\beta+\frac{1}{2})}\,y^u\,\mathrm{d}u,$$

where the contour has loops, if necessary, so that the poles of $\Gamma(u)$ and the poles of the function $\Gamma(-u-\alpha-\beta+\frac{1}{2})\Gamma(-u-\alpha+\beta+\frac{1}{2})$ are on opposite sides of it. For $k = \left(\begin{smallmatrix} cc\cos\theta & -\sin\theta \\ \sin\theta & \cos\theta \end{smallmatrix}\right) \in SO_2(\mathbb{R})$, we have (see [11])

$$W^{\pm}_{F,\mathbb{R}}\left(\begin{pmatrix} y & \\ & 1 \end{pmatrix} \cdot k\right) = e^{i\kappa\theta}\, W^{\pm}_{F,\mathbb{R}}\begin{pmatrix} y & \\ & 1 \end{pmatrix} = \mathrm{e}^{i\kappa\theta}\, W_{\pm\frac{\kappa}{2}, i\mu_F}(4\pi y), \qquad (y > 0),$$

if F is an eigenfunction of

$$\Delta_\kappa = y^2\left(\frac{\partial^2}{\partial x^2} + \frac{\partial^2}{\partial y^2}\right) - i\kappa y \frac{\partial}{\partial x}$$

with eigenvalue $\frac{1}{4} + \mu_F^2$. In (7.3) and (22), the Whittaker functions are determined by the signs of $-\ell_2$ and $-\ell_2'$, respectively. If F corresponds to a holomorphic, or anti-holomorphic, cuspform, there are no negative, or positive, respectively, terms in its Fourier expansion. We have

$$W^{+}_{F,\mathbb{R}}\left(\begin{pmatrix} y & \\ & 1 \end{pmatrix} \cdot k\right) = \mathrm{e}^{i\kappa\theta}\, W^{+}_{F,\mathbb{R}}\begin{pmatrix} y & \\ & 1 \end{pmatrix} = \mathrm{e}^{i\kappa\theta}\, W_{\frac{\kappa}{2}, \frac{\kappa_0-1}{2}}(4\pi y)$$

(for $\kappa \geq \kappa_0 \geq 12$, $y > 0$) for F corresponding to a holomorphic cuspform of weight κ_0.

Then, making the substitutions

$$t \longrightarrow \frac{t}{\ell_1}, \quad y \longrightarrow \frac{y}{|\ell_2|}, \quad t' \longrightarrow \frac{t'}{\ell_1'}, \quad y' \longrightarrow \frac{y'}{|\ell_2'|},$$

we can write (21) as

$$\begin{aligned} &\frac{\sqrt{|\ell_2|}\,\rho_F(-\ell_2)}{(\ell_1^2|\ell_2|)^{v-s}}\,\frac{\overline{\sqrt{|\ell_2'|}\,\rho_F(-\ell_2')}}{(\ell_1'^2|\ell_2'|)^{s}} \int_0^\infty\int_0^\infty\int_{H\cap K}\int_0^\infty\int_0^\infty\int_{H\cap K} (t^2y)^{v-s}\cdot(t'^2y')^s\,\mathcal{K}(h,m) \\ &\cdot W_\mu\begin{pmatrix} ty & & \\ & t & \\ & & 1 \end{pmatrix} W^{\pm}_{F,\mathbb{R}}\left(\begin{pmatrix} y & \\ & 1 \end{pmatrix}\cdot k\right)\cdot \overline{W}_\mu\begin{pmatrix} t'y' & & \\ & t' & \\ & & 1 \end{pmatrix} \overline{W}^{\pm}_{F,\mathbb{R}}\left(\begin{pmatrix} y' & \\ & 1 \end{pmatrix}\cdot k'\right) \\ &\cdot \mathrm{d}k\,\frac{\mathrm{d}y}{y^2}\,\frac{\mathrm{d}t}{t}\,\mathrm{d}k'\,\frac{\mathrm{d}y'}{y'^2}\,\frac{\mathrm{d}t'}{t'}, \end{aligned} \tag{23}$$

where

$$\mathcal{K}(h, m) = \int_U \varphi(u)\,\psi\left(huh^{-1}\right)\overline{\psi}(mum^{-1})\,\mathrm{d}u.$$

Recall that the Rankin–Selberg convolution $L(s, f \otimes F)$ is given by

$$L(s, f \otimes F) = L(s, f \otimes F_0) = \sum_{\ell_1, \ell_2 = 1}^{\infty} \frac{a(\ell_1, \ell_2)\lambda_{F_0}(\ell_2)}{(\ell_1^2 \ell_2)^s},$$

where F_0 is the basic ancestor of F and $\lambda_{F_0}(\ell)$ is the corresponding eigenvalue of the Hecke operator T_ℓ. Since $a(\ell_1, \ell_2) = a(\ell_1, -\ell_2)$, it follows from (18), (20), and (23) that

$$\begin{aligned} I &= \int\limits_{ZSL_3(\mathbb{Z})\backslash G} \text{Pé}(g)\, |f(g)|^2 \, \mathrm{d}g \\ &= \sum_{F \text{ in } GL_2} \frac{1}{2\pi i} \int\limits_{\Re(s)=\sigma} L(v+1-s,\ f \otimes F)\, L(s,\ \bar{f} \otimes \bar{F})\, \Gamma_\varphi(s)\, \mathrm{d}s, \end{aligned}$$

where

$$\begin{aligned} \Gamma_\varphi(s) &= \Gamma_\varphi(s,\ v,\ w,\ f,\ F) \\ &= \sum_{\pm} \rho_F(\pm 1)\overline{\rho_F(\pm 1)} \cdot \int_0^\infty \int_0^\infty \int_{H\cap K} \int_0^\infty \int_0^\infty \int_{H\cap K} (t^2 y)^{v-s+\frac{1}{2}} \cdot (t'^2 y')^{s-\frac{1}{2}}\, \mathcal{K}(h,\ m) \\ &\cdot W_\mu \begin{pmatrix} ty & & \\ & t & \\ & & 1 \end{pmatrix} W^{\pm}_{F,\mathbb{R}}\left(\begin{pmatrix} y & \\ & 1 \end{pmatrix} \cdot k \right) \cdot \overline{W}_\mu \begin{pmatrix} t'y' & & \\ & t' & \\ & & 1 \end{pmatrix} \overline{W}^{\pm}_{F,\mathbb{R}}\left(\begin{pmatrix} y' & \\ & 1 \end{pmatrix} \cdot k' \right) \\ &\cdot \mathrm{d}k\, \frac{\mathrm{d}y}{y^2}\, \frac{\mathrm{d}t}{t}\, \mathrm{d}k'\, \frac{\mathrm{d}y'}{y'^2}\, \frac{\mathrm{d}t'}{t'}, \end{aligned} \tag{24}$$

with all four possible sign choices in the sum. Note that we have also replaced s by $s - \frac{1}{2}$.

The kernel $\Gamma_\varphi(s)$ can be expressed as a Barnes-type (multiple) integral. To see this, note that

$$\psi\left(huh^{-1}\right) = \mathrm{e}^{2\pi i t (u_1 \sin\theta + u_2 \cos\theta)}, \qquad \overline{\psi}(mum^{-1}) = \mathrm{e}^{-2\pi i t'(u_1 \sin\theta' + u_2 \cos\theta')},$$

with $0 \le \theta,\ \theta' \le 2\pi$. Changing the variables $u_1 = r\cos\phi,\ u_2 = r\sin\phi$ ($r \ge 0$ and $0 \le \phi \le 2\pi$), one can write

$$\mathcal{K}(h,\ m) = \int_0^\infty \int_0^{2\pi} r^2 \varphi(r)\, \mathrm{e}^{2\pi i r t \sin(\theta+\phi)}\, \mathrm{e}^{-2\pi i r t' \sin(\theta'+\phi)}\, \mathrm{d}\phi\, \frac{\mathrm{d}r}{r}. \tag{25}$$

In (25), express the two exponentials using the Fourier expansion:

$$e^{iu\sin\theta} = \sum_{\ell=-\infty}^{\infty} J_\ell(u)\, e^{i\ell\theta}.$$

Recalling that

$$W^{\pm}_{F,\mathbb{R}}\left(\begin{pmatrix} y & \\ & 1\end{pmatrix}\cdot k\right) = \mathrm{e}^{i\kappa\theta}\, W^{\pm}_{F,\mathbb{R}}\begin{pmatrix} y & \\ & 1\end{pmatrix},$$

it follows that, up to a positive constant, $\Gamma_\varphi(s)$ is represented by

$$\sum_{\pm} \rho_F(\pm 1)\overline{\rho_F(\pm 1)}$$
$$\cdot \int_0^\infty\int_0^\infty\int_0^\infty\int_0^\infty (t^2y)^{v-s+\frac{1}{2}}\,(t'^2y')^{s-\frac{1}{2}} \cdot \int_0^\infty r^2\varphi(r)\, J_\kappa(2\pi r t)\, J_\kappa(2\pi r t')\,\frac{\mathrm{d}r}{r}$$
$$\cdot\, W_\mu\begin{pmatrix} ty & & \\ & t & \\ & & 1\end{pmatrix} W^{\pm}_{F,\mathbb{R}}\begin{pmatrix} y & \\ & 1\end{pmatrix} \overline{W}_\mu\begin{pmatrix} t'y' & & \\ & t' & \\ & & 1\end{pmatrix} \overline{W}^{\pm}_{F,\mathbb{R}}\begin{pmatrix} y' & \\ & 1\end{pmatrix} \frac{\mathrm{d}y}{y^2}\,\frac{\mathrm{d}t}{t}\,\frac{\mathrm{d}y'}{y'^2}\,\frac{\mathrm{d}t'}{t'}. \tag{26}$$

Here, we have also used the well-known identity $J_{-\kappa}(z) = (-1)^\kappa J_\kappa(z)$.

To continue the computation, express both $GL_3(\mathbb{R})$ Whittaker functions in (26) as (see [3])

$$W_\mu\begin{pmatrix} ty & & \\ & t & \\ & & 1\end{pmatrix} = \frac{1}{(2\pi i)^2}\int_{(\delta_1)}\int_{(\delta_2)} \pi^{-\xi_1-\xi_2}\, V(\xi_1,\xi_2)\, t^{1-\xi_1}\, y^{1-\xi_2}\,\mathrm{d}\xi_1\,\mathrm{d}\xi_2,$$

where

$$V(\xi_1,\xi_2) = \frac{1}{4}\,\frac{\Gamma\left(\frac{\xi_1+\alpha}{2}\right)\Gamma\left(\frac{\xi_1+\beta}{2}\right)\Gamma\left(\frac{\xi_1+\gamma}{2}\right)\Gamma\left(\frac{\xi_2-\alpha}{2}\right)\Gamma\left(\frac{\xi_2-\beta}{2}\right)\Gamma\left(\frac{\xi_2-\gamma}{2}\right)}{\Gamma\left(\frac{\xi_1+\xi_2}{2}\right)},$$

the vertical lines of integration being taken to the right of all poles of the integrand. We shall consider only the $(+,+)$ part of (26), assuming $\kappa \geq 0$ and

$$W^{+}_{F,\mathbb{R}}\begin{pmatrix} y & \\ & 1\end{pmatrix} = W_{\frac{\kappa}{2},\, i\mu_{F_0}}(4\pi y).$$

Interchanging the order of integration and applying standard integral formulas (see [15]), we write the integrals of the $(+,+)$ part of (26) corresponding to the above choice of $W^{+}_{F,\mathbb{R}}$ as

$$\frac{\pi^{-3(1+v)}}{128}\frac{1}{(2\pi i)^4}\int_{(\delta_1)}\int_{(\delta_2)}$$

$$\int_{(\delta_1')}\int_{(\delta_2')} V(\xi_1,\xi_2)\,\overline{V}(\xi_1',\xi_2')\,\frac{\Gamma\big(1+\frac{\kappa}{2}-s-\frac{\xi_1}{2}+v\big)\,\Gamma\big(\frac{\kappa}{2}+s-\frac{\xi_1'}{2}\big)}{\Gamma\big(\frac{\kappa}{2}+s+\frac{\xi_1}{2}-v\big)\,\Gamma\big(\frac{\kappa}{2}+1-s+\frac{\xi_1'}{2}\big)}$$

$$\cdot\,\Gamma\left(\frac{1-s-\xi_2+v-i\mu_{F_0}}{2}\right)\Gamma\left(\frac{1-s-\xi_2+v+i\mu_{F_0}}{2}\right)$$

$$\cdot\,\Gamma\left(\frac{s-\xi_2'-i\mu_{F_0}}{2}\right)\Gamma\left(\frac{s-\xi_2'+i\mu_{F_0}}{2}\right)$$

$$\cdot\,\frac{\Gamma\big(\frac{\xi_1+\xi_1'-2v}{2}\big)\,\Gamma\big(\frac{-\xi_1-\xi_1'+2v+w}{2}\big)}{\Gamma\big(\frac{w}{2}\big)}\,d\xi_2'\,d\xi_1'\,d\xi_2\,d\xi_1. \tag{27}$$

This representation holds, provided

$$\delta_1,\ \delta_2,\ \delta_1',\ \delta_2' > 0;$$
$$\Re(v)-\Re(s)-\delta_2 > -1;\quad \Re(s)-\delta_2' > 0;$$
$$\frac{3}{2} > 2\Re(s)-\delta_1' > 0;\quad -\frac{1}{2} > 2\Re(v)-2\Re(s)-\delta_1 > -2;$$
$$\Re(w) > \delta_1+\delta_1'-2\Re(v) > 0.$$

We remark that for all the other choices of $W^{\pm}_{F,\mathbb{R}}$, one obtains similar expressions.

For fixed F_0, a Maass cuspform of weight zero, or a classical holomorphic (or anti-holomorphic) cuspform of weight κ_0, the corresponding *Archimedean* sum over the K-types κ in the moment expansion can be evaluated using the effect of the Maass operators on F_0 given explicitly in [11] (see especially (4.70), (4.77), (4.78), and (4.83)).

We summarize the main result of this section in the following.

Theorem 1. *Let $P\acute{e}(g)$ defined in (11) be the Poincaré series associated to φ. Then, for s, v, $w \in \mathbb{C}$ with sufficiently large real parts, and f a cuspform on $SL_3(\mathbb{Z})$, we have*

$$\int_{ZSL_3(\mathbb{Z})\backslash G} P\acute{e}(g)\,|f(g)|^2\,dg$$

$$= \sum_{F\ in\ GL_2}\frac{1}{2\pi i}\int_{\Re(s)=\sigma} L(v+1-s,\ f\otimes F)\,L(s,\ \bar f\otimes \bar F)\,\Gamma_\varphi(s)\,ds,$$

where F runs over an orthonormal basis for all level-one cuspforms together with vertical integrals of all level-one Eisenstein series on $GL_2(\mathbb{Q})$, with no restriction on the right K-types. The weight function $\Gamma_\varphi(s)$ is given by

$$\Gamma_\varphi(s) = \sum_{\pm} \rho_F(\pm 1)\overline{\rho_F(\pm 1)}$$

$$\cdot \int_0^\infty\int_0^\infty\int_0^\infty\int_0^\infty (t^2 y)^{\nu-s+\frac{1}{2}}\,(t'^2 y')^{s-\frac{1}{2}} \cdot \int_0^\infty r^2\varphi(r)\,J_\kappa(2\pi r t)\,J_\kappa(2\pi r t')\frac{dr}{r}$$

$$\cdot\, W_\mu\begin{pmatrix} ty & & \\ & t & \\ & & 1\end{pmatrix} W^{\pm}_{F,\mathbb{R}}\begin{pmatrix} y & \\ & 1\end{pmatrix} \overline{W}_\mu\begin{pmatrix} t'y' & & \\ & t' & \\ & & 1\end{pmatrix} \overline{W}^{\pm}_{F,\mathbb{R}}\begin{pmatrix} y' & \\ & 1\end{pmatrix} \frac{dy}{y^2}\,\frac{dt}{t}\,\frac{dy'}{y'^2}\,\frac{dt'}{t'},$$

with all four possible sign choices in the sum.

7.4 Spectral Decomposition of Poincaré Series

We begin by showing that our Poincaré series Pé(g) is a degenerate GL_3 object (i.e., the cuspforms on $SL_3(\mathbb{Z})$ *do not* contribute to its spectral decomposition). We have the following.

Proposition 2. *The Poincaré series Pé(g) is orthogonal to the space of cuspforms on $SL_3(\mathbb{Z})$.*

Proof. Let f be a cuspform on $SL_3(\mathbb{Z})$ with Fourier expansion

$$f(g) \quad = \sum_{\gamma\in N(\mathbb{Z})\backslash H(\mathbb{Z})}\ \sum_{\ell_1=1}^{\infty}\ \sum_{\ell_2\neq 0} \frac{a(\ell_1,\ell_2)}{|\ell_1\ell_2|}\cdot W(L\gamma g).$$

Unwinding twice, it follows, as before, that

$$\int_{ZSL_3(\mathbb{Z})G} \text{Pé}(g)\,\bar{f}(g)\,dg \;= \sum_{\ell_1,\ell_2} \frac{\overline{a(\ell_1,\ell_2)}}{|\ell_1\ell_2|} \int_{ZN(\mathbb{Z})\backslash G/K} \varphi(g)\,\overline{W}(Lg)\,dg. \tag{28}$$

Now, write $g\in G$ in Iwasawa form,

$$g = \begin{pmatrix} 1 & x_1 & x_2\\ & 1 & x_3\\ & & 1\end{pmatrix}\begin{pmatrix} y_1y_2 & & \\ & y_1 & \\ & & 1\end{pmatrix}\begin{pmatrix} d & & \\ & d & \\ & & d\end{pmatrix} k \qquad (y_1,\,y_2>0,\; k\in K)$$

$$= \begin{pmatrix} y_1y_2\,d & & \\ & y_1 d & \\ & & d\end{pmatrix}\begin{pmatrix} 1 & x_1/y_2 & \\ & 1 & \\ & & 1\end{pmatrix}\begin{pmatrix} 1 & 0 & (x_2-x_1x_3)/y_1y_2\\ 0 & 1 & x_3/y_1\\ 0 & 0 & 1\end{pmatrix} k.$$

Then,

$$\varphi(g) = (y_1^2 y_2)^{\nu} \varphi \begin{pmatrix} 1 & 0 & (x_2 - x_1 x_3)/y_1 y_2 \\ 0 & 1 & x_3/y_1 \\ 0 & 0 & 1 \end{pmatrix} \tag{29}$$

and

$$W(Lg) = e^{2\pi i(\ell_2 x_1 + \ell_1 x_3)} \cdot W \begin{pmatrix} \ell_1 y_1 |\ell_2| y_2 & & \\ & \ell_1 y_1 & \\ & & 1 \end{pmatrix}. \tag{30}$$

Also, the integral in the right-hand side of (28) can be written explicitly as

$$\int\limits_{ZN(\mathbb{Z})\backslash G/K} \cdots \, dg = \int\limits_{y_2=0}^{\infty} \int\limits_{y_1=0}^{\infty} \int\limits_{x_3=-\infty}^{\infty} \int\limits_{x_2=-\infty}^{\infty} \int\limits_{x_1=0}^{1} \cdots \, dx_1 \, dx_2 \, dx_3 \, \frac{dy_1}{y_1^3} \frac{dy_2}{y_2^3}.$$

Letting

$$x_1 = t_1, \qquad x_2 = t_2 + t_1 t_3, \qquad x_3 = t_3,$$

the inner integral over t_1 is

$$\int\limits_0^1 e^{-2\pi i \ell_2 t_1} \, dt_1 = 0$$

(since $\ell_2 \neq 0$). Thus,

$$\int\limits_{ZSL_3(\mathbb{Z})\backslash G} \text{Pé}(g) \bar{f}(g) \, dg = 0.$$

□

Now, write the Poincaré series as

$$\text{Pé}(g) = \sum_{\gamma \in H(\mathbb{Z})\backslash SL_3(\mathbb{Z})} \varphi(\gamma g) = \sum_{\gamma \in P(\mathbb{Z})\backslash SL_3(\mathbb{Z})} \sum_{\beta \in U(\mathbb{Z})} \varphi(\beta \gamma g),$$

where $P(\mathbb{Z})$ denotes the subgroup of $SL_3(\mathbb{Z})$ with the bottom row $(0, 0, 1)$. By the Poisson summation formula, we have

$$\begin{aligned} \sum_{\beta \in U(\mathbb{Z})} \varphi(\beta g) &= \sum_{m_2, m_3 = -\infty}^{\infty} \varphi \left(\begin{pmatrix} 1 & & m_2 \\ & 1 & m_3 \\ & & 1 \end{pmatrix} \begin{pmatrix} 1 & x_1 & x_2 \\ & 1 & x_3 \\ & & 1 \end{pmatrix} \begin{pmatrix} y_1 y_2 & & \\ & y_1 & \\ & & 1 \end{pmatrix} \right) \\ &= \sum_{m_2, m_3 = -\infty}^{\infty} \varphi \left(\begin{pmatrix} 1 & x_1 & x_2 + m_2 \\ & 1 & x_3 + m_3 \\ & & 1 \end{pmatrix} \begin{pmatrix} y_1 y_2 & & \\ & y_1 & \\ & & 1 \end{pmatrix} \right) \\ &= \sum_{m_2, m_3 = -\infty}^{\infty} C_{\varphi}^{(m_2, m_3)}(x_1, y_1, y_2) \, e^{2\pi i (m_2 x_2 + m_3 x_3)}, \end{aligned}$$

where $C_\varphi^{(m_2,m_3)}(x_1, y_1, y_2)$ is given by

$$C_\varphi^{(m_2,m_3)}(x_1, y_1, y_2) = (y_1^2 y_2)^\nu \int_{\mathbb{R}^2} \varphi\begin{pmatrix} 1 & 0 & (u_2 - x_1u_3)/y_1y_2 \\ 0 & 1 & u_3/y_1 \\ 0 & 0 & 1 \end{pmatrix} e^{-2\pi i(m_2u_2+m_3u_3)}\, du_2\, du_3$$
$$= (y_1^2 y_2)^{\nu+1} \int_{\mathbb{R}^2} \varphi\begin{pmatrix} 1 & & t_2 \\ & 1 & t_3 \\ & & 1 \end{pmatrix} e^{-2\pi i[m_2y_1y_2t_2+(m_2x_1+m_3)y_1t_3]}\, dt_2\, dt_3. \qquad (31)$$

Therefore, denoting $C_\varphi^{(m_2,m_3)}(x_1, y_1, y_2)\, e^{2\pi i(m_2x_2+m_3x_3)}$ by $\widehat{\varphi}_g(m_2, m_3)$, we can write

$$\text{Pé}(g) = \sum_{\gamma\in P(\mathbb{Z})\backslash SL_3(\mathbb{Z})} \sum_{m_2,m_3=-\infty}^{\infty} \widehat{\varphi}_{\gamma g}(m_2, m_3).$$

Thus, by (31), we can decompose the Poincaré series Pé(g) as

$$\text{Pé}(g) = C(\varphi) \cdot E^{2,1}(g, \nu + 1) + \text{Pé}^*(g), \qquad (32)$$

where $E^{2,1}(g, \nu + 1)$ is the maximal parabolic Eisenstein series on $SL_3(\mathbb{Z})$ and

$$C(\varphi) = \int_{\mathbb{R}^2} \varphi\begin{pmatrix} 1 & & t_2 \\ & 1 & t_3 \\ & & 1 \end{pmatrix} dt_2\, dt_3. \qquad (33)$$

To obtain a spectral decomposition, we need to present the Poincaré series Pé(g) with the maximal parabolic Eisenstein series on $SL_3(\mathbb{Z})$ removed in a more useful way. To do so, we first write

$$\text{Pé}^*(g) = \sum_{\gamma\in P(\mathbb{Z})\backslash SL_3(\mathbb{Z})} \sum_{\substack{m_2,m_3=-\infty \\ (m_2,m_3)\neq(0,0)}}^{\infty} \widehat{\varphi}_{\gamma g}(m_2, m_3)$$
$$= \sum_{\gamma\in P(\mathbb{Z})\backslash SL_3(\mathbb{Z})} \sum_{\substack{\psi\in(U(\mathbb{Z})\backslash U(\mathbb{R}))^\frown \\ \psi\neq 1}} \widehat{\varphi}_{\gamma g}(\psi),$$

where

$$\widehat{\varphi}_g(\psi) = \int_U \varphi(ug)\overline{\psi(u)}\, du.$$

For $\beta \in H(\mathbb{Z})$, we observe that

$$\begin{aligned}\widehat{\varphi}_{\beta g}(\psi) &= \int_U \varphi(u\beta g)\overline{\psi(u)}\,\mathrm{d}u = \int_U \varphi(\beta\beta^{-1}u\beta g)\overline{\psi(u)}\,\mathrm{d}u \\ &= \int_U \varphi(\beta^{-1}u\beta g)\overline{\psi(u)}\,\mathrm{d}u = \int_U \varphi(ug)\overline{\psi(\beta u\beta^{-1})}\,\mathrm{d}u, \qquad (34)\end{aligned}$$

as $\varphi(\beta g) = \varphi(g)$ for $\beta \in H(\mathbb{Z})$ and $g \in G$. Setting $\psi^\beta(u) = \psi(\beta u \beta^{-1})$, the last integral in (34) is $\widehat{\varphi}_g(\psi^\beta)$.

Consider the characters on $U(\mathbb{Z})\backslash U(\mathbb{R})$:

$$\psi^m(u) = \mathrm{e}^{2\pi i m u_3}, \qquad \left(m \in \mathbb{Z}^\times \text{ and } u = \begin{pmatrix} 1 & & u_2 \\ & 1 & u_3 \\ & & 1 \end{pmatrix}\right).$$

Since every nontrivial character on $U(\mathbb{Z})\backslash U(\mathbb{R})$ is obtained as $(\psi^m)^\beta$, for unique $m \in \mathbb{Z}^\times$ and $\beta \in P^{1,1}(\mathbb{Z})\backslash H(\mathbb{Z})$, where $P^{1,1}(\mathbb{Z})$ is the parabolic subgroup of $H(\mathbb{Z})$, it follows from (34) that

$$\begin{aligned}\text{Pé}^*(g) &= \sum_{\gamma \in P(\mathbb{Z})\backslash SL_3(\mathbb{Z})} \ \sum_{\beta \in P^{1,1}(\mathbb{Z})\backslash H(\mathbb{Z})} \ \sum_{m \in \mathbb{Z}^\times} \widehat{\varphi}_{\beta\gamma g}(\psi^m) \\ &= \sum_{\gamma \in P^{1,1,1}(\mathbb{Z})\backslash SL_3(\mathbb{Z})} \ \sum_{m \in \mathbb{Z}^\times} \widehat{\varphi}_{\gamma g}(\psi^m).\end{aligned}$$

Let

$$\Theta = \left\{\begin{pmatrix} 1 & & \\ & * & * \\ & * & * \end{pmatrix}\right\}, \qquad U' = \left\{\begin{pmatrix} 1 & & * \\ & 1 & \\ & & 1 \end{pmatrix}\right\}, \qquad U'' = \left\{\begin{pmatrix} 1 & & \\ & 1 & * \\ & & 1 \end{pmatrix}\right\}.$$

Then, $\text{Pé}^*(g) =$

$$\sum_{\gamma \in P^{1,2}(\mathbb{Z})\backslash SL_3(\mathbb{Z})} \ \sum_{\beta \in P^{1,1}(\mathbb{Z})\backslash \Theta(\mathbb{Z})} \ \sum_{m \in \mathbb{Z}^\times} \int_{U''} \overline{\psi}^m(u'') \cdot \left(\int_{U'} \varphi(u'u''\beta\gamma g)\,\mathrm{d}u'\right)\mathrm{d}u''.$$

Setting

$$\widetilde{\varphi}(g) = \int_{U'} \varphi(u'g)\,\mathrm{d}u',$$

the last expression of $\text{Pé}^*(g)$ becomes

$$\text{Pé}^*(g) = \sum_{\gamma \in P^{1,2}(\mathbb{Z})\backslash SL_3(\mathbb{Z})} \ \sum_{\beta \in P^{1,1}(\mathbb{Z})\backslash \Theta(\mathbb{Z})} \ \sum_{m \in \mathbb{Z}^\times} \int_{U''} \overline{\psi}^m(u'')\widetilde{\varphi}(u''\beta\gamma g)\,\mathrm{d}u''. \qquad (35)$$

Let

$$\Phi(g) = \sum_{\beta \in P^{1,1}(\mathbb{Z})\backslash \Theta(\mathbb{Z})} \sum_{m \in \mathbb{Z}^\times} \int_{U''} \overline{\psi}^m(u'')\widetilde{\varphi}(u''\beta g)\, \mathrm{d}u''. \tag{36}$$

We need the following simple observation.

Lemma 1. *We have the equivariance*

$$\widetilde{\varphi}(pg) = |q|^{\nu+1} \cdot |a|^\nu \cdot |d|^{-2\nu-1} \cdot \widetilde{\varphi}(g), \qquad \left(\text{for } p = \begin{pmatrix} q & b & c \\ & a & \\ & & d \end{pmatrix} \in GL_3(\mathbb{R}) \right).$$

Proof. Indeed, since

$$\begin{pmatrix} 1 & & t \\ & 1 & \\ & & 1 \end{pmatrix}\begin{pmatrix} q & b & c \\ & a & \\ & & d \end{pmatrix} = \begin{pmatrix} q & b & td + c \\ & a & \\ & & d \end{pmatrix} = \begin{pmatrix} q & b & \\ & a & \\ & & d \end{pmatrix}\begin{pmatrix} 1 & & (td+c)/q \\ & 1 & \\ & & 1 \end{pmatrix},$$

we have

$$\widetilde{\varphi}(pg) = \int_{U'} \varphi(u'pg)\, \mathrm{d}u' = \left|\frac{qa}{d^2}\right|^\nu \cdot \int_{\mathbb{R}} \varphi\left(\begin{pmatrix} 1 & & (td+c)/q \\ & 1 & \\ & & 1 \end{pmatrix} g\right) \mathrm{d}t$$

$$= |q|^{\nu+1} \cdot |a|^\nu \cdot |d|^{-2\nu-1} \widetilde{\varphi}(g).$$

□

Assuming g of the form

$$g = \begin{pmatrix} a & * \\ & g' \end{pmatrix} \qquad (a \in \mathbb{R}^\times \text{ and } g' \in GL_2(\mathbb{R})),$$

(we can always do using the Iwasawa decomposition) and decomposing it as

$$g = \begin{pmatrix} a & * \\ & I_2 \end{pmatrix}\begin{pmatrix} 1 & \\ & g' \end{pmatrix},$$

we have

$$\widetilde{\varphi}(g) = |a|^{\nu+1} \cdot \widetilde{\varphi}\begin{pmatrix} 1 & \\ & g' \end{pmatrix}.$$

Since

$$\begin{pmatrix} 1 & \\ & D \end{pmatrix} g = \begin{pmatrix} a & * \\ & Dg' \end{pmatrix} \qquad (\text{for } D \in GL_2(\mathbb{R})),$$

it follows that $\Phi(g)$ defined in (36) descends to a GL_2 Poincaré series, with the corresponding Eisenstein series removed, of the type studied in [7–9]. Setting

$$\varphi^{(2)}\begin{pmatrix} 1 & x \\ & 1 \end{pmatrix} = \widetilde{\varphi}\begin{pmatrix} 1 & & \\ & 1 & x \\ & & 1 \end{pmatrix} \qquad (x \in \mathbb{R}),$$

and extending it to $GL_2(\mathbb{R})$ by

$$\varphi^{(2)}\left(\begin{pmatrix} a & \\ & d \end{pmatrix} gk\right) = \left|\frac{a}{d}\right|^{\frac{3v+1}{2}} \cdot \varphi^{(2)}(g), \qquad (g \in GL_2(\mathbb{R}),\ k \in O_2(\mathbb{R})),$$

we can write

$$\Phi\begin{pmatrix} a & * \\ & g' \end{pmatrix} = |a|^{v+1} \cdot |\det g'|^{-\frac{v+1}{2}} \cdot \sum_{\beta \in P^{1,1}(\mathbb{Z})\backslash SL_2(\mathbb{Z})} \sum_{m \in \mathbb{Z}^\times} \int_N \overline{\psi}^m(n)\, \varphi^{(2)}(n\beta g')\, dn, \tag{37}$$

with N the subgroup of upper-triangular unipotent elements in $GL_2(\mathbb{R})$. Note that, for

$$\varphi\begin{pmatrix} I_2 & u \\ & 1 \end{pmatrix} = \left(1 + ||u||^2\right)^{-\frac{w}{2}},$$

we have

$$\varphi^{(2)}\begin{pmatrix} 1 & x \\ & 1 \end{pmatrix} = \widetilde{\varphi}\begin{pmatrix} 1 & & \\ & 1 & x \\ & & 1 \end{pmatrix} = \int_{U'} \varphi\left(u' \begin{pmatrix} 1 & & \\ & 1 & x \\ & & 1 \end{pmatrix}\right) du'$$

$$= \int_{-\infty}^{\infty} \left(1 + u^2 + x^2\right)^{-\frac{w}{2}} du = \sqrt{\pi}\, \frac{\Gamma(\frac{w-1}{2})}{\Gamma(\frac{w}{2})} \cdot \left(1 + x^2\right)^{\frac{1-w}{2}}. \tag{38}$$

Then, by (2.2), (2.3), and (5.8) in [9], it follows that, for an orthonormal basis of Maass cuspforms which are simultaneous eigenfunctions of all the Hecke operators, we have the spectral decomposition

$$\Phi\begin{pmatrix} a & * \\ & g' \end{pmatrix}$$

$$= \frac{1}{2} \sum_{F-\text{even}} \overline{\rho_F(1)}\, L\Big(\frac{3v}{2} + 1, F\Big)\, \mathcal{G}\Big(\frac{1}{2} + i\mu_F;\ \frac{3v+1}{2}, w-1\Big)\, |a|^{v+1}\, |\det g'|^{-\frac{v+1}{2}}\, F(g')$$

$$+ \frac{1}{4\pi i} \int\limits_{\Re(s)=\frac{1}{2}} \frac{\zeta(\frac{3v}{2} + \frac{1}{2} + s)\, \zeta(\frac{3v}{2} + \frac{3}{2} - s)}{\pi^{-1+s}\, \Gamma(1-s)\, \zeta(2-2s)}$$

$$\cdot \mathcal{G}\Big(1 - s;\ \frac{3v+1}{2}, w-1\Big)\, |a|^{v+1}\, |\det g'|^{-\frac{v+1}{2}}\, E(g', s)\, ds,$$

where

$$\mathcal{G}(s;v,w) \;=\; \pi^{-v+\frac{1}{2}}\,\frac{\Gamma\left(\frac{-s+v+1}{2}\right)\Gamma\left(\frac{s+v}{2}\right)\Gamma\left(\frac{-s+v+w}{2}\right)\Gamma\left(\frac{s+v+w-1}{2}\right)}{\Gamma\left(\frac{w+1}{2}\right)\Gamma\left(v+\frac{w}{2}\right)}.$$

This decomposition holds provided $\Re(v)$ and $\Re(w)$ are sufficiently large. Hence, by (35) and (36), $\text{Pé}^*(g)$ has the induced spectral decomposition from GL_2,

$$\begin{aligned}
&\text{Pé}^*(g)\\
&= \frac{1}{2}\sum_{F-\text{even}} \overline{\rho_F(1)}\, L\left(\frac{3v}{2}+1,F\right)\mathcal{G}\left(\frac{1}{2}+i\mu_F;\frac{3v+1}{2},w-1\right)E_F^{1,2}(g,v+1)\\
&\quad+\frac{1}{4\pi i}\int\limits_{\Re(s)=\frac{1}{2}}\frac{\zeta(\frac{3v}{2}+\frac{1}{2}+s)\,\zeta(\frac{3v}{2}+\frac{3}{2}-s)}{\pi^{-1+s}\,\Gamma(1-s)\,\zeta(2-2s)}\,\mathcal{G}\left(1-s;\frac{3v+1}{2},w-1\right)\\
&\quad\cdot E^{1,1,1}\left(g,\frac{v+1}{2}-\frac{s}{3},\frac{2s}{3}\right)\,ds.
\end{aligned}$$

By Godement's criterion (see [2]), the minimal parabolic Eisenstein series $E^{1,1,1}$ inside the integral converges absolutely and uniformly on compact subsets of G/ZK for $\Re(v)$ sufficiently large. The meromorphic continuation of the Poincaré series $\text{Pé}(g)$ in $(v,w)\in\mathbb{C}^2$ follows by shifting the contour similarly to Sect. 5 of [9], or Theorem 4.17 in [7].

We summarize the main result of this section in the following theorem.

Theorem 2. *For $\Re(v)$ and $\Re(w)$ sufficiently large, the Poincaré series $\text{Pé}(g)$ associated to*

$$\varphi\begin{pmatrix} I_2 & u\\ & 1\end{pmatrix} \;=\; \left(1+||u||^2\right)^{-\frac{w}{2}}$$

has the spectral decomposition

$$\begin{aligned}
\text{Pé}(g) &= \frac{2\pi}{w-2}\cdot E^{2,1}(g,v+1)\\
&\quad+\frac{1}{2}\sum_{F-\text{even}} \overline{\rho_F(1)}\, L\left(\frac{3v}{2}+1,F\right)\mathcal{G}\left(\frac{1}{2}+i\mu_F;\frac{3v+1}{2},w-1\right)E_F^{1,2}(g,v+1)\\
&\quad+\frac{1}{4\pi i}\int\limits_{\Re(s)=\frac{1}{2}}\frac{\zeta(\frac{3v}{2}+\frac{1}{2}+s)\,\zeta(\frac{3v}{2}+\frac{3}{2}-s)}{\pi^{-1+s}\,\Gamma(1-s)\,\zeta(2-2s)}\,\mathcal{G}\left(1-s;\frac{3v+1}{2},w-1\right)\\
&\quad\cdot E^{1,1,1}\left(g,\frac{v+1}{2}-\frac{s}{3},\frac{2s}{3}\right)\,ds.
\end{aligned}$$

From Theorem 2, the meromorphic continuation of the Poincaré series follows by similar techniques to those used in [9] or [7]. However, we are not pursuing this spectral decomposition any further here, as for potential applications, one needs to use it together with asymptotic information for the weights appearing in the moment side, information which we do not have at the moment.

Remark 1. Let φ on U be defined by

$$\varphi\begin{pmatrix} I_2 & u \\ & 1 \end{pmatrix} = 2^{1-w}\sqrt{\pi}\,\frac{\Gamma(\frac{w}{2})\left(1+||u||^2\right)^{-\frac{w}{2}} F(\frac{w}{2},\frac{w}{2};w;\frac{1}{1+||u||^2})}{\Gamma(\frac{w-1}{2})},$$

and consider the Poincaré series $\text{Pé}(g)$ attached to this choice of φ. Representing the hypergeometric function by its power series,

$$F(\alpha,\beta;\gamma;z) = \frac{\Gamma(\gamma)}{\Gamma(\alpha)\Gamma(\beta)}\cdot\sum_{m=0}^{\infty}\frac{1}{m!}\frac{\Gamma(\alpha+m)\Gamma(\beta+m)}{\Gamma(\gamma+m)}z^m \qquad (|z|<1),$$

and using the last identity in (38), it follows, as in [5], Sect. 7.3, that the Poincaré series $\text{Pé}(g)$ with $v = 0$ satisfies a shifted functional equation (involving an Eisenstein series) as $w \longrightarrow 2-w$ (see also [9, 14]).

References

1. Jennifer Beineke and Daniel Bump. Moments of the Riemann zeta function and Eisenstein series. I. *J. Number Theory*, 105(1):150–174, 2004.
2. Armand Borel. Introduction to automorphic forms. In *Algebraic Groups and Discontinuous Subgroups (Proc. Sympos. Pure Math., Boulder, Colo., 1965)*, pages 199–210. Amer. Math. Soc., Providence, R.I., 1966.
3. Daniel Bump. *Automorphic forms on* GL(3, **R**), volume 1083 of *Lecture Notes in Mathematics*. Springer-Verlag, Berlin, 1984.
4. J. B. Conrey, D. W. Farmer, J. P. Keating, M. O. Rubinstein, and N. C. Snaith. Integral moments of L-functions. *Proc. London Math. Soc. (3)*, 91(1):33–104, 2005.
5. A. Diaconu and P. Garrett. Subconvexity bounds for automorphic L-functions. *J. Inst. Math. Jussieu*, 9(1):95–124, 2010.
6. A. Diaconu, P. Garrett, and D. Goldfeld. Moments for L-functions for $GL_r \times GL_{r-1}$. In preparation: http://www.math.umn.edu/garrett/m/v/.
7. Adrian Diaconu and Paul Garrett. Integral moments of automorphic L-functions. *J. Inst. Math. Jussieu*, 8(2):335–382, 2009.
8. Adrian Diaconu and Dorian Goldfeld. Second moments of quadratic Hecke L-series and multiple Dirichlet series. I. In *Multiple Dirichlet series, automorphic forms, and analytic number theory*, volume 75 of *Proc. Sympos. Pure Math.*, pages 59–89. Amer. Math. Soc., Providence, RI, 2006.
9. Adrian Diaconu and Dorian Goldfeld. Second moments of GL_2 automorphic L-functions. In *Analytic number theory*, volume 7 of *Clay Math. Proc.*, pages 77–105. Amer. Math. Soc., Providence, RI, 2007.

10. Adrian Diaconu, Dorian Goldfeld, and Jeffrey Hoffstein. Multiple Dirichlet series and moments of zeta and L-functions. *Compositio Math.*, 139(3):297–360, 2003.
11. W. Duke, J. B. Friedlander, and H. Iwaniec. The subconvexity problem for Artin L-functions. *Invent. Math.*, 149(3):489–577, 2002.
12. T. Estermann. On certain functions represented by Dirichlet series. *Proc. London Math. Soc.*, 27(435–448), 1928.
13. Dorian Goldfeld. *Automorphic forms and L-functions for the group* $\mathrm{GL}(n,\mathbf{R})$, volume 99 of *Cambridge Studies in Advanced Mathematics*. Cambridge University Press, Cambridge, 2006. With an appendix by Kevin A. Broughan.
14. A. Good. The convolution method for Dirichlet series. In *The Selberg trace formula and related topics (Brunswick, Maine, 1984)*, volume 53 of *Contemp. Math.*, pages 207–214. Amer. Math. Soc., Providence, RI, 1986.
15. I. S. Gradshteyn and I. M. Ryzhik. *Table of integrals, series, and products*. Academic Press Inc., Boston, MA, Fifth edition, 1994. Translation edited and with a preface by Alan Jeffrey.
16. A. Ivić. On the estimation of $\mathcal{Z}_2(s)$. In *Analytic and probabilistic methods in number theory (Palanga, 2001)*, pages 83–98. TEV, Vilnius, 2002.
17. Aleksandar Ivić, Matti Jutila, and Yoichi Motohashi. The Mellin transform of powers of the zeta-function. *Acta Arith.*, 95(4):305–342, 2000.
18. Matti Jutila. The Mellin transform of the fourth power of Riemann's zeta-function. In *Number theory*, volume 1 of *Ramanujan Math. Soc. Lect. Notes Ser.*, pages 15–29. Ramanujan Math. Soc., Mysore, 2005.
19. Nobushige Kurokawa. On the meromorphy of Euler products. I. *Proc. London Math. Soc. (3)*, 53(1):1–47, 1986.
20. Nobushige Kurokawa. On the meromorphy of Euler products. II. *Proc. London Math. Soc. (3)*, 53(2):209–236, 1986.
21. Yōichi Motohashi. A relation between the Riemann zeta-function and the hyperbolic Laplacian. *Ann. Scuola Norm. Sup. Pisa Cl. Sci. (4)*, 22(2):299–313, 1995.
22. Yoichi Motohashi. The Riemann zeta-function and the Hecke congruence subgroups. *Sūrikaisekikenkyūsho Kōkyūroku*, (958):166–177, 1996. Analytic number theory (Japanese) (Kyoto, 1994).
23. Yoichi Motohashi. *Spectral theory of the Riemann zeta-function*, volume 127 of *Cambridge Tracts in Mathematics*. Cambridge University Press, Cambridge, 1997.
24. Freydoon Shahidi. Third symmetric power L-functions for GL(2). *Compositio Math.*, 70(3):245–273, 1989.

Chapter 8
A Trace Formula of Special Values of Automorphic L-Functions

Bernhard Heim

Abstract Deligne introduced the concept of special values of automorphic L-functions. The arithmetic properties of these L-functions play a fundamental role in modern number theory. In this chapter, we prove a trace formula which relates special values of the Hecke, Rankin, and the central value of the Garrett triple L-function attached to primitive new forms. This type of trace formula is new and involves special values in the convergent and nonconvergent domain of the underlying L-functions.

Keywords Arithmetic trace formula • Garrett triple L-function • Critical values of L-functions • Saito-Kurokawa correspondence

8.1 Introduction and Statement of Results

The main result of this chapter is the discovery of an arithmetic trace formula. This formula relates special values of various kinds of automorphic L-functions. Our previous knowledge of the basic facts on the arithmetic nature of special values is built on the fundamental works of some of the pioneers in this field: Siegel [Si69], Klingen [Kl62], Shimura [Sh76], Zagier [Za77], Deligne [De79], and Garrett [Ga87].

Let $g \in S_k(SL_2(\mathbb{Z}))$ be a primitive (normalized Hecke eigenform) cusp form of integer weight k. Let $(f_j)_j \in S_{2k-2}$ and $(g_i)_i \in S_k$ be primitive eigenbases. The trace formula compares the weighted average $\sum_j$ of special values of the nontrivial piece of the triple L-function $L(f_j \otimes \mathrm{Sym}^2(g), c_k)$ evaluated at the central value c_k

B. Heim (✉)
German University of Technology (GUtech), PO Box 1816, Athaibah PC 130, Muscat, Sultanate of Oman
e-mail: bernhard.heim@gutech.edu.om

D. Bump et al. (eds.), *Multiple Dirichlet Series, L-functions and Automorphic Forms*, Progress in Mathematics 300, DOI 10.1007/978-0-8176-8334-4_8,

and the average $\sum_i$ of the triple L-function $L(g \otimes g \otimes g_i, 2k-2)$ and an error term expressed by special values related to the Rankin L-function attached to g. This special value $L(f_j \otimes \mathrm{Sym}^2(g), c_k)$ and the related triple L-function recently played a prominent role in the proof of the Gross-Prasad conjecture for Saito-Kurokawa lifts given by Ichino [Ich05]. More generally, Ikeda stated in [Ik06] a conjecture on the explicit value of a certain period which involves the central value of L-functions (Conjecture 5.1) of the type studied in this chapter. There, the nonvanishing of the central value is important. Recently, some progress has been obtained by Katsurada and Kawamura [KK06]. The focus of this chapter is the proof of the arithmetic trace formula and not applications. Nevertheless, we believe that there will be applications towards the problems proposed by Iwaniec and Sarnak in the survey article [IS02].

Before we go into more details, we put our results into a more general framework and give relations to other results.

Since the days of Euler (1707–1783), the analytic and arithmetic properties of infinite series of type

$$L(s) := \sum_{n=1}^{\infty} A(n)\, n^{-s} \quad (s \in \mathbb{C}) \tag{1}$$

at integral values $m = \ldots -2, -1, 0, 1, 2, \ldots$ have always revealed significant invariants and properties of the underlying *motivic* object related to the sequence $A(1), A(2), \ldots$ of complex numbers. Significant series arise when the function $A(n)$ is multiplicative and $L(s)$ converges absolutely and locally uniformly if $\mathrm{Re}(s)$ is large enough. These series are nowadays called L-functions.

Examples are given by the Dedekind zeta functions $\zeta_K(s)$, the Hasse-Weil zeta functions $Z_E(s)$, the Hecke L-function $L(f, s)$, and the Rankin L-functions $D(f, s)$ attached to algebraic number fields K, elliptic curves E, and primitive elliptic cusp forms f. They have a meromorphic continuation to the whole complex plane and satisfy a functional equation. Let us just recall some interesting properties. The Riemann zeta function $\zeta(s) := \zeta_{\mathbb{Q}}(s)$ has a single simple pole at $s = 1$. The nonvanishing at $\zeta(1+it)$ for $t \in \mathbb{R}$ directly leads to the prime number theorem. The Kronecker limit formula of ζ_K gives information on the regulator, class number, and other invariants of the number field K. From Euler, we know that

$$\zeta(2m) = \frac{(-1)^{m-1} 2^{2m-1} B_{2m}}{(2m)!} \pi^{2m} \text{ for } m \in \mathbb{N}. \tag{2}$$

Here, B_m denotes the mth Bernoulli number. Let $\Delta(z)$ be the Ramanujan function, the unique primitive cusp form of level 1 of weight 12, with Fourier coefficients $\tau(n)$. It is known that up to normalization, the values of the Rankin-type L-function $D(\Delta, s)$ at integral values within the "critical strip" are rational numbers, for example,

$$D(\Delta, 14) = \frac{\zeta(6)}{\zeta(3)} \sum_{n=1}^{\infty} \frac{\tau(n)^2}{n^{14}} = \frac{4^{14}}{14!} \pi^{17} \parallel \Delta \parallel^2. \tag{3}$$

Let $\langle\, ,\, \rangle$ be the Petersson scalar product and $\|\ \|$ the Petersson norm (see (14) for details). Then $\| \Delta \|^2 = 1.03536205679 \times 10^{-6}$ with 12-digit accuracy (see[Za77]).

The concept of critical values of a motivic L-function and conjectures on the arithmetic nature has been introduced by Deligne. Let $\widehat{L}(s) := \gamma(s)\, L(s)$ be the completion of $L(s)$ at infinity, i.e., $\gamma(s)$ is essentially a product of Γ-functions with functional equation $\widehat{L}(s) = \widehat{L}(w-s)$, $w \in \mathbb{N}$. Then $m \in \mathbb{Z}$ is a critical value if and only if $\gamma(m)$ and $\gamma(w-m)$ are finite. Deligne predicts that then $L(m) = \text{algebraic} \times \Omega_{\text{period}}$. Moreover, a certain functoriality of the action of the automorphism of the absolute Galois group over the involved number fields can be given.

Let $g \in S_k$ be primitive with Fourier coefficients $(a_n(g))_n$ and Satake parameter (see (15)) $\widetilde{\alpha}_p, \widetilde{\beta}_p$ for all finite prime numbers p. For simplification, we put

$$A_p(g) := \begin{pmatrix} \widetilde{\alpha}_p(g) & 0 \\ 0 & \widetilde{\beta}_p(g) \end{pmatrix}. \tag{4}$$

Then Hecke attached to g the L-function

$$L(g,s) := \prod_p \left\{\det\left(1_2 - A_p(g)\, p^{-s}\right)\right\}^{-1} \quad \text{for } \mathrm{Re}(s) > \frac{k}{2}. \tag{5}$$

With this notation, the Rankin L-function $D(g,s)$ and the triple L-function $L(f_1 \otimes f_2 \otimes f_3, s)$ are defined by

$$D(g,s) := \zeta(s-k+1)^{-1} \prod_p \left\{\det\left(1_4 - A_p(g) \otimes A_p(g)\, p^{-s}\right)\right\}^{-1} \quad \text{for } \mathrm{Re}(s) > k. \tag{6}$$

$$L(f_1 \otimes f_2 \otimes f_3, s) := \prod_p \left\{\det\left(1_8 - A_p(f_1) \otimes A_p(f_2) \otimes A_p(f_3)\, p^{-s}\right)\right\}^{-1} \quad \text{for } \mathrm{Re}(s) \gg 0. \tag{7}$$

Here, f_1, f_2, and f_3 are primitive elliptic cusp forms. Let $\widehat{L}(g,s)$, $\widehat{D}(g,s)$, etc., be the completed L-function, see (23)–(30). They all have a meromorphic continuation to the whole complex plane and satisfy certain functional equations. From this, the critical values can be explicitly determined. In contrast to the Rankin L-function, the center of the Hecke L-function and the triple L-function is always a critical point. The Hecke L-function vanishes in the center if the weight k is congruent to 2 modulo 4 and the triple L-function for the full modular group $SL_2(\mathbb{Z})$. This follows from the sign in the functional equation.

Recently, a piece of the triple L-function $L(f \otimes \mathrm{Sym}^2(g), s)$ attached to $g \in S_k$ and $f \in S_{2k-2}$ primitive (see (22) for an explicit definition) showed up in the proof of the Gross-Prasad conjecture of Saito-Kurokawa lifts. Among other things, Ichino [Ich05] showed that $L\left(f \otimes \mathrm{Sym}^2(g), 2k-2\right)$ is finite.

More precisely, we have the decomposition

$$L(f \otimes g \otimes g, s) = L\left(f \otimes \mathrm{Sym}^2(g), s\right) \cdot L(f, s-k+1). \tag{8}$$

Work of Deligne predicts that the unique critical value is given by $2k-2$ which matches with the center of the functional equation. Now, the vanishing of the triple L-function becomes obvious since the Hecke L-function of f vanishes at the center. So it remains an open question to study the arithmetic nature of $L(f \otimes \mathrm{Sym}^2(g), s)$. Ichino [Ich05] proved that the value is zero if and only if a certain period vanishes. Moreover, he described the transformation of the special value under any automorphism of $\mathbb{C}$. Recently, we have proven [Hei05]: Let g be given. Then there exists at least one f such that the value

$$L(f \otimes \mathrm{Sym}^2(g), 2k-2) \neq 0. \tag{9}$$

Starting with $f \in S_{2k-2}$ the property (9) is not always possible. Since, for example, in the case $k = 18$ we have $S_{10} = 0$. For general k, we cannot say much.

Theorem: Arithmetic Trace Formula. *Let k be an even positive integer. Let $g \in S_k$ be a primitive Hecke eigenform. Then we have*

$$\sum_{i=1}^{\dim S_{2k-2}} \frac{\widehat{L}(f_i, 2k-3)\ \widehat{L}\left(f_i \otimes Sym^2(g), 2k-2\right)}{\| f_i \|^2 \| g \|^4}$$

$$= (-1)^{k/2} \cdot 2^{k-2} \sum_{j=1}^{\dim S_k} \frac{\widehat{L}(g \otimes g \otimes g_j, 2k-2)}{\| g \|^4 \| g_j \|^2}$$

$$+\kappa_1 \left(\frac{\widehat{D}(g, 2k-2)}{\pi^{\frac{k}{2}-1} \| g \|^2} \right)^2 + \kappa_2 \frac{\widehat{D}(g, 2k-2)}{\pi^{\frac{k}{2}-1} \| g \|^2}. \tag{10}$$

Here, $(f_i)_i$ and $\left(g_j\right)_j$ are primitive Hecke eigenbases of S_{2k-2} and S_k, and the constants κ_1 and κ_2 can be explicitly given. We have

$$\kappa_1 = (-1)(-1)^{k/2} 2^4 \frac{\Gamma(k)^2}{(2k-2) B_{2k-2} \Gamma(k/2)^2}, \tag{11}$$

$$\kappa_2 = (-1)(-1)^{k/2} 2^{2k+1} \frac{\Gamma(k+1)}{(2k-2) B_k \Gamma(k/2)}. \tag{12}$$

Remark 1. We would like to note that in this chapter, we actually prove a more general trace formula. It involves the products of roots of L-values of type $\widehat{L}\left(f_i \otimes \mathrm{Sym}^2(g_{i_*}), 2k-2\right)$ on the left side and the more general triple L-function of type $\widehat{L}(g_{i_1} \otimes g_{i_2} \otimes g_j, 2k-2)$ on the right side (see (62)). Here, $i_* = i_1$ or i_2.

Remark 2. All the totally real algebraic numbers (see the Sects. 8.2.1 and 8.2.3 for more details)

$$\frac{\widehat{L}(f_i, 2k-3)}{\Omega_-(f_i)}, \quad \frac{\widehat{L}(g \otimes g \otimes g_j, 2k-2)}{\| g \|^2 \| g \|^2 \| g_j \|^2}, \text{ and } \frac{\widehat{D}(g, 2k-2)}{\pi^{\frac{k}{2}-1} \| g \|^2} \tag{13}$$

are given by evaluating an infinite product, which locally does not vanish in the domain of absolute and uniform convergence. Let $f, \ldots, \Phi$ be any Hecke eigenforms, then $K_{f,\ldots,\Phi}$ denotes the field over $\mathbb{Q}$ generated by the corresponding eigenvalues. We put K_k if we take all the eigenvalues of an Hecke eigenbasis of S_k. Then the values given in (13) are units in K_{f_i}, K_{g,g_j}, and K_g. This is not surprising. But new is the fact that these values can be explicitly used to study the central value of the L-function $L\left(f_i \otimes \mathrm{Sym}^2(g_{i*}), s\right)$ at the center of symmetry, at least on average.

8.2 Automorphic L-Functions

Let us recall some notation and basic facts on modular forms and L-functions. Moreover, we add some properties of Jacobi forms. For the general setting we refer the reader to Iwaniec [Iw97], Eichler and Zagier [EZ85], and Klingen [Kl90]. The overview article of van der Geer [Ge06] is also very useful.

8.2.1 Basics on L-Functions

Let $\mathbb{H}_g$ denote the Siegel upper half-space of genus g and let $\Gamma_g := Sp_g(\mathbb{Z})$ be the Siegel modular group of degree g. For k, an even nonnegative integer, let $M_k^{(g)}$ be the space of Siegel modular forms of weight k and genus g with respect to Γ_g. Let $S_k^{(g)}$ be the subspace of cusp forms. We recall the definition of the Petersson scalar product on S_k^g:

$$\langle F \,,\, G \rangle := \int_{\Gamma_g \backslash \mathbb{H}_g} F(Z)\, \overline{G(Z)} \det\left(\mathrm{Im}(Z)\right)^{k+g-1} \, \mathrm{d}Z. \tag{14}$$

Hence, $\| F \|^2 = \langle F \,,\, F \rangle$. To simplify notation, we drop the index g in the case $g = 1$. Examples of Siegel modular forms are given by Eisenstein series. Let $Z \in \mathbb{H}_g$ be an element of the Siegel upper half-space and let $k > g + 1$ be even. Then

$$E_k^g(Z) := \sum_{\left(\begin{smallmatrix} A & B \\ C & D \end{smallmatrix}\right) \in \Gamma_\infty \backslash \Gamma_g} \det(CZ + D)^{-k},$$

where $\Gamma_\infty := \left\{ \left(\begin{smallmatrix} A & B \\ 0 & D \end{smallmatrix}\right) \in \Gamma_g \right\}$. This series is absolutely and locally uniformly convergent on $\mathbb{H}_g$ and is an element of $M_k^{(g)}$. We denote its Fourier coefficients by $A^{E_k^g}(T)$, where $T \in \mathbb{A}_g$ runs through all half-integral symmetric semipositive matrices of size g. Here, $A^{E_k^g}(0) = 1$. It is useful to know that the coefficients $A^{E_k^g}(T)$ are rational and have bounded denominators.

Let $g \in S_k$ with Fourier coefficients $(a_n(g))_{n=1}^{\infty}$. The cusp form g is called primitive if g is a Hecke eigenform and if $a_1(g) = 1$. Let us assume that g is primitive. Then we attach to every prime number p the local parameters $\widetilde{\alpha}_p(g), \widetilde{\beta}_p(g) \in \mathbb{C}$ defined by the equations

$$\widetilde{\alpha}_p(g) + \widetilde{\beta}_p(g) = a_p(g) \text{ and } \widetilde{\alpha}_p(g) \cdot \widetilde{\beta}_p(g) = p^{k-1}. \tag{15}$$

Then the Satake parameters are given by

$$\alpha_p(g) := p^{-\frac{k-1}{2}} \widetilde{\alpha}_p(g) \text{ and } \beta_p(g) := p^{-\frac{k-1}{2}} \widetilde{\beta}_p(g). \tag{16}$$

By Deligne's proof of the Ramanujan–Petersson conjecture in [De71], we have $|\alpha_p(g)| = |\beta_p(g)| = 1$. For further simplification, we put

$$A_p(g) := \begin{pmatrix} \widetilde{\alpha}_p(g) & 0 \\ 0 & \widetilde{\beta}_p(g) \end{pmatrix}. \tag{17}$$

We begin now with the definition of the L-function $L(g, s)$ attached to g of Hecke type. We have the absolute convergent infinite product over all prime numbers

$$L(g, s) := \prod_p \left\{\det\left(1_2 - A_p(g)\, p^{-s}\right)\right\}^{-1} \quad \text{for } \mathrm{Re}(s) > \frac{k+1}{2}. \tag{18}$$

The standard L-function $D(g, s)$ or sometimes called the symmetric square L-function of g has been already defined in the introduction (6). We also have the identity

$$L(g, s) = \sum_{n=1}^{\infty} a_n(g)\, n^{-s}, \tag{19}$$

$$D(g, s) = \frac{\zeta(2s - 2k + 2)}{\zeta(s - k + 1)} \sum_{n=1}^{\infty} a_n(g)^2 n^{-s}. \tag{20}$$

Let now $f \in S_{2k-2}$ and $g \in S_k$ be primitive. Then, we put

$$S_p(g) := \begin{pmatrix} \widetilde{\alpha}_p(g)^2 & 0 & 0 \\ 0 & p^{k-1} & 0 \\ 0 & 0 & \widetilde{\beta}_p(g)^2 \end{pmatrix}. \tag{21}$$

The next L-function $L(f \otimes \mathrm{Sym}^2(g), s)$ is defined by

$$L(f \otimes \mathrm{Sym}^2(g), s) := \prod_p \left\{\det\left(1_6 - A_p(f) \otimes S_p(g)\, p^{-s}\right)\right\}^{-1} \quad \text{for } \mathrm{Re}(s) \gg 0. \tag{22}$$

Finally, we have the triple L-function (7) attached to primitive Hecke eigenforms $f_j \in S_{\nu(f_j)}$ for $j = 1, 2, 3$.

All these L-function have a meromorphic continuation to the whole complex s-plane. They also have a functional equation. This can be stated in the "right" way if we add the local factors corresponding to the Archimedean prime number. Let $\Gamma_{\mathbb{R}}(s) := \pi^{-\frac{s}{2}}\,\Gamma(s/2)$ and $\Gamma_{\mathbb{C}}(s) := 2\,(2\pi)^{-s}\,\Gamma(s)$ be the normalized Γ-function. Then we have for $g \in S_k$ primitive the completed L-functions

$$\widehat{L}(g, s) := \Gamma_{\mathbb{C}}(s)\, L(g, s), \tag{23}$$

$$\widehat{D}(g, s) := \Gamma_{\mathbb{R}}(s - k + 2)\Gamma_{\mathbb{C}}(s)\, D(g, s). \tag{24}$$

Then it is well known that $\widehat{L}(g, s)$ and $D(g, s)$ are entire function on the whole s-plane. They have the functional equation

$$\widehat{L}(g, s) = (-1)^{\frac{k}{2}}\, \widehat{L}(g, k - s) \tag{25}$$

and

$$\widehat{D}(g, s) = \widehat{D}(g, 2k - 1 - s). \tag{26}$$

In the setting of the triple L-function, we assume that $\nu(f_1) \geqslant \nu(f_2) \geqslant \nu(f_3)$. Since we are mainly interested in the balanced case, we assume that $\nu(f_2) + \nu(f_3) \geqslant \nu(f_1)$. Then

$$\begin{aligned}\widehat{L}\,(f_1 \otimes f_2 \otimes f_3, s) := \Gamma_{\mathbb{C}}\,(s)\,\Gamma_{\mathbb{C}}\,(s - \nu(f_1) + 1)\,\Gamma_{\mathbb{C}}\,(s - \nu(f_2) + 1) \\ \Gamma_{\mathbb{C}}\,(s - \nu(f_3) + 1)\, L\,(f_1 \otimes f_2 \otimes f_3, s).\end{aligned} \tag{27}$$

This function has a meromorphic continuation to the whole s-plane and satisfies the antisymmetric functional equation

$$\widehat{L}\,(f_1 \otimes f_2 \otimes f_3, s) = -\widehat{L}\,(f_1 \otimes f_2 \otimes f_3, \nu(f_1) + \nu(f_2) + \nu(f_3) - 2 - s)\,. \tag{28}$$

This L-function vanishes at the center $s_0 = \frac{\nu(f_1)+\nu(f_2)+\nu(f_3)}{2} - 1$. Moreover, let $f \in S_{2k-2}$ and $g \in S_k$ be primitive. Then we have by a straightforward calculation that

$$L(f \otimes g \otimes g, s) = L\left(f \otimes \mathrm{Sym}^2(g), s\right) \cdot L(f, s - k + 1). \tag{29}$$

We obtain the following completed L-function

$$\widehat{L}\left(f \otimes \mathrm{Sym}^2(g), s\right)$$
$$:= \Gamma_{\mathbb{C}}(s)\, \Gamma_{\mathbb{C}}(s-k+1)\, \Gamma_{\mathbb{C}}(s-2k+3)\, L(f \otimes \mathrm{Sym}^2(g), s). \quad (30)$$

It has a meromorphic continuation to the whole complex s-plane and has the functional equation $s \mapsto 4k-4-s$.

8.2.2 *Saito-Kurokawa Correspondance*

Let $M^+_{k-\frac{1}{2}}(\Gamma_0(4))$ be Kohnen's plus space. This is the space of modular forms of half-integral weight $k-\frac{1}{2}$ related to the group

$$\Gamma_0(4) := \left\{ \begin{pmatrix} a & b \\ c & d \end{pmatrix} \in SL_2(\mathbb{Z}) \,|\, c \equiv 0 \pmod 4 \right\},$$

where certain Fourier coefficients are zero. Let $S^+_{k-\frac{1}{2}}(\Gamma_0(4))$ be the subspace of cusp forms. Let $J_{k,1}$ be the space of Jacobi forms of weight k and index 1 and $J^{\mathrm{cusp}}_{k,1}$ the subspace of cusp forms. Jacobi forms are holomorphic functions on $\mathbb{H} \times \mathbb{C}$ which satisfy certain conditions (for details, see the standard reference [EZ85]).

Let $h_j \in S^+_{k-\frac{1}{2}}(\Gamma_0(4))$. Then there exists a Jacobi cusp form $\Phi_j \in J^{\mathrm{cusp}}_{k,1}$ via the isomorphism given in Theorem 5.4 in [EZ85]. This isomorphism is given on the level of Fourier coefficients and is compatible with the action of the Hecke algebra of Jacobi forms and modular forms of half-integral weight. Let $(\lambda(n))_n$ be the eigenvalues. Then $f(z) = \sum_n \lambda(n) e^{2\pi i n z} \in S_{2k-2}$ is a primitive Hecke eigenform. This is the Shimura isomorphism.

Moreover, these spaces are isomorphic to the (cuspidal) Maass Spezialschar, a certain subspace of $S_k^{(2)}$. Let further $\langle\, ,\, \rangle$, $\langle\, ,\, \rangle_J$, and $\langle\, ,\, \rangle_+$ denote the Petersson scalar products on $M_k^{(g)}$, the space of Jacobi forms, and the plus space. Moreover, let $\|\ \|_*$ be the related Petersson norm.

Let $g \in S_k$ be primitive. Then we denote by K_g the field generated by the eigenvalues of g. It is well known that K_g is a totally real number field. Let finitely many Hecke eigenforms $f_1, \ldots, f_l$ be given. They can be Siegel modular forms, Jacobi forms, or modular forms of half-integral weight. Then we denote by $K_{f_1,\ldots,f_l}$ the field generated by the eigenvalues. Let $f \in S_{2k-2}$ primitive be given. Then we can choose $h \in S^+_{k-\frac{1}{2}}(\Gamma_0(4))$ via the Shimura correspondance such that the Fourier coefficients are all contained in K_f. Similarly, we can choose the related Jacobi form Φ. Such h and Φ we call normalized.

8.2.3 *Algebraicity of Critical Values of Automorphic L-Functions*

The general philosophy of Deligne [De79] predicts for any motivic Dirichlet series $L(s)$ the structure of the arithmetic nature of certain "critical" values. The underlying assumption is that the Dirichlet series arise from some algebraic variety, Galois representation, or modular form and has a functional equation of the form

$$\widehat{L}(s) = \gamma(s)\, L(s) = \varepsilon \widehat{L}(w-s), \quad \varepsilon \text{ is root of unity, } w \text{ is a constant,} \tag{31}$$

and $\gamma(s)$ is a Γ-factor. Then all integers m for which $\gamma(m)$ and $\gamma(w-m)$ are finite are called critical values. It is expected that $L(m) = \text{algebraic} \times \Omega$, where Ω is a period "about which something nice can be said" (Don Zagier [[Za77], p. 118]).

(a) **Hecke L-function** $L(g,s)$
Let $g \in S_k$ be primitive. Then the critical values of the L-function $L(g,s)$ are given by the integers $m = 1, 2, \ldots, k-1$. We want also to remark that the center $m_0 = k/2$ is also a critical value and $L(g, m_0) = 0$ if $k \equiv 2 \pmod 4$. We know from the result of Eichler–Shimura–Manin that there exist two periods $\Omega_-(g), \Omega_+(g) \in \mathbb{R}$ such that for the critical values $m = \frac{k}{2}, \ldots, k-1$, we have

$$\frac{\widehat{L}(g,m)}{\Omega_{(-1)^m}(g)} \in K_g. \tag{32}$$

Here, we identify $(-1)^k$ with $+$ or $-$ in the obvious way. Analogous information for the other critical values follow directly from the functional equation (see also [Ge06], Sect. 26).

(b) **Rankin L-function** $D(g,s)$
Let $g \in S_k$ be primitive. Then the critical values of the Rankin-type L-function $D(g,s)$ are given by $m = 1, 3, \ldots, k-1$ and $k, k+2, \ldots, 2k-2$. Here, the center $m_0 = \frac{2k-2}{2}$ is not an integer and hence not a critical value. We have

$$\frac{D(g,m)}{\pi^{2m-k+1} \parallel g \parallel^2} = \left(2^{1-m}\Gamma\left(\frac{m-k+2}{2}\right)\Gamma(m)\right)^{-1} \pi^{\frac{k-m}{2}} \frac{\widehat{D}(g,m)}{\parallel g \parallel^2} \in K_g \tag{33}$$

for the even critical values. Supplementarily, we deduce from the functional equation that for the odd critical values, we have $D(g,m)/(\pi^m \parallel g \parallel^2) \in K_g$.

(c) **Triple L-function**
For the triple L-function $L(f_1 \otimes f_2 \otimes f_3, s)$ with $f_j \in S_{\nu_j}$, we fix the ordering $\nu(f_1) \geqslant \nu(f_2) \geqslant \nu(f_3)$ and assume that we are in the situation of the balanced case $\nu(f_2) + \nu(f_3) \geqslant \nu(f_1)$. Then the critical values m are given by

$$\nu(f_1) \leqslant m \leqslant \nu(f_2) + \nu(f_3) - 2. \tag{34}$$

Here, the center $m_0 = \frac{\nu(f_1)+\nu(f_2)+\nu(f_3)}{2} - 2$ is also a critical value. It can deduced from the functional equation and a finiteness theorem that the triple L-function vanishes in the center (see Orloff ([Or87]). Moreover, we have

$$\frac{L\,(f_1 \otimes f_2 \otimes f_3, m)}{\pi^{4m+A} \parallel f_1 \parallel^2 \parallel f_2 \parallel^2 \parallel f_3 \parallel^2} \in K_{f_1,f_2,f_2}, \tag{35}$$

with $A = 3 - \nu(f_1) - \nu(f_2) - \nu(f_3)$.

Examples:

(a) Let $f_1 = f \in S_{2k-2}$ and $f_2 = f_3 = g \in S_k$ be primitive. Then we have exactly one critical value $m = 2k - 2$. This is also equal to the center. Hence, we have $L\,(f \otimes g \otimes g, 2k-2) = 0$.

(b) Let $\nu(f_1) = \nu(f_2) = \nu(f_3) = k$. Then we have

$$\frac{\widehat{L}\,(f_1 \otimes f_2 \otimes f_3, 2k-2)}{\parallel f_1 \parallel^2 \parallel f_2 \parallel^2 \parallel f_3 \parallel^2} \in K_{f_1,f_2,f_2}. \tag{36}$$

(c) **L-function** $L(f \otimes \mathrm{Sym}^2(g), s)$
Let $f \in S_{2k-2}$ and $g \in S_k$ be primitive. Then there is one critical value, namely, $m = 2k - 2$, for the L-function $L(f \otimes \mathrm{Sym}^2(g), s)$. Moreover, we have

$$\frac{\widehat{L}\,\left(f \otimes \mathrm{Sym}^2(g), 2k-2\right)}{\Omega_+(f) \parallel g \parallel^4} \in K_{f,g}. \tag{37}$$

(see also Ichino [Ich05] for details).

8.3 Numerical Verification of the Trace Formula

We consider the arithmetic trace formula stated in the introduction for the weight $k = 12$. Let $\Delta \in S_{12}$ and $f \in S_{22}$ be the unique primitive Hecke eigenforms of weight 12 and 22. Then we have

$$\Delta(z) = q - 24q^2 + 252q^3 - 1472q^4 + 4830q^5 - 6048q^6 + \cdots = \sum_{n=1}^{\infty} \tau(n) q^n$$

$$f(z) = q - 288q^2 - 128844q^3 - 2014208q^4 + 21640950q^5 + \cdots = \sum_{n=1}^{\infty} b(n) q^n.$$

The Petersson norm a Hecke eigenform $g \in S_k$ can be identified with a special value of the standard zeta function $D(g, s)$ of , this is due to Rankin. The correspondance is given by

$$\| g \|^2 = \frac{(k-1)!}{2^{2k-1}\pi^{k+1}} D(g, k). \tag{38}$$

The special value $D(g, k)$ can be determined by meromorphic continuation. There is a useful program of Dokchister [Do04] to calculate such values. This leads to

$$\| \Delta \|^2 = 0.000001035362056804320922 34\ldots$$
$$\| f \|^2 = 0.0000200998183232743064523 1\ldots$$

Our first goal is to determine the numerical value of the left side of the trace formula. The value of $\widehat{L}\left(f \otimes \mathrm{Sym}^2(\Delta), 22\right)$ can again be determined with the program of Dokchister (see also Ichino [Ich05]). We have

$$L(f, 21) = 0.99984994142583825995245 16\ldots$$
$$\widehat{L}(f, 21) = 84.200021524454436595006560 1\ldots$$
$$\widehat{L}\left(f \otimes \mathrm{Sym}^2(\Delta), 22\right) = 0.75704862297802829562086575\ldots$$

Hence,

$$\sum_{i=1}^{\dim S_{2k-2}} \frac{\widehat{L}(f_i, 2k-3)\ \widehat{L}\left(f_i \otimes \mathrm{Sym}^2(g), 2k-2\right)}{\| f_i \|^2 \| g \|^4} \tag{39}$$

for $k = 12$ is equal to the numerical value

$$295841675765246464 3.22953541\ldots \tag{40}$$

This number has been obtained directly. From the proof of the trace formula, we know that this number should actually be a rational number (see (67)).

A careful analysis leads to the candidate

$$\frac{2^{56} \cdot 3^6 \cdot 5^4 \cdot 7}{131 \cdot 593} \tag{41}$$

which coincides with 2958416757652464643.22953541... in the range of precision.

On the right side, we first determine the value of

$$\frac{\widehat{D}(g, 2k-2)}{\pi^{\frac{k}{2}-1} \| g \|^2} \tag{42}$$

for $g = \Delta$. We obtain directly

$$D(\Delta, 22) = 0.99964571112477139783572962\ldots$$

and hence,

$$\frac{\widehat{D}(\Delta, 22)}{\pi^5 \parallel \Delta \parallel^2} = 110841.73409677216384571824 0\ldots\,.$$

The constants κ_0, κ_1, and κ_2

$$\kappa_0(k) = (-1)^{k/2} 2^{k-2},$$
$$\kappa_1(k) = (-1)(-1)^{k/2} 2^4 \frac{\Gamma(k)^2}{(2k-2)B_{2k-2}\Gamma(k/2)^2},$$
$$\kappa_2(k) = (-1)(-1)^{k/2} 2^{2k+1} \frac{\Gamma(k+1)}{(2k-2)B_k \Gamma(k/2)}$$

are explicitly given. Let $k = 12$. Then

$$\kappa_0 = 10240 = 2^{10}$$
$$\kappa_1 = -12995908.891263210741088\ldots = \frac{(-1)\cdot 2^{14}\cdot 3^7\cdot 5^2\cdot 7^2\cdot 23}{131\cdot 593}$$
$$\kappa_2 = 24052904584483.936324167872648\ldots = \frac{2^{32}\cdot 3^5\cdot 5^2\cdot 7^2\cdot 13}{691}.$$

We determine the special value of the triple L-function $L(\Delta \otimes \Delta \otimes \Delta, s)$ at $s = 22$ via the local factors of the Euler product by calculating the Satake parameters of Δ. Hence, we obtain

$$L(\Delta \otimes \Delta \otimes \Delta, 22) = 0.99602837097824593011931492\ldots\,. \tag{43}$$

Then we obtain for $k = 12$,

$$\sum_{j=1}^{\dim S_k} \frac{\widehat{L}(g \otimes g \otimes g_j, 2k-2)}{\parallel g \parallel^4 \parallel g_j \parallel^2}$$

is equal to

$$441423252695906.208342030317\ldots\,.$$

So, finally, we have for the expression

$$\kappa_0 \cdot \frac{\widehat{L}(\Delta \otimes \Delta \otimes \Delta, 22)}{\parallel \Delta \parallel^6} + \kappa_1 \cdot \left(\frac{\widehat{D}(\Delta, 22)}{\pi^5 \parallel \Delta \parallel^2}\right)^2 + \kappa_2 \cdot \frac{\widehat{D}(\Delta, 22)}{\pi^5 \parallel \Delta \parallel^2}$$

the explicit value

$$2958416757652464643.22111654\ldots. \tag{44}$$

This shows that the arithmetic trace formula for the weight $k = 12$ can be numerically verified.

8.4 Proof of the Arithmetic Trace Formula

This section is devoted to the arithmetic trace formula stated in the introduction. We give a proof which is constructive and explicit. Moreover, as already remarked, we give a more general formula which may be useful for further applications.

Proof. We prove the theorem with an extension of a technique related to the doubling method in the setting of modular and Jacobi forms. There, the so-called big cell plays a fundamental role. It is related to the unique nonnegligible orbit which leads to an integral representation of an automorphic L-function. For our purpose, it is not enough to know one orbit; we need them all. Actually, we need the whole pullback formula related to the orbits. What does this mean? Let us fix the diagonal embedding $\mathbb{H} \times \mathbb{H} \hookrightarrow \mathbb{H}_2$. Here,

$$(Z, W) \mapsto \begin{pmatrix} Z & 0 \\ 0 & W \end{pmatrix}. \tag{45}$$

Similarly, we will make use of the diagonal embedding $\mathbb{H}\times\mathbb{H}\times\mathbb{H}$ into $\mathbb{H}_3$. Let $\left(g_j\right)_j$ be a Hecke eigenbasis of S_k with $a_1(g_j) = 1$, i.e., g_j is assumed to be primitive. This always exists. Garrett [Ga84] has discovered the following beautiful formula:

$$E_k^{(2)}|_{\mathbb{H}\times\mathbb{H}} = E_k \otimes E_k + \sum_{j=1}^{\dim S_k} d_j \; g_j \otimes g_j. \tag{46}$$

It had been well-known since the time of Witt that the restriction of a modular form of genus n on blocks of size $n_1 + \cdots + n_l = n$ is an element of $M_k^{(n_1)} \otimes \cdots \otimes M_k^{(n_l)}$. That the image in the case $n = 2$ is contained in the "diagonal" of a Hecke eigenbasis was surprising. Most important is that the numbers d_j have a significant arithmetic meaning. They are related to a critical value of the Rankin L-function. From this, we can deduce that these numbers are elements of K_{g_j} and are not zero. They can be explicitly determined:

$$d_j = \frac{(-1)^{\frac{k}{2}} 2^{3-k} \pi D(g_j, 2k-2)}{(k-1)\zeta(k)\zeta(2k-2) \parallel g_j \parallel^2}. \tag{47}$$

The situation in the case $3 = 1 + 1 + 1$ is different. Garrett [Ga87] computed the scalar product of the restricted Eisenstein series with three elliptic cusp forms.

A detailed analysis and combination of the two papers of Garrett (see also [He99]) leads to the complete pullback formula. We obtain

$$E_k^{(3)}|_{\mathbb{H}\times\mathbb{H}\times\mathbb{H}} = E_k \otimes E_k \times E_k + \sum_{j=1}^{\dim S_k} d_j\, E_k \times g_j \otimes g_j$$
$$+ \sum_{j=1}^{\dim S_k} d_j\, g_j \otimes E_k \otimes g_j + \sum_{j=1}^{\dim S_k} d_j\, g_j \otimes g_j \otimes E_k$$
$$+ \sum_{i,j,m=1}^{\dim S_k} l_{i,j,m}\, g_i \otimes g_j \otimes g_m. \tag{48}$$

Here, we have $l_{i,j,m} \in K^{\times}_{g_i,g_j,g_m}$, the composition field of K_{g_i}, K_{g_j}, and K_{g_m}. These numbers are essentially critical values of the triple L-function in the sense of Deligne. They had been first explicitly determined by Garrett [Ga87] (see also Mizumoto [Mi97], page 192, and Heim [He99], page 236, for the explicit value of the constants and further explanation):

$$l_{i,j,m} = (-1)^{\frac{k}{2}} \cdot 2^{-5k+8} \frac{\Gamma(k-1)^3}{\Gamma(k)}$$
$$\times \frac{\pi^{3-2k}\, L(g_i \otimes g_j \otimes g_m, 2k-2)}{\zeta(2k-2)\,\zeta(k)\, \| g_i \|^2 \| g_j \|^2 \| g_m \|^2}. \tag{49}$$

Here, we would like to remark that all three cusp forms have the same weight. For a more general formula allowing different weights, one has to use differential operators. Moreover, the big cell is related to

$$\sum_{i,j,m=1}^{\dim S_k} l_{i,j,m}\, g_i \otimes g_j \otimes g_m.$$

But we will see immediately that one also needs one of the negligible orbits for the trace formula.

The next step is to extract the first coefficient of the Fourier expansion with respect to the third variable. It is important that this procedure is the same as starting with a Fourier-Jacobi expansion of the involved Siegel Eisenstein series, then extracting the first coefficient, and then restricting the domain $\mathbb{H}_2 \times \mathbb{C}^2$ to $\mathbb{H} \times \mathbb{H}$. Let B_k be the k-th Bernoulli number. Then we have

$$\frac{-2k}{B_k} E_k \otimes E_k + \sum_{j=1}^{\dim S_k} d_j\, E_k \otimes g_j + \sum_{j=1}^{\dim S_k} d_j\, g_j \otimes E_k$$
$$+ \frac{-2k}{B_k} \sum_{j=1}^{\dim S_k} d_j\, g_j \otimes g_j + \sum_{i,j,m=1}^{\dim S_k} l_{i,j,m}\, g_i \otimes g_j. \tag{50}$$

Here, we would like to mention that it turns out to be very convenient to have normalized our Siegel Eisenstein series, such that the 0th coefficient is always one since it is compatible with restricting the Eisenstein series to the diagonal.

Let $\delta_{i,j} = 1$ if $i = j$ and 0 otherwise. Then the coefficient of the basis element $g_i \otimes g_j \in S_k \otimes S_k$ is given by

$$\boxed{\delta_{ij}\, d_j \cdot \frac{-2k}{B_k} + \sum_{m=1}^{\dim S_k} l_{i,j,m}}. \tag{51}$$

Now, we do something which we have not found yet in the literature. We determine a second pullback formula of our Eisenstein series, with respect to a not obvious embedding of the Jacobi spaces $\mathbb{H}^J \times \mathbb{H}^J$ into $\mathbb{H}_3$ and obtain something new. Here, $\mathbb{H}^J := \mathbb{H} \times \mathbb{C}$. We start by looking directly at the Fourier-Jacobi expansion of the Eisenstein series of genus 3. It is convenient to parameterize elements of $\mathbb{H}_3$ in the following way:

$$Z = \begin{pmatrix} \tau_1 & z & z_1 \\ z & \tau_2 & z_2 \\ z_1 & z_2 & \tau_3 \end{pmatrix}. \tag{52}$$

We fix the diagonal embedding $\mathbb{H}^J \times \mathbb{H}^J \hookrightarrow \mathbb{H}_3$ given by

$$(\tau_1, z_1), (\tau_2, z_2) \mapsto \begin{pmatrix} \tau_1 & 0 & z_1 \\ 0 & \tau_2 & z_2 \\ z_1 & z_2 & \tau_3 \end{pmatrix}. \tag{53}$$

With this notation, the Fourier-Jacobi expansion of $E_k^{(3)}(Z)$ with respect to τ_3 is given by

$$E_k^{(3)}(Z) = \sum_{n=0}^{\infty} e_{k,n}^{(3)} \left(\left(\begin{smallmatrix} \tau_1 & z \\ z & \tau_2 \end{smallmatrix} \right), (z_1, z_2) \right) \mathrm{e}^{2\pi i n \tau_3}. \tag{54}$$

The Fourier-Jacobi coefficients are Jacobi forms on $\mathbb{H}_2 \times \mathbb{C}^2$ of weight k and index n. By switching to Jacobi Eisenstein series and having a "compatible" normalization, we normalize the Jacobi Eisenstein series in such a way that the 0-th Fourier coefficient is equal to 1. In this case, we have

$$E_{k,n}^{J,2} \left(\left(\begin{smallmatrix} \tau_1 & z \\ z & \tau_2 \end{smallmatrix} \right), (z_1, z_2) \right) = \frac{B_k}{-2k\, \sigma_{k-1}(n)}\, e_{k,n}^{(3)} \left(\left(\begin{smallmatrix} \tau_1 & z \\ z & \tau_2 \end{smallmatrix} \right), (z_1, z_2) \right). \tag{55}$$

Here, $\sigma_{k-1}(n) := \sum_{d|n} d^{k-1}$. Let $\left(\Phi_j\right)_j$ be a normalized Hecke eigenbasis of $J_{k,1}^{\text{cusp}}$, i.e., a Hecke eigenbasis such that all the Fourier coefficients are contained in the field K_{Φ_j} generated by all the eigenvalues. Let $f_j \in S_{2k-2}$ be primitive and correspond to Φ_j via the Shimura correspondance. Then $\left(f_j\right)_j$ is a Hecke eigenbasis of S_{2k-2} with the same eigenvalues. Obviously, we have $K_{f_j} = K_{\Phi_j}$. Arakawa [Ar94] found out that also in the setting of Jacobi forms the doubling method has a certain interpretation. But it turned out that the underlying Hecke-Jacobi theory is much more complicated as expected [AH98, He01]. But anyway, some results can be obtained. We deduce from [Ar94]

$$E_{k,1}^{J,2}|_{\mathbb{H}^J \times \mathbb{H}^J} = E_{k,1}^J \otimes E_{k,1}^J + \sum_{m=1}^{\dim J_{k,1}^{\text{cusp}}} \alpha_m \Phi_m \otimes \Phi_m. \tag{56}$$

Here, $E_{k,1}^J$ is the Jacobi-Eisenstein series of weight k and index 1 on $\mathbb{H} \times \mathbb{C}$ as introduced in [EZ85]. The numbers α_j are related to the critical values of the Hecke L-function attached to f_j. We have

$$\alpha_m = \frac{(-1)^{k/2}\pi\, 2^{1-k}}{(k-3/2)} \frac{L(f_m, 2k-3)}{\parallel \Phi_m \parallel^2 \zeta(2k-2)}. \tag{57}$$

For details, see [Ar94, He01]. Since up to normalization $E_{k,1}^J$ is the first Fourier-Jacobi coefficient of $E_k^{(2)}$ and these Eisenstein series are in the Maass Spezialschar, we have

$$E_{k,1}^J|_{\mathbb{H}} = E_k + \frac{B_k}{-2k} \sum_{j=1}^{\dim S_k} d_j\, g_j. \tag{58}$$

This formula can be deduced from the fact that the Siegel Eisenstein series of genus 2 is an element from the so-called Maass Spezialschar. It is then an easy exercise to obtain the formula. Further, we have formally that

$$\Phi_m|_{\mathbb{H}} = \sum_{j=1}^{\dim S_k} \gamma_j^m\, g_j. \tag{59}$$

From the arithmetic of the Fourier coefficients of the Jacobi form, we can deduce that γ_j^m are totally real algebraic numbers. Let now h_m be the modular form of half-integral weight directly related to the Jacobi form Φ_m via the isomorphism given in [EZ85], Theorem 5.4 (see also Sect. 8.2.2). Then we can combine Proposition 4.3 given in [He98] and the explicit description of Ichino [Ich05] of the square of the pullback of a Saito-Kurokawa lift. Again, by a straightforward calculation, we get

$$\left(\gamma_j^m\right)^2 = 2^{-k} \frac{\parallel h_m \parallel^2}{\parallel f_m \parallel^2 \parallel g_j \parallel^4} \widehat{L}\left(f_m \otimes \mathrm{Sym}^2(g_j), 2k-2\right). \tag{60}$$

Hence, we obtain the expression

$$\boxed{\frac{B_k}{-2k} d_i \cdot d_j + \frac{-2k}{B_k} \sum_{m=1}^{\dim S_{2k-2}} \alpha_m\, \gamma_i^m\, \gamma_j^m.} \tag{61}$$

for the coefficients of $g_i \otimes g_j$ in the pullback formula of $\frac{-2k}{B_k} E^{J,2}_{k,1}|_{\mathbb{H}\times\mathbb{H}}$. In the next step, we compare the two pullback formulas one in the setting of modular forms and the other deduced from the work of Arakawa in the setting of Jacobi forms. This leads to

$$\boxed{\delta_{ij}\, d_j \cdot \frac{-2k}{B_k} + \sum_{m=1}^{\dim S_k} l_{i,j,m} = \frac{B_k}{-2k}\, d_i \cdot d_j + \frac{-2k}{B_k} \sum_{m=1}^{\dim S_{2k-2}} \alpha_m\, \gamma_i^m\, \gamma_j^m .} \tag{62}$$

This formula is the heart of our approach. It contains much more information than we use at the moment. To prove the trace formula, we restrict ourself to the case $i = j$. We want to mention that if $i \neq j$, then on one side, the formula simplifies because the summand $\delta_{ij}\, d_j \cdot \frac{-2k}{B_k}$ disappears. But, on the other side, we only know the value of $\left(\gamma_i^m\right)^2$ which is a totally real algebraic number. So still, the delicate question of the sign of the root remains open. Nevertheless, we obtain from (24) and (47) the explicit formula

$$d_j = -\,2^{5-2k} \cdot \frac{\Gamma(k+1)}{\Gamma(k/2)} \cdot \frac{1}{B_k\, B_{2k-2}} \cdot \frac{\widehat{D}(g_j, 2k-2)}{\pi^{\frac{k}{2}-1} \parallel g_j \parallel^2}. \tag{63}$$

Moreover, from (27) and (49), we obtain

$$l_{j,j,m} = -\,2^{3-3k} \cdot \frac{k \cdot (2k-2)}{B_k\, B_{2k-2}} \frac{\widehat{L}(g_j \otimes g_j \otimes g_m, 2k-2)}{\parallel g_j \parallel^2 \parallel g_j \parallel^2 \parallel g_m \parallel^2}. \tag{64}$$

And from (23) and (57), we obtain

$$\alpha_j = (-1)^{\frac{k}{2}}\, 2^{1-k} \cdot \frac{2k-2}{B_{2k-2}} \frac{\widehat{L}(f_j, 2k-3)}{\parallel \Phi_j \parallel^2}. \tag{65}$$

Let $h_j \in S^+_{k-\frac{1}{2}}(\Gamma_0(4))$ be normalized and related to $\Phi_j \in J^{\mathrm{cusp}}_{k,1}$ via the isomorphism given in Theorem 5.4 in ([EZ85]. Then we obtain, for example, from ([KS89], Sect. 2), the transformation law for the square of the norms given by $\parallel \Phi_j \parallel^2 = 2^{2k-3} \parallel h_j \parallel^2$. This leads to

$$\alpha_j = (-1)^{\frac{k}{2}}\, 2^{4-3k} \cdot \frac{2k-2}{B_{2k-2}} \frac{\widehat{L}(f_j, 2k-3)}{\parallel h_j \parallel^2}. \tag{66}$$

Then we have

$$\alpha_j \left(\gamma_j^m\right)^2 = \kappa\, \frac{\widehat{L}(f_m, 2k-3)\widehat{L}\left(f_m \otimes \mathrm{Sym}^2(g_j), 2k-2\right)}{\parallel f_m \parallel^2 \cdot \parallel g_j \parallel^4}. \tag{67}$$

Here, $\kappa = (-1)^{\frac{k}{2}} 2^{4-4k} \frac{2k-2}{B_{2k-2}}$. Summarizing everything and plugging into (62), this leads to

$$-\frac{-2k}{B_k}\frac{\Gamma(k+1)}{\Gamma(k/2)} \cdot \frac{2^{5\text{-}2k}}{B_k B_{2k-2}} \frac{\widehat{D}(g_j, 2k-2)}{\pi^{\frac{k}{2}-1} \parallel g_j \parallel^2}$$

$$-\, 2^{3-3k} \frac{k \cdot (2k-2)}{B_k\, B_{2k-2}} \sum_{t=1}^{\dim S_k} \frac{\widehat{L}(g_j \otimes g_j \otimes g_t, 2k-2)}{\parallel g_j \parallel^2 \parallel g_j \parallel^2 \parallel g_t \parallel^2}$$

$$= \frac{B_k}{-2k} 2^{10-4k} \frac{\Gamma(k+1)^2}{\Gamma(k/2)^2} \cdot \frac{1}{B_k^2 B_{2k-2}^2} \left(\frac{\widehat{D}(g_j, 2k-2)}{\pi^{\frac{k}{2}-1} \parallel g_j \parallel^2} \right)^2$$

$$+ \, (-1)^{\frac{k}{2}} \frac{-2k}{B_k} 2^{4-4k} \frac{2k-2}{B_{2k-2}} \sum_{m=1}^{\dim S_{2k-2}} \frac{\widehat{L}(f_m, 2k-3) \widehat{L}\left(f_m \otimes \mathrm{Sym}^2(g_j), 2k-2\right)}{\parallel f_m \parallel^2 \cdot \parallel g_j \parallel^4}. \tag{68}$$

Finally, we obtain by a straightforward calculation the desired result.

References

[Ar94] T. Arakawa: *Jacobi Eisenstein series and a basis problem for Jacobi forms.* Comm. Mathematici Universitatis Sancti Pauli. **43** (1994), 181–216

[AH98] T. Arakawa, B. Heim: *Real analytic Jacobi Eisenstein series and Dirichlet series attached to three Jacobi forms.* Max-Planck-Institut Bonn. Series **66** (1998)

[De71] P. Deligne: *Formes modulaires et representations l-adic.* Lect. Notes Math. **179** (1971), Berlin-Heidelberg-New York, 139–172.

[De79] P. Deligne: *Valeurs de fonctions L et periode d'integrales.* Proc. Symposia Pure Math. **33** (1979), part 2, 313–346.

[Do04] T. Dokchitser: *Computing special values of motivic L-functions.* Exp. Math. **13** (2004), 137–149.

[EZ85] M. Eichler, D. Zagier: *The theory of Jacobi forms. Progress in Mathematics.* **Vol. 55**. Boston-Basel-Stuttgart: Birkhäuser (1985).

[Ga84] P. Garrett: *Pullbacks of Eisenstein series; applications.* Automorphic forms of several variables (Katata, 1983), 114–137, Progr. Math., **46** Birkhäuser Boston, Boston, MA, 1984.

[Ga87] P. Garrett: Decomposition of Eisenstein series: triple product L-functions. Ann. Math. **125** (1987), 209–235.

[Ge06] G. van der Geer: *Siegel modular forms.* manuscript on arXiv:math.AG/0605346 v1

[He98] B. Heim: *Über Poincare Reihen und Restriktionsabbildungen.* Abh. Math. Sem. Univ. Hamburg. **68** (1998), 79–89

[He99] B. Heim: *Pullbacks of Eisenstein series, Hecke-Jacobi theory and automorphic L-functions.* In: Automorphic Froms, Automorphic Representations and Arithmetic. Proceedings of Symposia of Pure Mathematics **66**, part 2 (1999).

[He01] B. Heim: *L-functions for Jacobi forms and the basis problem.* Manuscripta Math. **106** (2001), 489–503.

[Hei05] B. Heim: *On injectivity of the Satoh lifting of modular forms and the Taylor coefficients of Jacobi forms. Preprint* MPI-Bonn Series **68**, 2005

[Ich05] A. Ichino: *Pullbacks of Saito-Kurokawa Lifts.* Invent. Math. **162** (2005), 551–647.

[Ik01] T. Ikeda: *On the lifting of elliptic cusp forms to Siegel cusp forms of degree* $2n$. Ann. of Math. **154** no. 3 (2001), 641–681.

[Ik06] T. Ikeda: *Pullback of the lifting of elliptic cusp forms and Miyawakis conjecture.* Duke Math. Journal, **131** no. 3 (2006), 469–497.

[Iw97] H. Iwaniec: *Topics in classical automorphic forms.* **Vol. 17** Graduate Studies in mathematics, AMS (1997).

[IS02] H. Iwaniec and P. Sarnak: *Perspectives on the analytic theory of L-functions.* GAFA 2000 (Tel Aviv, 1999), Geom. Funct. Anal. (2000), Special Volume, Part II, 705–741.

[KK06] H. Katsurada, Hisa-aki Kawamura: *A certain Dirichlet series of Rankin-Selberg type associated with the Ikeda lifting.* Department of Mathemtics Hokkaido University, series 2006, number 806

[Kl62] H. Klingen: *Über die Werte der Dedekindschen Zetafunktion.* Math. Ann. **145** (1962), 265–272.

[Kl90] H. Klingen: *Introductory lectures on Siegel modular forms.* Cambridge Studies in Advanced Mathematics, **20**. Cambridge University Press, Cambridge, 1990.

[Ko80] W. Kohnen: *Modular forms of half-integral weight on* $\Gamma_0(4)$. Math. Ann. **248** (1980), 249–266.

[KS89] W. Kohnen, N.-P. Skoruppa: *A certain Dirichlet series attached to Siegel modular forms of degree two.* Invent. Math. **95** (1989), 449–476

[Ma79] H. Maass: *Über eine Spezialschar von Modulformen zweiten Grades I,II,III.* Invent. Math. **52, 53, 53** (1979), 95–104, 249–253, 255–265.

[Mi97] S. Mizumoto: *Nearly Holomorphic Eisenstein Liftings.* Abh. Math. Sem. Univ. Hamburg **67** (1997), 173–194,

[Or87] T. Orloff: *Special values and mixed weight triple products (with an appendix by Don Blasius).* Invent. Math. **90** (1987).

[Sh73] G. Shimura: *On modular forms of half-integral weight.* Ann. of Math. **97** (1973), 440–481.

[Sh76] G. Shimura: *The special values of the zeta functions associated with cusp forms.* Comm. pure appl. Math. **29** (1976), 783–804.

[Sh95] G. Shimura: *Eisenstein series and zeta functions on symplectic groups.* Inv. Math. **119** (1995), 539–584.

[Si69] C.L. Siegel: *Berechnung von Zetafunktionen an ganzzahligen Stellen.* Nachr. Akad. Wiss. Göttingen (1969), 87–102.

[Za77] D. Zagier: *Modular forms whose Fourier coefficients involve zeta-functions of quadratic fields.* Modular functions of one variable, VI (Proc. Second Internat. Conf., Univ. Bonn, Bonn, 1976), pp. 105–169. Lecture Notes in Math., **Vol. 627**, Springer, Berlin, 1977.

[Za80] D. Zagier: *Sur la conjecture de Saito-Kurokawa (d'après H. Maass).* Sém. Delange-Pisot-Poitou 1979/1980, Progress in Math. **12** (1980), 371–394

Chapter 9
The Adjoint L-Function of $SU_{2,1}$

Joseph Hundley

To the memory of my grandfather, Harold H. Hensold, Jr.

Abstract We modify Ginzburg's construction for the adjoint L-function of $GL(3)$ to accommodate quasi-split unitary groups.

Keywords Adjoint L-function of SU(2,1) • Rankin–Selberg method • Exceptional group G_2

In these notes, we give a construction for a certain L-function attached to a globally generic automorphic representation of the quasi-split unitary group in 3 variables associated to a quadratic extension E/F of number fields. Recall that the finite Galois form of the L-group of this group is a semidirect product of $GL_3(\mathbb{C})$ and $\mathrm{Gal}(E/F)$. The representation we consider has the property that when restricted to $GL_3(\mathbb{C})$, it is the adjoint representation of this group. For this reason, we refer to the associated L-function as the adjoint L-function. In fact, there are two representations of $GL_3(\mathbb{C}) \rtimes \mathrm{Gal}(E/F)$ with the above property—related to one another by twisting by the unique nontrivial one-dimensional representation of $\mathrm{Gal}(E/F)$. We pin down precisely which one we are talking about in Sect. 9.1.1. Let us mention that a small modification of this construction gives the other.

The construction is a slight modification of that given in [3].

J. Hundley (✉)
Department of Mathematics, Mailcode 4408, Southern Illinois University, 1245 Lincoln Drive, Carbondale, IL 62901, USA
e-mail: jhundley@math.siu.edu

D. Bump et al. (eds.), *Multiple Dirichlet Series, L-functions and Automorphic Forms*, Progress in Mathematics 300, DOI 10.1007/978-0-8176-8334-4_9,

9.1 Notation

Let F be a global field, and $\mathbb{A}$ its ring of adèles. Let $E = F(\tau)$ be a quadratic extension, such that $\rho := \tau^2 \in F$. Let $J = \begin{pmatrix} & & 1 \\ & 1 & \\ 1 & & \end{pmatrix}$. Abusing notation, we will also denote by J the analogous matrix of any size with points in any ring (with unity). The F points of our special unitary group may be thought of as the set of 3×3 matrices with determinant 1 with entries in E such that $gJ\,{}^t\bar{g} = J$. Here, $\bar{}$ denotes conjugation by the nontrivial element of $\mathrm{Gal}(E/F)$. Presently, we shall also identify this group with a group of matrices having entries in F. We denote this group by $SU_{2,1}$.

We consider also the split exceptional group of type G_2 defined over F, which we denote simply by G_2. We recall a few facts about this group (cf. [4], pp. 350–57). First, it may be realized as the identity component of the group of automorphisms of a seven-dimensional vector space which preserve a general skew-symmetric trilinear form. Second, this seven-dimensional "standard" representation of G_2 is orthogonal: the image also preserves a symmetric bilinear form. We wish now to pin things down explicitly. It will be convenient to realize G_2 as a subgroup of SO_8.

Thus, we consider $SO_8 = \{g \in GL_8 : gJ\,{}^tg = J\}$. Let

$$v_0 = {}^t(0,0,0,1,-1,0,0,0).$$

By SO_7, we mean the stabilizer of v_0 in SO_8. Let V_0 denote the orthogonal complement of v_0, defined relative to J. To fix an embedding of G_2, into SO_7, we fix a trilinear form of V_0 in general position, namely:

$$\begin{aligned} T := e_7^* \wedge (e_4^* + e_5^*) \wedge e_2^* + e_1^* \wedge (e_4^* + e_5^*) \wedge e_8^* \\ + e_6^* \wedge (e_4^* + e_5^*) \wedge e_3^* + 2e_3^* \wedge e_2^* \wedge e_8^* - 2e_6^* \wedge e_7^* \wedge e_1^* \end{aligned}$$

(which is obtained from the form written down on p. 357 of [4] via suitable identifications). The identity component of the stabilizer of T in $GL(V_0)$ is a group of type G_2, defined and split over F, and contained in SO_7 as defined above.

Now, let $v_\rho = {}^t(0,0,1,0,0,\rho,0,0)$, and let H_ρ denote the stabilizer of v_ρ in G_2.

Lemma 1 $H_\rho \cong SU_{2,1}$.

Remarks 2 *1. This is essentially the same embedding of $SU_{2,1}$ into G_2 described on p. 371 of [1].*

2. One may also obtain this embedding by making the following identifications between an F-basis of E^3 and one for the orthogonal complement of $\langle v_0, v_\rho\rangle$ in F^8.

$$\begin{array}{lll} (1,0,0) \leftrightarrow e_1 & (-\tau^{-1},0,0) \leftrightarrow e_2 & (0,-2\tau,0) \leftrightarrow e_3 - \rho e_6 \\ (0,-2,0) \leftrightarrow e_4 + e_5 & (0,0,2\tau) \leftrightarrow e_7 & (0,0,2) \leftrightarrow e_8 \end{array}$$

Proof. On the one hand, we know from [6], pp. 808–810 that the stabilizer of a vector in this representation having nonzero length (relative to J) is isomorphic to either SL_3 or $SU(Q)$ for a suitable Q. On the other hand, H_ρ is clearly contained in the group of automorphisms of the six-dimensional complement of v_ρ in V_0 which preserve both the original symmetric bilinear form and the skew-symmetric form obtained by plugging in v_ρ as one of the arguments of T. This latter group is isomorphic to $U_{2,1}$ (with an isomorphism being given by the identification of bases above). The result follows. □

To aid in visualizing these groups and checking various assertions below, we write down a general element of each of their Lie algebras.

$$G_2 : \begin{pmatrix} T_1 & a & c & d & d & e & f & 0 \\ g & T_2-T_1 & b & -c & -c & d & 0 & -f \\ h & l & 2T_1-T_2 & a & a & 0 & -d & -e \\ i & -h & g & 0 & 0 & -a & c & -d \\ i & -h & g & 0 & 0 & -a & c & -d \\ j & i & 0 & -g & -g & T_2-2T_1 & -b & -c \\ k & 0 & -i & h & h & -l & T_1-T_2 & -a \\ 0 & -k & -j & -i & -i & -h & -g & -T_1 \end{pmatrix} \tag{1}$$

$$SU_{2,1} : \begin{pmatrix} T_1 & a & -\rho e & d & d & e & f & 0 \\ \rho a & T_1 & -\rho d & \rho e & \rho e & d & 0 & -f \\ h & l & 0 & a & a & 0 & -d & -e \\ -\rho l & -h & \rho a & 0 & 0 & -a & -\rho e & -d \\ -\rho l & -h & \rho a & 0 & 0 & -a & -\rho e & -d \\ -\rho h & -\rho l & 0 & -\rho a & -\rho a & 0 & \rho d & \rho e \\ k & 0 & \rho l & h & h & -l & -T_1 & -a \\ 0 & -k & \rho h & \rho l & \rho l & -h & -\rho a & -T_1 \end{pmatrix} \tag{2}$$

The set of upper triangular matrices in G_2 is a Borel subgroup B_{G_2}, and the set of diagonal matrices in G_2 is a maximal torus T_{G_2}. We use this torus and Borel to define notions of "standard" for parabolics and Levis. We also fix a maximal compact subgroup $K = \prod_v K_v$ of $G_2(\mathbb{A})$ such that $G_2(F_v) = B_{G_2}(F_v)K_v$ for all v and $K_v = G_2(\mathfrak{o}_v)$ for almost all finite v. (Here, $\mathfrak{o}_v$ denotes the ring of integers of F_v.)

For any matrix A, we let ${}_tA$ denote the "other transpose" $J\,{}^tAJ$, obtained by reflecting A over the diagonal that runs from upper right to lower left. Finally, if H is any F group, then $H(F\backslash\mathbb{A}) := H(F)\backslash H(\mathbb{A})$.

9.1.1 The Representation r

Let us now describe explicitly the representation r which appears in the Langlands L-function we will construct. We first describe the L-group we consider, which is the finite Galois form of the L-group of $U_{2,1}(E/F)$. Let Fr denote the nontrivial element of Gal(E/F). Our L-group is $GL_3(\mathbb{C}) \rtimes \mathrm{Gal}(E/F)$, where the semidirect product structure is such that

$$\mathrm{Fr} \cdot g \cdot \mathrm{Fr} = {}_t g^{-1}. \tag{3}$$

Now, consider the eight-dimensional complex vector space of 3×3 traceless matrices, with an action of $GL_3(\mathbb{C})$ by conjugation. The definition

$$\mathrm{Fr} \cdot X = {}_t X \tag{4}$$

extends this to a well-defined action of $GL_3 \rtimes \mathrm{Gal}(E/F)$. This is our representation r.

It is not difficult to check that there is only one other way to define the action of Fr which is compatible with (3) and (4), namely $\mathrm{Fr} \cdot X = -{}_t X$. Now, it is part of the L-group formalism that the parameter of π at an unramified place v is in the identity component iff ρ is a square in the completion of F at v. Hence, if we let r' denote the representation corresponding to the action $\mathrm{Fr} \cdot X = -{}_t X$, then $L^S(s, \pi, r')$ is the twist of $L^S(s, \pi, r)$ by the quadratic character corresponding to the extension E/F. An integral for this L-function may be obtained from the one considered in this chapter by inserting this character into the induction data for the Eisenstein series.

9.1.2 *Eisenstein Series*

We shall make use of the same Eisenstein series on $G_2(F\backslash\mathbb{A})$ as in [3]. We recall the definition. Let P denote the standard maximal parabolic of G_2 such that the short simple root of G_2 is a root of the Levi factor of P. Take f a K-finite flat section of the fiber bundle of representations $\mathrm{Ind}_{P(\mathbb{A})}^{G_2(\mathbb{A})} |\delta_P|^s$ (nonnormalized induction). Thus, for each s, $f(g,s)$ is a function $G_2(\mathbb{A}) \to \mathbb{C}$ such that $f(pg,s) = |\delta_P(p)|^s f(g,s)$ for all $p \in P(\mathbb{A}), g \in G_2(\mathbb{A})$, and for $k \in K$, the value of $f(k,s)$ is independent of s. The associated Eisenstein series is defined by the formula

$$E(g,s) = \sum_{\gamma \in P(F)\backslash G_2(F)} f(\gamma g, s)$$

for $\Re(s)$ sufficiently large, and by meromorphic continuation elsewhere.

9.2 Unfolding

Lemma 3 *The space $P(F)\backslash G_2(F)/SU_{2,1}(F)$ has two elements, represented by the identity and (any representative in $G_2(F)$ for) the simple reflection in the Weyl group of G_2 associated to the long simple root, which we denote w_2.*

Proof. This follows easily from our characterization of $SU_{2,1}$ as a stabilizer. Indeed, $P(F)\backslash G_2(F)/SU_{2,1}(F)$ may be identified with the set of $P(F)$ orbits in the $G_2(F)$ orbit of v_ρ. Write $v \in G_2(F)v_\rho$ as ${}^t(v_1, v_2, v_3)$ with $v_1, v_3 \in F^2$ and $v_2 \in F^4$. Either $v_3 = 0$ or not. This distinction clearly separates $P(F)$ orbits, and in particular separates the $P(F)$ orbit of the identity from that of w_2. On the other hand, an element of the $G_2(F)$ orbit of v_ρ is certainly in V_0 and of norm 2ρ. It is not hard to check that $P(F)$ permutes the set of such elements with $v_3 = 0$ and $v_3 \neq 0$ each transitively. □

Let φ_π be a cusp form in the space of an irreducible automorphic cuspidal representations π of $SU_{2,1}(\mathbb{A})$. Let N denote the maximal unipotent subgroup of $SU_{2,1}$

$$\left\{\begin{pmatrix} 1 & x & y \\ & 1 & -\bar{x} \\ & & 1 \end{pmatrix} : x, y \in E, \operatorname{Tr} y + \operatorname{Norm} x = 0\right\}.$$

Fix a nontrivial additive character ψ of $(F\backslash\mathbb{A})$, and let ψ_N denote the character of $N(\mathbb{A})$ with coordinates as above to $\psi\left(\frac{1}{2}\operatorname{Tr} x\right)$. (The $\frac{1}{2}$ is for convenience: it cancels the 2 that arises when we take the trace of an element of F.)

We assume that the integral

$$W_{\varphi_\pi}(g) := \int_{N(F\backslash\mathbb{A})} \varphi_\pi(ng)\psi_N(n)\, dn \tag{5}$$

does not vanish identically. (And hence, that π is generic.) We consider the integral

$$I(\varphi_\pi, f, s) := \int_{SU_{2,1}(F\backslash\mathbb{A})} \varphi_\pi(g)E(g, s).$$

Theorem 4 (The Unfolding) *Let N_2 denote the two-dimensional unipotent subgroup of $SU_{2,1}$ corresponding to the coordinates e and f in* (2). *Then for $\Re(s)$ sufficiently large,*

$$I(\varphi_\pi, f, s) = \int_{N_2(\mathbb{A})\backslash SU_{2,1}(\mathbb{A})} W_{\varphi_\pi}(g) f(w_2 g, s) dg. \tag{6}$$

Proof. By the lemma, we find that $I(\varphi_\pi, f, s)$ is equal to

$$\int_{(SU_{2,1}\cap P)(F)\backslash SU_{2,1}(\mathbb{A})} \varphi_\pi(g) f(g, s) dg$$

$$+ \int_{(SU_{2,1}\cap w_2 P w_2)(F)\backslash SU_{2,1}(\mathbb{A})} \varphi_\pi(g) f(w_2 g, s) dg.$$

The first integral vanishes by the cuspidality of π.

The group $SU_{2,1} \cap w_2 P w_2$ consists of the one-dimensional F-split torus and the two-dimensional unipotent group N_2. Incidentally, when written as elements of $GL_3(E)$, this unipotent group is

$$\left\{ \begin{pmatrix} 1 & r\tau & t\tau + \frac{r^2\rho}{2} \\ & 1 & r\tau \\ & & 1 \end{pmatrix} : r, t \in F \right\}.$$

We now expand φ_π along the subgroup of elements of the form

$$\begin{pmatrix} 1 & s & -\frac{s^2}{2} \\ & 1 & -s \\ & & 1 \end{pmatrix}.$$

The constant term vanishes by cuspidality. The remaining terms are permuted simply transitively by the action of the F-split torus. The term corresponding to 1 yields the integral (5). □

9.3 Unramified Computations

We now consider the value of the local analogue of (6) at a place where all data is unramified. Thus, let F be a nonarchimedean local field. We denote the nonarchimedean valuation on F by v and the cardinality of the residue field by q. We keep the definitions of all the algebraic groups above. However, we now allow the possibility that ρ is a square in F. In this case, the group $SU_{2,1}$ defined by the equations above is isomorphic to SL_3 over F, and what we are proving is essentially the local result proved in [3]. We assume that ρ and 2 are both units in F. In this section, we encounter only the F points of algebraic groups, so we suppress the "(F)."

Let f be the spherical vector in the induced representation $\mathrm{Ind}_P^{G_2} |\delta_P|^s$, and let W denote the normalized spherical vector in the Whittaker model of an unramified local representation π of GL_3. The integral we consider is

$$I(s,\pi) = \int_{N_2 \backslash SU_{2,1}} W(g) f(w_2 g, s) \mathrm{d}g.$$

The main result of this section is the following:

Proposition 5 *For $\Re(s)$ sufficiently large,*

$$I(s,\pi) = \frac{L(3s-1,\pi,r)}{\zeta(3s)\zeta(6s-2)\zeta(3s-9)}.$$

Here, all zeta and L-functions are local. Thus, if q is the number of elements in the residue field of F, then $\zeta(3s) = (1 - q^{-3s})^{-1}$, etc.

Proof. We begin with some computations which are valid regardless of whether or not ρ is a square in F. The one-dimensional subgroup of $SU_{2,1}$ corresponding to the variable d in (2) maps isomorphically onto the quotient $N_2\backslash N$. An element of this group may also be expressed as $x_{\alpha_2}(\rho u)x_{2\alpha_1+\alpha_2}(-u)$, where x_{α_2} and $x_{2\alpha_1+\alpha_2}$ are maps of $\mathbb{G}_a$ onto the one-parameter unipotent subgroups of G_2 corresponding to the indicated roots. These subgroups correspond to the variables b and d in (1). Using the Iwasawa decomposition, we express the integral over $N_2\backslash SU_{2,1}$ as integrals over the maximal compact K, the torus T, and this one-dimensional subgroup. Since W and f are spherical, and the volume of K is 1, we may erase the integral over K. Also $w_2 x_{2\alpha_1+\alpha_2}(u) \in P$. Hence, we find that

$$I(s,\pi) = \int_T \int_F f(w_2 x_{\alpha_2}(\rho u)t, s)\psi_N(u)\mathrm{d}u W(t)\delta_{\mathrm{B}}^{-1}(t)\mathrm{d}t.$$

Here, δ_{B} denotes the modular quasi-character of the Borel subgroup of $SU_{2,1}$. Now, an element of T may be visualized as an element of $SL_3(F(\sqrt{\rho}))$ of the form

$$\begin{pmatrix} a+b\sqrt{\rho} & & \\ & \frac{a-b\sqrt{\rho}}{a+b\sqrt{\rho}} & \\ & & \frac{1}{a-b\sqrt{\rho}} \end{pmatrix}.$$

The corresponding 8×8 matrix is

$$\begin{pmatrix} a & -b & & & & & & \\ -b\rho & a & & & & & & \\ & & \frac{a^2}{N} & -\frac{ab}{N} & -\frac{ab}{N} & -\frac{b^2}{N} & & \\ & & -\frac{ab\rho}{N} & \frac{a^2}{N} & \frac{b^2\rho}{N} & \frac{ab}{N} & & \\ & & -\frac{ab\rho}{N} & \frac{b^2\rho}{N} & \frac{a^2}{N} & \frac{ab}{N} & & \\ & & -\frac{b^2\rho^2}{N} & \frac{ab\rho}{N} & \frac{ab\rho}{N} & \frac{a^2}{N} & & \\ & & & & & & \frac{a}{N} & \frac{b}{N} \\ & & & & & & \frac{b\rho}{N} & \frac{a}{N} \end{pmatrix}, \quad \text{where } N := a^2 - b^2\rho.$$

We now write the Iwasawa decomposition for this as an element of G_2. First, suppose that $|b\rho| \leq |a|$. Then the decomposition is

$$
\begin{pmatrix}
1 & -\frac{b}{a} & & & & & & \\
 & 1 & & & & & & \\
 & & 1 & -\frac{b}{a} & -\frac{b}{a} & -\frac{b^2}{a^2} & & \\
 & & & 1 & & \frac{b}{a} & & \\
 & & & & 1 & \frac{b}{a} & & \\
 & & & & & 1 & & \\
 & & & & & & 1 & \frac{b}{a} \\
 & & & & & & & 1
\end{pmatrix}
\begin{pmatrix}
\frac{N}{a} & & & & & & & \\
 & a & & & & & & \\
 & & \frac{N}{a^2} & & & & & \\
 & & & 1 & & & & \\
 & & & & 1 & & & \\
 & & & & & \frac{a^2}{N} & & \\
 & & & & & & \frac{1}{a} & \\
 & & & & & & & \frac{a}{N}
\end{pmatrix}
\begin{pmatrix}
1 & & & & & & & \\
\frac{b\rho}{a} & 1 & & & & & & \\
 & & 1 & & & & & \\
 & & \frac{b\rho}{a} & 1 & & & & \\
 & & \frac{b\rho}{a} & & 1 & & & \\
 & & -\frac{b^2\rho^2}{a^2} & -\frac{b\rho}{a} & -\frac{b\rho}{a} & 1 & & \\
 & & & & & & 1 & \\
 & & & & & & -\frac{b\rho}{a} & 1
\end{pmatrix}.
$$

If, $|b\rho| > |a|$, it is

$$
\begin{pmatrix}
1 & -\frac{a}{b\rho} & & & & & & \\
 & 1 & & & & & & \\
 & & 1 & -\frac{a}{b\rho} & -\frac{a}{b\rho} & -\left(\frac{a}{b\rho}\right)^2 & & \\
 & & & 1 & & \frac{a}{b\rho} & & \\
 & & & & 1 & \frac{a}{b\rho} & & \\
 & & & & & 1 & & \\
 & & & & & & 1 & \frac{a}{b\rho} \\
 & & & & & & & 1
\end{pmatrix}
\begin{pmatrix}
\frac{N}{b\rho} & & & & & & & \\
 & b\rho & & & & & & \\
 & & \frac{N}{b^2\rho^2} & & & & & \\
 & & & 1 & & & & \\
 & & & & 1 & & & \\
 & & & & & \frac{b^2\rho^2}{N} & & \\
 & & & & & & \frac{1}{b\rho} & \\
 & & & & & & & \frac{b\rho}{N}
\end{pmatrix}
\times
$$

$$
\times
\begin{pmatrix}
 & 1 & & & & & & \\
-1 & \frac{a}{b\rho} & & & & & & \\
 & & & & & -1 & & \\
 & & & & -1 & -\frac{a}{b\rho} & & \\
 & & & -1 & & -\frac{a}{b\rho} & & \\
 & & -1 & \frac{a}{b\rho} & \frac{a}{b\rho} & \frac{a^2}{b^2\rho^2} & & \\
 & & & & & & & -1 \\
 & & & & & & 1 & \frac{a}{b\rho}
\end{pmatrix}.
$$

Let us denote the three factors by u', t', and k', respectively. Then u' has the property that $w_2 u' w_2^{-1}$ and $w_2[x_{\alpha_2}(u), u']w_2^{-1}$ (where $[\,,\,]$ denotes the commutator) are both in P. Thus, $f(w_2 x_{\alpha_2}(u)t, s) = f(w_2 x_{\alpha_2}(u)t', s)$. We have

$$
I(s,\pi) = \int_T \left(\int_F f(w_2 x_{\alpha_2}(u), s)\psi(\alpha_2(t')u)\mathrm{d}u \right) K(t)\delta_{\mathrm{B}}^{-\frac{1}{2}}(t)\delta_P^s(w_2 t' w_2)|\alpha_2(t')|\mathrm{d}t,
$$

where t' is as above, and $K(t) := W(t)\delta_{\mathrm{B}}(t)^{-\frac{1}{2}}$. We find that

$$
\delta_{\mathrm{B}}^{-\frac{1}{2}}(t) = |N|^{-1}, \quad \delta_P(w_2 t' w_2) = \frac{|N|^2}{\max(|a|,|b|)^3}, \quad |\alpha_2(t')| = \frac{\max(|a|,|b|)^3}{|N|}.
$$

Lemma 6

$$
\int_F f(w_2 x_{\alpha_2}(u), s)\psi(cu)\mathrm{d}u = (1-q^{-3s})\frac{(1-q^{(-3s+1)(v(c)+1)})}{(1-q^{-3s+1})}.
$$

Proof. There is an embedding j of SL_2 into G_2 such that $j\begin{pmatrix} & 1 \\ -1 & \end{pmatrix} = w_2$ and $j\begin{pmatrix} 1 & u \\ & 1 \end{pmatrix} = x_{\alpha_2}(u)$. The lemma is a well-known computation from SL_2 applied to this copy of SL_2. One has only to check that $f\left(j\begin{pmatrix} t & \\ & t^{-1} \end{pmatrix}, s\right) = t^{-3s}$. □

Let $x = q^{-3s+1}$. Then the above reads

$$(1-q^{-1}x)\frac{(1-x^{v(c)+1})}{(1-x)}.$$

To complete the argument, we must consider the two cases ($SU_{2,1}$ splits over F or does not) separately.

9.3.1 Split Case

In this case, put $t_1 = a + b\sqrt{\rho}$ and $t_2 = a - b\sqrt{\rho}$. Then t_1 and t_2 are just two independent variables ranging over $F^{\times}$. The quantity called "N" above us equal to t_1t_2. Since ρ and 2 are units, $\max(|a|,|b|) = \max(|t_1|,|t_2|)$. Let β_1, β_2 denote the simple roots of SL_3. We get

$$\delta_P(w_2t'w_2) = \frac{|t_1t_2|^2}{\max(|t_1|,|t_2|)^3} = \min(|t_1^{-1}t_2^2|,|t_1^2t_2^{-1}|) = \min(|\beta_1(t)|,|\beta_2(t)|),$$

$$|\alpha_2(t')| = \frac{\max(|t_1|,|t_2|)^3}{|t_1t_2|} = \max(|\beta_1(t)|,|\beta_2(t)|).$$

Now, define two integer-valued variables, depending on t by $m_i = v(\beta_i(t)),\ i = 1,2$. As t ranges over the torus of SL_3, the pair (m_1,m_2) ranges over

$$\{(m_1,m_2)\in\mathbb{Z}^2 : m_1 - m_2 \text{ is divisible by } 3\}.$$

Every part of our local integral can now be expressed in terms of m_1 and m_2. First, we consider the function $K(t)$. This is evaluated using the Casselman–Shalika formula [2]. It is equal to zero unless m_1 and m_2 are both nonnegative. If m_1 and m_2 are both nonnegative, then the pair corresponds to a dominant weight for the group $PGL_3(\mathbb{C})$. Let Γ_{m_1,m_2} denote the corresponding irreducible finite-dimensional representation, which we may also regard as a representation of $GL_3(\mathbb{C})$. Then we have

$$K(t) = \operatorname{Tr}\Gamma_{m_1,m_2}(\tilde{t}_\pi),$$

where $\tilde{t}_\pi$ is the conjugacy class of $GL_3(\mathbb{C})$ associated to the local representation π. Also $\delta_P(w_2t'w_2) = q^{-\max(m_1,m_2)}$, $|\alpha_2(t')| = q^{-\min(m_1,m_2)}$, and $\delta_B^{-\frac{1}{2}}(t) = q^{-m_1-m_2}$. Thus, we consider,

$$(1-q^{-1}x)\sum_{m_1,m_2}\frac{1-x^{\min(m_1,m_2)+1}}{1-x}x^{\max(m_1,m_2)}\operatorname{Tr}\Gamma_{m_1,m_2}(\tilde{t}_\pi),$$

where the sum is over m_1, m_2 both nonnegative, such that $m_1 - m_2$ is divisible by 3.

We now make use of the relationship between local Langlands L-functions and the Poincaré series of certain graded algebras. We first review some definitions. Fix $N \in \mathbb{N}$, and

$$A = \bigoplus_{i_1,\dots,i_N \in \mathbb{N}} A_{i_1,\dots,i_N}$$

a graded algebra over a field k. The Poincaré series of A is a power series in N indeterminates

$$\sum_{i_1,\dots,i_N=0}^{\infty} \dim(A_{i_1,\dots,i_N}) T_1^{i_1} \dots T_N^{i_N}.$$

The graded algebra which is relevant for consideration of Langlands L-functions is described as follows. Let ${}^L G$ be a semisimple complex Lie group and (r, V) a finite-dimensional representation. Inside the symmetric algebra $\mathrm{Sym}^*(V)$, we consider the subalgebra $\mathrm{Sym}^*(V)^{{}^L U}$ of ${}^L U$ invariants. (Here, ${}^L U$ is a maximal unipotent subgroup of ${}^L G$.) This subalgebra contains the highest weight vectors of each of the irreducible components of $\mathrm{Sym}^*(V)$ and is graded by the semigroup of dominant weights of ${}^L G$ as well as by degree.

Let us use a slightly different notation from that above. We use X for the indeterminate associated to the grading by degree and $T_1, \dots, T_N$ for the grading by weight. Let π be an unramified representation of $G(F)$ where F is a nonarchimedean local field and G is a semisimple algebraic group such that ${}^L G$ is the L-group. Let $\tilde{t}_\pi$ be the semisimple conjugacy class in ${}^L G$ corresponding to π. Then it follows from the definitions that the local Langlands L-function $L(s, \pi, r)$ may be obtained from the Poincaré series of $\mathrm{Sym}^*(V)^{{}^L U}$ by substituting q^{-s} for X and $\mathrm{Tr}\, \Gamma_{k_1\varpi_1+\dots+k_N\varpi_N}(\tilde{t}_\pi)$ for $T_1^{k_1} \dots T_N^{k_N}$. Here, $\Gamma_{k_1\varpi_1+\dots+k_N\varpi_N}$ denotes the irreducible finite-dimensional representation of ${}^L G$ with highest weight $k_1\varpi_1 + \dots + k_N\varpi_N$.

In cases when ${}^L G$ is reductive but not semisimple, this discussion must be adapted, as the choice of maximal unipotent ${}^L U$ does not by itself pin down a basis for the weight lattice which may be used to define the grading. In the case at hand, one needs only to observe that, since the adjoint representation of $GL_3(\mathbb{C})$ factors through the projection to $PGL_3(\mathbb{C})$ (which is semisimple), each of the representations appearing in the decomposition of the symmetric algebra must as well. Alternatively, one may define the grading using the weights of the derived group.

Proposition 5 in this case is reduced to the claim that the Poincaré series of the ${}^L U$ invariants for the adjoint representation of $GL_3(\mathbb{A})$ is given by

$$\frac{1}{(1-X^2)(1-X^3)} \sum_{\substack{m_1,m_2=0 \\ 3|(m_1-m_2)}}^{\infty} \frac{1-X^{\min(m_1,m_2)+1}}{1-X} X^{\max(m_1,m_2)} T_1^{m_1} T_2^{m_2}. \tag{7}$$

This may be deduced from the computations in Sect. 3 of [3]. It also follows readily from example a) on p. 14 of [5].

Remark 7 *The Poincaré series* (7) *of* $\mathrm{Sym}^*(V)^{^LU}$ *may be summed, with the result being the rational function:*

$$\frac{1 - T_1^3 T_2^3 X^6}{(1 - T_1 T_2 X)(1 - T_1 T_2 X^2)(1 - T_1^3 X^3)(1 - T_2^3 X^3)(1 - X^2)(1 - X^3)}.$$

9.3.2 NonSplit Case

Suppose now that ρ is not a square in the local field F. Then it is part of the L-group formalism that the semisimple conjugacy class $\tilde{t}_\pi$ in ${}^LG = GL_3(\mathbb{C}) \rtimes \mathrm{Gal}(E/F)$ associated to π is in the coset corresponding to the nontrivial element of $\mathrm{Gal}(E/F)$, which we denote Fr . Each such conjugacy class contains an element of the form

$$\left(\begin{pmatrix} \mu & & \\ & \pm 1 & \\ & & \mu^{-1} \end{pmatrix}, \mathrm{Fr}\right).$$

Adjusting by an element of the center, we may assume the sign in the middle is plus.

Referring to Sect. 9.1.1, we see that our eight-dimensional representation decomposes into a five-dimensional $+1$ eigenspace for Fr, on which $\begin{pmatrix} \mu & & \\ & 1 & \\ & & \mu^{-1} \end{pmatrix}$ acts with eigenvalues $\mu^2, \mu, 1, \mu^{-1}, \mu^{-2}$, and a three-dimensional -1 eigenspace for Fr, on which $\begin{pmatrix} \mu & & \\ & 1 & \\ & & \mu^{-1} \end{pmatrix}$ acts with eigenvalues $\mu, 1, \mu^{-1}$. Hence, the local L-function is

$$\frac{1}{(1 - \mu^2 x)(1 - \mu^2 x^2)(1 - x^2)(1 - \mu^{-2} x)(1 - \mu^{-2} x^2)},$$

where $x = q^{-3s+1}$ as before. This may also be written as

$$\frac{1}{(1 - x^2)} \sum_{k_1, k_2 = 0}^{\infty} \mathrm{Tr}(\Gamma_{k_1} \otimes \Gamma_{k_2}) \begin{pmatrix} \mu^2 & \\ & \mu^{-2} \end{pmatrix} x^{k_1 + 2k_2}.$$

Turning to the local integral, we find that in this case we have $m_1 = m_2$. Let us therefore denote this quantity simply "m."

Lemma 8 *With the notation as above, we have*

$$K(t) = \mathrm{Tr}\, \Gamma_m \begin{pmatrix} \mu^2 & \\ & \mu^{-2} \end{pmatrix}.$$

Proof. This can be verified either by direct computation or by a close reading of [2]. In either method, it is necessary first to identify the precise unramified character of the torus of $SU_{2,1}$ corresponding to $\left(\begin{pmatrix} \mu & & \\ & \pm 1 & \\ & & \mu^{-1} \end{pmatrix}, \mathrm{Fr}\right)$. For the convenience of the reader, we record that it is the map sending the torus element with coordinates a and b as above to $\mu^{v(a^2-b^2\rho)}$. (Here, v is again the discrete valuation on the field F.)

This case of the proposition now follows from the identity

$$\sum_{k_1,k_2=0}^{\infty} \sum_{i=0}^{\min(k_1,k_2)} X^{k_1+2k_2} T^{k_1+k_2-2i} = \frac{1}{(1-X^3)(1-TX)(1-TX^2)}$$

$$= \frac{1}{1-X^3} \sum_{m=0}^{\infty} X^m \frac{1-X^{m+1}}{1-X} T^m,$$

which is straightforward to verify. □

This also completes the proof of Proposition 5. □

References

1. A. Ben-Artzi and D. Ginzburg, A tower of liftings for $\mathrm{SU}_{2,1}$. J. Ramanujan Math. Soc. 18 (2003), no. 4, 369–383.
2. W. Casselman and J. Shalika, The unramified principal series of p-adic groups. II. The Whittaker function. Compositio Math. 41 (1980), no. 2, 207–231.
3. D. Ginzburg, Rankin-Selberg integral for the adjoint representation of GL_3. Invent. Math. 105 (1991), no. 3, 571–588.
4. W. Fulton, J. Harris, *Representation theory. A first course.* Graduate Texts in Mathematics, 129. Readings in Mathematics. Springer-Verlag, New York, 1991. xvi+551 pp. ISBN: 0–387–97527–6; 0–387–97495–4
5. Kirillov, Anatol N. Decomposition of symmetric and exterior powers of the adjoint representation of $\mathfrak{gl}_N$. I. Unimodality of principal specialization of the internal product of the Schur functions, in *Infinite analysis, Part A, B (Kyoto, 1991)*, 545–579, Adv. Ser. Math. Phys., 16, World Sci. Publ., River Edge, NJ, 1992.
6. S. Rallis, G. Schiffmann, Theta correspondence associated to G_2. Amer. J. Math. 111 (1989), no. 5, 801–849

Chapter 10
Symplectic Ice

Dmitriy Ivanov

Abstract In this chapter, we construct an ice model (a six-vertex model) whose partition function equals the product of a deformation of Weyl's denominator and an irreducible character of the symplectic group $Sp(2n,\mathbb{C})$. Similar results have been obtained by Brubaker et al. (Schur polynomials and the Yang–Baxter equation. Comm. Math. Phys. 308(2):281–301, 2011) (for the general linear group) and by Hamel and King (Symplectic shifted tableaux and deformations of Weyl's denominator formula for sp$(2n)$, J. Algebraic Combin. **16**, 2002, no. 3, 269–300, 2003) (for the symplectic group). The difference between our result and that of (Hamel and King, Symplectic shifted tableaux and deformations of Weyl's denominator formula for sp$(2n)$, J. Algebraic Combin. **16**, 2002, no. 3, 269–300, 2003) is that our Boltzmann weights for cap vertices are different from those in (Hamel and King, Symplectic shifted tableaux and deformations of Weyl's denominator formula for sp$(2n)$, J. Algebraic Combin. **16**, 2002, no. 3, 269–300, 2003). Also, our proof uses the Yang–Baxter equation, while that of Hamel and King does not.

Keywords Yang–Baxter equation • Weyl character formula • Partition function • Ice model

10.1 Introduction

Hamel and King [5] showed how characters of irreducible representations times a deformed Weyl denominator equal partition functions of certain ice models, and Brubaker et al. [4] showed how to use the Yang–Baxter equation to investigate these ice models. In this chapter, we use the Yang–Baxter equation to investigate a certain

D. Ivanov (✉)
e-mail: dmitriy.ivanov.2@gmail.com

D. Bump et al. (eds.), *Multiple Dirichlet Series, L-functions and Automorphic Forms*, Progress in Mathematics 300, DOI 10.1007/978-0-8176-8334-4_10,

ice model (a 6-vertex model). We consider a graph in the shape of a grid with caps on the right end. A *state* of this system consists of assignment of signs $+$ or $-$ to each edge. To each vertex, we assign a number (called *Boltzmann weight*); the Boltzmann weight of a vertex depends on the signs adjacent to it. A Boltzmann weight of a state is equal to the product of the Boltzmann weights of all vertices, and the *partition function* is the sum of the Boltzmann weights of all possible states. We show that the partition function of this ice model equals the product of an irreducible character of the symplectic group $Sp(2n, \mathbb{C})$ with a deformation of the Weyl denominator. A similar result was originally proved by Hamel and King [5], but the Boltzmann weights (for the vertices at the caps) here are different from theirs. Also, our proof uses the Yang–Baxter equation, whereas the proof of Hamel and King does not. Our proof is similar to that of an analogous result for the general linear group in Brubaker et al. [4]. Our proof also uses Proctor patterns; a good reference for this topic is Beineke et al. [1, 2]. This gives us 6-vertex models for the characters of $Sp(2n, \mathbb{C})$. Our result can also be interpreted as an example of an *exactly solved model* in the sense of statistical mechanics, that is, an ice model whose partition function can be computed explicitly.

Ice models can also be used to describe Whittaker functions. Let F be a locally compact field and G a split reductive group over F. Let T denote a split maximal torus of G, B the positive Borel subgroup, and U the unipotent radical of $B = TU$. Let ψ denote a nontrivial character of U. A Whittaker model of an irreducible representation (π, V) of G is a space $\mathcal{W}_\pi$ of functions W on G satisfying $W(ug) = \psi(u)W(g)$ which is closed under right translations and such that W_π is isomorphic to V (as G-modules). It is known that a Whittaker model is unique (if it exists). The Casselman–Shalika formula relates the characters of the L-group with values of the Whittaker functions on a p-adic group. In some cases, a Whittaker function can be described as the partition function of a statistical system in the 6-vertex model. For example, Brubaker et al. [4] used a certain statistical system in the 6-vertex model and Yang–Baxter equation to study p-adic Whittaker functions of type A.

The ice model studied here is similar to the U-turn models used by Kuperberg [6] to enumerate classes of alternating sign matrices. After the work in this chapter was done, Brubaker et al. [3] followed our arguments (using the same ice model but with different Boltzmann weights) to represent a Whittaker function on the metaplectic double cover of $Sp(2n, F)$ where F is a non-archimedean local field.

I would like to thank Daniel Bump for suggesting this problem and for his encouragement and guidance. This work was supported in part by NSF grant DMS-0652817.

10.2 The Yang–Baxter Equation

We shall give a new proof of a result due to Hamel and King [5] based on the Yang–Baxter equation. It represents the product of the character of the symplectic group times a deformation of the Weyl denominator as a partition function of a certain

Table 10.1 Boltzmann weights for gamma and delta ice

Γ vertex	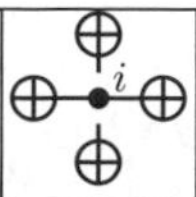	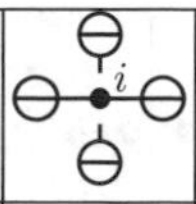	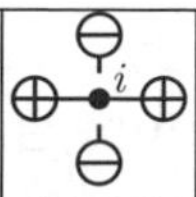	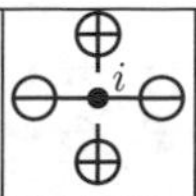	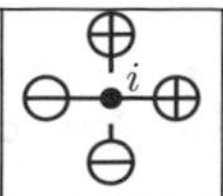	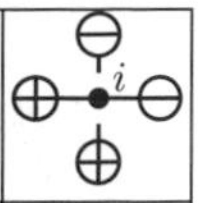
Boltzmann weight	1	z_i^{-1}	t	z_i^{-1}	$z_i^{-1}(t+1)$	1
Δ vertex	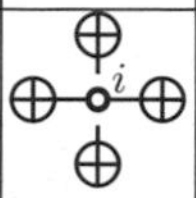	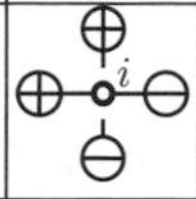	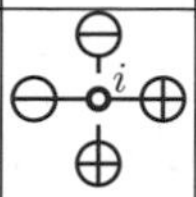	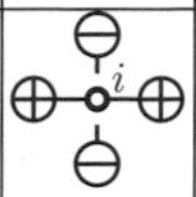	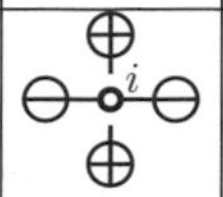	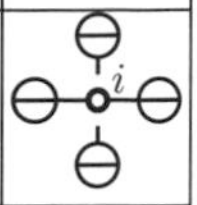
Boltzmann weight	z_i	$z_i(t+1)$	1	$z_i t$	1	1

type of ice. We begin by explaining the notation we will be using later. By *ice* we shall mean a lattice (or a more general graph), to each edge of which we may assign a sign (either + or −). We shall consider six different types of ice, called *delta* (Δ) *ice, gamma* (Γ) *ice, delta–delta* ($\Delta\Delta$) *ice, delta–gamma ice* ($\Delta\Gamma$), *gamma–delta* ($\Gamma\Delta$) *ice*, and *gamma–gamma* ($\Gamma\Gamma$) *ice*. To each vertex, we assign a weight (called *Boltzmann weight*); for each type of ice, the weight depends on signs + or − that will be assigned to the four adjacent edges. Let $z_1, \ldots, z_n, t$ be complex numbers with $z_i \neq 0$ for all i. We shall refer to z_i as *spectral parameters* and t *a deformation parameter*. In Tables 10.1 and 10.2, we give Boltzmann weights for each type of ice. The labels i and j are integers between 1 and r. They refer to the spectral parameters and depend on the row of the matrix.

We will consider planar graphs. Each vertex will have four (or occasionally only two) adjacent edges. Edges on the interior will join two vertices, but edges on the boundary will only have a single adjacent vertex. Each vertex will be assigned a set of Boltzmann weights from one of the six types of ice. Each boundary edge will be assigned a fixed sign + or −. Interior edges will also be assigned signs, but these will not be fixed.

By an *admissible configuration* (or admissible state), we shall mean a labeling of the edges of the graph by signs + or − such that each vertex on the graph is listed in the table above. The *Boltzmann weight* of an admissible state is simply the product of the Boltzmann weights of all vertices in this state. The *partition function* of an ice is the sum of Boltzmann weights of all admissible states. (This definition of the partition function comes from statistical mechanics.)

The Yang–Baxter equation will be an important tool in our proof. Here is the version of the Yang–Baxter equation that we will use:

Lemma 1. *Let $X, Y \in \{\Gamma, \Delta\}$, and let S be a vertex of type X, T of type Y and R of type XY. We use the spectral parameter z_i at the vertex S, z_j at the vertex T and*

Table 10.2 Auxiliary Boltzmann weights

Type						
$\Gamma\Delta$	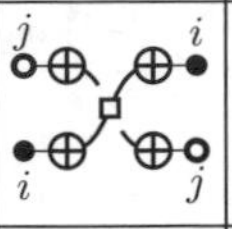	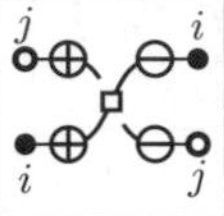	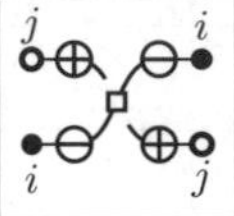	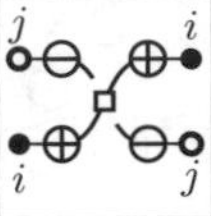	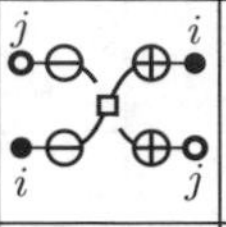	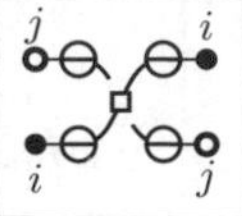
Boltz. weight	$t^2 z_j - z_i^{-1}$	$(t+1)z_j$	$tz_j + z_i^{-1}$	$tz_j + z_i^{-1}$	$(t+1)z_i^{-1}$	$z_i^{-1} - z_j$
$\Delta\Delta$	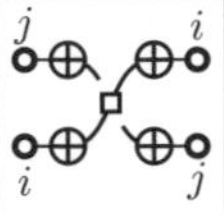	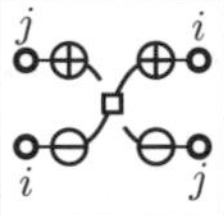	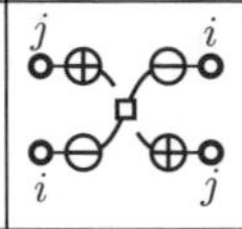	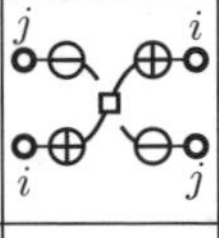	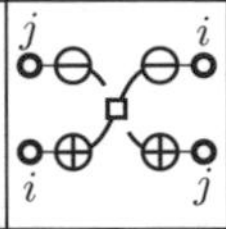	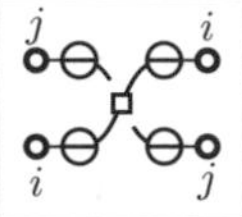
Boltz. weight	$tz_i + z_j$	$z_j(t+1)$	$tz_j - tz_i$	$z_i - z_j$	$(t+1)z_i$	$z_i + tz_j$
$\Gamma\Gamma$	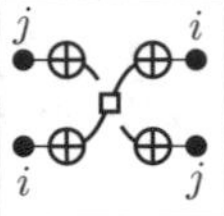	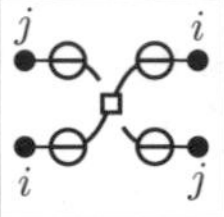	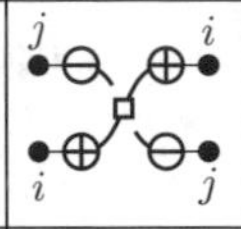	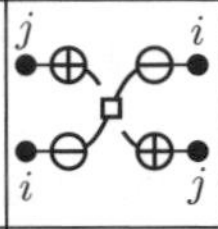	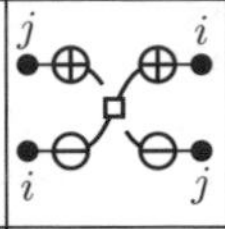	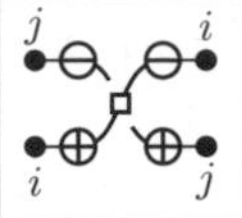
Boltz. weight	$tz_i^{-1} + z_j^{-1}$	$tz_j^{-1} + z_i^{-1}$	$tz_j^{-1} - tz_i^{-1}$	$z_i^{-1} - z_j^{-1}$	$(t+1)z_i^{-1}$	$(t+1)z_j^{-1}$
$\Delta\Gamma$	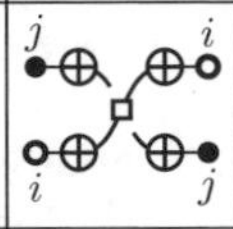	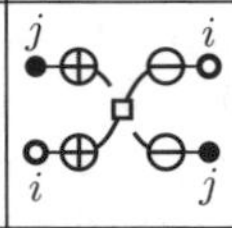	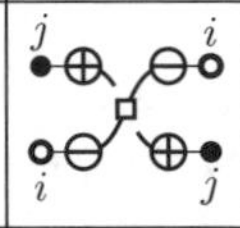	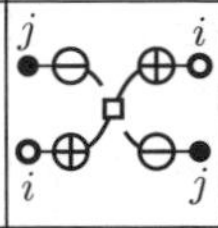	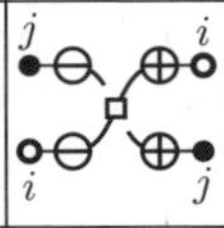	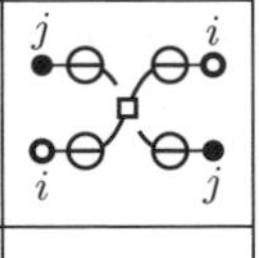
Boltz. weight	$z_i - z_j^{-1}$	$(t+1)z_i$	$tz_i + z_j^{-1}$	$tz_i + z_j^{-1}$	$(t+1)z_j^{-1}$	$-t^2 z_i + z_j^{-1}$

the spectral parameters z_i, z_j at the vertex R. The Boltzmann weights are given in the table on the previous page. Then the partition functions of

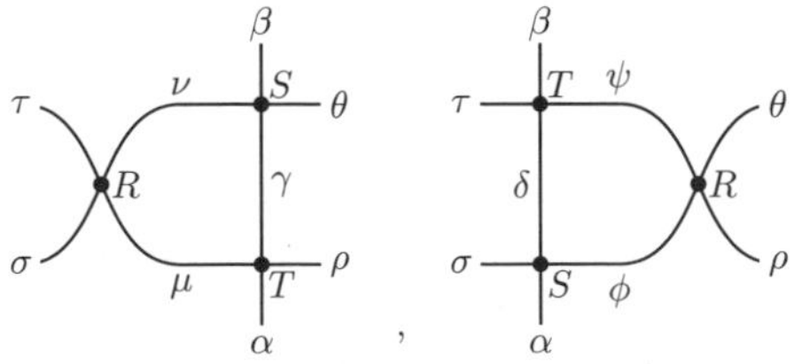

are equal for every fixed combination of signs $\sigma, \tau, \alpha, \beta, \rho, \theta$.

Note that by definition in the partition functions, $\sigma, \tau, \alpha, \beta, \rho, \theta$ are fixed, but the signs ν, μ, γ and ψ, ϕ, δ of the interior edges are summed.

We refer the reader to Brubaker et al. [4] for the proof of this proposition. Indeed, the case $X = Y = \Gamma$ follows from Theorem 3 of [4], and a similar argument shows that the proposition is true in the other three cases. Informally, this means that if the signs on the outer edges are fixed, then we can push the braid (i.e., a vertex of type XY) to the right without changing the partition function. We emphasize that the vertices R, S, T have the same Boltzmann weights in both pictures. A consequence of this proposition is that it allows us to switch the spectral parameters without changing the partition function. We see that z_i corresponds to the top row and z_j corresponds to the bottom row while in the second picture z_i corresponds to the bottom row and z_j corresponds to the top one.

Let us use the Yang–Baxter equation to show that the following two configurations have equal partition functions:

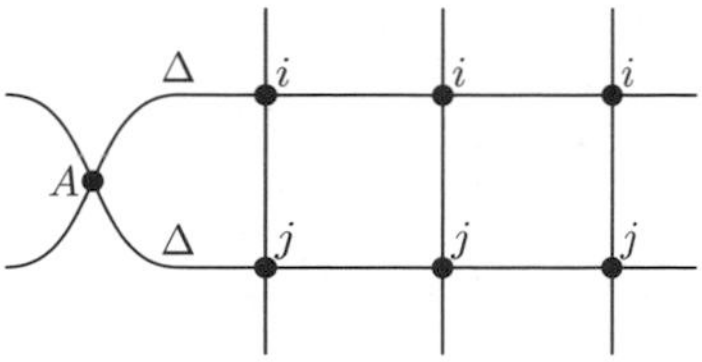

(1)

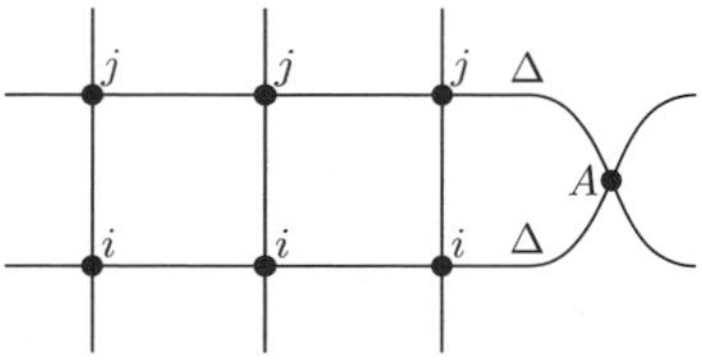

(2)

Here it is assumed that the six boundary edges are given some fixed assignment of signs $\pm$ and that these are the same for the two configurations. Note that the braid in this example connects two rows of delta ice (in both cases), so the vertex A is a vertex of delta–delta ice. The spectral parameters corresponding to these rows have been switched; in other words, if the spectral parameters of the rows of the first ice are z_i (top row) and z_j (bottom row), then after applying the Yang–Baxter equation, z_j will correspond to the top row and z_i will correspond to the bottom row.

Indeed, using Lemma 1 for delta ice (i.e., with $X = Y = \Delta$), we see that the partition functions of (1) equal the partition function of

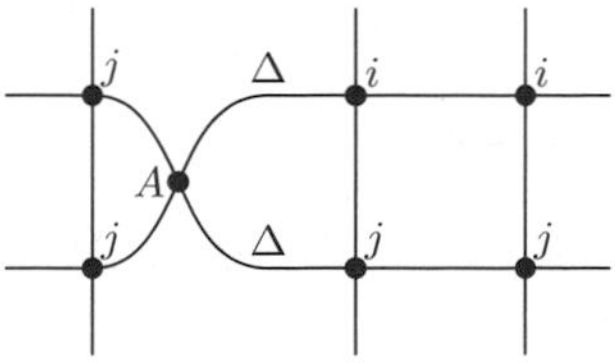

Repeating this process two more times, we see that this equals the partition function of (2).

The Yang–Baxter equation can even be applied to ice which contains rows of both delta ice and gamma ice; in this case, the braid will have more than one vertex. Let us show that the partition functions of the following two systems are equal:

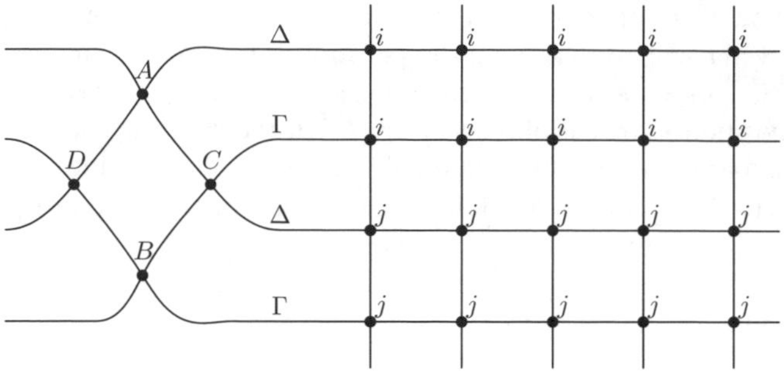

(3)

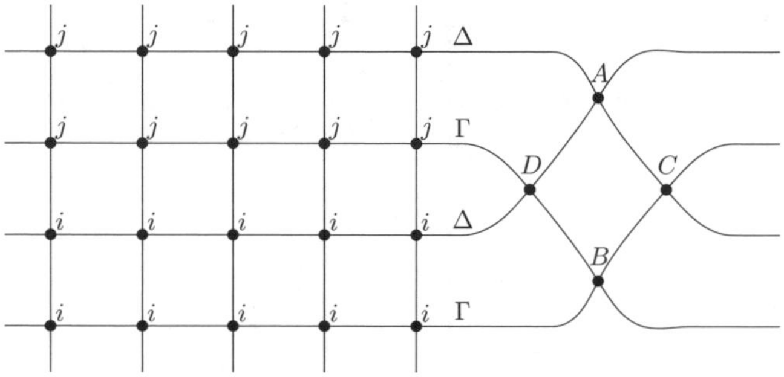

(4)

Here the vertex A is a vertex of delta–delta ice, the vertex B is a vertex of gamma–gamma ice, the vertex C is a vertex of gamma–delta ice, and the vertex D is a vertex of delta–gamma ice. The vertices in the rows are of Δ or Γ type as labeled. In the second system, the spectral parameters of the delta rows and of the gamma rows have been switched. This means that if the spectral parameters in the first system corresponding to the delta rows are z_i (upper delta row) and z_j (lower delta row), then in the second system, the spectral parameter corresponding to the upper delta row is z_j and the spectral parameter corresponding to the lower row is z_i. The same is true for the rows of gamma ice. This example will be used later in our proof.

To prove this, using the Yang–Baxter equation in the form of Lemma 1 with $X, Y = \Gamma, \Delta$, we see that the partition function of (3) equals the partition function of:

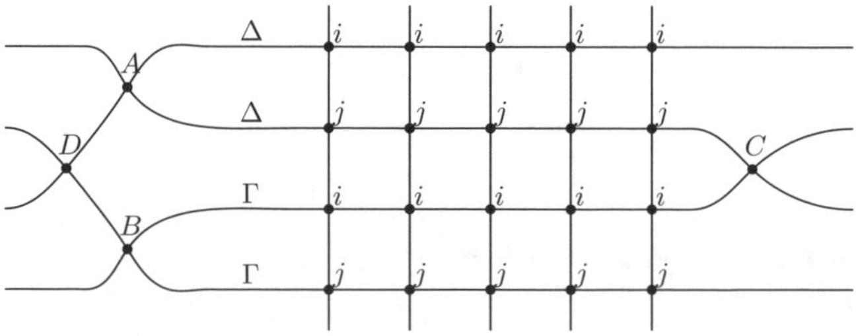

Then applying this again with $X, Y = \Delta, \Delta$, this equals the partition function of:

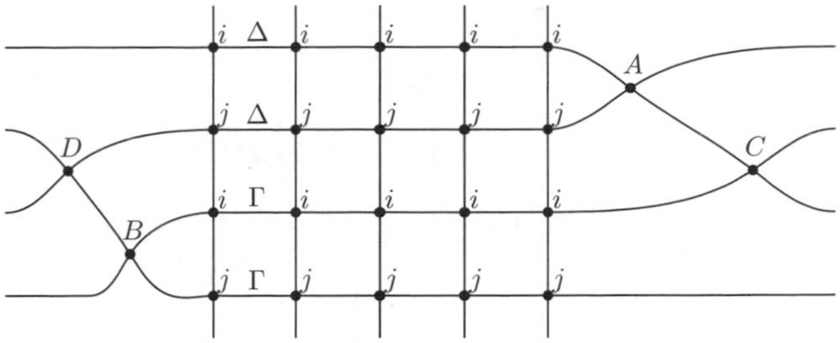

Next using Lemma 1 with $X = Y = \Gamma$ moves the B vertex across (not shown), then finally using the lemma with $X = \Delta, Y = \Gamma$ shows that this partition function equals that of (4).

Yang–Baxter equation will be a useful tool later when we prove that the partition function of certain ice is invariant under interchanging spectral parameters.

Let $\lambda = (\lambda_1, \lambda_2, \ldots, \lambda_n) \in \mathbb{Z}^n$; we assume that $\lambda_1 \geq \lambda_2 \geq \lambda_3, \ldots \geq \lambda_n \geq 0$. We may view λ as a dominant weight of the symplectic group $Sp(2n, \mathbb{C})$. We recall that $Sp(2n, \mathbb{C})$ consists of all $2n$ by $2n$ invertible matrices g satisfying $gJg^T = J$ where $J = \begin{pmatrix} 0 & -I \\ I & 0 \end{pmatrix}$; here 0 denotes the zero matrix and I denotes the identity matrix. Let χ_λ be the character of the irreducible representation of $Sp(2n, \mathbb{C})$ with highest weight λ. By the Weyl character formula, there exists a Laurent polynomial in n variables $s_\lambda^{sp}(x_1, x_2, .., x_n) \in \mathbb{C}[x_1, x_1^{-1}, x_2, x_2^{-1}, .., x_n, x_n^{-1}]$ such that $\chi_\lambda(g) = s_\lambda^{sp}(z_1, z_2, .., z_n)$ where

$$z_1, z_2, .., z_n, z_1^{-1}, z_2^{-1}, .., z_n^{-1}$$

are the eigenvalues of $g \in Sp(2n, \mathbb{C})$. Define

$$D(z_1, z_2, .., z_n, t) = \prod_i (1 + tz_i^2) \cdot \prod_{i<j} (1 + tz_iz_j)(1 + tz_iz_j^{-1}) \cdot z^{-\rho},$$

where $z^{-\rho} = z_1^{-n} z_2^{-n+1} \ldots z_n^{-1}$. We will show that $s_\lambda^{sp} \cdot D$ equals the partition function of a six-vertex model system. We recall that the Weyl character formula states that

$$\chi_\lambda(g) = \frac{\sum_{w \in W} (-1)^{l(w)} z^{w(\lambda+\rho)}}{\prod_{\alpha \in \Phi^+} (z^{\alpha/2} - z^{-\alpha/2})}.$$

In this formula, W is the Weyl group, $l(w)$ is the length function, and Φ^+ is the set of positive roots. Here $z^{(\lambda+\rho)}$ means $z_1^{\lambda_1+n} z_2^{\lambda_2+n-1} \ldots z_n^{\lambda_n+1}$, and $z^{w(\lambda+\rho)}$ means $(w(z_1))^{\lambda_1+n}(w(z_2))^{\lambda_2+n-1} \ldots (w(z_n))^{\lambda_n+1}$, We now observe that the denominator in this formula can be written as

$$\prod_{\alpha\in\Phi+}(z^{\alpha/2}-z^{-\alpha/2})=\prod_{\alpha\in\Phi+}(z^{-\alpha/2}-z^{\alpha/2})$$

$$=z^{-\rho}\prod_{\alpha\in\Phi+}(1-z^{\alpha})=z^{-\rho}\prod_{i}(1-z_i^2)\cdot\prod_{i<j}(1-z_iz_j)(1-z_iz_j^{-1}).$$

Hence, we see that $D(z_1, z_2, .., z_n, -1)$ is the denominator in the Weyl character formula, so this result is a deformation of the Weyl character formula.

10.3 The Character as a Partition Function

We consider the following problem. We consider a domain which consists of a rectangular lattice with $2n$ rows; each odd numbered row is a row of delta ice, and each even numbered row is a row of gamma ice. Hence, there are n rows of delta ice and n rows of gamma ice. Let z_i be the spectral parameter corresponding to the ith row of delta ice and also the ith row of gamma ice. We assume that $\lambda = (\lambda_1, .., \lambda_n) \in \mathbb{Z}^n$ be a partition (this means that $\lambda_1 \geq \lambda_2 \geq \cdots \geq \lambda_n \geq 0$). We assign signs to each edge as follows: on the left column, we assign $-$ to each row of delta ice (i.e., to each odd numbered row), and a $+$ to each even numbered row. On the bottom, we assign $+$ to each edge. On the top, we assign $-$ to each column labeled $\lambda_i + n + 1 - i$ (for $1 \leq i \leq n$), and we assign $+$ to each remaining column. The columns of this lattice are numbered in descending order from $\lambda_1 + n$ to 1. On the right side, we connect each row of delta ice with the following row of gamma ice with a "cap," to be described below.

If $n = 2$ and $\lambda = (3, 1)$, we thus have the following configuration:

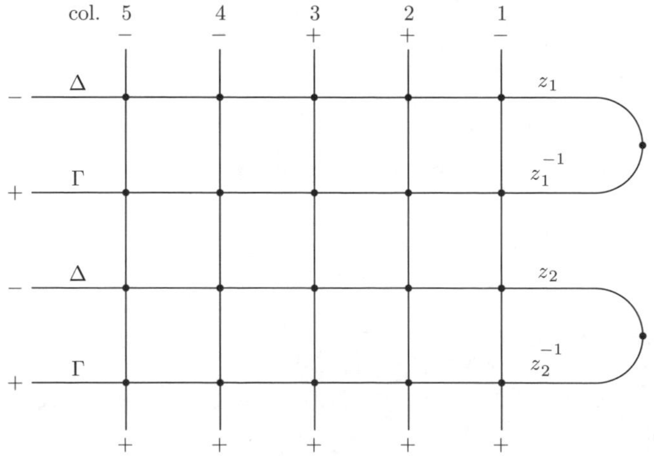

(5)

Each cap includes a new vertex whose Boltzmann weights for the cap are as follows:

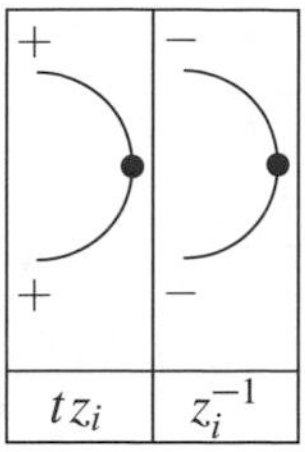

Other combinations of signs are not allowed. Thus, the signs at the ends of a row of delta ice and the row of gamma ice right below it must be the same (thus, we can have either two + signs or two − signs).

We consider the partition function of the ice described above. Our result is the following theorem:

Theorem 1. *Let $\lambda = (\lambda_1, \ldots, \lambda_n) \in \mathbb{Z}^n$ be a partition. Then the partition function of the ice described above is given by $D \cdot s_\lambda^{sp}(z_1, z_2, \ldots, z_n)$.*

Proof. Let I_1 denote the given ice (5). Let $Z(I)$ denote the partition function of the ice I. Consider $Z(I_1)/D$. We first show that this quotient is invariant under the action of the Weyl group (which is generated by permutations of $z_1, .., z_n$ and also by the functions that take z_i to z_i^{-1} (for some i)).

Lemma 2. *$Z(I_1)/D$ is invariant under any permutation of $z_1, \ldots, z_n$.*

Proof. Since the group S_n is generated by transpositions of the form $(k, k+1)$, it suffices to show that $Z(I_1)/D$ is invariant under the interchange $i \leftrightarrow i+1$ (for any i).

We attach a braid to I_1. (The picture below is the part of the ice which corresponds to spectral parameters with indices i and $i+1$, namely, two rows of delta ice with spectral parameters z_i and z_{i+1} and two rows of gamma ice with spectral parameters z_i^{-1} and z_{i+1}^{-1}. The picture (ice) extends above and below this part.)

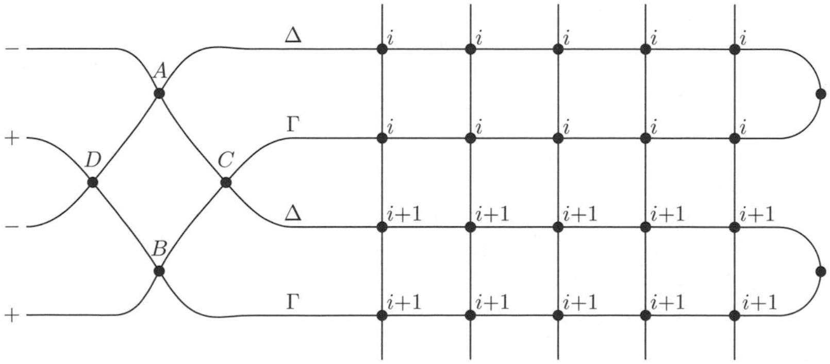

(6)

Let us denote this new ice by I_2. The only admissible configuration of spins here is shown in the picture below.

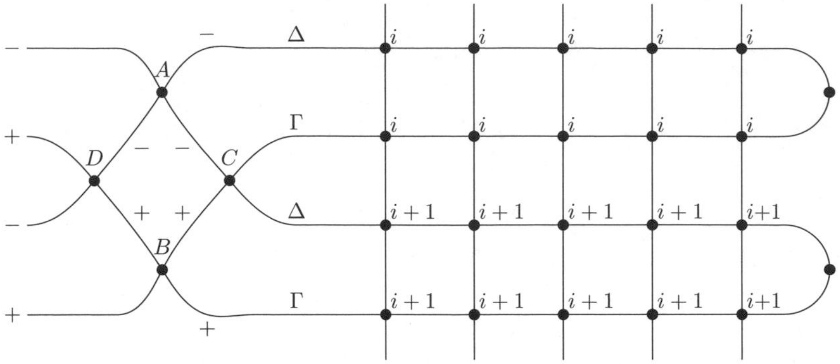

(7)

Therefore, every state of this new boundary problem (with ice I_2) determines a unique state of the original problem (with ice I_1). Hence, the partition function of I_2 is the partition function of the ice I_1 times the partition function of the braid. The partition function of the braid is equal to $(z_i + tz_{i+1})(tz_i^{-1} + z_{i+1}^{-1})(tz_i + z_{i+1}^{-1})(tz_{i+1} + z_i^{-1})$, and therefore,

$$Z(I_2) = (z_i + tz_{i+1})(tz_i^{-1} + z_{i+1}^{-1})(tz_i + z_{i+1}^{-1})(tz_{i+1} + z_i^{-1})Z(I_1).$$

An application of the Yang–Baxter equation tells us that the partition function of I_2 is equal to the partition of the ice I_3 (see picture below) with z_i, z_{i+1} switched. We will denote this by $Z^*(I_3)$; $*$ means that z_i, z_{i+1} are switched.

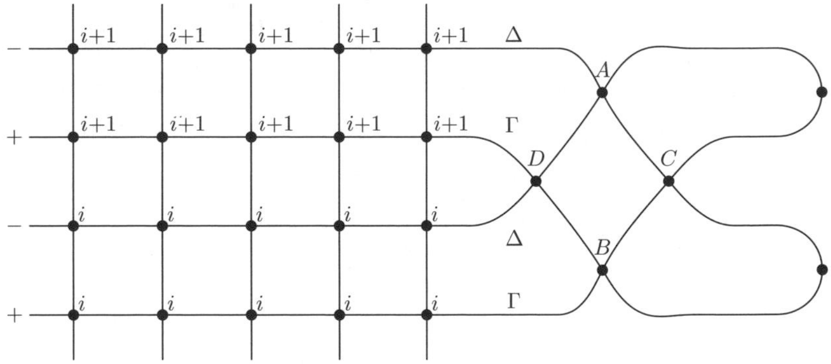

Note that $Z^*(I_3) = \sum_{e_i} Z^*(I_4)(e_1, e_2, e_3, e_4) \cdot Z(I_5)(e_1, e_2, e_3, e_4)$ where the sum is taken over all possible spins e_1, e_2, e_3, e_4, and the pictures of I_4, I_5 are shown below. Here are the ices I_4 (left) and I_5:

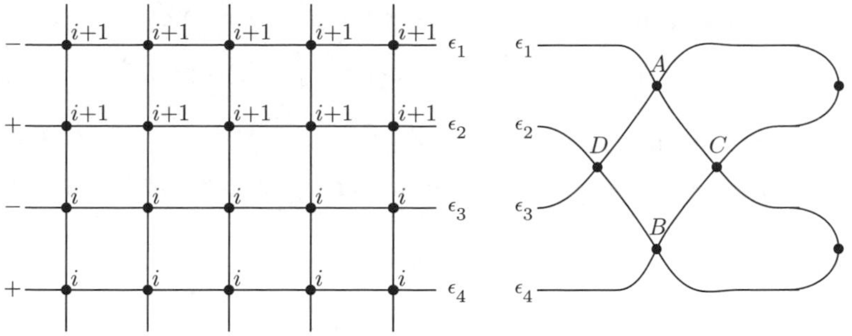

At this point, we used the computer program SAGE [7] to check that the following equality holds.

Lemma 3. *For every choice of $e_1, e_2, e_3, e_4 \in \{\pm\}$, we have*

$$Z(I_5)(e_1, e_2, e_3, e_4) = z_i^{-2} z_{i+1}^{-2} \cdot Z^*(I_6)(e_1, e_2, e_3, e_4) \cdot (tz_i + z_{i+1})(tz_{i+1} + z_i)(tz_i z_{i+1} + 1)^2.$$

Here I_6 is the ice that consists of two caps (see picture).

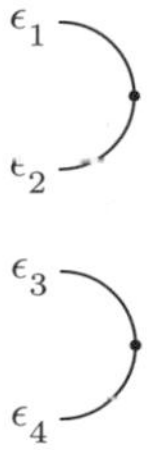

Note, in particular, that the expression $z_i^{-2} z_{i+1}^{-2} \cdot (tz_i + z_{i+1})(tz_{i+1} + z_i)(tz_i z_{i+1} + 1)^2$ does not depend on the choice of spins e_1, e_2, e_3, e_4, and therefore, we get $Z(I_2) = Z^*(I_3) = \sum_{e_i} Z^*(I_4)(e_1, e_2, e_3, e_4) \cdot z_i^{-2} z_{i+1}^{-2} \cdot Z^*(I_6) \cdot (tz_i + z_{i+1})(tz_{i+1} + z_i)(tz_i z_{i+1} + 1)^2$, or, equivalently,

$$\begin{aligned}(z_i + tz_{i+1})(tz_i^{-1} + z_{i+1}^{-1})(tz_i + z_{i+1}^{-1})(tz_{i+1} + z_i^{-1})Z(I_1) \\ = \sum_{e_i} Z^*(I_4)(e_1, e_2, e_3, e_4) \cdot z_i^{-2} z_{i+1}^{-2} \cdot Z^*(I_6) \\ \cdot (tz_i + z_{i+1})(tz_{i+1} + z_i)(tz_i z_{i+1} + 1)^2.\end{aligned}$$

We observe that $\sum_{e_i} Z^*(I_4) \cdot Z^*(I_6) = Z^*(I_1)$, and hence, we obtain the following equality:

$$(z_i + tz_{i+1})(tz_i^{-1} + z_{i+1}^{-1})(tz_i + z_{i+1}^{-1})(tz_{i+1} + z_i^{-1})Z(I_1)$$
$$= Z^*(I_1) \cdot z_i^{-2} z_{i+1}^{-2} \cdot (tz_i + z_{i+1})(tz_{i+1} + z_i)(tz_i z_{i+1} + 1)^2.$$

After simplifying it, we obtain

$$(tz_i + z_{i+1})Z(I_1) = (tz_{i+1} + z_i)Z^*(I_1),$$

which means that

$$Z(I_1) = \frac{(tz_{i+1} + z_i)}{(tz_i + z_{i+1})} Z(I_1)^*.$$

Recall that we defined D as

$$z^{-\rho} \cdot \prod_i (1 + tz_i^2) \cdot \prod_{i<j} (1 + tz_i z_j)(1 + tz_i z_j^{-1}).$$

If we let D^* denote the same expression, but with z_i and z_{i+1} switched, i.e.

$$D^*(z_1, z_2, \ldots, z_i, z_{i+1}, \ldots, z_n) = D(z_1, z_2, \ldots, z_{i+1}, z_i, \ldots, z_n),$$

then it is easy to check that $\frac{D}{D^*} = \frac{(tz_{i+1}+z_i)}{(tz_i+z_{i+1})}$, and therefore, we conclude that

$$Z(I_1) = \frac{D}{D^*} Z(I_1)^*,$$

which is to say that

$$Z(I_1)/D = Z(I_1)^*/D^*,$$

which is exactly what we needed to show. □

Lemma 4. *$Z(I_1)/D$ is invariant under the change $z_i \leftrightarrow z_i^{-1}$.*

Proof. Since we have already shown that this expression is invariant under $z_i \leftrightarrow z_{i+1}$ for any i, it suffices to show that $Z(I_1)/D$ is invariant under $z_n \leftrightarrow z_n^{-1}$. We transform the last row of the given ice (the row of gamma ice corresponding to spectral parameter z_n^{-1}) into a row of delta ice as follows: we change the sign on the left edge from $+$ to $-$ and we also change the signs of all the entries on the edges in the last row. We observe that in the last row of horizontal edges, only the following types of Gamma ice can appear:

Gamma Ice	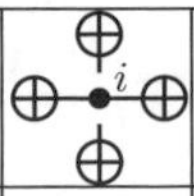	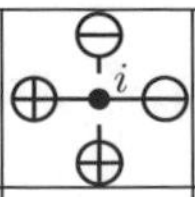	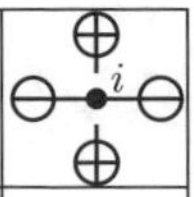
Boltzmann weight	1	1	z_i

These change to:

Delta Ice			
Boltzmann weight	1	1	z_i

We observe that the Boltzmann weights are unchanged, and therefore, the partition function is also unchanged by this transformation. Let us denote this new ice by I_7; hence, $Z(I_7) = Z(I_1)$. (We note that this will work only in the last row because it is essential that there be no $-$ signs on the bottom edges.) We also define a new cap which connects two rows of delta ice. The signs on the ends of this new cap must be opposite (thus, one of them must be a $+$ and the other one must be a $-$), and the Boltzmann weights for this cap are shown in the picture below.

+	−
−	+
tz_i	z_i^{-1}

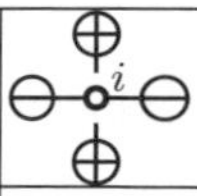

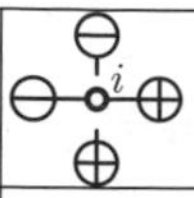

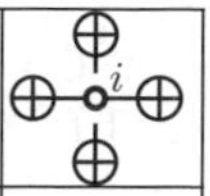

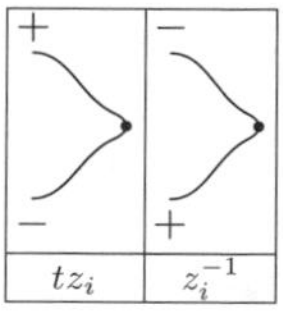

So now, the last two rows are rows of delta ice. We attach a braid (delta–delta vertex) to this ice, as shown in the picture below. We denote this new ice by I_8.

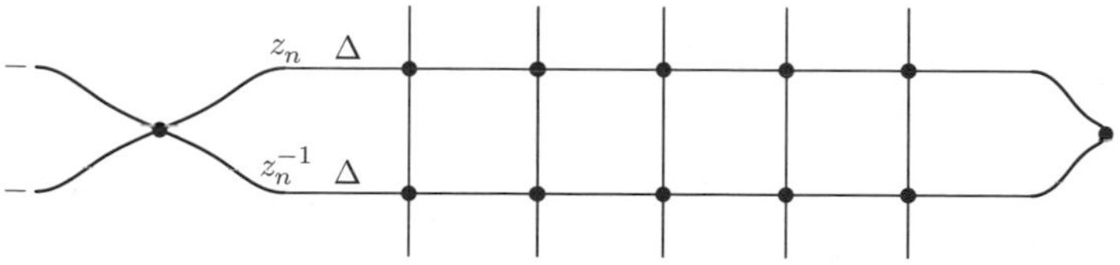

The only admissible configuration of spins for the delta–delta vertex has all four adjacent edges $-$. Therefore, every state of this new boundary problem (with ice I_8) determines a unique state of the original problem (with ice I_7). Hence, the partition function of each state of I_8 is the partition function of the corresponding state of ice I_7 times the partition function of the braid. The partition function of the braid is equal to $(z_n + tz_n^{-1})$.

An application of the Yang–Baxter equation tells us that the partition function of I_8 is equal to the partition of the ice I_9 (see picture below) with z_n and z_n^{-1} switched.

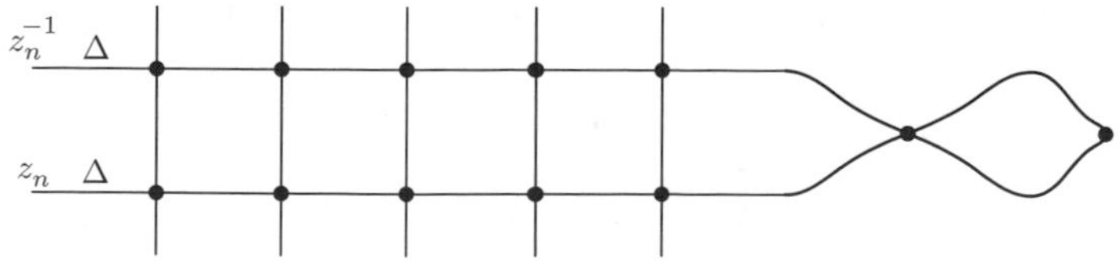

We will denote this by $Z^*(I_9)$; $*$ means that z_n and z_n^{-1} are switched. We have $Z^*(I_9) = \sum_{e_1,e_2} Z^*(I_{10})(e_1,e_2) \cdot Z^*(I_{11})(e_1,e_2)$ where the sum is taken over all possible spins e_1, e_2, and the pictures of I_{10} and I_{11} are shown below.

Here are pictures of I_{10} (left) and I_{11} (right):

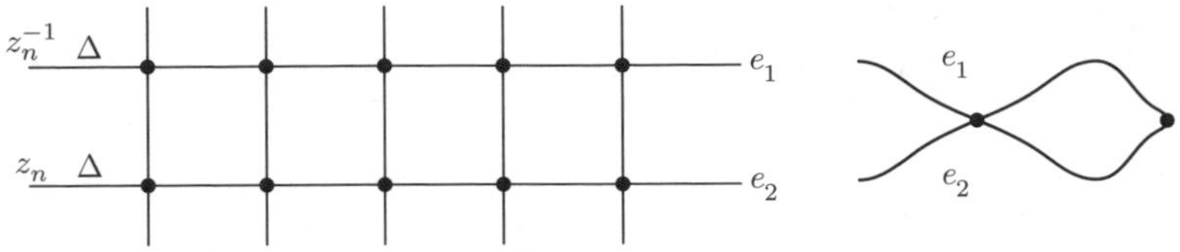

We checked that the following equality holds:

$$Z^*(I_{11}) = (tz_n + z_n^{-1})Z^*(I_{12}),$$

where I_{12} is the ice that only consists of the new cap (see picture)

This result is very similar to the fish equation (Lemma 9 on page 14 of Kuperberg [6]), although the Boltzmann weights that are used in the fish equation and our Boltzmann weights are different. Note in particular that the factor $(tz_n + z_n^{-1})$ does not depend on the choice of spins e_1, e_2, and therefore, we get

$$\begin{aligned} Z^*(I_9) &= \sum_{e_1,e_2} Z^*(I_{10})(e_1,e_2) \cdot Z^*(I_{11})(e_1,e_2) \\ &= (tz_n + z_n^{-1}) \cdot \sum_{e_1,e_2} Z^*(I_9)(e_1,e_2) Z^*(I_{12})(e_1,e_2) = (tz_n + z_n^{-1})Z^*(I_7). \end{aligned}$$

Now we may change the last row back to the gamma ice. As mentioned before, the partition function is not affected by this change; this means that $Z^*(I_7) = Z^*(I_1)$. Hence, we obtain the following equality:

$$(z_n + tz_n^{-1})Z(I_1) = (tz_n + z_n^{-1})Z^*(I_1).$$

From this, we conclude that $\frac{Z(I_1)}{Z(I_1)^*} = \frac{(tz_n+z_n^{-1})}{(z_n+tz_n^{-1})}$. But we also have $\frac{D}{D^*} = \frac{(tz_n+z_n^{-1})}{(z_n+tz_n^{-1})}$, so that $\frac{Z(I_1)}{Z(I_1)^*} = \frac{D}{D^*}$, and so this proves that $Z(I_1)/D$ is invariant under the change $z_i \leftrightarrow z_i^{-1}$. □

Lemma 5. *$Z(I_1)/D$ is a polynomial in t (with coefficients in*

$$\mathbb{C}[z_1, z_2, \ldots, z_n, z_1^{-1}, z_2^{-1}, \ldots, z_n^{-1}]),$$

and it is independent of t.

Proof. We observe that we may write $D = z^{-\rho} \cdot \prod_{\alpha \in \Phi^+} (1 + t z^{\alpha})$, where Φ^+ denotes the set of positive roots. Let $D' = z^{\rho} \cdot \prod_{\alpha \in \Phi^-} (1 + t z^{\alpha})$.

We see that DD' is invariant under the action of the Weyl group (since the Weyl group simply permutes all the roots), and therefore, $Z(I)D' = (Z(I_1)/D) \cdot (DD')$ is also invariant under the action of the Weyl group.

But $Z(I_1)D'$ is clearly divisible by D' (here we view both terms as polynomials in t), and therefore, since $Z(I_1)D'$ is invariant under the action of the Weyl group, it must also be divisible by D. Now, since D and D' are relatively prime in a unique factorization domain $\mathbb{C}[z_1, z_1^{-1}, z_2, z_2^{-1}, \ldots, z_n, z_n^{-1}][t]$ (note that this ring is a localization of a UFD $\mathbb{C}[z_1, z_2, \ldots, z_n,][t]$, so it is indeed a UFD), it follows that D must divide $Z(I_1)$, as desired.

To show that $Z(I_1)/D$ is independent of t, we observe that D has degree n^2 (as a polynomial in t). Hence, it suffices to show that $Z(I_1)$ has degree n^2 as a polynomial in t. In fact, since we already know that $Z(I_1)/D$ is a polynomial in t, it suffices to show that the degree of $Z(I_1))$ is $\leq n^2$. $Z(I_1)$ is the sum of Boltzmann weights of each state, so it suffices to show that the partition function of each state has degree $\leq n^2$.

So consider an admissible state. We observe that the only (possible) powers of t come from the following types of vertices:

Gamma Ice	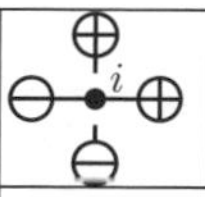	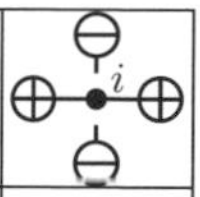
Boltzmann weight	$z_i(t+1)$	t
Delta Ice	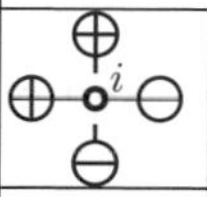	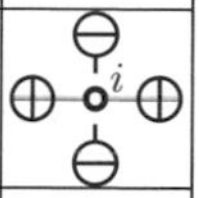
Boltzmann weight	$z_i(t+1)$	$z_i t$

We see that we need to count the number of $-$ signs in the rows of edges between the rows of delta ice and gamma ice.

We recall Lemma 5 of Brubaker et al. [4], which counts the number of $-$ signs in the rows of edges above and below a row of gamma ice. We state this lemma here for the convenience of the reader:

Lemma 6. *Suppose we have a row of gamma ice and that the sign of the left edge is* $+$. *Let* m *be the number of* $-$ *signs in the vertical row of edges above this row of gamma ice and* m' *be the number of* $-$ *signs in the row of vertical edges below this row. Then* $m = m'$ *if and only if the sign of the right edge is* $-$ *and* $m = m' - 1$ *if and only if the sign of the right edge is* $+$. *Moreover, if* $\alpha_1, \alpha_2, \ldots,$ *are the locations of the* $-$ *signs in the row of edges above this row of gamma ice and* $\beta_1, \beta_2, \ldots$ *are the locations of* $-$ *signs in the row of edges below this row, the sequences* $\alpha_1, \alpha_2, \ldots$ *and* $\beta_1, \beta_2, \ldots$ *interleave.*

By symmetry of the delta ice and gamma ice, we obtain the following lemma:

Lemma 7. *Suppose we have a row of delta ice and that the sign of the left edge is* $-$*. Let* m *be the number of* $-$ *signs in the row of edges above this row of delta ice and* m' *be the number of* $-$ *signs in the row of edges below this row. Then* $m = m'$ *if and only if the sign of the right edge is* $-$ *and* $m = m' - 1$ *if and only if the sign of the right edge is* $+$*. Moreover, if* $\alpha_1, \alpha_2, \ldots,$ *are the locations of the* $-$ *signs in the row of edges above this row of delta ice and* $\beta_1, \beta_2, \ldots$ *are the locations of* $-$ *signs in the row of edges below this row, the sequences* $\alpha_1, \alpha_2, \ldots$ *and* $\beta_1, \beta_2, \ldots$ *interleave.*

Since the number of $-$ signs in the top row of edges is n, we see that the number of $-$ signs in the rows of edges between the rows of delta ice and gamma ice is $\leq n(n-1)$.

Moreover, since there are n caps, the degree of t coming from the Boltzmann weights of these caps is $\leq n$. We see that if the signs on the cap are $++$, then the Boltzmann weight is tz_i, so it will contribute only one power of t to the Boltzmann weight of a state; if the signs are $--$, then the Boltzmann weight is z_i^{-1}, which does not contribute any powers of t.

Hence, the degree of the partition function of each state is $\leq n^2$, and as explained above, this means that $Z(I_1)/D$ is independent of t. This completes the proof of Lemma 4. □

10.4 Proctor Patterns

By a *Proctor pattern*, we shall mean a sequence of rows of nonnegative integers $a_{i,j}$ and $b_{i,j}$

$$\begin{array}{ccccccccc}
a_{0,1} & & a_{0,2} & & \cdots & a_{0,r} & & \\
 & b_{1,1} & & b_{1,2} & \cdots & & b_{1,r} & \\
 & & a_{1,2} & & \cdots & a_{1,r} & \vdots & \\
 & & & \ddots & \cdots & & & \\
 & & & & & & b_{r,r} &
\end{array}$$

such that the rows interleave. This last condition means that

$$\min\{a_{(i-1,j)}, a_{(i,j)}\} \geq b_{i,j} \geq \max\{a_{(i-1,j+1)}, a_{(i,j+1)}\}$$

and also

$$\min\{b_{(i+1,j-1)}, b_{(i,j-1)}\} \geq a_{i,j} \geq \max\{b_{(i+1,j)}, b_{(i,j)}\}.$$

We now would like to establish a bijection between the set of all admissible states and the Proctor patterns. Consider a row of delta ice and a row of gamma ice right below it. Let a_i denote the locations of $-$ signs in the top row of vertical edges, b_i

denote the locations of $-$ signs of middle row of vertical edges, and c_i denote the locations of $-$ signs of bottom row of vertical edges. We consider two cases:

Case I: the signs on the cap that connects these two rows are $--$. Then (using Lemmas 6 and 7) we have the following inequalities:

$$a_1 \geq b_1 \geq a_2 \geq \cdots \geq b_m$$

$$b_1 \geq c_1 \geq b_2 \cdots \geq b_m.$$

Case II: the signs on the cap that connects these two rows are $++$. Then we have the following inequalities:

$$a_1 \geq b_1 \geq a_2 \geq \cdots \geq b_{m-1} \geq a_m$$

$$b_1 \geq c_1 \geq b_2 \cdots \geq b_{m-1} \geq c_{m-1}.$$

In this case, we define $b_m = 0$.

We see that the locations of $-$ signs in the rows of vertical edges gives us a Proctor pattern. Here are three rows of a Proctor pattern:

$$\begin{array}{ccccccc} a_1 & & a_2 & \ldots & a_m & & \\ & b_1 & & b_2 & \ldots & b_m & \\ & & c_1 & \ldots & c_{m-1}. & & \end{array}$$

Moreover, we see that this is a one-to-one correspondence, meaning that given any Proctor pattern, there exists an admissible state such that the locations of $-$ signs are exactly as described by this pattern. This also follows from the Lemmas 6 and 7 above.

Since $Z(I_1)/D$ is independent of t, we may choose $t = -1$. This way, the Boltzmann weight of the vertices

Gamma ice	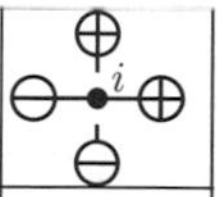
Boltzmann weight	$z_i(t+1)$
Delta ice	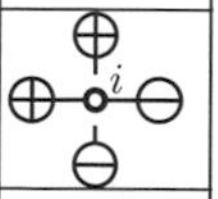
Boltzmann weight	$z_i(t+1)$

is equal to 0, so we consider all possible states which omit both of these vertices.

We observe that there exists a bijection between the Weyl group and the set of all Proctor patterns of the particular type, in which each entry in the pattern (except for entries in the top row) is equal to one of the entries above it (or to 0 in the case of the last entry in an even numbered row). This bijection is given as follows:

given a Proctor pattern, let d_i denote the sum of the entries of the $(2i-1)$th row; in other words, it denotes the sum of the entries in each odd numbered row; here $1 \le i \le n$. Also let $d_{n+1} = 0$. Note that the numbers $d_i - d_{i+1}$ for $1 \le i \le n$, all distinct. In fact, since each entry in the pattern (except for entries in the top row) is equal to one of the entries above it (or to 0 in the case of the last entry in an even numbered row), it follows that $d_i - d_{i+1}$ is equal to $\lambda_j + \rho_j$ for some j $(1 \le j \le n)$. In other words, we get a permutation of n distinct elements $\lambda_1 + \rho_1, \lambda_2 + \rho_2, \dots, \lambda_n + \rho_n$. And for any such permutation and any choice of the signs on the caps, we can construct a Proctor pattern corresponding to it. We see that $w(z_1), w(z_2) \dots w(z_n)$ is equal to $z_{\sigma(1)}^{\epsilon_1}, z_{\sigma(2)}^{\epsilon_2}, z_{\sigma(n)}^{\epsilon_n}$ for some $\sigma \in S_n$ and each $\epsilon_i \in \{\pm 1\}$; the choice of ϵ_i depends on whether an admissible state corresponds to case I or case II described above. Hence, we see that we indeed get a bijection between the Weyl group and the set of all Proctor patterns of the particular type described above. Since we've already observed that there is a bijection between the set of all Proctor patterns of the particular type and the set of all admissible states, by composing these two bijections, we obtain a bijection between the set of all admissible states and the Weyl group. An explicit formula for this bijection is given by $w \to (-1)^{l(w)} z^{w(\lambda+\rho)}$, where $(-1)^{l(w)} z^{w(\lambda+\rho)}$ represents the Boltzmann weight of the corresponding admissible state. Here $z^{(\lambda+\rho)}$ means $z_1^{\lambda_1+n} z_2^{\lambda_2+n-1} \dots z_n^{\lambda_n+1}$ and $z^{w(\lambda+\rho)}$ means $(w(z_1))^{\lambda_1+n}(w(z_2))^{\lambda_2+n-1} \dots (w(z_n))^{\lambda_n+1}$.

Note that the partition function is equal to $\sum_{w \in W} (-1)^{l(w)} z^{w(\lambda+\rho)}$, which is equal to the numerator in the Weyl character formula. Also note that with $t = -1$, we have $D = \prod_i (1 - z_i^2) \cdot \prod_{i<j} (1 - z_i z_j)(1 - t z_i z_j^{-1}) \cdot z^{-\rho}$, which is the denominator in the Weyl character formula. Hence, we have obtained a deformation of the Weyl character formula (which, as we mentioned before, is due to Hamel and King [5]), and the proof of the theorem is complete. □

References

1. J. Beineke, B. Brubaker, and S. Frechette, *A crystal description for symplectic multiple Dirichlet series*, in this volume.
2. J. Beineke, B. Brubaker, and S. Frechette, *Weyl group multiple Dirichlet series of Type C*, Pacific J. Math. (to appear).
3. B. Brubaker, D. Bump, G. Chinta, and P. Gunnells, *Metaplectic Whittaker functions and crystals of type B*, in this volume.
4. B. Brubaker, D. Bump, and S. Friedberg. Schur polynomials and the Yang-Baxter equation. *Comm. Math. Phys.*, 308(2):281–301, 2011.
5. A. M. Hamel and R. C. King, *Symplectic shifted tableaux and deformations of Weyl's denominator formula for* sp(2n), J. Algebraic Combin. **16** (2002), no. 3, 269–300 (2003).
6. Greg Kuperberg, *Symmetry classes of alternating-sign matrices under one roof*, Ann. of Math. (2) **156** (2002), no. 3, 835–866.
7. W. A. Stein et al., *Sage Mathematics Software (Version 4.1)*, The Sage Development Team, 2009, http://www.sagemath.org.

Chapter 11
On Witten Multiple Zeta-Functions Associated with Semisimple Lie Algebras III

Yasushi Komori, Kohji Matsumoto, and Hirofumi Tsumura

Abstract We prove certain general forms of functional relations among Witten multiple zeta-functions in several variables (or zeta-functions of root systems). The structural background of these functional relations is given by the symmetry with respect to Weyl groups. From these relations, we can deduce explicit expressions of values of Witten zeta-functions at positive even integers, which are written in terms of generalized Bernoulli numbers of root systems. Furthermore, we introduce generating functions of Bernoulli numbers of root systems, using which we can give an algorithm of calculating Bernoulli numbers of root systems.

Keywords Witten zeta-functions • Root systems • Lie algebras • Bernoulli polynomials • Weyl groups

11.1 Introduction

Let $\mathfrak{g}$ be a complex semisimple Lie algebra with rank r. The Witten zeta-function associated with $\mathfrak{g}$ is defined by

Y. Komori
Department of Mathematics, Rikkyo University, Nishi-Ikebukuro, Toshima-ku, Tokyo 171-8501, Japan
e-mail: komori@rikkyo.ac.jp

K. Matsumoto (✉)
Graduate School of Mathematics, Nagoya University, Chikusa-ku, Nagoya 464-8602, Japan
e-mail: kohjimat@math.nagoya-u.ac.jp

H. Tsumura
Department of Mathematics and Information Sciences, Tokyo Metropolitan University, 1-1, Minami-Ohsawa, Hachioji, Tokyo 192-0397, Japan
e-mail: tsumura@tmu.ac.jp

D. Bump et al. (eds.), *Multiple Dirichlet Series, L-functions and Automorphic Forms*, Progress in Mathematics 300, DOI 10.1007/978-0-8176-8334-4_11,

$$\zeta_W(s;\mathfrak{g}) = \sum_{\varphi} (\dim \varphi)^{-s}, \tag{1}$$

where the summation runs over all finite-dimensional irreducible representations φ of $\mathfrak{g}$.

Let $\mathbb{N}$ be the set of positive integers, $\mathbb{N}_0 = \mathbb{N} \cup \{0\}$, $\mathbb{Z}$ the ring of rational integers, $\mathbb{Q}$ the rational number field, $\mathbb{R}$ the real number field, and $\mathbb{C}$ the complex number field, respectively. Witten's motivation [38] for introducing the above zeta-function is to express the volumes of certain moduli spaces in terms of special values of (1). This expression is called Witten's volume formula, which especially implies that

$$\zeta_W(2k;\mathfrak{g}) = C_W(2k,\mathfrak{g})\pi^{2kn} \tag{2}$$

for any $k \in \mathbb{N}$, where n is the number of all positive roots and $C_W(2k,\mathfrak{g}) \in \mathbb{Q}$ (Witten [38], Zagier [39]).

When $\mathfrak{g} = \mathfrak{sl}(2)$, the corresponding Witten zeta-function is nothing but the Riemann zeta-function $\zeta(s)$. It is classically known that

$$\zeta(2k) = \frac{-1}{2}\frac{(2\pi\sqrt{-1})^{2k}}{(2k)!}B_{2k} \quad (k \in \mathbb{N}), \tag{3}$$

where B_{2k} is the $2k$-th Bernoulli number. Formula (2) is a generalization of (3), but the values $C_W(2k,\mathfrak{g})$ are not explicitly determined in the work of Witten and of Zagier. In this chapter, we introduce a root-system theoretic generalization of Bernoulli numbers and periodic Bernoulli functions, and express $C_W(2k,\mathfrak{g})$ explicitly in terms of generalized periodic Bernoulli functions $P(\mathbf{k},\mathbf{y};\Delta)$ (defined in Sect. 11.4). This result will be given in Theorem 8. Note that Szenes [31,32] also studied generalizations of Bernoulli polynomials from the viewpoint of the theory of arrangement of hyperplanes, which include $P(\mathbf{k},\mathbf{y};\Delta)$ mentioned above. However, our root-system theoretic approach enables us to show that our $P(\mathbf{k},\mathbf{y};\Delta)$ and its generating function $F(\mathbf{k},\mathbf{y};\Delta)$ are quite natural extensions of the classical ones (see Theorems proved in Sect. 11.6).

Our explicit expression of $C_W(2k,\mathfrak{g})$ is obtained as a special case of a general family of functional relations, which is another main result of this chapter. To explain this, we first introduce the multivariable version of Witten zeta-functions.

Let Δ be the set of all roots of $\mathfrak{g}$, Δ_+ the set of all positive roots of $\mathfrak{g}$, $\Psi = \{\alpha_1,\ldots,\alpha_r\}$ the fundamental system of Δ, and $\alpha_j^\vee$ the coroot associated with α_j $(1 \le j \le r)$. Let $\lambda_1,\ldots,\lambda_r$ be the fundamental weights satisfying $\langle\alpha_i^\vee,\lambda_j\rangle = \lambda_j(\alpha_i^\vee) = \delta_{ij}$ (Kronecker's delta). A more explicit form of $\zeta_W(s;\mathfrak{g})$ can be written down in terms of roots and weights by using Weyl's dimension formula (see (1.4) of [16]). Inspired by that form, we introduced in [16] the multivariable version of Witten zeta-function

$$\zeta_r(\mathbf{s};\mathfrak{g}) = \sum_{m_1=1}^{\infty}\cdots\sum_{m_r=1}^{\infty}\prod_{\alpha\in\Delta_+}\langle\alpha^\vee, m_1\lambda_1+\cdots+m_r\lambda_r\rangle^{-s_\alpha}, \tag{4}$$

where $\mathbf{s} = (s_\alpha)_{\alpha\in\Delta_+} \in \mathbb{C}^n$. In the case that $\mathfrak{g}$ is of type X_r, we call (4) the zeta-function of the root system of type X_r and also denote it by $\zeta_r(\mathbf{s}; X_r)$, where $X = A, B, C, D, E, F, G$. Note that from (1.5) and (1.7) in [16], we have

$$\zeta_W(s; \mathfrak{g}) = K(\mathfrak{g})^s \zeta_r(s, \ldots, s; \mathfrak{g}), \tag{5}$$

where

$$K(\mathfrak{g}) = \prod_{\alpha\in\Delta_+} \langle \alpha^\vee, \lambda_1 + \cdots + \lambda_r \rangle. \tag{6}$$

More generally, in [16], we introduced multiple zeta-functions associated with sets of roots. The main body of [16] is devoted to the study of recursive structures in the family of those zeta-functions that can be described in terms of Dynkin diagrams of underlying root systems.

The Euler–Zagier r-fold sum is defined by the multiple series

$$\begin{aligned}\zeta_r(s_1, \ldots, s_r) = \sum_{m_1=1}^{\infty} \cdots \sum_{m_r=1}^{\infty} m_1^{-s_1} (m_1 + m_2)^{-s_2} \times \\ \times \cdots \times (m_1 + \cdots + m_r)^{-s_r}.\end{aligned} \tag{7}$$

The harmonic product formula

$$\zeta(s_1)\zeta(s_2) = \zeta(s_1 + s_2) + \zeta_2(s_1, s_2) + \zeta_2(s_2, s_1), \tag{8}$$

due to L. Euler (where $\zeta(s)$ denotes the Riemann zeta-function), and its r-ple analogue are classical examples of functional relations (cf. Bradley [3]). However, no other functional relations among multiple zeta-functions have been discovered for a long time.

Let

$$\zeta_{MT,2}(s_1, s_2, s_3) = \sum_{m=1}^{\infty} \sum_{n=1}^{\infty} m^{-s_1} n^{-s_2} (m + n)^{-s_3}. \tag{9}$$

This series is called Tornheim's harmonic double sum or the Mordell–Tornheim double zeta-function, after the work of Tornheim [33] and Mordell [27] in 1950s. But it is to be noted that this sum actually coincides with the Witten zeta-function (4) for $\mathfrak{g} = \mathfrak{sl}(3)$, that is, the simple Lie algebra of type A_2. Recently, the third-named author [36] proved that there are certain functional relations among $\zeta_{MT,2}(s_1, s_2, s_3) = \zeta_2(s_1, s_2, s_3; A_2)$ and the Riemann zeta-function. Moreover, he obtained the same type of functional relations for various relatives of $\zeta_{MT,2}(s_1, s_2, s_3)$ (see [34, 35]). The method in these papers can be called the "u-method," because an auxiliary parameter $u > 1$ was introduced to ensure the absolute convergence in the argument.

In [24], by the same "u-method," the second- and the third-named authors proved certain functional relations among $\zeta_3(\mathbf{s}; A_3)$, $\zeta_2(\mathbf{s}; A_2)$, and the Riemann zeta-function. The papers [22, 23, 25, 26] are also devoted to the study of some new functional relations for certain (mainly) double and triple zeta-functions and their relatives.

The above papers give many examples of functional relations. Therefore, it is timely to investigate the structural reason underlying these functional relations. The first hint on this question was supplied by Nakamura's paper [28], in which he presented a new simple proof of the result of the third-named author [36]. (Nakamura's method has then been applied in [22, 23].)

It can be observed from Nakamura's proof that the functional relations proved in [28, 36] are connected with the symmetry with respect to the symmetric group $\mathfrak{S}_3$, which is the Weyl group of the Lie algebra of type A_2. This suggests the formulation of general functional relations using Weyl groups.

One of the main purposes of this chapter is to show that such general forms of functional relations can indeed be proved. In Sect. 11.2, we prepare some notation and preliminary results about root systems, Weyl groups, and convex polytopes, which play essential roles in the study of the structural background of functional relations. We will give general forms of functional relations in Sects. 11.3 and 11.4. The most general form of functional relations is Theorem 3, which is specialized to the case of "Weyl group symmetric" linear combinations $S(\mathbf{s}, \mathbf{y}; I; \Delta)$ (defined by (110)) of zeta-functions of root systems with exponential factors in Theorems 5 and 6. A naïve form of Theorem 6 has been announced in Sect. 3 of [11], but we consider the generalized form with exponential factors because this form can be applied to evaluations of L-functions of root systems (see [17] for the details).

The theorems mentioned above give expressions of linear combinations of zeta-functions in terms of certain multiple integrals involving Lerch zeta-functions. Since the values of Lerch zeta-functions at positive integers (≥ 2) can be written as Bernoulli polynomials, we can show more explicit forms of those multiple integrals in some special cases. We will carry out this procedure using generating functions (Theorem 4).

In particular, we find that the value $S(\mathbf{s}, \mathbf{y}; \Delta) = S(\mathbf{s}, \mathbf{y}; \emptyset; \Delta)$ at $\mathbf{s} = \mathbf{k} = (k_\alpha)$, where all k_α's are positive integers (≥ 2), is essentially a generalization $P(\mathbf{k}, \mathbf{y}; \Delta)$ of periodic Bernoulli functions. The generating function $F(\mathbf{t}, \mathbf{y}; \Delta)$ of $P(\mathbf{k}, \mathbf{y}; \Delta)$ will be evaluated in Theorem 7. Consequently, we can prove a generalization of Witten's volume formula (2) (Theorem 8).

In Sect. 11.5, we will show the Weyl group symmetry of $S(\mathbf{s}, \mathbf{y}; \Delta)$, $F(\mathbf{t}, \mathbf{y}; \Delta)$, and $P(\mathbf{k}, \mathbf{y}; \Delta)$. For our purpose, it is not sufficient to consider the usual Weyl group only, and hence, we will introduce a certain extension $\widehat{W}$ of the affine Weyl group and will prove the symmetry with respect to $\widehat{W}$ (Theorems 9, 11, and 12). These results ensure that the existence of functional relations is indeed based on the symmetry with respect to $\widehat{W}$. Although the symmetry by affine translations is itself trivial in the case of periodic Bernoulli functions, it plays an important role to construct the action of Weyl groups on a generalization of Bernoulli polynomials introduced in Sect. 11.6. (See Theorem 16 for details.) It is to be noted that Weyl groups already played a role in Zagier's sketch [39] of the proof of (2) and also in Gunnells–Sczech's computation [6].

In Sect. 11.6, we will prove that $P(\mathbf{k},\mathbf{y};\Delta)$ can be continued to polynomials in $\mathbf{y}$ (Theorem 13). This may be regarded as a (root-system theoretic) generalization of Bernoulli polynomials. Since $P(\mathbf{k},\mathbf{y};\Delta)$ is essentially the same as $S(\mathbf{k},\mathbf{y};\Delta)$, this gives explicit relations among special values of zeta-functions of root systems. Moreover, in the same theorem, we will give the continuation of $F(\mathbf{t},\mathbf{y};\Delta)$.

As examples, in Sect. 11.7, we will calculate P and F explicitly in the cases of A_1, A_2, A_3, and B_2. In particular, from the explicit expansion of generating functions F, we will determine the value of $C_W(2,A_2)$, $C_W(2,A_3)$, and $C_W(2,B_2)$ in (2). The case of A_2 type is included in the results of Mordell [27], Subbarao et al. [29], and Zagier [39]. Furthermore, Gunnells–Sczech [6] studied general cases and, in particular, gave explicit examples for $C_W(2k,A_3)$. Here we can deduce explicit value relations and special values of multivariable zeta-functions of any simple algebra $\mathfrak{g}$ by the same argument, at least in principle, though the actual procedure will become quite complicated when the rank of $\mathfrak{g}$ becomes higher. We also provide an example of a functional relation in the A_2 case.

Some parts of the contents of [16] and this chapter have been already announced briefly in [11–13, 15].

11.2 Root Systems, Weyl Groups, and Convex Polytopes

In this preparatory section, we first fix notation and summarize basic facts about root systems and Weyl groups. See [2, 8–10] for the details. Let V be an r-dimensional real vector space equipped with an inner product $\langle\cdot,\cdot\rangle$. We denote the norm of $v\in V$ by $\|v\|=\langle v,v\rangle^{1/2}$. The dual space V^* is identified with V via the inner product of V. Let Δ be a finite reduced root system in V and $\Psi=\{\alpha_1,\ldots,\alpha_r\}$ its fundamental system. Let Δ_+ and Δ_- be the set of all positive and negative roots, respectively. Then we have a decomposition of the root system $\Delta=\Delta_+\coprod\Delta_-$. Let $Q^\vee$ be the coroot lattice, P the weight lattice, P_+ the set of integral dominant weights, and P_{++} the set of integral strongly dominant weights, respectively, defined by

$$Q^\vee=\bigoplus_{i=1}^r\mathbb{Z}\alpha_i^\vee,\quad P=\bigoplus_{i=1}^r\mathbb{Z}\lambda_i,\quad P_+=\bigoplus_{i=1}^r\mathbb{N}_0\lambda_i,\quad P_{++}=\bigoplus_{i=1}^r\mathbb{N}\lambda_i,\tag{10}$$

where the fundamental weights $\{\lambda_j\}_{j=1}^r$ are a basis dual to $\Psi^\vee$ satisfying $\langle\alpha_i^\vee,\lambda_j\rangle=\delta_{ij}$. Let

$$\rho=\frac{1}{2}\sum_{\alpha\in\Delta_+}\alpha=\sum_{j=1}^r\lambda_j\tag{11}$$

be the lowest strongly dominant weight. Then $P_{++}=P_++\rho$.

We define the reflection σ_α with respect to a root $\alpha \in \Delta$ as

$$\sigma_\alpha : V \to V, \quad \sigma_\alpha : v \mapsto v - \langle \alpha^\vee, v \rangle \alpha, \tag{12}$$

and for a subset $A \subset \Delta$, let $W(A)$ be the group generated by reflections σ_α for $\alpha \in A$. Let $W = W(\Delta)$ be the Weyl group. Then $\sigma_j = \sigma_{\alpha_j}$ $(1 \leq j \leq r)$ generates W. Namely, we have $W = W(\Psi)$. Any two fundamental systems Ψ, Ψ' are conjugate under W.

We denote the fundamental domain called the fundamental Weyl chamber by

$$C = \{v \in V \mid \langle \Psi^\vee, v \rangle \geq 0\}, \tag{13}$$

where $\langle \Psi^\vee, v \rangle$ means any of $\langle \alpha^\vee, v \rangle$ for $\alpha^\vee \in \Psi^\vee$. Then W acts on the set of Weyl chambers $WC = \{wC \mid w \in W\}$ simply transitively. Moreover, if $wx = y$ for $x, y \in C$, then $x = y$ holds. The stabilizer W_x of a point $x \in V$ is generated by the reflections which stabilize x. We see that $P_+ = P \cap C$.

Let $\mathrm{Aut}(\Delta)$ be the subgroup of all the automorphisms $\mathrm{GL}(V)$ which stabilizes Δ (see [8, Sect. 12.2]). Then the Weyl group W is a normal subgroup of $\mathrm{Aut}(\Delta)$, and there exists a subgroup $\Omega \subset \mathrm{Aut}(\Delta)$ such that

$$\mathrm{Aut}(\Delta) = \Omega \ltimes W. \tag{14}$$

The subgroup Ω is isomorphic to the group $\mathrm{Aut}(\Gamma)$ of automorphisms of the Dynkin diagram Γ. For $w \in \mathrm{Aut}(\Delta)$, we set

$$\Delta_w = \Delta_+ \cap w^{-1} \Delta_- \tag{15}$$

and the length function $\ell(w) = |\Delta_w|$ (see [9, Sect. 1.6]). The subgroup Ω is characterized as $w \in \Omega$ if and only if $\ell(w) = 0$. Note that $w\Delta_w = \Delta_- \cap w\Delta_+ = -\Delta_{w^{-1}}$ and $\ell(w) = \ell(w^{-1})$.

For $u \in V$, let $\tau(u)$ be the translation by u, that is,

$$\tau(u) : V \to V, \quad \tau(u) : v \mapsto v + u. \tag{16}$$

Since $\mathrm{Aut}(\Delta)$ stabilizes the coroot lattice $Q^\vee$, we can define

$$\widehat{W} = \mathrm{Aut}(\Delta) \ltimes \tau(Q^\vee). \tag{17}$$

Then $\widehat{W} = (\Omega \ltimes W) \ltimes \tau(Q^\vee) \simeq \Omega \ltimes (W \ltimes \tau(Q^\vee))$. It should be noted that $\widehat{W}$ is an extension of the affine Weyl group $W \ltimes \tau(Q^\vee)$ different from the extended affine Weyl group $W \ltimes \tau(P^\vee)$ (see [2, 10] for the details of affine Weyl groups).

Let $\widehat{V} = V \times \mathbb{R}$ and $\delta = (0, 1) \in \widehat{V}$. We embed V in $\widehat{V}$ and we have $\widehat{V} = V \oplus \mathbb{R}\,\delta$. For $\gamma = \eta + c\delta \in \widehat{V}$ with $\eta \in V$ and $c \in \mathbb{R}$, we associate an affine linear functional on V as $\gamma(v) = \langle \eta, v \rangle + c$. Let $\widehat{Q}^\vee$ be the affine coroot lattice defined by

$$\widehat{Q}^\vee = Q^\vee \oplus \mathbb{Z}\,\delta \tag{18}$$

(see [10]).

For a set X, let $\mathfrak{F}(X)$ be the set of all functions $f : X \to \mathbb{C}$. For a function $f \in \mathfrak{F}(P)$, we define a subset

$$H_f = \{\lambda \in P \mid f(\lambda) = 0\} \tag{19}$$

and for a subset A of $\mathfrak{F}(P)$, define $H_A = \bigcup_{f \in A} H_f$. It is noted that we typically work with linear functions, and in such cases, H_A is the intersection of P and the union of some hyperplanes.

One sees that an action of W is induced on $\mathfrak{F}(P)$ as $(wf)(\lambda) = f(w^{-1}\lambda)$. Note that $V \subset \widehat{V} \subset \mathfrak{F}(P)$, where the second inclusion is given by the associated functional mentioned above.

Let $I \subset \{1, \ldots, r\}$ and $\Psi_I = \{\alpha_i \mid i \in I\} \subset \Psi$. Let V_I be the linear subspace spanned by Ψ_I. Then $\Delta_I = \Delta \cap V_I$ is a root system in V_I whose fundamental system is Ψ_I. For the root system Δ_I, we denote the corresponding coroot lattice (resp. weight lattice, etc.) by $Q_I^\vee = \bigoplus_{i \in I} \mathbb{Z}\alpha_i^\vee$ (resp. $P_I = \bigoplus_{i \in I} \mathbb{Z}\lambda_i$, etc.). We define

$$C_I = \{v \in C \mid \langle \Psi_{I^c}^\vee, v\rangle = 0, \quad \langle \Psi_I^\vee, v\rangle > 0\}, \tag{20}$$

where I^c is the complement of I. Then the dimension of the linear span of C_I is $|I|$, and we have a disjoint union

$$C = \coprod_{I \subset \{1,\ldots,r\}} C_I \tag{21}$$

and the collection of all sets wC_I for $w \in W$ and $I \subset \{1, \ldots, r\}$ is called the Coxeter complex (see [9, Sect. 1.15]; it should be noted that we use a little different notation), which partitions V and we have a decomposition

$$P_+ = \coprod_{I \subset \{1,\ldots,r\}} P_{I++}, \tag{22}$$

where

$$P_{I++} = P_+ \cap C_I. \tag{23}$$

In particular, $P_{\emptyset++} = \{0\}$ and $P_{\{1,\ldots,r\}++} = P_{++}$.

The natural embedding $\iota : Q_I^\vee \to Q^\vee$ induces the projection $\iota^* : P \to P_I$. Namely, for $\lambda \in P$, $\iota^*(\lambda)$ is defined as a unique element of P_I satisfying $\langle \iota(q), \lambda\rangle = \langle q, \iota^*(\lambda)\rangle$ for all $q \in Q_I^\vee$. Let

$$W^I = \{w \in W \mid \Delta_{I+}^\vee \subset w\Delta_+^\vee\}. \tag{24}$$

Then we have the following key lemmas to functional relations among zeta-functions. Note that the statements hold trivially in the case $I = \emptyset$, and hence, we deal with $I \neq \emptyset$ in their proofs.

Lemma 1. *The subset W^I coincides with the minimal (right) coset representatives $\{w \in W \mid \ell(\sigma_i w) > \ell(w)$ for all $i \in I\}$ of the parabolic subgroup $W(\Delta_I)$ (see [9, Sect. 1.10]). Therefore, $|W^I| = \bigl(W(\Delta) : W(\Delta_I)\bigr)$.*

Proof. Let $w \in W^I$. Then $\Delta^\vee_{I+} \subset w\Delta^\vee_+$, which implies $\Delta^\vee_{I+} \cap w\Delta^\vee_- = \emptyset$. In particular, $\alpha_i^\vee \notin w\Delta^\vee_-$ for $i \in I$, which yields $\Delta^\vee_+ \cap w\Delta^\vee_- = (\Delta^\vee_+ \setminus \{\alpha_i^\vee\}) \cap w\Delta^\vee_-$. Therefore,

$$\sigma_i(\Delta^\vee_+ \cap w\Delta^\vee_-) = \sigma_i\bigl((\Delta^\vee_+ \setminus \{\alpha_i^\vee\}) \cap w\Delta^\vee_-\bigr) = (\Delta^\vee_+ \setminus \{\alpha_i^\vee\}) \cap \sigma_i w\Delta^\vee_- \subset \Delta^\vee_+ \cap \sigma_i w\Delta^\vee_- \tag{25}$$

and $\ell(\sigma_i w) \geq \ell(w)$. Since $\ell(\sigma_i w) = \ell(w) \pm 1$, we have $\ell(\sigma_i w) = \ell(w) + 1$ and w is a minimal coset representative.

Assume that $w \in W$ satisfies $\ell(\sigma_i w) > \ell(w)$ for all $i \in I$. Then we have

$$\sigma_i(\Delta^\vee_+ \cap w\Delta^\vee_-) = \bigl((\Delta^\vee_+ \setminus \{\alpha_i^\vee\}) \cap \sigma_i w\Delta^\vee_-\bigr) \cup \bigl(\{-\alpha_i\} \cap \sigma_i w\Delta^\vee_-\bigr). \tag{26}$$

Since $|\sigma_i(\Delta^\vee_+ \cap w\Delta^\vee_-)| = \ell(w)$ and $|(\Delta^\vee_+ \setminus \{\alpha_i^\vee\}) \cap \sigma_i w\Delta^\vee_-| \geq \ell(\sigma_i w) - 1 = \ell(w)$, we have $|\{-\alpha_i\} \cap \sigma_i w\Delta^\vee_-| = 0$. It implies that no element of $\Delta^\vee_+ \cap w\Delta^\vee_-$ is sent to $\Delta^\vee_-$ by σ_i for $i \in I$, and hence, $\Psi_I^\vee \cap w\Delta^\vee_- = \emptyset$. Since $0 \notin \langle \Delta^\vee, \rho\rangle$ and $\alpha^\vee \in w\Delta^\vee_-$ if and only if $\langle \alpha^\vee, w\rho\rangle < 0$, we have $\langle \Psi_I^\vee, w\rho\rangle > 0$ and hence $\langle \Delta^\vee_{I+}, w\rho\rangle > 0$. It follows that $\Delta^\vee_{I+} \cap w\Delta^\vee_- = \emptyset$ and $w \in W^I$. □

Lemma 2.

$$\iota^{*-1}(P_{I+}) = P_{I+} \oplus P_{I^c} = \bigcup_{w \in W^I} wP_+. \tag{27}$$

Proof. The first equality is clear. We prove the second equality.

Assume $w \in W^I$. Then for $\lambda \in P_+$, we have $\langle \Delta^\vee_{I+}, w\lambda\rangle = \langle w^{-1}\Delta^\vee_{I+}, \lambda\rangle \subset \langle \Delta^\vee_+, \lambda\rangle \geq 0$. Hence, $wP_+ \subset \iota^{*-1}(P_{I+})$.

Conversely, assume $\lambda \in \iota^{*-1}(P_{I+})$. Since $|\Delta^\vee| < \infty$, it is possible to fix a sufficiently small constant $c > 0$ such that $0 < |\langle \Delta^\vee, c\rho\rangle| < 1$. Then we see that $\lambda + c\rho$ is regular (see [8, Sect. 10.1]), that is, $0 \notin \langle \Delta^\vee, \lambda + c\rho\rangle$ and the signs of $\langle \alpha^\vee, \lambda\rangle$ and $\langle \alpha^\vee, \lambda + c\rho\rangle$ coincide if $\langle \alpha^\vee, \lambda\rangle \neq 0$, because $\langle \Delta^\vee, \lambda\rangle \subset \mathbb{Z}$. Let $\tilde{\Delta}^\vee_+ = \{\alpha^\vee \in \Delta^\vee \mid \langle \alpha^\vee, \lambda + c\rho\rangle > 0\}$. Then $\tilde{\Delta}^\vee_+$ is a positive system, and hence, there exists an element $w \in W$ such that $\tilde{\Delta}^\vee_+ = w\Delta^\vee_+$. Since $\lambda \in \iota^{*-1}(P_{I+})$, we have $\Delta^\vee_{I+} \subset \tilde{\Delta}^\vee_+$. Hence, $\Delta^\vee_{I+} \subset w\Delta^\vee_+$, that is, $w \in W^I$. Moreover, $\langle \tilde{\Delta}^\vee_+, \lambda + c\rho\rangle > 0$ implies $\langle \Delta^\vee_+, w^{-1}(\lambda + c\rho)\rangle > 0$ and $\langle \Delta^\vee_+, w^{-1}\lambda\rangle \geq 0$ again due to the integrality. Therefore, $\lambda \in wP_+$. □

Lemma 3. *For $\lambda \in \iota^{*-1}(P_{I+})$, an element $w \in W^I$ satisfying $\lambda \in wP_+$ (whose existence is assured by Lemma 2) is unique if and only if $\lambda \notin H_{\Delta^\vee \setminus \Delta_I^\vee}$.*

Proof. Assume $\alpha^\vee \in \Delta^\vee \setminus \Delta_I^\vee$ and $\lambda \in \iota^{*-1}(P_{I+}) \cap H_{\alpha^\vee}$. Let $w \in W^I$ satisfy $\lambda \in wP_+$. Then $\sigma_\alpha\lambda = \lambda \in wP_+$ and hence $w^{-1}\lambda = \sigma_{w^{-1}\alpha}w^{-1}\lambda \in P_+$, which further implies $w^{-1}\alpha^\vee \in \Delta'^\vee$, where $\Delta'^\vee$ is a coroot system orthogonal to $w^{-1}\lambda$ whose fundamental system is given by $\Psi'^\vee = \{\alpha_i^\vee \in \Psi^\vee \mid \langle \alpha_i^\vee, w^{-1}\lambda\rangle = 0\}$

(see [9, Sect. 1.12]). If $\Psi'^{\vee} \subset w^{-1}\Delta_I^{\vee}$, then $W(\Psi'^{\vee})\Psi'^{\vee} = \Delta'^{\vee} \subset w^{-1}\Delta_I^{\vee}$, and hence $w^{-1}\alpha^{\vee} \in w^{-1}\Delta_I^{\vee}$, which contradicts to the assumption $\alpha^{\vee} \notin \Delta_I^{\vee}$. Therefore, there exists a fundamental coroot $\alpha_i^{\vee} \in \Psi'^{\vee} \setminus w^{-1}\Delta_I^{\vee}$, which satisfies $\sigma_i w^{-1}\lambda = w^{-1}\lambda \in P_+$ by construction. Since $w \in W^I$, we have $w^{-1}\Delta_{I+}^{\vee} \subset \Delta_+^{\vee} \setminus \{\alpha_i^{\vee}\}$. Hence, $\sigma_i w^{-1}\Delta_{I+}^{\vee} \subset \Delta_+^{\vee}$, because $\sigma_i(\Delta_+^{\vee} \setminus \{\alpha_i^{\vee}\}) \subset \Delta_+^{\vee}$. Then putting $w' = w\sigma_i$, we have $W^I \ni w' \neq w$ such that $\lambda \in wP_+ \cap w'P_+$.

Conversely, assume that there exist $w, w' \in W^I$ such that $w \neq w'$ and $\lambda \in wP_+ \cap w'P_+$. This implies that $w^{-1}\lambda = w'^{-1}\lambda$ is on a wall of C and hence $\lambda \in H_{\Delta^{\vee}}$. Let $\Delta''^{\vee} = \{\alpha^{\vee} \in \Delta^{\vee} \mid \lambda \in H_{\alpha^{\vee}}\}$ be a coroot system orthogonal to λ so that $\lambda \in H_{\Delta''^{\vee}}$. Assume $\Delta''^{\vee} \subset \Delta_I^{\vee}$. Then by $\lambda = ww'^{-1}\lambda$, we have $ww'^{-1} \in W_{\lambda}$ and hence $ww'^{-1} \in W(\Delta_I)$, because $W_{\lambda} = W(\Delta''^{\vee}) \subset W(\Delta_I^{\vee})$ by the assumption. Since $\mathrm{id} \neq ww'^{-1} \in W(\Delta_I)$, there exists a coroot $\alpha^{\vee} \in \Delta_{I+}^{\vee}$ such that $\beta^{\vee} = ww'^{-1}\alpha^{\vee} \in \Delta_{I-}^{\vee}$. Then, since $w^{-1}(ww'^{-1})\Delta_{I+}^{\vee} \subset \Delta_+^{\vee}$ and $w^{-1}\Delta_{I+}^{\vee} \subset \Delta_+^{\vee}$, we have $w^{-1}\beta^{\vee} \in \Delta_+^{\vee}$ from the first inclusion and $w^{-1}(-\beta^{\vee}) \in \Delta_+^{\vee}$ from the second one, which leads to the contradiction. Therefore, $\lambda \in H_{\alpha^{\vee}}$ for $\alpha \in \Delta''^{\vee} \setminus \Delta_I^{\vee}$. □

Next, we give some definitions and facts about convex polytopes (see [7,40]) and their triangulations. For a subset $X \subset \mathbb{R}^N$, we denote by $\mathrm{Conv}(X)$ the convex hull of X. A subset $\mathcal{P} \subset \mathbb{R}^N$ is called a convex polytope if $\mathcal{P} = \mathrm{Conv}(X)$ for some finite subset $X \subset \mathbb{R}^N$. Let $\mathcal{P}$ be a d-dimensional polytope. Let $\mathcal{H}$ be a hyperplane in $\mathbb{R}^N$. Then $\mathcal{H}$ divides $\mathbb{R}^N$ into two half-spaces. If $\mathcal{P}$ is entirely contained in one of the two closed half-spaces and $\mathcal{P} \cap \mathcal{H} \neq \emptyset$, then $\mathcal{H}$ is called a supporting hyperplane. For a supporting hyperplane $\mathcal{H}$, a subset $\mathcal{F} = \mathcal{P} \cap \mathcal{H} \neq \emptyset$ is called a face of the polytope $\mathcal{P}$. If the dimension of a face $\mathcal{F}$ is j, then we call it a j-face $\mathcal{F}$. A 0-face is called a vertex and a $(d-1)$-face a facet. For convenience, we regard $\mathcal{P}$ itself as its unique d-face. Let $\mathrm{Vert}(\mathcal{P})$ be the set of the vertices of $\mathcal{P}$. Then

$$\mathcal{F} = \mathrm{Conv}(\mathrm{Vert}(\mathcal{P}) \cap \mathcal{F}), \tag{28}$$

for a face $\mathcal{F}$.

A triangulation of a polytope is a partition of it into simplexes that intersect each other in entire faces. It is known that a convex polytope can be triangulated without adding any vertices. Here we give an explicit procedure of a triangulation of $\mathcal{P}$. Number all the vertices of $\mathcal{P}$ as $\mathbf{p}_1, \ldots, \mathbf{p}_k$. For a face $\mathcal{F}$, by $\mathcal{N}(\mathcal{F})$, we mean the vertex $\mathbf{p}_j$ whose index j is the smallest in the vertices belonging to $\mathcal{F}$. A full flag Φ is defined by the sequence

$$\Phi : \mathcal{F}_0 \subset \mathcal{F}_1 \subset \cdots \subset \mathcal{F}_{d-1} \subset \mathcal{F}_d = \mathcal{P}, \tag{29}$$

with j-faces $\mathcal{F}_j$ such that $\mathcal{N}(\mathcal{F}_j) \notin \mathcal{F}_{j-1}$.

Theorem 1 ([30]). *All the collection of the simplexes with vertices $\mathcal{N}(\mathcal{F}_0), \ldots, \mathcal{N}(\mathcal{F}_{d-1}), \mathcal{N}(\mathcal{F}_d)$ associated with full flags gives a triangulation.*

Remark 1. This procedure only depends on the face poset structure of $\mathcal{P}$ (see [7, Sect. 5]).

For $\mathbf{a} = {}^t(a_1, \ldots, a_N), \mathbf{b} = {}^t(b_1, \ldots, b_N) \in \mathbb{C}^N$, we define $\mathbf{a} \cdot \mathbf{b} = a_1 b_1 + \cdots + a_N b_N$. The definition of polytopes above is that of "V-polytopes." We mainly deal with another representation of polytopes, "H-polytopes," instead. Namely, consider a bounded subset of the form

$$\mathcal{P} = \bigcap_{i \in I} \mathcal{H}_i^+ \subset \mathbb{R}^N, \tag{30}$$

where $|I| < \infty$ and $\mathcal{H}_i^+ = \{\mathbf{x} \in \mathbb{R}^N \mid \mathbf{a}_i \cdot \mathbf{x} \geq h_i\}$ with $\mathbf{a}_i \in \mathbb{R}^N$ and $h_i \in \mathbb{R}$. The following theorem is intuitively clear but nontrivial (see, e.g., [40, Theorem 1.1]).

Theorem 2 (Weyl–Minkowski). *H-polytopes are V-polytopes and vice versa.*

We have a representation of k-faces in terms of hyperplanes $\mathcal{H}_i = \{\mathbf{x} \in \mathbb{R}^N \mid \mathbf{a}_i \cdot \mathbf{x} = h_i\}$.

Proposition 1. *Let $J \subset I$. Assume that $\mathcal{F} = \mathcal{P} \cap \bigcap_{j \in J} \mathcal{H}_j \neq \emptyset$. Then $\mathcal{F}$ is a face.*

Proof. Let $\mathbf{x} \in \mathcal{P}$. Then $\mathbf{x} \in \bigcap_{j \in J} \mathcal{H}_j^+$ and hence $\mathbf{a}_j \cdot \mathbf{x} \geq h_j$ for all $j \in J$. Set $\mathbf{a} = \sum_{j \in J} \mathbf{a}_j$ and $h = \sum_{j \in J} h_j$. Let $\mathcal{H}$ be a hyperplane defined by $\{\mathbf{x} \in \mathbb{R}^N \mid \mathbf{a} \cdot \mathbf{x} = h\}$. Then $\mathbf{a} \cdot \mathbf{x} \geq h$ for $\mathbf{x} \in \mathcal{P}$ and $\mathcal{P} \subset \mathcal{H}^+ = \{\mathbf{x} \in \mathbb{R}^N \mid \mathbf{a} \cdot \mathbf{x} \geq h\}$.

Let $\mathbf{x} \in \mathcal{P} \cap \mathcal{H}$. Then $\mathbf{a} \cdot \mathbf{x} = h$. Since $\mathbf{a}_j \cdot \mathbf{x} = h_j + c_j \geq h_j$ with some $c_j \geq 0$ for all $j \in J$, we have $c_j = 0$ and $\mathbf{x} \in \mathcal{H}_j$ for all $j \in J$. Thus, $\mathbf{x} \in \mathcal{F}$. It is easily seen that $\mathbf{x} \in \mathcal{F}$ satisfies $\mathbf{x} \in \mathcal{P} \cap \mathcal{H}$. Therefore, $\mathcal{F} = \mathcal{P} \cap \mathcal{H}$ and $\mathcal{H}$ is a supporting hyperplane. □

Proposition 2. *Let $\mathcal{H}$ be a supporting hyperplane and $\mathcal{F} = \mathcal{P} \cap \mathcal{H}$ a k-face. Then there exists a set of indices $J \subset I$ such that $|J| = (\dim \mathcal{P}) - k$ and $\mathcal{F} = \mathcal{P} \cap \bigcap_{j \in J} \mathcal{H}_j$.*

Proof. Assume $d = N$ without loss of generality. Let $\mathbf{x} \in \mathcal{F}$. Then $\mathbf{x} \in \mathcal{H}$ and hence $\mathbf{x} \in \partial \mathcal{P}$ since $\mathcal{P} \subset \mathcal{H}^+$. If $\mathbf{x} \in \mathcal{H}_i^+ \setminus \mathcal{H}_i$ for all $i \in I$, then $\mathbf{x}$ is in the interior of $\mathcal{P}$, which contradicts to the above. Thus, $\mathbf{x} \in \mathcal{H}_j$ for some $j \in I$.

First we assume that $\mathcal{F} = \mathcal{P} \cap \mathcal{H}$ is a facet. Then there exists a subset $\{\mathbf{x}_1, \ldots, \mathbf{x}_N\} \subset \mathcal{F}$ such that $\mathbf{x}_2 - \mathbf{x}_1, \ldots, \mathbf{x}_N - \mathbf{x}_1$ are linearly independent. Let $\mathcal{C}$ be the convex hull of $\{\mathbf{x}_1, \ldots, \mathbf{x}_N\}$. We consider that $\mathcal{C} \subset \mathcal{F}$ is equipped with the relative topology. Note that for $\mathbf{x} \in \mathcal{C}$, we have $\mathbf{x} \in \mathcal{H}_{i(\mathbf{x})}$, where $i(\mathbf{x}) \in I$. We show that there exists an open subset $\mathcal{U} \subset \mathcal{C}$ and $i \in I$ such that $\mathcal{H}_i \cap \mathcal{U}$ is dense in $\mathcal{U}$. Fix an order $\{i_1, i_2, \ldots\} = I$. If $\mathcal{H}_{i_1} \cap \mathcal{C}$ is dense in $\mathcal{C}$, then we have done. Hence, we assume that it is false. Then there exists an open subset $\mathcal{U}_1 \subset \mathcal{C}$ such that $\mathcal{H}_{i_1} \cap \mathcal{U}_1 = \emptyset$. Similarly, we see that $\mathcal{H}_{i_2} \cap \mathcal{U}_1$ is dense in $\mathcal{U}_1$ unless there exists an open subset $\mathcal{U}_2 \subset \mathcal{U}_1$ such that $\mathcal{H}_{i_2} \cap \mathcal{U}_2 = \emptyset$. Since $|I| < \infty$, repeated application of this argument yields the assertion. Thus, there exists a subset $\{\mathbf{x}'_1, \ldots, \mathbf{x}'_N\} \subset \mathcal{H}_i \cap \mathcal{U}$ such that $\mathbf{x}'_2 - \mathbf{x}'_1, \ldots, \mathbf{x}'_N - \mathbf{x}'_1$ are linearly independent. Hence, we have $\mathcal{F} \subset \mathcal{H}_i$ and $\mathcal{H} = \mathcal{H}_i$.

For any k-face $\mathcal{F}$, there exists a sequence of faces such that

$$\mathcal{F} = \mathcal{F}_k \subset \mathcal{F}_{k+1} \subset \cdots \subset \mathcal{F}_{N-1} \subset \mathcal{P}, \tag{31}$$

where $\mathcal{F}_j$ $(k \le j \le N-1)$ is a j-face. Since $\mathcal{F}_j$ is a facet of $\mathcal{F}_{j+1}$, by the induction on dimensions, we have $\mathcal{F} = \mathcal{P} \cap \bigcap_{j\in J} \mathcal{H}_j$ for some $J \subset I$ with $|J| = N-k$. □

Lemma 4. *Let* $\mathbf{a} = {}^t(a_1,\dots,a_N) \in \mathbb{C}^N$ *and* σ *be a simplex with vertices* $\mathbf{p}_0,\dots,\mathbf{p}_N \in \mathbb{R}^N$ *in general position. Then*

$$\int_\sigma \mathrm{e}^{\mathbf{a}\cdot\mathbf{x}}\mathrm{d}\mathbf{x} = N!\mathrm{Vol}(\sigma)\sum_{m=0}^{N} \frac{\mathrm{e}^{\mathbf{a}\cdot\mathbf{p}_m}}{\prod_{j\neq m}\mathbf{a}\cdot(\mathbf{p}_m-\mathbf{p}_j)} = N!\mathrm{Vol}(\sigma)T(\mathbf{a}\cdot\mathbf{p}_0,\dots,\mathbf{a}\cdot\mathbf{p}_N), \tag{32}$$

where

$$T(x_0,\dots,x_N) = \det\begin{pmatrix} 1 & \cdots & 1 \\ x_0 & \cdots & x_N \\ \vdots & \ddots & \vdots \\ x_0^{N-1} & \cdots & x_N^{N-1} \\ \mathrm{e}^{x_0} & \cdots & \mathrm{e}^{x_N} \end{pmatrix} \Bigg/ \det\begin{pmatrix} 1 & \cdots & 1 \\ x_0 & \cdots & x_N \\ \vdots & \ddots & \vdots \\ x_0^{N-1} & \cdots & x_N^{N-1} \\ x_0^{N} & \cdots & x_N^{N} \end{pmatrix}, \tag{33}$$

$$\mathrm{Vol}(\sigma) = \int_\sigma 1\mathrm{d}\mathbf{x} = \frac{1}{N!}|\det \tilde{P}|, \tag{34}$$

and

$$\tilde{P} = \begin{pmatrix} 1 & \cdots & 1 \\ \mathbf{p}_0 & \cdots & \mathbf{p}_N \end{pmatrix}, \tag{35}$$

with $\mathbf{p}_j$ *regarded as column vectors.*

Proof. Let τ be the simplex whose vertices are ${}^t(0,\dots,\overset{j}{1},\dots,0)$, $1 \le j \le N$, and the origin. Then by changing variables from $\mathbf{x}$ to $\mathbf{y}$ as $\mathbf{x} = \mathbf{p}_0 + P\mathbf{y}$ with the $N \times N$ matrix $P = (\mathbf{p}_1 - \mathbf{p}_0,\dots,\mathbf{p}_N - \mathbf{p}_0)$, we have

$$\int_\sigma \mathrm{e}^{\mathbf{a}\cdot\mathbf{x}}\mathrm{d}\mathbf{x} = \mathrm{e}^{\mathbf{a}\cdot\mathbf{p}_0}|\det P|\int_\tau \mathrm{e}^{\mathbf{b}\cdot\mathbf{y}}\mathrm{d}\mathbf{y}, \tag{36}$$

where ${}^t\mathbf{b} = (b_1,\dots,b_N) = {}^t\mathbf{a}P$. Since

$$\int_0^{1-y_1-\cdots-y_{m-1}} \mathrm{e}^{c+(b_1-c)y_1+\cdots+(b_m-c)y_m}\mathrm{d}y_m = \frac{1}{b_m-c}\left(\mathrm{e}^{b_m+(b_1-b_m)y_1+\cdots+(b_{m-1}-b_m)y_{m-1}} - \mathrm{e}^{c+(b_1-c)y_1+\cdots+(b_{m-1}-c)y_{m-1}}\right), \tag{37}$$

we see that

$$\int_\tau \mathrm{e}^{\mathbf{b}\cdot\mathbf{y}}\mathrm{d}\mathbf{y} = q_0(\mathbf{b}) + \sum_{m=1}^{N} q_m(\mathbf{b})\mathrm{e}^{b_m}, \tag{38}$$

where $q_j(\mathbf{b})$ for $0 \leq j \leq N$ are rational functions in $b_1, \ldots, b_N$. In particular, we have

$$q_0(\mathbf{b}) = \frac{(-1)^N}{b_1 \cdots b_N} \tag{39}$$

and hence

$$\int_\sigma \mathrm{e}^{\mathbf{a}\cdot\mathbf{x}} d\mathbf{x} = \frac{|\det(\mathbf{p}_0 - \mathbf{p}_1, \ldots, \mathbf{p}_0 - \mathbf{p}_N)|}{(\mathbf{a}\cdot(\mathbf{p}_0 - \mathbf{p}_1)) \cdots (\mathbf{a}\cdot(\mathbf{p}_0 - \mathbf{p}_N))} \mathrm{e}^{\mathbf{a}\cdot\mathbf{p}_0} + \sum_{m=1}^{N} \tilde{q}_m(\mathbf{a}) \mathrm{e}^{\mathbf{a}\cdot\mathbf{p}_m}, \tag{40}$$

where $\tilde{q}_m(\mathbf{a})$ are certain rational functions in $a_1, \ldots, a_N$. Since $\mathrm{e}^{\mathbf{a}\cdot\mathbf{p}_j}$ for $0 \leq j \leq N$ are linearly independent over the field of rational functions in $a_1, \ldots, a_N$, exchanging the roles of the indices 0 and j in the change of variables in (36) yields

$$\begin{aligned} \int_\sigma \mathrm{e}^{\mathbf{a}\cdot\mathbf{x}} d\mathbf{x} &= \sum_{m=0}^{N} \frac{|\det(\mathbf{p}_m - \mathbf{p}_0, \ldots, \mathbf{p}_m - \mathbf{p}_{m-1}, \mathbf{p}_m - \mathbf{p}_{m+1}, \ldots, \mathbf{p}_m - \mathbf{p}_N)|}{\prod_{j \neq m} \mathbf{a}\cdot(\mathbf{p}_m - \mathbf{p}_j)} \mathrm{e}^{\mathbf{a}\cdot\mathbf{p}_m} \\ &= N!\mathrm{Vol}(\sigma) \sum_{m=0}^{N} \frac{\mathrm{e}^{\mathbf{a}\cdot\mathbf{p}_m}}{\prod_{j \neq m} \mathbf{a}\cdot(\mathbf{p}_m - \mathbf{p}_j)}. \end{aligned} \tag{41}$$

Generally, we have

$$\sum_{m=0}^{N} \frac{\mathrm{e}^{x_m}}{\prod_{j \neq m}(x_m - x_j)} = T(x_0, \ldots, x_N) \tag{42}$$

by the Laplace expansion of the numerator of the right-hand side of (33) with respect to the last row and hence the result (32). □

Although the following lemma is a direct consequence of the second expression of (32) with the definition of Schur polynomials and the Jacobi–Trudi formula (see [19]), we give a direct proof for convenience.

Lemma 5. *Let* $\mathbf{a} \in \mathbb{C}^N$, σ *and* $\mathbf{p}_0, \ldots, \mathbf{p}_N \in \mathbb{R}^N$ *are the same as in Lemma 4. Then the Taylor expansion with respect to* $\mathbf{a}$ *is given by*

$$\begin{aligned} \int_\sigma \mathrm{e}^{\mathbf{a}\cdot\mathbf{x}} d\mathbf{x} = \mathrm{Vol}(\sigma) \Bigg(& 1 + \frac{1}{N+1} \sum_{0 \leq i \leq N} \mathbf{a}\cdot\mathbf{p}_i + \cdots \\ & + \frac{N!}{(N+k)!} \sum_{\substack{k_0, \ldots, k_N \geq 0 \\ k_0 + \cdots + k_N = k}} (\mathbf{a}\cdot\mathbf{p}_0)^{k_0} \cdots (\mathbf{a}\cdot\mathbf{p}_N)^{k_N} + \cdots \Bigg). \end{aligned} \tag{43}$$

Proof. We recall Dirichlet's integral (see [37, Chap. XII, Sect. 12.5]) for nonnegative integers k_j and a continuous function g, that is,

$$\begin{aligned}&\int_{\tilde{\tau}} y_0^{k_0}\cdots y_N^{k_N} g(y_0+\cdots+y_N)\mathrm{d}y_0\cdots\mathrm{d}y_N\\&=\frac{k_0!\cdots k_N!}{(N+k_0+\cdots+k_N)!}\int_0^1 g(t)t^{k_0+\cdots+k_N+N}\mathrm{d}t,\end{aligned} \tag{44}$$

where $\tilde{\tau}$ is the $(N+1)$-dimensional simplex with their vertices ${}^t(0,\ldots,\overset{j}{1},\ldots,0)$, $0 \leq j \leq N$, and the origin. This formula is easily obtained by repeated application of the beta integral.

We calculate

$$f(\mathbf{a}) = \frac{1}{k!}\int_{\sigma} (\mathbf{a}\cdot\mathbf{x})^k \mathrm{d}\mathbf{x}. \tag{45}$$

By multiplying

$$1 = (k+1)\int_0^1 s^k \mathrm{d}s, \tag{46}$$

and changing variables as $\mathbf{x}' = s\mathbf{x}$, we obtain

$$f(\mathbf{a}) = \frac{k+1}{k!}\int_{\tilde{\sigma}} (\mathbf{a}\cdot\mathbf{x}')^k s^{-N}\mathrm{d}s\mathrm{d}\mathbf{x}', \tag{47}$$

where $\tilde{\sigma} = \bigcup_{0\leq s\leq 1}\begin{pmatrix} s \\ s\sigma \end{pmatrix}$ is an $(N+1)$-dimensional simplex. Again, we change variables as $\tilde{P}\tilde{\mathbf{y}} = \tilde{\mathbf{x}} = \begin{pmatrix} s \\ \mathbf{x}' \end{pmatrix}$. Then

$$\begin{aligned}f(\mathbf{a}) &= \frac{k+1}{k!}|\det\tilde{P}|\int_{\tilde{\tau}}\left(\sum_{j=0}^N(\mathbf{a}\cdot\mathbf{p}_j)y_j\right)^k\left(\sum_{j=0}^N y_j\right)^{-N}\mathrm{d}\tilde{\mathbf{y}}\\&=\frac{k+1}{k!}|\det\tilde{P}|\sum_{\substack{k_0,\ldots,k_N\geq 0\\k_0+\cdots+k_N=k}}(\mathbf{a}\cdot\mathbf{p}_0)^{k_0}\cdots(\mathbf{a}\cdot\mathbf{p}_N)^{k_N}\frac{k!}{k_0!\cdots k_N!}\\&\quad\times\int_{\tilde{\tau}} y_0^{k_0}\cdots y_N^{k_N}\left(\sum_{j=0}^N y_j\right)^{-N}\mathrm{d}\tilde{\mathbf{y}},\end{aligned} \tag{48}$$

where $\tilde{\mathbf{y}} = {}^t(y_0, \ldots, y_N)$. Hence applying (44), we obtain

$$f(\mathbf{a}) = \frac{k+1}{k!}|\det \tilde{P}| \sum_{\substack{k_0,\ldots,k_N \geq 0 \\ k_0+\cdots+k_N=k}} (\mathbf{a}\cdot\mathbf{p}_0)^{k_0}\cdots(\mathbf{a}\cdot\mathbf{p}_N)^{k_N} \frac{k!}{(N+k)!}\frac{1}{k+1}$$

$$= \mathrm{Vol}(\sigma)\frac{N!}{(N+k)!} \sum_{\substack{k_0,\ldots,k_N \geq 0 \\ k_0+\cdots+k_N=k}} (\mathbf{a}\cdot\mathbf{p}_0)^{k_0}\cdots(\mathbf{a}\cdot\mathbf{p}_N)^{k_N}. \tag{49}$$

□

It should be remarked that $\mathbf{a}\cdot\mathbf{p}_j$ for $0 \leq j \leq N$ are not linearly independent. Thus, the coefficients with respect to them are not unique. Lemma 5 is a special and exact case of the following lemma.

Lemma 6. *Let* $\mathbf{a} \in \mathbb{C}^N$, σ *and* $\mathbf{p}_0, \ldots, \mathbf{p}_N \in \mathbb{R}^N$ *are the same as in Lemma 4. Then for* $\mathbf{b} \in \mathbb{C}^N$, *the coefficients of total degree* k *with respect to* $\mathbf{a}$ *of the Taylor expansion of*

$$\int_\sigma \mathrm{e}^{(\mathbf{a}+\mathbf{b})\cdot\mathbf{x}}d\mathbf{x} \tag{50}$$

are holomorphic functions in $\mathbf{b}$ *of the form*

$$\mathrm{Vol}(\sigma)\sum_{m=0}^{N} \sum_{\substack{k_0,\ldots,k_N \geq 0 \\ k_0+\cdots+k_N=k}} c_{m,k_0,\ldots,k_N} \frac{\mathrm{e}^{\mathbf{b}\cdot\mathbf{p}_m}}{\prod_{j\neq m}\left(\mathbf{b}\cdot(\mathbf{p}_m-\mathbf{p}_j)\right)^{k_j+1}}, \tag{51}$$

where $c_{m,k_0,\ldots,k_N} \in \mathbb{Q}$.

Proof. We assume $|\mathbf{a}\cdot(\mathbf{p}_m-\mathbf{p}_j)| < |\mathbf{b}\cdot(\mathbf{p}_m-\mathbf{p}_j)|$ for all $j \neq m$. Then we have

$$\frac{\mathrm{e}^{(\mathbf{a}+\mathbf{b})\cdot\mathbf{p}_m}}{\prod_{j\neq m}(\mathbf{a}+\mathbf{b})\cdot(\mathbf{p}_m-\mathbf{p}_j)}$$

$$= \frac{\mathrm{e}^{\mathbf{b}\cdot\mathbf{p}_m}}{\prod_{j\neq m}\mathbf{b}\cdot(\mathbf{p}_m-\mathbf{p}_j)} \frac{\mathrm{e}^{\mathbf{a}\cdot\mathbf{p}_m}}{\prod_{j\neq m}\left(1+\frac{\mathbf{a}\cdot(\mathbf{p}_m-\mathbf{p}_j)}{\mathbf{b}\cdot(\mathbf{p}_m-\mathbf{p}_j)}\right)}$$

$$= \frac{\mathrm{e}^{\mathbf{b}\cdot\mathbf{p}_m}}{\prod_{j\neq m}\mathbf{b}\cdot(\mathbf{p}_m-\mathbf{p}_j)} \mathrm{e}^{\mathbf{a}\cdot\mathbf{p}_m} \prod_{j\neq m}\sum_{k_j=0}^{\infty}\left(-\frac{\mathbf{a}\cdot(\mathbf{p}_m-\mathbf{p}_j)}{\mathbf{b}\cdot(\mathbf{p}_m-\mathbf{p}_j)}\right)^{k_j}$$

$$= \sum_{k=0}^{\infty} \sum_{\substack{k_0,\ldots,k_N \geq 0 \\ k_0+\cdots+k_N=k}} \frac{\mathrm{e}^{\mathbf{b}\cdot\mathbf{p}_m}}{\prod_{j\neq m}\left(\mathbf{b}\cdot(\mathbf{p}_m-\mathbf{p}_j)\right)^{k_j+1}} \frac{(\mathbf{a}\cdot\mathbf{p}_m)^{k_m}}{k_m!} \prod_{j\neq m}\left(\mathbf{a}\cdot(\mathbf{p}_j-\mathbf{p}_m)\right)^{k_j}. \tag{52}$$

Applying this result to the second member of (32) with $\mathbf{a}$ replaced by $\mathbf{a}+\mathbf{b}$, we see that the coefficients of total degree k are of the form (51). The holomorphy with respect to $\mathbf{a}$ and $\mathbf{b}$ follows from the original integral form (50). Therefore, (51) is valid for the whole space with removable singularities. □

11.3 General Functional Relations

The purpose of this section is to give a very general formulation of functional relations. For $f, g \in \mathfrak{F}(P)$ and $I, J \subset \{1, \dots, r\}$, we define

$$\zeta(f, g; J; \Delta) = \sum_{\lambda \in P_{J++} \setminus H_g} \frac{f(\lambda)}{g(\lambda)}, \tag{53}$$

and

$$S(f, g; I; \Delta) = \sum_{\lambda \in \iota^{*-1}(P_{I+}) \setminus H_g} \frac{f(\lambda)}{g(\lambda)}. \tag{54}$$

We assume that (54) is absolutely convergent for a fixed I. By (24) and Lemma 2, we have $\mathrm{id} \in W^I$ and $P_+ \subset \iota^{*-1}(P_{I+})$. Hence, (53) is also absolutely convergent for all J since $P_{J++} \subset P_+$ by (22).

For $s \in \mathbb{C}$, $\Re s > 1$ and $x, c \in \mathbb{R}$, let

$$L_s(x, c) = -\frac{\Gamma(s+1)}{(2\pi\sqrt{-1})^s} \sum_{\substack{n \in \mathbb{Z} \\ n+c \neq 0}} \frac{\mathbf{e}((n+c)x)}{(n+c)^s}. \tag{55}$$

Here and hereafter, we use the standard notation $\mathbf{e}(x) = \mathrm{e}^{2\pi\sqrt{-1}x}$ for simplicity. Let D_{I^c} be a finite subset of $(Q^\vee \setminus \{0\}) \oplus \mathbb{R}\delta \subset \widehat{V}$. Then any element of $\gamma \in D_{I^c}$ can be written as $\gamma = \eta_\gamma + c_\gamma \delta$ ($\eta_\gamma \in Q^\vee \setminus \{0\}$, $c_\gamma \in \mathbb{R}$). We assume that D_{I^c} contains $B_{I^c} = \{\gamma_i\}_{i \in I^c}$ where $\gamma_i = \eta_i + c_i \delta$ for $i \in I^c$ such that $\{\eta_i\}_{i \in I^c}$ forms a basis of $Q^\vee_{I^c}$ and $c_i \in \mathbb{R}$. Let $\{\mu_i\}_{i \in I^c} \subset P_{I^c}$ be a basis dual to $\{\eta_i\}_{i \in I^c}$.

Theorem 3. *Let $s_\gamma \in \mathbb{C}$ with $\Re s_\gamma > 1$ for $\gamma \in D_{I^c}$ and let $\mathbf{y} \in V_{I^c}$. We assume that*

$$f(\lambda + \mu) = f(\lambda)\mathbf{e}(\langle \mathbf{y}, \mu \rangle), \tag{56}$$

$$g(\lambda + \mu) = g^\sharp(\lambda) \prod_{\gamma \in D_{I^c}} \gamma(\lambda + \mu)^{s_\gamma}, \tag{57}$$

for any $\lambda \in P_{I+}$ and any $\mu \in P_{I^c}$, where $g^\sharp \in \mathfrak{F}(P_I)$. (Hence f depends on $\mathbf{y}$, and g depends on s_γ's.) Then

(i) We have

$$S(f,g;I;\Delta)=\sum_{w\in W^I}\sum_{J\subset\{1,\ldots,r\}}\frac{1}{N_{w,J}}\zeta(w^{-1}f,w^{-1}g;J;\Delta)$$
$$=\sum_{J\subset I}\zeta(f\cdot f^{\sharp},g^{\sharp};J;\Delta), \tag{58}$$

where

$$N_{w,J}=|wW(\Delta_{J^c})\cap W^I| \tag{59}$$

and $f^{\sharp}\in\mathfrak{F}(P_I)$ *is defined by*

$$f^{\sharp}(\lambda)=(-1)^{|D_{I^c}|}\mathbf{e}(-\langle\mathbf{y},\nu\rangle)\left(\prod_{\gamma\in D_{I^c}}\frac{(2\pi\sqrt{-1})^{s_\gamma}}{\Gamma(s_\gamma+1)}\right)$$
$$\times\int_0^1\cdots\int_0^1\mathbf{e}\left(-\sum_{\gamma\in D_{I^c}\setminus B_{I^c}}\gamma(\lambda-\nu)x_\gamma\right)\left(\prod_{\gamma\in D_{I^c}\setminus B_{I^c}}L_{s_\gamma}(x_\gamma,c_\gamma)\right)$$
$$\times\left(\prod_{i\in I^c}L_{s_{\gamma_i}}\left(\langle\mathbf{y},\mu_i\rangle-\sum_{\gamma\in D_{I^c}\setminus B_{I^c}}x_\gamma\langle\eta_\gamma,\mu_i\rangle,c_i\right)\right)\prod_{\gamma\in D_{I^c}\setminus B_{I^c}}\mathrm{d}x_\gamma, \tag{60}$$

and

$$\nu=\sum_{i\in I^c}c_i\mu_i\in P_{I^c}\otimes\mathbb{R}. \tag{61}$$

(ii) The second member of (58) *consists of* $2^r\big(W(\Delta):W(\Delta_I)\big)$ *terms.*
(iii) If $H_{\Delta^\vee\setminus\Delta_I^\vee}\subset H_g$*, then* $\zeta(w^{-1}f,w^{-1}g;J;\Delta)=0$ *unless* $N_{w,J}=1$.

Proof. First, we claim that for $w,w'\in W^I$ and $\lambda\in wP_{J++}$; we have $\lambda\in w'P_{J++}$ if and only if $w'\in wW(\Delta_{J^c})$. In fact, $\lambda\in wP_{J++}\cap w'P_{J++}$ implies $w'^{-1}\lambda\in P_{J++}$ and $(w^{-1}w')w'^{-1}\lambda\in P_{J++}$, and hence, $w^{-1}w'$ stabilizes $P_{J++}=P_+\cap C_J$. Therefore, $w^{-1}w'\in W(\Delta_{J^c})$. The converse statement is shown by reversing the arguments, and we have the claim. By using this claim, Lemma 2, and the decomposition (22), we have

$$S(f,g;I;\Delta)=\sum_{\lambda\in\iota^{*-1}(P_{I+})\setminus H_g}\frac{f(\lambda)}{g(\lambda)}$$
$$=\sum_{w\in W^I}\sum_{J\subset\{1,\ldots,r\}}\frac{1}{N_{w,J}}\sum_{\lambda\in wP_{J++}\setminus H_g}\frac{f(\lambda)}{g(\lambda)}. \tag{62}$$

Therefore,

$$
\begin{aligned}
S(f,g;I;\Delta) &= \sum_{w\in W^I}\sum_{J\subset\{1,\dots,r\}}\frac{1}{N_{w,J}}\sum_{\lambda\in P_{J++}\setminus w^{-1}H_g}\frac{f(w\lambda)}{g(w\lambda)} \\
&= \sum_{w\in W^I}\sum_{J\subset\{1,\dots,r\}}\frac{1}{N_{w,J}}\sum_{\lambda\in P_{J++}\setminus H_{w^{-1}g}}\frac{(w^{-1}f)(\lambda)}{(w^{-1}g)(\lambda)} \\
&= \sum_{w\in W^I}\sum_{J\subset\{1,\dots,r\}}\frac{1}{N_{w,J}}\zeta(w^{-1}f,w^{-1}g;J;\Delta),
\end{aligned}
\tag{63}
$$

where the last member consists of $2^r\big(W(\Delta):W(\Delta_I)\big)$ terms since the cardinality of the power set of $\{1,\dots,r\}$ is 2^r and $|W^I|=\big(W(\Delta):W(\Delta_I)\big)$ by Lemma 1, which implies the statement (ii).

Assume that $\gamma=\eta_\gamma+c_\gamma\delta\in(Q^\vee\setminus\{0\})\oplus\mathbb{R}\delta$ and $\lambda\in P$. Then $\gamma(\lambda)=\langle\eta_\gamma,\lambda\rangle+c_\gamma\in\mathbb{Z}+c_\gamma$ and for $\gamma(\lambda)\neq 0$, we have

$$
\begin{aligned}
\frac{1}{\gamma(\lambda)^{s_\gamma}} &= \sum_{\substack{n_\gamma\in\mathbb{Z}\\ n_\gamma+c_\gamma\neq 0}}\int_0^1\frac{\mathbf{e}((n_\gamma+c_\gamma-\gamma(\lambda))x_\gamma)}{(n_\gamma+c_\gamma)^{s_\gamma}}\mathrm{d}x_\gamma \\
&= \int_0^1\sum_{\substack{n_\gamma\in\mathbb{Z}\\ n_\gamma+c_\gamma\neq 0}}\frac{\mathbf{e}((n_\gamma+c_\gamma-\gamma(\lambda))x_\gamma)}{(n_\gamma+c_\gamma)^{s_\gamma}}\mathrm{d}x_\gamma \\
&= -\frac{(2\pi\sqrt{-1})^{s_\gamma}}{\Gamma(s_\gamma+1)}\int_0^1 L_{s_\gamma}(x_\gamma,c_\gamma)\mathbf{e}(-\gamma(\lambda)x_\gamma)\mathrm{d}x_\gamma,
\end{aligned}
\tag{64}
$$

where we have used the absolute convergence because of $\Re s_\gamma>1$.

By using (57), (56), and Lemma 2, we have

$$
S(f,g;I;\Delta)=\sum_{\lambda\in P_{I+}}\sum_{\substack{\mu\in P_{I^c}\\ \lambda+\mu\notin H_g}}\frac{f(\lambda)}{g^\sharp(\lambda)}\mathbf{e}(\langle\mathbf{y},\mu\rangle)\left(\prod_{\gamma\in D_{I^c}}\frac{1}{\gamma(\lambda+\mu)^{s_\gamma}}\right). \tag{65}
$$

Applying (64) to the right-hand side of the above, we obtain

$$
\begin{aligned}
S(f,g;I;\Delta) = \sum_{\lambda\in P_{I+}}\sum_{\substack{\mu\in P_{I^c}\\ \lambda+\mu\notin H_g}}&\frac{f(\lambda)}{g^\sharp(\lambda)}\mathbf{e}(\langle\mathbf{y},\mu\rangle)\left(\prod_{\gamma\in B_{I^c}}\frac{1}{\gamma(\lambda+\mu)^{s_\gamma}}\right) \\
&\times\left(\prod_{\gamma\in D_{I^c}\setminus B_{I^c}}-\frac{(2\pi\sqrt{-1})^{s_\gamma}}{\Gamma(s_\gamma+1)}\int_0^1 L_{s_\gamma}(x_\gamma,c_\gamma)\mathbf{e}(-\gamma(\lambda+\mu)x_\gamma)\mathrm{d}x_\gamma\right).
\end{aligned}
\tag{66}
$$

Note that if $\gamma(\lambda)=0$, then $c_\gamma\in\mathbb{Z}$ and the last member of (64) vanishes. Hence, we may add the case $\gamma(\lambda+\mu)=0$ for $\gamma\in D_{I^c}\setminus B_{I^c}$ in the above. Therefore, by using $H_g=H_{g^\sharp}\cup H_{D_{I^c}\setminus B_{I^c}}\cup H_{B_{I^c}}$ and putting $\mu=\sum_{i\in I^c}n_i\mu_i$ $(n_i\in\mathbb{Z})$, we have

$$
\begin{aligned}
&S(f,g;I;\Delta)\\
&=\left(\prod_{\gamma\in D_{I^c}\setminus B_{I^c}}-\frac{(2\pi\sqrt{-1})^{s_\gamma}}{\Gamma(s_\gamma+1)}\right)\sum_{\lambda\in P_{I+}\setminus H_{g^\sharp}}\frac{f(\lambda)}{g^\sharp(\lambda)}\\
&\quad\times\sum_{\substack{n_i\in\mathbb{Z}\\ n_i+c_i\neq 0\\ i\in I^c}}\int_0^1\cdots\int_0^1\mathbf{e}\left(-\sum_{\gamma\in D_{I^c}\setminus B_{I^c}}\gamma(\lambda)x_\gamma\right)\left(\prod_{\gamma\in D_{I^c}\setminus B_{I^c}}L_{s_\gamma}(x_\gamma,c_\gamma)\right)\\
&\quad\times\prod_{i\in I^c}\frac{\mathbf{e}\Big(\big(\langle\mathbf{y},\mu_i\rangle-\sum_{\gamma\in D_{I^c}\setminus B_{I^c}}x_\gamma\langle\eta_\gamma,\mu_i\rangle\big)n_i\Big)}{(n_i+c_i)^{s_{\gamma_i}}}\prod_{\gamma\in D_{I^c}\setminus B_{I^c}}\mathrm{d}x_\gamma\\
&=(-1)^{|D_{I^c}|}\mathbf{e}(-\langle\mathbf{y},\nu\rangle)\left(\prod_{\gamma\in D_{I^c}}\frac{(2\pi\sqrt{-1})^{s_\gamma}}{\Gamma(s_\gamma+1)}\right)\sum_{\lambda\in P_{I+}\setminus H_{g^\sharp}}\frac{f(\lambda)}{g^\sharp(\lambda)}\\
&\quad\times\int_0^1\cdots\int_0^1\mathbf{e}\left(-\sum_{\gamma\in D_{I^c}\setminus B_{I^c}}\gamma(\lambda-\nu)x_\gamma\right)\left(\prod_{\gamma\in D_{I^c}\setminus B_{I^c}}L_{s_\gamma}(x_\gamma,c_\gamma)\right)\\
&\quad\times\left(\prod_{i\in I^c}L_{s_{\gamma_i}}\Big(\langle\mathbf{y},\mu_i\rangle-\sum_{\gamma\in D_{I^c}\setminus B_{I^c}}x_\gamma\langle\eta_\gamma,\mu_i\rangle,c_i\Big)\right)\prod_{\gamma\in D_{I^c}\setminus B_{I^c}}\mathrm{d}x_\gamma,
\end{aligned}\tag{67}
$$

which is equal to the third member of (58), because $P_{I+}=\bigcup_{J\subset I}P_{J++}$. Hence, the statement (i).

As for the last claim (iii) of the theorem, we first note that $N_{w,J}>1$ if and only if $wP_{J++}\subset H_{\Delta^\vee\setminus\Delta_I^\vee}$. This follows from Lemma 3 and the definition of $N_{w,J}$, with noting the claim proved on the first line of the present proof. Now assume $H_{\Delta^\vee\setminus\Delta_I^\vee}\subset H_g$. Then, if $N_{w,J}>1$, we have $wP_{J++}\subset H_g$. This implies $\zeta(w^{-1}f,w^{-1}g;J;\Delta)=0$, because the definition (53) is an empty sum in this case. □

In this chapter, we mainly discuss the case when $D_{I^c}\subset Q^\vee$. Nevertheless, we give the above generalized form of Theorem 3, by which we can treat the case of zeta-functions of Hurwitz type.

For a real number x, let $\{x\}$ denote its fractional part $x-[x]$. If $s=k\geq 2$ and c are integers, then it is known (see [1]) that

$$\begin{aligned} L_k(x,c) &= -\frac{k!}{(2\pi\sqrt{-1})^k} \sum_{n\in\mathbb{Z}\setminus\{0\}} \frac{\mathbf{e}(nx)}{n^k} \\ &= B_k(\{x\}), \end{aligned} \tag{68}$$

where $B_k(x)$ is the kth Bernoulli polynomial. Thus if all $L_s(x,c)$ in the integrand on the right-hand side of (60) are written in terms of Bernoulli polynomials, then we have a chance to obtain an explicit form of the right-hand side of (60). We calculate the integral in question through a generating function instead of a direct calculation. The result will be stated in Theorem 4 below.

Remark 2. When $s_\gamma = 1$, the argument (64) is not valid, because the series is conditionally convergent. Hence, on the right-hand side of (67), $s_{\gamma_i} = 1$ for $i \in I^{\mathrm{c}}$ is not allowed. However, for $n \in \mathbb{Z} \setminus \{0\}$, we have

$$\frac{1}{n} = (-2\pi\sqrt{-1}) \int_0^1 B_1(x)\mathbf{e}(-nx)\mathrm{d}x, \tag{69}$$

where the right-hand side vanishes if $n = 0$. Thus the value $s_\gamma = 1$ is also allowed for $\gamma \in (D_{I^{\mathrm{c}}} \setminus B_{I^{\mathrm{c}}}) \cap \widehat{Q}^\vee$ on the right-hand side of (66) and hence in Theorem 3.

Let $M, N \in \mathbb{N}$, $\mathbf{k} = (k_l)_{1\le l\le M+N} \in \mathbb{N}_0^{M+N}$, $\mathbf{y} = (y_i)_{1\le i\le M} \in \mathbb{R}^M$ and $b_j \in \mathbb{C}$, $c_{ij} \in \mathbb{R}$ for $1 \le i \le M$ and $1 \le j \le N$. Let

$$P(\mathbf{k},\mathbf{y}) = \int_0^1 \cdots \int_0^1 \exp\left(\sum_{j=1}^N b_j x_j\right) \left(\prod_{j=1}^N B_{k_j}(x_j)\right) \left(\prod_{i=1}^M B_{k_{N+i}}\left(\left\{y_i - \sum_{j=1}^N c_{ij}x_j\right\}\right)\right) \prod_{j=1}^N \mathrm{d}x_j. \tag{70}$$

For $\mathbf{t} = (t_l)_{1\le l\le M+N} \in \mathbb{C}^{M+N}$, we define a generating function of $P(\mathbf{k},\mathbf{y})$ by

$$F(\mathbf{t},\mathbf{y}) = \sum_{\mathbf{k}\in\mathbb{N}_0^{M+N}} P(\mathbf{k},\mathbf{y}) \prod_{j=1}^{M+N} \frac{t_j^{k_j}}{k_j!}. \tag{71}$$

Lemma 7. *(i) The generating function $F(\mathbf{t},\mathbf{y})$ is absolutely convergent, uniformly on $\mathcal{D}_R \times \mathbb{R}^M$ where $\mathcal{D}_R = \{t \in \mathbb{C} \mid |t| \le R\}^{M+N}$ with $0 < R < 2\pi$.*

(ii) The function $F(\cdot,\mathbf{y})$ is analytically continued to a meromorphic function in $\mathbf{t}$ on the whole space $\mathbb{C}^{M+N}$, and we have

$$\begin{aligned} F(\mathbf{t},\mathbf{y}) = &\left(\prod_{j=1}^{M+N} \frac{t_j}{\mathrm{e}^{t_j}-1}\right) \int_0^1 \cdots \int_0^1 \left(\prod_{j=1}^N \exp((b_j+t_j)x_j)\right) \\ &\times \left(\prod_{i=1}^M \exp\left(t_{N+i}\left\{y_i - \sum_{j=1}^N c_{ij}x_j\right\}\right)\right) \prod_{j=1}^N \mathrm{d}x_j \end{aligned}$$

$$= \left(\prod_{j=1}^{M+N} \frac{t_j}{e^{t_j} - 1} \right) \sum_{m_1 = c_1^-}^{c_1^+} \cdots \sum_{m_M = c_M^-}^{c_M^+} \exp\left(\sum_{i=1}^{M} t_{N+i}(\{y_i\} + m_i) \right)$$

$$\times \int_{\mathcal{P}_{\mathbf{m},\mathbf{y}}} \exp\left(\sum_{j=1}^{N} (a_j + b_j) x_j \right) \prod_{j=1}^{N} dx_j, \tag{72}$$

where c_i^+ *is the minimum integer satisfying* $c_i^+ \geq \sum_{\substack{1 \leq j \leq N \\ c_{ij} > 0}} c_{ij}$ *and* c_i^- *is the maximum integer satisfying* $c_i^- \leq \sum_{\substack{1 \leq j \leq N \\ c_{ij} < 0}} c_{ij}$ *and*

$$a_j = t_j - \sum_{i=1}^{M} t_{N+i} c_{ij}, \tag{73}$$

$$\mathcal{P}_{\mathbf{m},\mathbf{y}} = \left\{ \mathbf{x} = (x_j)_{1 \leq j \leq N} \;\middle|\; \begin{array}{c} 0 \leq x_j \leq 1, \quad (1 \leq j \leq N) \\ \{y_i\} + m_i - 1 \leq \sum_{j=1}^{N} c_{ij} x_j \leq \{y_i\} + m_i, \\ (1 \leq i \leq M) \end{array} \right\}, \tag{74}$$

which is a convex polytope.

Proof. (i) Fix $R' \in \mathbb{R}$ such that $R < R' < 2\pi$. Then by the Cauchy integral formula

$$\frac{B_k(x)}{k!} = \frac{1}{2\pi\sqrt{-1}} \int_{|z|=R'} \frac{z e^{zx}}{e^z - 1} \frac{dz}{z^{k+1}}, \tag{75}$$

we have for $0 \leq x \leq 1$

$$\left| \frac{B_k(x)}{k!} \right| \leq \frac{1}{2\pi} \int_{|z|=R'} \left| \frac{z e^{zx}}{e^z - 1} \right| \frac{|dz|}{R'^{k+1}} \leq \frac{C_{R'}}{R'^k}, \tag{76}$$

where

$$C_{R'} = \max\left\{ \left| \frac{z e^{zx}}{e^z - 1} \right| \;\middle|\; |z| = R',\ 0 \leq x \leq 1 \right\}. \tag{77}$$

Therefore,

$$
\begin{aligned}
\left| P(\mathbf{k},\mathbf{y}) \prod_{j=1}^{M+N} \frac{t_j^{k_j}}{k_j!} \right| &\le \left(\prod_{j=1}^{M+N} |t_j|^{k_j} \right) \int_0^1 \cdots \int_0^1 \left| \exp\left(\sum_{j=1}^{N} b_j x_j \right) \left(\prod_{j=1}^{N} \frac{B_{k_j}(x_j)}{k_j!} \right) \right. \\
&\quad \times \left. \left(\prod_{i=1}^{M} \frac{B_{k_{N+i}}\left(\left\{ y_i - \sum_{j=1}^{N} c_{ij} x_j \right\} \right)}{k_{N+i}!} \right) \right| \prod_{j=1}^{N} \mathrm{d}x_j \\
&\le C \left(\prod_{j=1}^{M+N} R^{k_j} \right) \int_0^1 \cdots \int_0^1 C_{R'}^{M+N} \prod_{j=1}^{M+N} \frac{1}{R'^{k_j}} \prod_{j=1}^{N} \mathrm{d}x_j \\
&= C C_{R'}^{M+N} \prod_{j=1}^{M+N} \left(\frac{R}{R'} \right)^{k_j},
\end{aligned} \tag{78}
$$

where $C = \exp\left(\sum_{j=1}^{N} |\Re b_j| \right)$. Since

$$
\sum_{\mathbf{k}\in\mathbb{N}_0^{M+N}} C C_{R'}^{M+N} \prod_{j=1}^{M+N} \left(\frac{R}{R'} \right)^{k_j} = C C_{R'}^{M+N} \prod_{j=1}^{M+N} \frac{1}{1 - R/R'} < \infty, \tag{79}
$$

we have the uniform and absolute convergence of $F(\mathbf{t},\mathbf{y})$.

(ii) Noting the absolute convergence shown in (i), we obtain

$$
\begin{aligned}
F(\mathbf{t},\mathbf{y}) &= \sum_{\mathbf{k}\in\mathbb{N}_0^{M+N}} P(\mathbf{k},\mathbf{y}) \prod_{j=1}^{M+N} \frac{t_j^{k_j}}{k_j!} \\
&= \sum_{\mathbf{k}\in\mathbb{N}_0^{M+N}} \int_0^1 \cdots \int_0^1 \exp\left(\sum_{j=1}^{N} b_j x_j \right) \left(\prod_{j=1}^{N} B_{k_j}(x_j) \frac{t_j^{k_j}}{k_j!} \right) \\
&\quad \times \left(\prod_{i=1}^{M} B_{k_{N+i}}\left(\left\{ y_i - \sum_{j=1}^{N} c_{ij} x_j \right\} \right) \frac{t_{N+i}^{k_{N+i}}}{k_{N+i}!} \right) \prod_{j=1}^{N} \mathrm{d}x_j \\
&= \left(\prod_{j=1}^{M+N} \frac{t_j}{\mathrm{e}^{t_j} - 1} \right) \int_0^1 \cdots \int_0^1 \exp\left(\sum_{j=1}^{N} b_j x_j \right) \left(\prod_{j=1}^{N} \exp(t_j x_j) \right) \\
&\quad \times \left(\prod_{i=1}^{M} \exp\left(t_{N+i} \left\{ y_i - \sum_{j=1}^{N} c_{ij} x_j \right\} \right) \right) \prod_{j=1}^{N} \mathrm{d}x_j
\end{aligned}
$$

$$= \left(\prod_{j=1}^{M+N} \frac{t_j}{e^{t_j}-1}\right) \int_0^1 \cdots \int_0^1 \exp\Bigl(\sum_{j=1}^{N}\Bigl(b_j + t_j - \sum_{i=1}^{M} t_{N+i} c_{ij}\Bigr)x_j\Bigr)$$

$$\times \exp\Bigl(\sum_{i=1}^{M} t_{N+i}\Bigl(y_i - \Bigl[y_i - \sum_{j=1}^{N} c_{ij} x_j\Bigr]\Bigr)\Bigr) \prod_{j=1}^{N} dx_j. \tag{80}$$

Here the inequality

$$y_i - c_i^+ \le y_i - \sum_{j=1}^{N} c_{ij} x_j \le y_i - c_i^- \tag{81}$$

implies

$$\{y_i\} + c_i^- \le y_i - \left[y_i - \sum_{j=1}^{N} c_{ij} x_j \right] \le \{y_i\} + c_i^+, \tag{82}$$

and for $m_i \in \mathbb{Z}$, the region of $\mathbf{x}$ satisfying

$$y_i - \left[y_i - \sum_{j=1}^{N} c_{ij} x_j \right] = \{y_i\} + m_i \tag{83}$$

is given by

$$\{y_i\} + m_i - 1 < \sum_{j=1}^{N} c_{ij} x_j \le \{y_i\} + m_i. \tag{84}$$

Therefore,

$$F(\mathbf{t}, \mathbf{y}) = \left(\prod_{j=1}^{M+N} \frac{t_j}{e^{t_j}-1}\right) \sum_{m_1=c_1^-}^{c_1^+} \cdots \sum_{m_M=c_M^-}^{c_M^+} \exp\left(\sum_{i=1}^{M} t_{N+i}(\{y_i\} + m_i)\right)$$

$$\times \int_{\mathcal{P}_{\mathbf{m},\mathbf{y}}} \exp\Bigl(\sum_{j=1}^{N}(a_j + b_j)x_j\Bigr) \prod_{j=1}^{N} dx_j, \tag{85}$$

which is a meromorphic function in $\mathbf{t}$ on the whole space $\mathbb{C}^{M+N}$. □

Remark 3. When the polytope $\mathcal{P}_{\mathbf{m},\mathbf{y}}$ is simple, then the corresponding integral is calculated by Proposition 3.10 in [4]. Generally, the polytopes we treat may not be simple. In [17], we avoid such difficulty using a certain limiting process.

Lemma 8. *The function $P(\mathbf{k}, \mathbf{y})$ is continuous with respect to $\mathbf{y}$ on $\mathbb{R}^M$. The function $F(\mathbf{t}, \mathbf{y})$ is continuous on $\{t \in \mathbb{C} \mid |t| < 2\pi\}^{M+N} \times \mathbb{R}^M$ and holomorphic in $\mathbf{t}$ for a fixed $\mathbf{y} \in \mathbb{R}^M$.*

Proof. For $\mathbf{y} \in \mathbb{R}^M$ and $\mathbf{x} = (x_j)_{1\le j\le N} \in \mathbb{R}^N$, let $h(\mathbf{y},\mathbf{x})$ be the integrand of (70). Let $\{\mathbf{y}_n\}_{n=1}^{\infty}$ be a sequence in $\mathbb{R}^M$ with $\lim_{n\to\infty}\mathbf{y}_n = \mathbf{y}_\infty$, and we set $h_n(\mathbf{x}) = h(\mathbf{y}_n,\mathbf{x})$ and $h_\infty(\mathbf{x}) = h(\mathbf{y}_\infty,\mathbf{x})$. Then for $\mathbf{x} \in [0,1]^N$,

$$\lim_{n\to\infty} h_n(\mathbf{x}) = h_\infty(\mathbf{x}) \tag{86}$$

holds if $\mathbf{x}$ satisfies

$$(\mathbf{y}_\infty)_i - \sum_{j=1}^{N} c_{ij}x_j \notin \mathbb{Z}, \tag{87}$$

for all $1 \le i \le M$. Hence, (86) holds almost everywhere, and we have

$$\begin{aligned}\lim_{n\to\infty} P(\mathbf{k},\mathbf{y}_n) &= \lim_{n\to\infty}\int_{[0,1]^N} h_n(\mathbf{x})\prod_{j=1}^{N}\mathrm{d}x_j\\ &= \int_{[0,1]^N}\lim_{n\to\infty} h_n(\mathbf{x})\prod_{j=1}^{N}\mathrm{d}x_j\\ &= \int_{[0,1]^N} h_\infty(\mathbf{x})\prod_{j=1}^{N}\mathrm{d}x_j\\ &= P(\mathbf{k},\mathbf{y}_\infty),\end{aligned} \tag{88}$$

where we have used the uniform boundedness $h_n(\mathbf{x}) \le C$ for some $C > 0$ and for all $n \in \mathbb{N}$ and $\mathbf{x} \in [0,1]^N$.

Combining the continuity of $P(\mathbf{k},\mathbf{y})$, definition (71), and Lemma 7, we obtain the continuity and the holomorphy of $F(\mathbf{t},\mathbf{y})$. □

Now we return to the situation $\mathbf{y} \in V_{I^{\mathrm{c}}}$. Let $y_i = \langle \mathbf{y}, \mu_i\rangle$ for $i \in I^{\mathrm{c}}$ and we identify $\mathbf{y}$ with $(y_i)_{i\in I^{\mathrm{c}}} \in \mathbb{R}^{|I^{\mathrm{c}}|}$. We set $\mathbb{Q}[\mathbf{y}] = \mathbb{Q}[(y_i)_{i\in I^{\mathrm{c}}}]$, $A(\mathbf{y}) = \sum_{i\in I^{\mathrm{c}}}\mathbb{Z}\, y_i + \mathbb{Z}$ and $|\mathbf{k}| = \sum_{\gamma\in D_{I^{\mathrm{c}}}} k_\gamma$. Let

$$\overline{D} = \{\gamma\circ\tau(-\nu) \mid \gamma \in D_{I^{\mathrm{c}}} \setminus B_{I^{\mathrm{c}}}\}, \qquad \nu = \sum_{i\in I^{\mathrm{c}}} c_i\mu_i \in P_{I^{\mathrm{c}}}. \tag{89}$$

Theorem 4. *Assume the same condition as in Theorem 3. Moreover, we assume that $D_{I^{\mathrm{c}}} \subset \widehat{Q}^{\vee}$ and $s_\gamma = k_\gamma$ are integers for all $\gamma \in D_{I^{\mathrm{c}}}$ such that $k_\gamma \ge 2$ for $\gamma \in B_{I^{\mathrm{c}}}$ and $k_\gamma \ge 1$ otherwise. Then $f^\sharp \in \mathfrak{F}(P_I)$ in (60) is of the form*

$$f^\sharp(\lambda) = \mathbf{e}(-\langle \mathbf{y},\nu\rangle)\sum_{k=0}^{|\mathbf{k}|}(\pi\sqrt{-1})^{|\mathbf{k}|-(k+N)}\sum_{\eta}\frac{f_\eta^{(k)}(\lambda)}{g_\eta^{(k)}(\lambda)}, \tag{90}$$

where η runs over a certain finite set of indices, $N = |D_{I^c} \setminus B_{I^c}|$ and

$$f_\eta^{(k)} \in \mathbb{Q}[\mathbf{y}]\mathbf{e}(\mathbb{Q} \cdot A(\mathbf{y}) \cdot \overline{D}), \quad g_\eta^{(k)} \in \big(A(\mathbf{y}) \cdot \overline{D}\big)^{k+N}. \tag{91}$$

Proof. From (60), we have

$$f^\sharp(\lambda) = (-1)^{|D_{I^c}|}\mathbf{e}(-\langle \mathbf{y}, \nu\rangle)\Bigg(\prod_{\gamma\in D_{I^c}} \frac{(2\pi\sqrt{-1})^{k_\gamma}}{k_\gamma!}\Bigg)P(\mathbf{k}, \mathbf{y}, \lambda), \tag{92}$$

where for $\mathbf{k} = (k_\gamma)_{\gamma\in D_{I^c}}$,

$$\begin{aligned} P(\mathbf{k}, \mathbf{y}, \lambda) = \int_0^1 \cdots \int_0^1 \mathbf{e}\Bigg(-\sum_{\gamma\in D_{I^c}\setminus B_{I^c}} \gamma(\lambda-\nu)x_\gamma\Bigg)\Bigg(\prod_{\gamma\in D_{I^c}\setminus B_{I^c}} B_{k_\gamma}(x_\gamma)\Bigg) \\ \times \Bigg(\prod_{i\in I^c} B_{k_{\gamma_i}}\Bigg(\Bigg\{y_i - \sum_{\gamma\in D_{I^c}\setminus B_{I^c}} x_\gamma\langle \eta_\gamma, \mu_i\rangle\Bigg\}\Bigg)\Bigg) \prod_{\gamma\in D_{I^c}\setminus B_{I^c}} \mathrm{d}x_\gamma. \end{aligned} \tag{93}$$

Hence, $P(\mathbf{k}, \mathbf{y}, \lambda)$ is of the form (70). Therefore, by applying Lemma 7, we find that $P(\mathbf{k}, \mathbf{y}, \lambda)$ is obtained as the coefficient of the term

$$\prod_{\gamma\in D_{I^c}} t_\gamma^{k_\gamma} \tag{94}$$

in the generating function

$$\begin{aligned} F(\mathbf{t}, \mathbf{y}, \lambda) &= \sum_{\mathbf{k}\in\mathbb{N}_0^{|D_{I^c}|}} P(\mathbf{k}, \mathbf{y}, \lambda) \prod_{\gamma\in D_{I^c}} \frac{t_\gamma^{k_\gamma}}{k_\gamma!} \\ &= \Bigg(\prod_{\gamma\in D_{I^c}} \frac{t_\gamma}{\mathrm{e}^{t_\gamma}-1}\Bigg) \sum_{\substack{m_i=0\\ i\in I^c}}^{c_i^+} \exp\Bigg(\sum_{i\in I^c} t_{\gamma_i}(\{y_i\}+m_i)\Bigg) \\ &\quad \times \int_{\mathcal{P}_{\mathbf{m},\mathbf{y}}} \exp\big((\mathbf{a}+\mathbf{b})\cdot\mathbf{x}\big) \prod_{\gamma\in D_{I^c}\setminus B_{I^c}} \mathrm{d}x_\gamma, \end{aligned} \tag{95}$$

where $\mathbf{a} = (a_\gamma)_{\gamma\in D_{I^c}\setminus B_{I^c}} \in \mathbb{R}^N$, $\mathbf{b} = (b_\gamma)_{\gamma\in D_{I^c}\setminus B_{I^c}} \in \mathbb{C}^N$ with

$$a_\gamma = t_\gamma - \sum_{i\in I^c} t_{\gamma_i}\langle \eta_\gamma, \mu_i\rangle, \tag{96}$$

$$b_\gamma = -2\pi\sqrt{-1}\gamma(\lambda-\nu). \tag{97}$$

Since $\langle \eta_\gamma, \mu_i \rangle \in \mathbb{Z}$, any vertex $\mathbf{p}_j$ of $\mathcal{P}_{\mathbf{m},\mathbf{y}}$ satisfies

$$(\mathbf{p}_j)_\gamma \in \sum_{i \in I^c} \mathbb{Q}\, y_i + \mathbb{Q} = \mathbb{Q} \cdot A(\mathbf{y}). \tag{98}$$

The first two factors of the last member of (95) are expanded as

$$\Big(\prod_{\gamma \in D_{I^c}} \frac{t_\gamma}{e^{t_\gamma} - 1}\Big)\Big(\prod_{i \in I^c} \exp\big(t_{\gamma_i}(\{y_i\} + m_i)\big)\Big) = \sum_{\mathbf{k}' \in \mathbb{N}_0^{|D_{I^c}|}} P_{\mathbf{k}'}(\mathbf{y}) \prod_{\gamma \in D_{I^c}} t_\gamma^{k'_\gamma}, \tag{99}$$

where $P_{\mathbf{k}'}(\mathbf{y}) \in \mathbb{Q}[\mathbf{y}]$ is of total degree at most $|\mathbf{k}'|$.

Next, we calculate the contribution of $\mathbf{t}$ of the integral part. By a triangulation $\mathcal{P}_{\mathbf{m},\mathbf{y}} = \bigcup_{l=1}^{L(\mathbf{m})} \sigma_{l,\mathbf{m},\mathbf{y}}$ in Theorem 1, the integral on $\mathcal{P}_{\mathbf{m},\mathbf{y}}$ is reduced to those on $\sigma_{l,\mathbf{m},\mathbf{y}}$. Since (94) is of total degree $|\mathbf{k}|$, and $\mathbf{a}$ is of the same degree as $\mathbf{t}$ by (96), the contribution of $\mathbf{t}$ comes from terms of total degree $\kappa \le |\mathbf{k}|$ with respect to $\mathbf{a}$. By Lemma 6, we see that these terms are calculated as the special values of the functions $h(\mathbf{b}')$ at $\mathbf{b}$, where $h(\mathbf{b}')$ is a holomorphic function on $\mathbb{C}^N$ of the form

$$\mathrm{Vol}(\sigma_{l,\mathbf{m},\mathbf{y}}) \sum_{q=0}^{N} \sum_{\substack{\kappa_0,\ldots,\kappa_N \ge 0 \\ \kappa_0 + \cdots + \kappa_N = \kappa}} c_{q,\kappa_0,\ldots,\kappa_N} \frac{e^{\mathbf{b}' \cdot \mathbf{p}_{j_q}}}{\prod_{\substack{q'=0 \\ q' \ne q}}^{N} \big(\mathbf{b}' \cdot (\mathbf{p}_{j_q} - \mathbf{p}_{j_{q'}})\big)^{\kappa_{q'}+1}}, \tag{100}$$

and $\mathbf{p}_{j_q}$'s are the vertices of $\sigma_{l,\mathbf{m},\mathbf{y}}$ and $\mathrm{Vol}(\sigma_{l,\mathbf{m},\mathbf{y}}) \in \mathbb{Q}[\mathbf{y}]$ is of total degree at most N due to (98). Since

$$\mathbf{b} \cdot \mathbf{p}_j = -2\pi\sqrt{-1} \sum_{\gamma \in D_{I^c} \setminus B_{I^c}} \gamma(\lambda - \nu)(\mathbf{p}_j)_\gamma \in \pi\sqrt{-1}\mathbb{Q} \cdot A(\mathbf{y}) \cdot \overline{D}, \tag{101}$$

each term of (100) is an element of

$$\mathbb{Q}[\mathbf{y}] \frac{\mathbf{e}(\mathbb{Q} \cdot A(\mathbf{y}) \cdot \overline{D})}{(\pi\sqrt{-1}\mathbb{Q} \cdot A(\mathbf{y}) \cdot \overline{D})^{\kappa'+N}} = \frac{1}{(\pi\sqrt{-1})^{\kappa'+N}} \mathbb{Q}[\mathbf{y}] \frac{\mathbf{e}(\mathbb{Q} \cdot A(\mathbf{y}) \cdot \overline{D})}{\big(A(\mathbf{y}) \cdot \overline{D}\big)^{\kappa'+N}}, \tag{102}$$

where $0 \le \kappa' \le \kappa \le |\mathbf{k}|$.

Combining (99) and (100) for all $\mathbf{m}$ and $l \in L(\mathbf{m})$ appearing in the sum, we see that the coefficient of (94) is of the form (90). □

Remark 4. It may happen that the denominator of (90) vanishes. However, the original form (60) implies that $f^\sharp$ is well defined on P_I. In fact, the values can be obtained by use of analytic continuation in (100).

Remark 5. It should be noted that in Theorems 3 and 4, we have treated $\mathbf{y} \in V_{I^c}$ as a fixed parameter. In general, as a function of $\mathbf{y} \in V_{I^c}$, (90) is not a real analytic function on the whole space V_{I^c}. We study this fact in a special case in Sect. 11.6.

We conclude this section with the following proposition, whose proof is a direct generalization of that of (2.1) in [16].

Proposition 3. *Let $f, g \in \mathfrak{F}(P)$ and $J \subset \{1, \ldots, r\}$. Assume that $J = J_1 \coprod J_2$ and that f and g are decomposed as*

$$f(\lambda_1 + \lambda_2) = f_1(\lambda_1) f_2(\lambda_2), \qquad g(\lambda_1 + \lambda_2) = g_1(\lambda_1) g_2(\lambda_2), \tag{103}$$

for $f_1, g_1 \in \mathfrak{F}(P_{J_1})$, $f_2, g_2 \in \mathfrak{F}(P_{J_2})$, $\lambda_1 \in P_{J_1}$, $\lambda_2 \in P_{J_2}$. Then we have

$$\zeta(f, g; J; \Delta) = \zeta(f_1, g_1; J_1; \Delta) \zeta(f_2, g_2; J_2; \Delta). \tag{104}$$

11.4 Functional Relations, Value Relations, and Generating Functions

Hereafter, we deal with the special case

$$f(\lambda) = \mathbf{e}(\langle \mathbf{y}, \lambda \rangle), \tag{105}$$

$$g(\lambda) = \prod_{\alpha \in \Delta_+} \langle \alpha^\vee, \lambda \rangle^{s_\alpha}, \tag{106}$$

for $\mathbf{y} \in V$ and $\mathbf{s} = (s_\alpha)_{\alpha \in \overline{\Delta}} \in \mathbb{C}^{|\Delta_+|}$, where $\overline{\Delta}$ is the quotient of Δ obtained by identifying α and $-\alpha$. We define an action of $\mathrm{Aut}(\Delta)$ by

$$(w\mathbf{s})_\alpha = s_{w^{-1}\alpha}, \tag{107}$$

and let

$$\zeta_r(\mathbf{s}, \mathbf{y}; \Delta) = \sum_{\lambda \in P_{++}} \mathbf{e}(\langle \mathbf{y}, \lambda \rangle) \prod_{\alpha \in \Delta_+} \frac{1}{\langle \alpha^\vee, \lambda \rangle^{s_\alpha}}. \tag{108}$$

Then we have

$$\zeta(f, g; J; \Delta) = \begin{cases} \zeta_r(\mathbf{s}, \mathbf{y}; \Delta), & \text{if } J = \{1, \ldots, r\}, \\ 0, & \text{otherwise}, \end{cases} \tag{109}$$

because for $J \neq \{1, \ldots, r\}$, we have $P_{J++} \subset H_{\Delta^\vee} = H_g$. When $\mathbf{y} = 0$, the function $\zeta_r(\mathbf{s}; \Delta) = \zeta_r(\mathbf{s}, 0; \Delta)$ coincides with the zeta-function of the root system Δ, defined by (3.1) of [16]. Therefore, (108) is the Lerch-type generalization of zeta-functions of root systems. Also we have

$$S(f, g; I; \Delta) = \sum_{\lambda \in \iota^{*-1}(P_{I+}) \setminus H_{\Delta^\vee}} \mathbf{e}(\langle \mathbf{y}, \lambda \rangle) \prod_{\alpha \in \Delta_+} \frac{1}{\langle \alpha^\vee, \lambda \rangle^{s_\alpha}}, \tag{110}$$

which we denote by $S(\mathbf{s}, \mathbf{y}; I; \Delta)$. Note that $H_{\Delta^\vee}$ coincides with the set of all walls of Weyl chambers. Let

$$\mathcal{S} = \{\mathbf{s} = (s_\alpha) \in \mathbb{C}^{|\Delta_+|} \mid \Re s_\alpha > 1 \quad \text{for } \alpha \in \Delta_+\}. \tag{111}$$

Note that $\mathcal{S}$ is an $\mathrm{Aut}(\Delta)$-invariant set.

Lemma 9. *$\zeta_r(\mathbf{s},\mathbf{y};\Delta)$ and $S(\mathbf{s},\mathbf{y};I;\Delta)$ are absolutely convergent, uniformly on $\mathcal{D}\times V$ where $\mathcal{D}$ is any compact subset of the set $\mathcal{S}$. Hence, they are continuous on $\mathcal{S}\times V$, and holomorphic in $\mathbf{s}$ for a fixed $\mathbf{y}\in V$.*

Proof. Since for $\mathbf{s}\in\mathcal{D}$, $\alpha\in\Delta_+$ and $\lambda\in P\setminus H_{\Delta^\vee}$,

$$\left|\frac{1}{\langle\alpha^\vee,\lambda\rangle^{s_\alpha}}\right|\leq\frac{e^{\pi|\Im s_\alpha|}}{|\langle\alpha^\vee,\lambda\rangle|^{\Re s_\alpha}} \tag{112}$$

(the factor $e^{\pi|\Im s_\alpha|}$ appears when $\langle\alpha^\vee,\lambda\rangle$ is negative), we have

$$\begin{aligned}\left|\prod_{\alpha\in\Delta_+}\frac{1}{\langle\alpha^\vee,\lambda\rangle^{s_\alpha}}\right|&\leq h(\mathbf{s})\prod_{\alpha\in\Delta_+}\frac{1}{|\langle\alpha^\vee,\lambda\rangle|^{\Re s_\alpha}}\leq h(\mathbf{s})\prod_{\alpha\in\Psi}\frac{1}{|\langle\alpha^\vee,\lambda\rangle|^{\Re s_\alpha}}\\&\leq A\prod_{\alpha\in\Psi}\frac{1}{|\langle\alpha^\vee,\lambda\rangle|^{B}}\end{aligned} \tag{113}$$

where $h(\mathbf{s})=\prod_{\alpha\in\Delta_+}e^{\pi|\Im s_\alpha|}$ and $A=\max_{\mathbf{s}\in\mathcal{D}}h(\mathbf{s})$, $B=\min_{\alpha\in\Psi}(\min_{\mathbf{s}\in\mathcal{D}}\Re s_\alpha)>1$. It follows that

$$\sum_{\lambda\in P\setminus H_{\Delta^\vee}}\prod_{\alpha\in\Psi}\frac{1}{|\langle\alpha^\vee,\lambda\rangle|^{B}}=|W|\sum_{\lambda\in P_{++}}\prod_{\alpha\in\Psi}\frac{1}{|\langle\alpha^\vee,\lambda\rangle|^{B}}=|W|(\zeta(B))^r<\infty, \tag{114}$$

and hence the uniform and absolute convergence on $\mathcal{D}\times V$. □

Remark 6. Although the statements in Lemma 9 and in the rest of this chapter hold for larger regions than $\mathcal{S}$, we work with $\mathcal{S}$ for simplicity. For instance, the above proof of Lemma 9 holds for the region $\{\mathbf{s}=(s_\alpha)\in\mathbb{C}^{|\Delta_+|}\mid\Re s_\alpha>1\text{for }\alpha\in\Psi,\ \Re s_\alpha>0\quad\text{otherwise}\}$. (See also Remark 2.) In more general cases, we may need to specify an order of summation because the convergence is conditional. One way to do this is the Q-limit procedure treated in [6].

First, we apply Theorem 3 to the case $I\neq\emptyset$. Then Theorem 3 implies the following theorem:

Theorem 5. *When $I\neq\emptyset$, for $\mathbf{s}\in\mathcal{S}$ and $\mathbf{y}\in V$, we have*

$$\begin{aligned}&S(\mathbf{s},\mathbf{y};I;\Delta)\\&=\sum_{w\in W^I}\left(\prod_{\alpha\in\Delta_{w^{-1}}}(-1)^{-s_\alpha}\right)\zeta_r(w^{-1}\mathbf{s},w^{-1}\mathbf{y};\Delta)\\&=(-1)^{|\Delta_+\setminus\Delta_{I+}|}\left(\prod_{\alpha\in\Delta_+\setminus\Delta_{I+}}\frac{(2\pi\sqrt{-1})^{s_\alpha}}{\Gamma(s_\alpha+1)}\right)\sum_{\lambda\in P_{I++}}\mathbf{e}(\langle\mathbf{y},\lambda\rangle)\prod_{\alpha\in\Delta_{I+}}\frac{1}{\langle\alpha^\vee,\lambda\rangle^{s_\alpha}}\end{aligned}$$

$$\times \int_0^1 \cdots \int_0^1 \mathbf{e}\left(-\sum_{\alpha\in\Delta_+\setminus(\Delta_{I+}\cup\Psi)} x_\alpha\langle\alpha^\vee,\lambda\rangle\right)\left(\prod_{\alpha\in\Delta_+\setminus(\Delta_{I+}\cup\Psi)} L_{s_\alpha}(x_\alpha,0)\right)$$

$$\times\left(\prod_{i\in I^c} L_{s_{\alpha_i}}\left(\langle\mathbf{y},\lambda_i\rangle-\sum_{\alpha\in\Delta_+\setminus(\Delta_{I+}\cup\Psi)} x_\alpha\langle\alpha^\vee,\lambda_i\rangle,0\right)\right)\prod_{\alpha\in\Delta_+\setminus(\Delta_{I+}\cup\Psi)} dx_\alpha. \tag{115}$$

The second member consists of $\left(W(\Delta):W(\Delta_I)\right)$ *terms.*

Proof. It is easy to check (56) and (57) for (105) and (106), with $D_{I^c}=\Delta_+\setminus\Delta_{I+}$, $B_{I^c}=\Psi_{I^c}$ (hence $c_\gamma=0$ for all $\gamma\in D_{I^c}$), and $g^\sharp(\lambda)=\prod_{\alpha\in\Delta_{I+}}\langle\alpha^\vee,\lambda\rangle^{s_\alpha}$ for $\lambda\in P_{I+}$. Since $P_{J++}\subset H_{g^\sharp}$ for all $J\subsetneq I$, we have the second equality.

Next, we check the first equality. From (58), (109), and Theorem 3 (iii), we have

$$S(\mathbf{s},\mathbf{y};I;\Delta)=\sum_{w\in W^I}\zeta(w^{-1}f,w^{-1}g;\{1,\ldots,r\};\Delta). \tag{116}$$

Further,

$$\begin{aligned}\zeta(w^{-1}f;w^{-1}g;\{1,\ldots,r\};\Delta)&=\sum_{\lambda\in P_{++}}\mathbf{e}(\langle\mathbf{y},w\lambda\rangle)\prod_{\alpha\in\Delta_+}\frac{1}{\langle\alpha^\vee,w\lambda\rangle^{s_\alpha}}\\&=\sum_{\lambda\in P_{++}}\mathbf{e}(\langle w^{-1}\mathbf{y},\lambda\rangle)\prod_{\alpha\in\Delta_+}\frac{1}{\langle w^{-1}\alpha^\vee,\lambda\rangle^{s_\alpha}}\\&=\sum_{\lambda\in P_{++}}\mathbf{e}(\langle w^{-1}\mathbf{y},\lambda\rangle)\prod_{\alpha\in w^{-1}\Delta_+}\frac{1}{\langle\alpha^\vee,\lambda\rangle^{s_{w\alpha}}},\end{aligned} \tag{117}$$

by rewriting α as $w\alpha$. When

$$\alpha\in w^{-1}\Delta_+\cap\Delta_-=-(\Delta_+\cap w^{-1}\Delta_-)=-\Delta_w, \tag{118}$$

we further replace α by $-\alpha$. Then we have

$$\begin{aligned}&\zeta(w^{-1}f;w^{-1}g;\{1,\ldots,r\};\Delta)\\&=\left(\prod_{\alpha\in\Delta_w}(-1)^{-s_{w\alpha}}\right)\sum_{\lambda\in P_{++}}\mathbf{e}(\langle w^{-1}\mathbf{y},\lambda\rangle)\prod_{\alpha\in\Delta_+}\frac{1}{\langle\alpha^\vee,\lambda\rangle^{s_{w\alpha}}}\\&=\left(\prod_{\alpha\in\Delta_{w^{-1}}}(-1)^{-s_\alpha}\right)\zeta_r(w^{-1}\mathbf{s},w^{-1}\mathbf{y};\Delta),\end{aligned} \tag{119}$$

where we have used the fact that $w\Delta_w=-\Delta_{w^{-1}}$. Hence, the first equality follows.

Lemma 1 implies that the second member of (115) consists of $\big(W(\Delta):W(\Delta_I)\big)$ terms. □

Next, we deal with the case $I=\emptyset$. Let $S(\mathbf{s},\mathbf{y};\Delta)=S(\mathbf{s},\mathbf{y};\emptyset;\Delta)$. Then we have the following theorem by Theorem 3.

Theorem 6. *For* $\mathbf{s}\in\mathcal{S}$ *and* $\mathbf{y}\in V$,

$$\begin{aligned} S(\mathbf{s},\mathbf{y};\Delta) &= \sum_{w\in W}\left(\prod_{\alpha\in\Delta_{w^{-1}}}(-1)^{-s_\alpha}\right)\zeta_r(w^{-1}\mathbf{s},w^{-1}\mathbf{y};\Delta) \\ &= (-1)^{|\Delta_+|}\bigg(\prod_{\alpha\in\Delta_+}\frac{(2\pi\sqrt{-1})^{s_\alpha}}{\Gamma(s_\alpha+1)}\bigg)\int_0^1\cdots\int_0^1\bigg(\prod_{\alpha\in\Delta_+\setminus\Psi}L_{s_\alpha}(x_\alpha,0)\bigg) \\ &\quad\times\left(\prod_{i=1}^{r}L_{s_{\alpha_i}}\left(\langle\mathbf{y},\lambda_i\rangle-\sum_{\alpha\in\Delta_+\setminus\Psi}x_\alpha\langle\alpha^\vee,\lambda_i\rangle,0\right)\right)\prod_{\alpha\in\Delta_+\setminus\Psi}\mathrm{d}x_\alpha. \end{aligned} \tag{120}$$

The above two theorems are general functional relations among zeta-functions of root systems with exponential factors. In some cases, it is possible to deduce, from these theorems, more explicit functional relations among zeta-functions (see Example 5). However, in general, it is not easy to deduce explicit forms of functional relations from (115) by direct calculations. Therefore, in our forthcoming paper [17], we will consider some structural background of our technique more deeply and will present much improved versions of Theorems 7 and 13. In fact, by using these results, we will give explicit forms of other concrete examples which we do not treat in this chapter. On the other hand, in [14], we will introduce another technique of deducing explicit forms. This can be regarded as a certain refinement of the "u-method" developed in our previous papers. By using this technique, we give explicit functional relations among zeta-functions associated with root systems of types A_3, $C_2(\simeq B_2)$, B_3, and C_3.

Now we study special values of $S(\mathbf{s},\mathbf{y};\Delta)$. We recall that by (68),

$$L_k(x,0)=B_k(\{x\}) \tag{121}$$

for a real number x. Motivated by this observation, for $\mathbf{k}=(k_\alpha)_{\alpha\in\overline{\Delta}}\in\mathbb{N}_0^{|\Delta_+|}$ and $\mathbf{y}\in V$, we define

$$\begin{aligned} P(\mathbf{k},\mathbf{y};\Delta) &= \int_0^1\cdots\int_0^1\bigg(\prod_{\alpha\in\Delta_+\setminus\Psi}B_{k_\alpha}(x_\alpha)\bigg) \\ &\quad\times\left(\prod_{i=1}^{r}B_{k_{\alpha_i}}\left(\left\{\langle\mathbf{y},\lambda_i\rangle-\sum_{\alpha\in\Delta_+\setminus\Psi}x_\alpha\langle\alpha^\vee,\lambda_i\rangle\right\}\right)\right)\prod_{\alpha\in\Delta_+\setminus\Psi}\mathrm{d}x_\alpha, \end{aligned} \tag{122}$$

so that

$$S(\mathbf{k},\mathbf{y};\Delta)=(-1)^{|\Delta_+|}\Bigl(\prod_{\alpha\in\Delta_+}\frac{(2\pi\sqrt{-1})^{k_\alpha}}{k_\alpha!}\Bigr)P(\mathbf{k},\mathbf{y};\Delta) \tag{123}$$

for $\mathbf{k}\in\mathcal{S}\cap\mathbb{N}_0^{|\Delta_+|}$. This function $P(\mathbf{k},\mathbf{y};\Delta)$ may be regarded as a generalization of the periodic Bernoulli functions and $B_{\mathbf{k}}(\Delta)=P(\mathbf{k},0;\Delta)$ the Bernoulli numbers (see [1]). We define generating functions of $P(\mathbf{k},\mathbf{y};\Delta)$ and $B_{\mathbf{k}}(\Delta)$ as

$$F(\mathbf{t},\mathbf{y};\Delta)=\sum_{\mathbf{k}\in\mathbb{N}_0^{|\Delta_+|}}P(\mathbf{k},\mathbf{y};\Delta)\prod_{\alpha\in\Delta_+}\frac{t_\alpha^{k_\alpha}}{k_\alpha!}, \tag{124}$$

$$F(\mathbf{t};\Delta)=\sum_{\mathbf{k}\in\mathbb{N}_0^{|\Delta_+|}}B_{\mathbf{k}}(\Delta)\prod_{\alpha\in\Delta_+}\frac{t_\alpha^{k_\alpha}}{k_\alpha!}, \tag{125}$$

where $\mathbf{t}=(t_\alpha)_{\alpha\in\overline{\Delta}}$ with $|t_\alpha|<2\pi$. Assume Δ is irreducible and not of type A_1. Then by Lemma 8, we see that $P(\mathbf{k},\mathbf{y};\Delta)$ is continuous in $\mathbf{y}$ on V and $F(\mathbf{t},\mathbf{y};\Delta)$ is continuous on $\{t\in\mathbb{C}\mid |t|<2\pi\}^{|\Delta_+|}\times V$ and holomorphic in $\mathbf{t}$ for a fixed $\mathbf{y}\in V$. Further, by Lemma 7(ii), we see that for a fixed $\mathbf{y}\in V$, $F(\mathbf{t},\mathbf{y};\Delta)$ is analytically continued to a meromorphic function in $\mathbf{t}$ on the whole space $\mathbb{C}^{|\Delta_+|}$. For explicit examples, see (250) for $P((2,2,2),\mathbf{y};A_2)$ in the region $0<y_2<y_1<1$ and [17, Example 3] for $P((2,2,2,2),\mathbf{y};C_2)$.

Theorem 7. *We have*

$$\begin{aligned}F(\mathbf{t},\mathbf{y};\Delta)&=\Bigl(\prod_{\alpha\in\Delta_+}\frac{t_\alpha}{e^{t_\alpha}-1}\Bigr)\int_0^1\cdots\int_0^1\Bigl(\prod_{\alpha\in\Delta_+\setminus\Psi}\exp(t_\alpha x_\alpha)\Bigr)\\&\quad\times\Bigl(\prod_{i=1}^r\exp\Bigl(t_{\alpha_i}\Bigl\{\langle\mathbf{y},\lambda_i\rangle-\sum_{\alpha\in\Delta_+\setminus\Psi}x_\alpha\langle\alpha^\vee,\lambda_i\rangle\Bigr\}\Bigr)\Bigr)\prod_{\alpha\in\Delta_+\setminus\Psi}dx_\alpha\\&=\Bigl(\prod_{\alpha\in\Delta_+}\frac{t_\alpha}{e^{t_\alpha}-1}\Bigr)\sum_{m_1=0}^{2\langle\rho^\vee,\lambda_1\rangle-1}\cdots\sum_{m_r=0}^{2\langle\rho^\vee,\lambda_r\rangle-1}\exp\Bigl(\sum_{i=1}^r t_{\alpha_i}(\{\langle\mathbf{y},\lambda_i\rangle\}+m_i)\Bigr)\\&\quad\times\int_{\mathcal{P}_{\mathbf{m},\mathbf{y}}}\exp\Bigl(\sum_{\alpha\in\Delta_+\setminus\Psi}t_\alpha^*x_\alpha\Bigr)\prod_{\alpha\in\Delta_+\setminus\Psi}dx_\alpha,\end{aligned} \tag{126}$$

where $\rho^\vee=\frac{1}{2}\sum_{\alpha\in\Delta_+}\alpha^\vee$, $t_\alpha^*=t_\alpha-\sum_{i=1}^r t_{\alpha_i}\langle\alpha^\vee,\lambda_i\rangle$, $\mathbf{m}=(m_1,\ldots,m_r)$ *and*

$$\mathcal{P}_{\mathbf{m},\mathbf{y}}=\left\{\mathbf{x}=(x_\alpha)_{\alpha\in\Delta_+\setminus\Psi}\;\middle|\;\begin{aligned}&0\le x_\alpha\le 1,\quad(\alpha\in\Delta_+\setminus\Psi)\\&\{\langle\mathbf{y},\lambda_i\rangle\}+m_i-1\le\sum_{\alpha\in\Delta_+\setminus\Psi}x_\alpha\langle\alpha^\vee,\lambda_i\rangle\\&\qquad\le\{\langle\mathbf{y},\lambda_i\rangle\}+m_i,\quad(1\le i\le r)\end{aligned}\right\} \tag{127}$$

is a convex polytope. In particular, we have

$$F(\mathbf{t};\Delta)=\Bigl(\prod_{\alpha\in\Delta_+}\frac{t_\alpha}{e^{t_\alpha}-1}\Bigr)\sum_{m_1=1}^{2\langle\rho^\vee,\lambda_1\rangle-1}\cdots\sum_{m_r=1}^{2\langle\rho^\vee,\lambda_r\rangle-1}\exp\Bigl(\sum_{i=1}^{r}t_{\alpha_i}m_i\Bigr)$$
$$\times\int_{\mathcal{P}_{\mathbf{m}}}\exp\Bigl(\sum_{\alpha\in\Delta_+\setminus\Psi}t_\alpha^* x_\alpha\Bigr)\prod_{\alpha\in\Delta_+\setminus\Psi}\mathrm{d}x_\alpha, \tag{128}$$

where

$$\mathcal{P}_{\mathbf{m}}=\mathcal{P}_{\mathbf{m},0}$$
$$=\left\{\mathbf{x}=(x_\alpha)_{\alpha\in\Delta_+\setminus\Psi}\;\middle|\;\begin{array}{l}0\le x_\alpha\le 1,\quad(\alpha\in\Delta_+\setminus\Psi)\\ m_i-1\le\sum_{\alpha\in\Delta_+\setminus\Psi}x_\alpha\langle\alpha^\vee,\lambda_i\rangle\le m_i,\quad(1\le i\le r)\end{array}\right\}. \tag{129}$$

Proof. Applying Lemma 7 to the case $N=|\Delta_+\setminus\Psi|$, $M=|\Psi|$, $b_j=0$, $k_j=k_\alpha$, $y_i=\langle\mathbf{y},\lambda_i\rangle$, and $c_{ij}=\langle\alpha^\vee,\lambda_i\rangle$, we obtain (126). □

From the above theorem, we can deduce the following formula. In the case when all k_α's are the same, this formula gives a refinement of Witten's formula (2). In other words, it gives a multiple generalization of the classical formula (3). In [38], Witten showed that the volume of certain moduli spaces can be written in terms of special values of series (1). Moreover, he remarked that the volume is rational in the orientable case, which implies (2). Zagier [39] gives a brief sketch of a more number-theoretic demonstration of (2). Szenes [31] provides an algorithm of the evaluations by use of iterated residues. In our method, the rational number $C_W(2k,\mathfrak{g})$ is expressed in terms of generalized Bernoulli numbers, which can be calculated by use of the generating functions.

Theorem 8. *Assume that Δ is an irreducible root system. Let $k_\alpha=k_{\|\alpha\|}\in\mathbb{N}$ and $\mathbf{k}=(k_\alpha)_{\alpha\in\overline{\Delta}}$. Then we have*

$$\zeta_r(2\mathbf{k};\Delta)=\frac{(-1)^{|\Delta_+|}}{|W|}\Bigl(\prod_{\alpha\in\Delta_+}\frac{(2\pi\sqrt{-1})^{2k_\alpha}}{(2k_\alpha)!}\Bigr)B_{2\mathbf{k}}(\Delta)\in\mathbb{Q}\,\pi^{2\sum_l k_l|(\Delta_+)_l|}, \tag{130}$$

where l runs over the lengths of roots and $(\Delta_+)_l=\{\alpha\in\Delta_+\mid\|\alpha\|=l\}$.

Proof. Since the vertices $\mathbf{p}_j$ of $\mathcal{P}_{\mathbf{m}}$ satisfy $(\mathbf{p}_j)_\alpha\in\mathbb{Q}$, by Theorem 7 and Lemma 5, we have

$$B_{2\mathbf{k}}(\Delta)=P(2\mathbf{k},0;\Delta)\in\mathbb{Q}, \tag{131}$$

and hence by (123),

$$S(2\mathbf{k},0;\Delta) = (-1)^{|\Delta_+|}\Bigl(\prod_{\alpha\in\Delta_+}\frac{(2\pi\sqrt{-1})^{2k_\alpha}}{(2k_\alpha)!}\Bigr)B_{2\mathbf{k}}(\Delta) \in \mathbb{Q}\,\pi^{2\sum_l k_l|(\Delta_+)_l|}. \tag{132}$$

On the other hand, by Theorem 6,

$$S(2\mathbf{k},0;\Delta) = |W|\zeta_r(2\mathbf{k};\Delta), \tag{133}$$

since roots of the same length form a single orbit. Therefore, we have (130). □

Remark 7. The assumption of the irreducibility of Δ in Theorem 8 is not essential. Since a reducible root system is decomposed into a direct sum of some irreducible root systems, this assumption can be removed by use of Proposition 3.

Remark 8. It is also to be stressed that our formula covers the case when some of the k_α's are not the same. For example, let $X_r = C_2(\simeq B_2)$. Then we can take the positive roots as $\{\alpha_1,\alpha_2,2\alpha_1+\alpha_2,\alpha_1+\alpha_2\}$ with $\|\alpha_1^\vee\| = \|\alpha_1^\vee + 2\alpha_2^\vee\|$, $\|\alpha_2^\vee\| = \|\alpha_1^\vee + \alpha_2^\vee\|$. We see that

$$\begin{aligned}\zeta_2(2,4,4,2;C_2) &= \sum_{m=1}^\infty\sum_{n=1}^\infty \frac{1}{m^2n^4(m+n)^4(m+2n)^2}\\ &= \frac{53}{6810804000}\pi^{12}.\end{aligned} \tag{134}$$

Explicit forms of generating functions can be calculated with the aid of Theorem 1 and Lemma 4. We give some more explicit examples in Sect. 11.7.

11.5 Actions of $\widehat{W}$

In this section, we study the action of $\widehat{W}$ on $S(\mathbf{s},\mathbf{y};\Delta)$, $F(\mathbf{t},\mathbf{y};\Delta)$, and $P(\mathbf{k},\mathbf{y};\Delta)$. First, consider the action of $\mathrm{Aut}(\Delta)\subset\widehat{W}$. Note that $P\setminus H_{\Delta^\vee}$ is an $\mathrm{Aut}(\Delta)$-invariant set, because $H_{\Delta^\vee}$ is $\mathrm{Aut}(\Delta)$ invariant. An action of $\mathrm{Aut}(\Delta)$ is naturally induced on any function f in $\mathbf{s}$ and $\mathbf{y}$ as follows: For $w\in\mathrm{Aut}(\Delta)$,

$$(wf)(\mathbf{s},\mathbf{y}) = f(w^{-1}\mathbf{s},w^{-1}\mathbf{y}). \tag{135}$$

Theorem 9. *For* $\mathbf{s}\in\mathcal{S}$ *and* $\mathbf{y}\in V$, *and for* $w\in\mathrm{Aut}(\Delta)$, *we have*

$$(wS)(\mathbf{s},\mathbf{y};\Delta) = \left(\prod_{\alpha\in\Delta_{w^{-1}}}(-1)^{-s_\alpha}\right)S(\mathbf{s},\mathbf{y};\Delta), \tag{136}$$

if $s_\alpha\in\mathbb{Z}$ *for* $\alpha\in\Delta_{w^{-1}}$.

Proof. From (110), we have

$$(wS)(\mathbf{s},\mathbf{y};\Delta)=\sum_{\lambda\in P\setminus H_{\Delta^\vee}}\mathbf{e}(\langle w^{-1}\mathbf{y},\lambda\rangle)\prod_{\alpha\in\Delta_+}\frac{1}{\langle\alpha^\vee,\lambda\rangle^{s_{w\alpha}}}. \tag{137}$$

Rewriting λ as $w^{-1}\lambda$ and noting that $P\setminus H_{\Delta^\vee}$ is $\mathrm{Aut}(\Delta)$ invariant, we have

$$\begin{aligned}(wS)(\mathbf{s},\mathbf{y};\Delta)&=\sum_{\lambda\in P\setminus H_{\Delta^\vee}}\mathbf{e}(\langle\mathbf{y},\lambda\rangle)\prod_{\alpha\in\Delta_+}\frac{1}{\langle w\alpha^\vee,\lambda\rangle^{s_{w\alpha}}}\\&=\sum_{\lambda\in P\setminus H_{\Delta^\vee}}\mathbf{e}(\langle\mathbf{y},\lambda\rangle)\prod_{w^{-1}\alpha\in\Delta_+}\frac{1}{\langle\alpha^\vee,\lambda\rangle^{s_\alpha}}\\&=\Bigl(\prod_{\alpha\in\Delta_{w^{-1}}}(-1)^{-s_\alpha}\Bigr)S(\mathbf{s},\mathbf{y};\Delta).\end{aligned} \tag{138}$$

Thus, we have □

Theorem 10. *For* $\mathbf{s}\in\mathcal{S}$ *and* $\mathbf{y}\in V$*, we have* $S(\mathbf{s},\mathbf{y};\Delta)=0$ *if there exists an element* $w\in\mathrm{Aut}(\Delta)_{\mathbf{s}}\cap\mathrm{Aut}(\Delta)_{\mathbf{y}}$ *such that* $s_\alpha\in\mathbb{Z}$ *for* $\alpha\in\Delta_{w^{-1}}$ *and*

$$\sum_{\alpha\in\Delta_{w^{-1}}}s_\alpha\notin 2\mathbb{Z}, \tag{139}$$

where $\mathrm{Aut}(\Delta)_{\mathbf{s}}$ *and* $\mathrm{Aut}(\Delta)_{\mathbf{y}}$ *are the stabilizers of* $\mathbf{s}$ *and* $\mathbf{y}$ *respectively by regarding* $\mathbf{y}\in V/Q^\vee$.

Proof. Assume (139). Then by Theorem 9,

$$\Bigl(1-\Bigl(\prod_{\alpha\in\Delta_{w^{-1}}}(-1)^{-s_\alpha}\Bigr)\Bigr)S(\mathbf{s},\mathbf{y};\Delta)=0, \tag{140}$$

which implies $S(\mathbf{s},\mathbf{y};\Delta)=0$. □

Lemma 10. *The group* $\mathrm{Aut}(\Delta)$ *acts on* $\mathbb{C}^{|\Delta_+|}$ *by*

$$(w\mathbf{t})_\alpha:=\begin{cases}t_{w^{-1}\alpha}, & \text{if } \alpha\in\Delta_+\setminus\Delta_{w^{-1}},\\ -t_{w^{-1}\alpha}, & \text{if } \alpha\in\Delta_{w^{-1}},\end{cases} \tag{141}$$

where $\mathbf{t}=(t_\alpha)_{\alpha\in\overline{\Delta}}\in\mathbb{C}^{|\Delta_+|}$ *and the representative* α *runs over* Δ_+.

Proof. What we have to check is that the definition (141) indeed defines an action. Since

$$\bigl(v(w\mathbf{t})\bigr)_\alpha=\begin{cases}(w\mathbf{t})_{v^{-1}\alpha}, & \text{if } \alpha\in\Delta_+\setminus\Delta_{v^{-1}},\\ -(w\mathbf{t})_{v^{-1}\alpha}, & \text{if } \alpha\in\Delta_{v^{-1}},\end{cases} \tag{142}$$

we have

$$(v(w\mathbf{t}))_\alpha = t_{(vw)^{-1}\alpha}, \tag{143}$$

if and only if either

1. $\alpha \in \Delta_+ \setminus \Delta_{v^{-1}}$ and $v^{-1}\alpha \in \Delta_+ \setminus \Delta_{w^{-1}}$

or

2. $\alpha \in \Delta_{v^{-1}}$ and $-v^{-1}\alpha \in \Delta_{w^{-1}}$

holds. Here, the minus sign in the second case is caused by the fact that if $\alpha \in \Delta_{v^{-1}}$, then $v^{-1}\alpha \in v^{-1}\Delta_{v^{-1}} = -\Delta_v \subset \Delta_-$. Therefore, (143) is valid if and only if $\alpha \in \Delta_+$ and

$$\begin{aligned}\alpha &\in \bigl(v(\Delta_+ \setminus \Delta_{w^{-1}}) \cap v\Delta_+\bigr) \cup \bigl(v(-\Delta_{w^{-1}}) \cap v\Delta_-\bigr)\\ &= (v\Delta_+ \setminus vw\Delta_-) \cup (v\Delta_- \cap vw\Delta_+)\\ &= (v\Delta_+ \cap vw\Delta_+) \cup (v\Delta_- \cap vw\Delta_+)\\ &= vw\Delta_+.\end{aligned} \tag{144}$$

This condition is equivalent to $\alpha \in \Delta_+ \setminus \Delta_{(vw)^{-1}}$. This implies $v(w\mathbf{t}) = (vw)(\mathbf{t})$. □

Note that we defined two types of actions of $\mathrm{Aut}(\Delta)$ on $\mathbb{C}^{|\Delta_+|}$, that is, (107) and (141). The action (141) is used only on the variable $\mathbf{t}$ and should not be confused with the action (107).

If Δ is of type A_1, then $F(\mathbf{t}, \mathbf{y}; A_1) = te^{t\{y\}}/(e^t - 1)$ (see Example 1) is an even or, in other words, $\mathrm{Aut}(\Delta)$-invariant function except for $y \in \mathbb{Z}$. In the multiple cases, $F(\mathbf{t}, \mathbf{y}; \Delta)$ is revealed to be really an $\mathrm{Aut}(\Delta)$-invariant function. To show it, we need some notation and facts. Fix $1 \le m \le r$. Note that $\sigma_m\Delta_+ = (\Delta_+ \setminus \{\alpha_m\}) \coprod \{-\alpha_m\}$. Let $\Delta_1 = (\Delta_+ \setminus \Psi) \cap \sigma_m(\Delta_+ \setminus \Psi)$ and $\Psi_1 = \Psi \cap \sigma_m(\Delta_+ \setminus \Psi)$ so that $\sigma_m(\Delta_+ \setminus \Psi) = \Delta_1 \coprod \Psi_1$. Let $\Delta_2 = (\Delta_+ \setminus \Psi) \cap \sigma_m\Psi$ and $\Psi_2 = \Psi \cap \sigma_m\Psi$. Then we have $\Delta_+ \setminus \Psi = \Delta_1 \coprod \Delta_2$ and $\Psi = \Psi_1 \coprod \Psi_2 \coprod \{\alpha_m\}$. Moreover, we see that σ_m fixes Ψ_2 pointwise and $\Psi_1 = \sigma_m\Delta_2$.

Lemma 11.

$$\sum_{\alpha_i \in \Psi_1} \lambda_i \langle \alpha_i^\vee, \alpha_m \rangle = \alpha_m - 2\lambda_m. \tag{145}$$

Proof. Note that $\alpha_i \in \Psi_1$ if and only if $\langle \alpha_i^\vee, \alpha_m \rangle \neq 0$ and $\alpha_i \neq \alpha_m$. Let v be the left-hand side. Then we have

$$\langle \alpha_k^\vee, v \rangle = \begin{cases} \langle \alpha_k^\vee, \alpha_m \rangle, & \text{if } \alpha_k \in \Psi_1,\\ 0, & \text{if } \alpha_k \in \Psi_2 \cup \{\alpha_m\}, \end{cases} \tag{146}$$

which determines the right-hand side uniquely. □

An action of $\mathrm{Aut}(\Delta)$ is naturally induced on any function f in $\mathbf{t}$ and $\mathbf{y}$ as follows: For $w \in \mathrm{Aut}(\Delta)$,

$$(wf)(\mathbf{t},\mathbf{y}) = f(w^{-1}\mathbf{t}, w^{-1}\mathbf{y}). \tag{147}$$

Theorem 11. *Assume that Δ is an irreducible root system. If Δ is not of type A_1, then*

$$(wF)(\mathbf{t},\mathbf{y};\Delta) = F(\mathbf{t},\mathbf{y};\Delta) \tag{148}$$

for $\mathbf{t} \in \mathbb{C}^{|\Delta_+|}$ and $\mathbf{y} \in V$, and for $w \in \mathrm{Aut}(\Delta)$. Hence, for $\mathbf{k} \in \mathbb{N}_0^{|\Delta_+|}$ and $\mathbf{y} \in V$,

$$(wP)(\mathbf{k},\mathbf{y};\Delta) = \Bigl(\prod_{\alpha\in\Delta_{w^{-1}}} (-1)^{-k_\alpha}\Bigr) P(\mathbf{k},\mathbf{y};\Delta). \tag{149}$$

Remark 9. If $\mathbf{k}$ is in the region $\mathcal{S}$ of absolute convergence with respect to $\mathbf{s}$, the relation (123) and Theorem 9 immediately imply (149), while if $\mathbf{k} \notin \mathcal{S}$, it should be proved independently.

Remark 10. The assumption of the irreducibility is not essential by the same reason as in Remark 7.

Proof. It is sufficient to show (148) for the cases $w = \sigma_m \in W$ and $w = \omega \in \Omega$ because $\mathrm{Aut}(\Delta)$ is generated by simple reflections and the subgroup Ω. Applying the simple reflection σ_m to the second member of (126), we have

$$\begin{aligned}
&(\sigma_m F)(\mathbf{t},\mathbf{y};\Delta)\\
&= \Biggl(\prod_{\alpha\in\Delta_+} \frac{t_\alpha}{e^{t_\alpha}-1}\Biggr) \int_0^1 \cdots \int_0^1 \Biggl(\prod_{\alpha\in\Delta_+\setminus\Psi} \exp(t_{\sigma_m\alpha} x_\alpha)\Biggr)\\
&\quad \times \exp\Biggl(t_{\alpha_m}\Biggl(1 - \Biggl\{\langle\sigma_m\mathbf{y},\lambda_m\rangle - \sum_{\alpha\in\Delta_+\setminus\Psi} x_\alpha\langle\alpha^\vee,\lambda_m\rangle\Biggr\}\Biggr)\Biggr)\\
&\quad \times \Biggl(\prod_{\substack{i=1\\ i\neq m}}^{r} \exp\Biggl(t_{\sigma_m\alpha_i}\Biggl\{\langle\sigma_m\mathbf{y},\lambda_i\rangle - \sum_{\alpha\in\Delta_+\setminus\Psi} x_\alpha\langle\alpha^\vee,\lambda_i\rangle\Biggr\}\Biggr)\Biggr) \prod_{\alpha\in\Delta_+\setminus\Psi} dx_\alpha,
\end{aligned} \tag{150}$$

where we have used the fact that by the action of σ_m, the factor $\prod_{\alpha\in\Delta_+} t_\alpha/(e^{t_\alpha}-1)$ is sent to

$$\begin{aligned}
\frac{-t_{\sigma_m\alpha_m}}{e^{-t_{\sigma_m\alpha_m}}-1} \prod_{\alpha\in\Delta_+\setminus\{\alpha_m\}} \frac{t_{\sigma_m\alpha}}{e^{t_{\sigma_m\alpha}}-1} &= \frac{t_{\sigma_m\alpha_m} e^{t_{\sigma_m\alpha_m}}}{e^{t_{\sigma_m\alpha_m}}-1} \prod_{\alpha\in\Delta_+\setminus\{\alpha_m\}} \frac{t_{\sigma_m\alpha}}{e^{t_{\sigma_m\alpha}}-1}\\
&= e^{t_{\alpha_m}} \prod_{\alpha\in\Delta_+} \frac{t_\alpha}{e^{t_\alpha}-1}.
\end{aligned} \tag{151}$$

Therefore, rewriting x_α as $x_{\sigma_m\alpha}$, we have

$$
\begin{aligned}
(\sigma_m F)(\mathbf{t},\mathbf{y};\Delta) &= \Bigl(\prod_{\alpha\in\Delta_+}\frac{t_\alpha}{e^{t_\alpha}-1}\Bigr)\int_0^1\cdots\int_0^1\Bigl(\prod_{\alpha\in\Delta_+\setminus\Psi}\exp(t_{\sigma_m\alpha}x_{\sigma_m\alpha})\Bigr)\\
&\quad\times\exp\Bigl(t_{\alpha_m}\Bigl(1-\Bigl\{\langle\mathbf{y},\sigma_m\lambda_m\rangle-\sum_{\alpha\in\Delta_+\setminus\Psi}x_{\sigma_m\alpha}\langle\sigma_m\alpha^\vee,\sigma_m\lambda_m\rangle\Bigr\}\Bigr)\Bigr)\\
&\quad\times\prod_{\substack{i=1\\ i\neq m}}^{r}\exp\Bigl(t_{\sigma_m\alpha_i}\Bigl\{\langle\mathbf{y},\sigma_m\lambda_i\rangle-\sum_{\alpha\in\Delta_+\setminus\Psi}x_{\sigma_m\alpha}\langle\sigma_m\alpha^\vee,\sigma_m\lambda_i\rangle\Bigr\}\Bigr)\prod_{\alpha\in\Delta_+\setminus\Psi}\mathrm{d}x_{\sigma_m\alpha}\\
&= \Bigl(\prod_{\alpha\in\Delta_+}\frac{t_\alpha}{e^{t_\alpha}-1}\Bigr)\int_0^1\cdots\int_0^1\Bigl(\prod_{\alpha\in\sigma_m(\Delta_+\setminus\Psi)}\exp(t_\alpha x_\alpha)\Bigr)\\
&\quad\times\exp\Bigl(t_{\alpha_m}\Bigl(1-\Bigl\{\langle\mathbf{y},\lambda_m-\alpha_m\rangle-\sum_{\alpha\in\sigma_m(\Delta_+\setminus\Psi)}x_\alpha\langle\alpha^\vee,\lambda_m-\alpha_m\rangle\Bigr\}\Bigr)\Bigr)\\
&\quad\times\prod_{\alpha_i\in\Psi\setminus\{\alpha_m\}}\exp\Bigl(t_{\sigma_m\alpha_i}\Bigl\{\langle\mathbf{y},\lambda_i\rangle-\sum_{\alpha\in\sigma_m(\Delta_+\setminus\Psi)}x_\alpha\langle\alpha^\vee,\lambda_i\rangle\Bigr\}\Bigr)\prod_{\alpha\in\sigma_m(\Delta_+\setminus\Psi)}\mathrm{d}x_\alpha\\
&= \Bigl(\prod_{\alpha\in\Delta_+}\frac{t_\alpha}{e^{t_\alpha}-1}\Bigr)\int_0^1\cdots\int_0^1\prod_{\alpha\in\Delta_1}\exp(t_\alpha\{x_\alpha\})\prod_{\alpha_i\in\Psi_1}\exp(t_{\alpha_i}\{x_{\alpha_i}\})\\
&\quad\times\exp\Bigl(t_{\alpha_m}\Bigl(1-\Bigl\{\langle\mathbf{y},\lambda_m-\alpha_m\rangle-\sum_{\alpha\in\Delta_1\cup\Psi_1}x_\alpha\langle\alpha^\vee,\lambda_m-\alpha_m\rangle\Bigr\}\Bigr)\Bigr)\\
&\quad\times\Bigl(\prod_{\alpha_j\in\Psi_2}\exp\Bigl(t_{\alpha_j}\Bigl\{\langle\mathbf{y},\lambda_j\rangle-\sum_{\alpha\in\Delta_1\cup\Psi_1}x_\alpha\langle\alpha^\vee,\lambda_j\rangle\Bigr\}\Bigr)\Bigr)\\
&\quad\times\Bigl(\prod_{\alpha_i\in\Psi_1}\exp\Bigl(t_{\sigma_m\alpha_i}\Bigl\{\langle\mathbf{y},\lambda_i\rangle-\sum_{\alpha\in\Delta_1\cup\Psi_1}x_\alpha\langle\alpha^\vee,\lambda_i\rangle\Bigr\}\Bigr)\Bigr)\prod_{\alpha\in\Delta_1\cup\Psi_1}\mathrm{d}x_\alpha.
\end{aligned}
\tag{152}
$$

Here we change variables from $\mathbf{x}=(x_\alpha)_{\alpha\in\Delta_1\cup\Psi_1}$ to $\mathbf{z}=(z_\alpha)_{\alpha\in\Delta_1\cup\Delta_2}$ as

$$
z_\alpha=\begin{cases} x_\alpha, & \text{if } \alpha\in\Delta_1,\\ \langle\mathbf{y},\lambda_i\rangle-\sum_{\beta\in\Delta_1\cup\Psi_1}x_\beta\langle\beta^\vee,\lambda_i\rangle, & \text{if } \alpha=\sigma_m\alpha_i\in\Delta_2,\end{cases}
\tag{153}
$$

so that the Jacobian matrix is calculated as

$$
\frac{\partial\mathbf{z}}{\partial\mathbf{x}}=\begin{pmatrix} I_{|\Delta_1|} & 0\\ * & -I_{|\Delta_2|}\end{pmatrix},
\tag{154}
$$

where I_p is the $p \times p$ identity matrix, since

$$\begin{aligned} z_{\sigma_m \alpha_i} &= \langle \mathbf{y}, \lambda_i \rangle - \sum_{\alpha \in \Delta_1 \cup \Psi_1} x_\alpha \langle \alpha^\vee, \lambda_i \rangle \\ &= \langle \mathbf{y}, \lambda_i \rangle - \sum_{\alpha \in \Delta_1} x_\alpha \langle \alpha^\vee, \lambda_i \rangle - x_{\alpha_i}. \end{aligned} \tag{155}$$

Thus, we have $|\det \partial \mathbf{x}/\partial \mathbf{z}| = 1$. For $\alpha = \sigma_m \alpha_k \in \Delta_2$ and $\alpha_i \in \Psi_1$, we have

$$\langle \alpha^\vee, \lambda_i \rangle = \langle \sigma_m \alpha_k^\vee, \lambda_i \rangle = \langle \alpha_k^\vee, \sigma_m \lambda_i \rangle = \langle \alpha_k^\vee, \lambda_i \rangle = \delta_{ki}, \tag{156}$$

and hence,

$$\begin{aligned} x_{\alpha_i} &= \langle \mathbf{y}, \lambda_i \rangle - \sum_{\alpha \in \Delta_1} z_\alpha \langle \alpha^\vee, \lambda_i \rangle - z_{\sigma_m \alpha_i} \\ &= \langle \mathbf{y}, \lambda_i \rangle - \sum_{\alpha \in \Delta_+ \setminus \Psi} z_\alpha \langle \alpha^\vee, \lambda_i \rangle. \end{aligned} \tag{157}$$

For the fourth factor of the last integral in (152), we have

$$\langle \alpha^\vee, \lambda_j \rangle = \langle \sigma_m \alpha_k^\vee, \lambda_j \rangle = \langle \alpha_k^\vee, \sigma_m \lambda_j \rangle = \langle \alpha_k^\vee, \lambda_j \rangle = 0, \tag{158}$$

for $\alpha = \sigma_m \alpha_k \in \Delta_2$ and $\alpha_j \in \Psi_2$, and hence

$$\begin{aligned} \langle \mathbf{y}, \lambda_j \rangle - \sum_{\alpha \in \Delta_1 \cup \Psi_1} x_\alpha \langle \alpha^\vee, \lambda_j \rangle &= \langle \mathbf{y}, \lambda_j \rangle - \sum_{\alpha \in \Delta_1} z_\alpha \langle \alpha^\vee, \lambda_j \rangle \\ &= \langle \mathbf{y}, \lambda_j \rangle - \sum_{\alpha \in \Delta_+ \setminus \Psi} z_\alpha \langle \alpha^\vee, \lambda_j \rangle. \end{aligned} \tag{159}$$

For the third factor, we have

$$\begin{aligned} &\langle \mathbf{y}, \lambda_m - \alpha_m \rangle - \sum_{\alpha \in \Delta_1 \cup \Psi_1} x_\alpha \langle \alpha^\vee, \lambda_m - \alpha_m \rangle \\ &= \langle \mathbf{y}, \lambda_m - \alpha_m \rangle - \sum_{\alpha \in \Delta_1} x_\alpha \langle \alpha^\vee, \lambda_m - \alpha_m \rangle - \sum_{\alpha_i \in \Psi_1} x_{\alpha_i} \langle \alpha_i^\vee, \lambda_m - \alpha_m \rangle \\ &= \langle \mathbf{y}, \lambda_m - \alpha_m \rangle - \sum_{\alpha \in \Delta_1} x_\alpha \langle \alpha^\vee, \lambda_m - \alpha_m \rangle + \sum_{\alpha_i \in \Psi_1} x_{\alpha_i} \langle \alpha_i^\vee, \alpha_m \rangle \\ &= \langle \mathbf{y}, \lambda_m - \alpha_m \rangle - \sum_{\alpha \in \Delta_1} z_\alpha \langle \alpha^\vee, \lambda_m - \alpha_m \rangle \\ &\quad + \sum_{\alpha_i \in \Psi_1} \Bigl(\langle \mathbf{y}, \lambda_i \rangle - \sum_{\alpha \in \Delta_+ \setminus \Psi} z_\alpha \langle \alpha^\vee, \lambda_i \rangle \Bigr) \langle \alpha_i^\vee, \alpha_m \rangle \end{aligned} \tag{160}$$

by using (157). Hence, we have

$$\begin{aligned}
&\langle \mathbf{y}, \lambda_m - \alpha_m \rangle - \sum_{\alpha \in \Delta_1 \cup \Psi_1} x_\alpha \langle \alpha^\vee, \lambda_m - \alpha_m \rangle \\
&= \langle \mathbf{y}, \lambda_m - \alpha_m \rangle - \sum_{\alpha \in \Delta_1} z_\alpha \langle \alpha^\vee, \lambda_m - \alpha_m \rangle \\
&\quad + \sum_{\alpha_i \in \Psi_1} \langle \mathbf{y}, \lambda_i \rangle \langle \alpha_i^\vee, \alpha_m \rangle - \sum_{\alpha \in \Delta_+ \setminus \Psi} \sum_{\alpha_i \in \Psi_1} z_\alpha \langle \alpha^\vee, \lambda_i \rangle \langle \alpha_i^\vee, \alpha_m \rangle \\
&= -\langle \mathbf{y}, \lambda_m \rangle + \sum_{\alpha \in \Delta_+ \setminus \Psi} z_\alpha \langle \alpha^\vee, \lambda_m \rangle,
\end{aligned} \tag{161}$$

where in the last line, we have used Lemma 11 and the fact that for $\alpha = \sigma_m \alpha_k \in \Delta_2$, we have

$$\langle \alpha^\vee, \lambda_m - \alpha_m \rangle = \langle \sigma_m \alpha_k^\vee, \sigma_m \lambda_m \rangle = \langle \alpha_k^\vee, \lambda_m \rangle = 0. \tag{162}$$

Since all the factors of the integrand of the right-hand side of (152) are periodic functions with period 1, we integrate the interval [0, 1] with respect to z_α. Therefore, using (153), (157), (159), and (161), we have

$$\begin{aligned}
&(\sigma_m F)(\mathbf{t}, \mathbf{y}; \Delta) \\
&= \Bigl(\prod_{\alpha \in \Delta_+} \frac{t_\alpha}{e^{t_\alpha} - 1} \Bigr) \int_0^1 \cdots \int_0^1 \Bigl(\prod_{\alpha \in \Delta_1} \exp(t_\alpha \{z_\alpha\}) \Bigr) \\
&\quad \times \Bigl(\prod_{\alpha_i \in \Psi_1} \exp\Bigl(t_{\alpha_i} \Bigl\{ \langle \mathbf{y}, \lambda_i \rangle - \sum_{\alpha \in \Delta_+ \setminus \Psi} z_\alpha \langle \alpha^\vee, \lambda_i \rangle \Bigr\} \Bigr) \Bigr) \\
&\quad \times \exp\Bigl(t_{\alpha_m} \Bigl(1 - \Bigl\{ -\langle \mathbf{y}, \lambda_m \rangle + \sum_{\alpha \in \Delta_+ \setminus \Psi} z_\alpha \langle \alpha^\vee, \lambda_m \rangle \Bigr\} \Bigr) \Bigr) \\
&\quad \times \Bigl(\prod_{\alpha_j \in \Psi_2} \exp\Bigl(t_{\alpha_j} \Bigl\{ \langle \mathbf{y}, \lambda_j \rangle - \sum_{\alpha \in \Delta_+ \setminus \Psi} z_\alpha \langle \alpha^\vee, \lambda_j \rangle \Bigr\} \Bigr) \Bigr) \\
&\quad \times \Bigl(\prod_{\alpha_i \in \Psi_1} \exp(t_{\sigma_m \alpha_i} \{z_{\sigma_m \alpha_i}\}) \Bigr) \prod_{\alpha \in \Delta_+ \setminus \Psi} dz_\alpha \\
&= \Bigl(\prod_{\alpha \in \Delta_+} \frac{t_\alpha}{e^{t_\alpha} - 1} \Bigr) \int_0^1 \cdots \int_0^1 \Bigl(\prod_{\alpha \in \Delta_+ \setminus \Psi} \exp(t_\alpha z_\alpha) \Bigr) \\
&\quad \times \Bigl(\prod_{\alpha_i \in \Psi \setminus \{\alpha_m\}} \exp\Bigl(t_{\alpha_i} \Bigl\{ \langle \mathbf{y}, \lambda_i \rangle - \sum_{\alpha \in \Delta_+ \setminus \Psi} z_\alpha \langle \alpha^\vee, \lambda_i \rangle \Bigr\} \Bigr) \Bigr)
\end{aligned}$$

$$\times \exp\Bigl(t_{\alpha_m}\Bigl(1-\Bigl\{-\langle \mathbf{y},\lambda_m\rangle + \sum_{\alpha\in\Delta_+\setminus\Psi} z_\alpha\langle\alpha^\vee,\lambda_m\rangle\Bigr\}\Bigr)\Bigr) \prod_{\alpha\in\Delta_+\setminus\Psi} \mathrm{d}z_\alpha$$
$$= F(\mathbf{t},\mathbf{y};\Delta), \tag{163}$$

where in the last line, we have used the fact that for any $\alpha_m \in \Psi$, there exists a root $\alpha \in \Delta_+ \setminus \Psi$ such that $\langle\alpha^\vee,\lambda_m\rangle \neq 0$ and thus $1-\{-x\} = \{x\}$ for $x \in \mathbb{R}\setminus\mathbb{Z}$ implies that the integrand coincides with that of $F(\mathbf{t},\mathbf{y};\Delta)$ almost everywhere.

Lastly, we check the invariance with respect to $\omega \in \Omega$. Since $\omega \in \Omega$ permutes Ψ and leaves Δ_+ and hence $\Delta_+ \setminus \Psi$ invariant, we have

$$\begin{aligned}
&(\omega F)(\mathbf{t},\mathbf{y};\Delta)\\
&= \Bigl(\prod_{\alpha\in\Delta_+}\frac{t_{\omega\alpha}}{\mathrm{e}^{t_{\omega\alpha}}-1}\Bigr)\int_0^1\cdots\int_0^1\Bigl(\prod_{\alpha\in\Delta_+\setminus\Psi}\exp(t_{\omega\alpha}x_{\omega\alpha})\Bigr)\\
&\quad\times\Bigl(\prod_{i=1}^r \exp\Bigl(t_{\omega\alpha_i}\Bigl\{\langle\omega^{-1}\mathbf{y},\lambda_i\rangle - \sum_{\alpha\in\Delta_+\setminus\Psi} x_{\omega\alpha}\langle\omega\alpha^\vee,\omega\lambda_i\rangle\Bigr\}\Bigr)\Bigr)\prod_{\alpha\in\Delta_+\setminus\Psi}\mathrm{d}x_{\omega\alpha}\\
&= \Bigl(\prod_{\alpha\in\Delta_+}\frac{t_{\alpha}}{\mathrm{e}^{t_{\alpha}}-1}\Bigr)\int_0^1\cdots\int_0^1\Bigl(\prod_{\alpha\in\Delta_+\setminus\Psi}\exp(t_{\alpha}x_{\alpha})\Bigr)\\
&\quad\times\Bigl(\prod_{i=1}^r \exp\Bigl(t_{\omega\alpha_i}\Bigl\{\langle\mathbf{y},\omega\lambda_i\rangle - \sum_{\alpha\in\Delta_+\setminus\Psi} x_{\alpha}\langle\alpha^\vee,\omega\lambda_i\rangle\Bigr\}\Bigr)\Bigr)\prod_{\alpha\in\Delta_+\setminus\Psi}\mathrm{d}x_{\alpha}\\
&= F(\mathbf{t},\mathbf{y};\Delta).
\end{aligned} \tag{164}$$

□

It is possible to extend the action of $\mathrm{Aut}(\Delta)$ to that of $\widehat{W}$ as follows: For $q \in Q^\vee$,

$$\begin{aligned}
(\tau(q)\mathbf{s})_\alpha &= s_\alpha,\\
(\tau(q)\mathbf{t})_\alpha &= t_\alpha,\\
\tau(q)\mathbf{y} &= \mathbf{y}+q.
\end{aligned} \tag{165}$$

We can observe the periodicity of S, F, and P with respect to $\mathbf{y}$ from (110), (122), and the first line of (126). From this periodicity, we have

Theorem 12. *The action of* $\mathrm{Aut}(\Delta)$ *is extended to that of* $\widehat{W}$ *and is given by*

$$\begin{aligned}
(\tau(q)S)(\mathbf{s},\mathbf{y};\Delta) &= S(\mathbf{s},\mathbf{y};\Delta),\\
(\tau(q)F)(\mathbf{t},\mathbf{y};\Delta) &= F(\mathbf{t},\mathbf{y};\Delta),\\
(\tau(q)P)(\mathbf{k},\mathbf{y};\Delta) &= P(\mathbf{k},\mathbf{y};\Delta),
\end{aligned} \tag{166}$$

for $q \in Q^\vee$.

Remark 11. Some statements related with $S(\mathbf{s},\mathbf{y};\Delta)$ or $\zeta_r(\mathbf{s},\mathbf{y};\Delta)$ in Sects. 11.4 and 11.5 hold on any regions in $\mathbf{s}$ to which these functions are analytically continued. In particular, in the case $\mathbf{y}=0$, as we noticed at the beginning of Sect. 11.4, the function $\zeta_r(\mathbf{s};\Delta)=\zeta_r(\mathbf{s},0;\Delta)$ coincides with the zeta-function defined in [16], and its analytic continuation is given in [16, Theorem 6.1] or by Essouabri's theory [5].

11.6 Generalization of Bernoulli Polynomials

In the previous sections, we have investigated $P(\mathbf{k},\mathbf{y};\Delta)$ as a continuous function in $\mathbf{y}$. In fact, this function is not real analytic in $\mathbf{y}$ in general. However, they are piecewise real analytic, and each piece is actually a polynomial in $\mathbf{y}$. In this section, we will prove this fact and will discuss basic properties of those polynomials.

Let $\mathfrak{D}=\{\mathbf{y}\in V \mid 0\leq\langle\mathbf{y},\lambda_i\rangle\leq 1,\ (1\leq i\leq r)\}$ be a period-parallelotope of $F(\mathbf{t},\cdot;\Delta)$ with its interior. Let $\mathscr{R}$ be the set of all linearly independent subsets $\mathbf{R}=\{\beta_1,\ldots,\beta_{r-1}\}\subset\Delta$, $\mathfrak{H}_{\mathbf{R}^\vee}=\bigoplus_{i=1}^{r-1}\mathbb{R}\beta_i^\vee$ the hyperplane passing through $\mathbf{R}^\vee\cup\{0\}$ and

$$\mathfrak{H}_{\mathscr{R}}:=\bigcup_{\substack{\mathbf{R}\in\mathscr{R}\\ q\in Q^\vee}}(\mathfrak{H}_{\mathbf{R}^\vee}+q). \tag{167}$$

Lemma 12. *We have*

$$\mathfrak{H}_{\mathscr{R}}=\bigcup_{w\in W}w\left(\bigcup_{j=1}^{r}(\mathfrak{H}_{\Psi^\vee\setminus\{\alpha_j^\vee\}}+\mathbb{Z}\alpha_j^\vee)\right). \tag{168}$$

The set $\{\mathfrak{H}_{\mathbf{R}^\vee}+q \mid \mathbf{R}\in\mathscr{R}, q\in Q^\vee\}$ is locally finite, that is, for any $\mathbf{y}\in V$, there exists a neighborhood $U(\mathbf{y})$ such that $U(\mathbf{y})$ intersects finitely many of these hyperplanes.

Proof. Fix $\mathbf{R}\in\mathscr{R}$. Then $\tilde{\Delta}^\vee=\Delta^\vee\cap\mathfrak{H}_{\mathbf{R}^\vee}$ is a coroot system so that $\mathbf{R}^\vee\subset\tilde{\Delta}^\vee$. Let μ be a nonzero vector normal to $\mathfrak{H}_{\mathbf{R}^\vee}$. Then there exists an element $w\in W$ such that $w^{-1}\mu\in C$. Put $w^{-1}\mu=\sum_{j=1}^r c_j\lambda_j$ with $c_j\geq 0$. Then $\alpha^\vee=\sum_{j=1}^r a_j\alpha_j^\vee\in\Delta^\vee$ orthogonal to $w^{-1}\mu$ should satisfy $\sum_{j=1}^r a_jc_j=0$. Since a_j are all nonpositive or nonnegative, we have $a_j=0$ for j such that $c_j\neq 0$. Hence, $c_j=0$ except for only one j, because $w^{-1}\tilde{\Delta}^\vee\subset\Delta^\vee$ is orthogonal to $w^{-1}\mu$ with codimension 1. That is, $w^{-1}\mu=c\lambda_j$ for some $c>0$. Therefore, $w(\Psi^\vee\setminus\{\alpha_j^\vee\})$ is a fundamental system of $\tilde{\Delta}^\vee$ and $\mathfrak{H}_{\mathbf{R}^\vee}=\mathfrak{H}_{w(\Psi^\vee\setminus\{\alpha_j^\vee\})}=w\mathfrak{H}_{\Psi^\vee\setminus\{\alpha_j^\vee\}}$. Moreover, $Q^\vee=wQ^\vee=\bigoplus_{i=1}^r\mathbb{Z}w\alpha_i^\vee$, which implies

$$\mathfrak{H}_{\mathbf{R}^\vee}+Q^\vee=w\mathfrak{H}_{\Psi^\vee\setminus\{\alpha_j^\vee\}}+\mathbb{Z}w\alpha_j^\vee, \tag{169}$$

since $\bigoplus_{i=1,i\neq j}^{r}\mathbb{Z}\alpha_i^\vee \subset \mathfrak{H}_{\Psi^\vee\setminus\{\alpha_j^\vee\}}$. This shows that $\mathfrak{H}_{\mathscr{R}}$ is contained in the right-hand side of (168). The opposite inclusion is clear. The local finiteness follows from the expression (168) and $|W| < \infty$. □

Due to the local finiteness shown in Lemma 12 and $\partial\mathfrak{D} \subset \mathfrak{H}_{\mathscr{R}}$, we denote by $\mathfrak{D}^{(\nu)}$ each open connected component of $\mathfrak{D} \setminus \mathfrak{H}_{\mathscr{R}}$ so that

$$\mathfrak{D} \setminus \mathfrak{H}_{\mathscr{R}} = \coprod_{\nu\in\mathfrak{J}} \mathfrak{D}^{(\nu)}, \tag{170}$$

where $\mathfrak{J}$ is a set of indices. Let $\mathscr{V}$ be the set of all linearly independent subsets $\mathbf{V} = \{\beta_1, \dots, \beta_r\} \subset \Delta_+$ and $\mathscr{A} = \{0,1\}^{n-r}$, where $n = |\Delta_+|$. Let $\mathscr{W} = \mathscr{V} \times \mathscr{A}$. For subsets $u = \{u_1, \dots, u_k\}, v = \{v_1, \dots, v_{k+1}\} \subset V$, and $c = \{c_1, \dots, c_{k+1}\} \subset \mathbb{R}$, let

$$H(\mathbf{y}; u, v, c) = \det\begin{pmatrix} \langle u_1, v_1\rangle & \cdots & \langle u_1, v_{k+1}\rangle \\ \vdots & \ddots & \vdots \\ \langle u_k, v_1\rangle & \cdots & \langle u_k, v_{k+1}\rangle \\ \langle \mathbf{y}, v_1\rangle + c_1 & \cdots & \langle \mathbf{y}, v_{k+1}\rangle + c_{k+1} \end{pmatrix}. \tag{171}$$

We give a simple description of the polytopes $\mathcal{P}_{\mathbf{m},\mathbf{y}}$ defined by (127). For $\gamma \in \Delta_+$, $a \in \{0,1\}$, and $\mathbf{y} \in V$, we define $\mathbf{u}(\gamma, a) \in \mathbb{R}^{n-r}$ by

$$\mathbf{u}(\gamma, a)_\alpha = \begin{cases} (-1)^{1-a}\langle \alpha^\vee, \lambda_i\rangle, & \text{if } \gamma = \alpha_i \in \Psi, \\ (-1)^a \delta_{\alpha\gamma}, & \text{if } \gamma \notin \Psi, \end{cases} \tag{172}$$

where α runs over $\Delta_+ \setminus \Psi$, and define $v(\gamma, a; \mathbf{y}) \in \mathbb{R}$ by

$$v(\gamma, a; \mathbf{y}) = \begin{cases} (-1)^{1-a}(\{\langle \mathbf{y}, \lambda_i\rangle\} + m_i - a), & \text{if } \gamma = \alpha_i \in \Psi, \\ (-1)^a a = -a, & \text{if } \gamma \notin \Psi. \end{cases} \tag{173}$$

Further, we define

$$\mathcal{H}_{\gamma,a}(\mathbf{y}) = \{\mathbf{x} = (x_\alpha)_{\alpha\in\Delta_+\setminus\Psi} \in \mathbb{R}^{n-r} \mid \mathbf{u}(\gamma, a)\cdot\mathbf{x} = v(\gamma, a; \mathbf{y})\}, \tag{174}$$

and

$$\mathcal{H}^+_{\gamma,a}(\mathbf{y}) = \{\mathbf{x} = (x_\alpha)_{\alpha\in\Delta_+\setminus\Psi} \in \mathbb{R}^{n-r} \mid \mathbf{u}(\gamma, a)\cdot\mathbf{x} \geq v(\gamma, a; \mathbf{y})\}, \tag{175}$$

where for $\mathbf{w} = (w_\alpha), \mathbf{x} = (x_\alpha) \in \mathbb{C}^{n-r}$, we have set

$$\mathbf{w}\cdot\mathbf{x} = \sum_{\alpha\in\Delta_+\setminus\Psi} w_\alpha x_\alpha. \tag{176}$$

Then we have

$$\mathcal{P}_{\mathbf{m},\mathbf{y}} = \bigcap_{\substack{\gamma\in\Delta_+ \\ a\in\{0,1\}}} \mathcal{H}^+_{\gamma,a}(\mathbf{y}). \tag{177}$$

We use the identification $\mathbb{C}\otimes V \simeq \mathbb{C}^r$ through $\mathbf{y} \mapsto (y_i)_{i=1}^r$ where $y_i = \langle \mathbf{y}, \lambda_i\rangle$ with $\langle\cdot,\cdot\rangle$ bilinearly extended over $\mathbb{C}$. For $\mathbf{k} = (k_\alpha)_{\alpha\in\overline{\Delta}} \in \mathbb{N}_0^n$, we set $|\mathbf{k}| = \sum_{\alpha\in\overline{\Delta}} k_\alpha$.

Theorem 13. *In each $\mathfrak{D}^{(\nu)}$, the functions $F(\mathbf{t},\mathbf{y};\Delta)$ and $P(\mathbf{k},\mathbf{y};\Delta)$ are real analytic in $\mathbf{y}$. Moreover, $F(\mathbf{t},\mathbf{y};\Delta)$ is analytically continued to a meromorphic function $F^{(\nu)}(\mathbf{t},\mathbf{y};\Delta)$ from each $\mathbb{C}^n \times \mathfrak{D}^{(\nu)}$ to the whole space $\mathbb{C}^n \times (\mathbb{C}\otimes V)$. Similarly, $P(\mathbf{k},\mathbf{y};\Delta)$ is analytically continued to a polynomial function $B^{(\nu)}_{\mathbf{k}}(\mathbf{y};\Delta) \in \mathbb{Q}[\mathbf{y}]$ from each $\mathfrak{D}^{(\nu)}$ to the whole space $\mathbb{C}\otimes V$ with its total degree at most $|\mathbf{k}| + n - r$.*

Proof. Throughout this proof, we fix an index $\nu \in \mathfrak{J}$. Note that $\{\langle \mathbf{y},\lambda_i\rangle\} = \langle \mathbf{y},\lambda_i\rangle$ holds for $\mathbf{y} \in \mathring{\mathfrak{D}}$. We show this statement by several steps. In the first three steps, we investigate the dependence of vertices of $\mathcal{P}_{\mathbf{m},\mathbf{y}}$ on $\mathbf{y} \in \mathfrak{D}^{(\nu)}$, and in the last two steps, by use of this result and triangulation, we show the analyticity of the generating function. We fix $\mathbf{m} \in \mathbb{N}_0^r$ except in the last step.

(Step 1.) Let $\mathbf{V} = \{\beta_1,\ldots,\beta_r\} \subset \Delta_+$ and $a_\gamma \in \{0,1\}$ for $\gamma \in \Delta_+ \setminus \mathbf{V}$. Consider the intersection of $|\Delta_+ \setminus \mathbf{V}|(= n-r)$ hyperplanes

$$\bigcap_{\gamma\in\Delta_+\setminus\mathbf{V}} \mathcal{H}_{\gamma,a_\gamma}(\mathbf{y}) = \{\mathbf{x} = (x_\alpha) \mid \mathbf{u}(\gamma,a_\gamma)\cdot\mathbf{x} = v(\gamma,a_\gamma;\mathbf{y}) \text{ for } \gamma \in \Delta_+\setminus\mathbf{V}\}. \tag{178}$$

Then this set consists of the solutions of the system of the $(n-r)$ linear equations

$$\begin{cases} \sum_{\alpha\in\Delta_+\setminus\Psi} x_\alpha \langle \alpha^\vee,\lambda_j\rangle = \langle \mathbf{y},\lambda_j\rangle + m_j - a_{\alpha_j}, & \text{for } \gamma = \alpha_j \in \Psi\setminus\mathbf{V}, \\ x_\gamma = a_\gamma, & \text{for } \gamma \in \Delta_+ \setminus (\Psi\cup\mathbf{V}). \end{cases} \tag{179}$$

Let $I = \{i \mid \beta_i \in \mathbf{V}\setminus\Psi\}$ and $J = \{j \mid \alpha_j \in \Psi\setminus\mathbf{V}\}$. Note that $|I| = |J| =: k$ and $\{\beta_i \mid i \in I^c\} = \{\alpha_j \mid j \in J^c\}$. The system of the linear equations (179) has a unique solution if and only if

$$\det(\langle \beta_i^\vee,\lambda_j\rangle)_{j\in J}^{i\in I} \neq 0, \tag{180}$$

and also if and only if

$$\mathbf{V} \in \mathscr{V}, \tag{181}$$

since

$$\begin{aligned}
\left|\det(\langle\beta_i^\vee,\lambda_j\rangle)_{1\le j\le r}^{1\le i\le r}\right| &= \left|\det\begin{pmatrix}(\langle\beta_i^\vee,\lambda_j\rangle)_{j\in J}^{i\in I} & (\langle\beta_i^\vee,\lambda_j\rangle)_{j\in J^c}^{i\in I}\\ (\langle\alpha_i^\vee,\lambda_j\rangle)_{j\in J}^{i\in J^c} & (\langle\alpha_i^\vee,\lambda_j\rangle)_{j\in J^c}^{i\in J^c}\end{pmatrix}\right| \\
&= \left|\det\begin{pmatrix}(\langle\beta_i^\vee,\lambda_j\rangle)_{j\in J}^{i\in I} & * \\ 0 & I_{|J^c|}\end{pmatrix}\right| \\
&= \left|\det(\langle\beta_i^\vee,\lambda_j\rangle)_{j\in J}^{i\in I}\right|,
\end{aligned} \tag{182}$$

where I_p is the $p\times p$ identity matrix. We assume (181) and denote by $\mathbf{p}(\mathbf{y};\mathbf{W})$ the unique solution, where $\mathbf{W}=(\mathbf{V},\mathbf{A})\in\mathscr{W}$ with the sequence $\mathbf{A}=(a_\gamma)_{\gamma\in\Delta_+\setminus\mathbf{V}}$ regarded as an element of $\mathscr{A}$. We see that $\mathbf{p}(\mathbf{y};\mathbf{W})$ depends on $\mathbf{y}$ affine linearly.

(Step 2.) We define $\xi_{\mathbf{y}}:\mathscr{W}\to\mathbb{R}^{n-r}$ by

$$\xi_{\mathbf{y}}:\mathbf{W}\mapsto\mathbf{p}(\mathbf{y};\mathbf{W}). \tag{183}$$

Any vertex (i.e., 0-face) of $\mathcal{P}_{\mathbf{m},\mathbf{y}}$ is defined by the intersection of $(n-r)$ hyperplanes by Proposition 2. Hence $\mathrm{Vert}(\mathcal{P}_{\mathbf{m},\mathbf{y}})\subset\xi_{\mathbf{y}}(\mathscr{W})$. On the other hand, $\mathcal{P}_{\mathbf{m},\mathbf{y}}$ is defined by n pairs of inequalities in (127). The point $\mathbf{p}(\mathbf{y};\mathbf{W})$ is a vertex of $\mathcal{P}_{\mathbf{m},\mathbf{y}}$ if all of those inequalities hold. We see that $(n-r)$ pairs among them are satisfied, because

$$\mathbf{p}(\mathbf{y};\mathbf{W})\in\bigcap_{\gamma\in\Delta_+\setminus\mathbf{V}}\mathcal{H}_{\gamma,a_\gamma}(\mathbf{y}), \tag{184}$$

and also it is easy to check

$$\mathbf{p}(\mathbf{y};\mathbf{W})\in\bigcap_{\gamma\in\Delta_+\setminus\mathbf{V}}\big(\mathcal{H}^+_{\gamma,1-a_\gamma}(\mathbf{y})\setminus\mathcal{H}_{\gamma,1-a_\gamma}(\mathbf{y})\big). \tag{185}$$

Therefore, $\mathbf{p}(\mathbf{y};\mathbf{W})\in\mathrm{Vert}(\mathcal{P}_{\mathbf{m},\mathbf{y}})$ if and only if the remaining r pairs of inequalities are satisfied, that is,

$$\begin{aligned}
\mathbf{p}(\mathbf{y};\mathbf{W}) &\in \bigcap_{\substack{\beta\in\mathbf{V}\\ a\in\{0,1\}}}\mathcal{H}^+_{\beta,a}(\mathbf{y}) \\
&= \{\mathbf{x}=(x_\alpha)\mid \mathbf{u}(\beta,a)\cdot\mathbf{x}\ge v(\beta,a;\mathbf{y})\text{ for }\beta\in\mathbf{V},a\in\{0,1\}\},
\end{aligned} \tag{186}$$

or equivalently $\mathbf{x}=\mathbf{p}(\mathbf{y};\mathbf{W})$ satisfies r pairs of the linear inequalities

$$\begin{cases}\langle\mathbf{y},\lambda_l\rangle+m_l-1\le\sum_{\alpha\in\Delta_+\setminus\Psi}x_\alpha\langle\alpha^\vee,\lambda_l\rangle\le\langle\mathbf{y},\lambda_l\rangle+m_l, \\ \qquad\qquad\text{for }\beta=\alpha_l\in\mathbf{V}\cap\Psi, \\ 0\le x_\beta\le 1, \qquad\qquad\text{for }\beta\in\mathbf{V}\setminus\Psi.\end{cases} \tag{187}$$

We see that it depends on $\mathbf{y}$ whether $\mathbf{p}(\mathbf{y};\mathbf{W})$ is a vertex, or in other words, whether the solution of (179) satisfies (187). We will show in the next step that $\mathbf{p}(\mathbf{y};\mathbf{W}) \in \mathcal{H}(\mathbf{y};\mathbf{W})$ implies $\mathbf{y} \in \mathfrak{H}_{\mathscr{R}}$, where

$$\mathcal{H}(\mathbf{y};\mathbf{W}) = \bigcup_{\substack{\beta\in\mathbf{V}\\ a\in\{0,1\}}} \mathcal{H}_{\beta,a}(\mathbf{y}). \tag{188}$$

Then for $\mathbf{y} \in \mathfrak{D} \setminus \mathfrak{H}_{\mathscr{R}}$, we can uniquely determine the $(n-r)$ hyperplanes on which the point $\mathbf{p}(\mathbf{y};\mathbf{W})$ lies; they are $\{\mathcal{H}_{\gamma,a_\gamma}(\mathbf{y})\}_{\gamma\in\Delta_+\setminus\mathbf{V}}$. Therefore, $\xi_{\mathbf{y}}$ is an injection.

For $\beta \in \mathbf{V}$ and $a \in \{0,1\}$, we define $f_{\beta,a} : \mathring{\mathfrak{D}} \to \mathbb{R}$ by

$$f_{\beta,a} : \mathbf{y} \mapsto \mathbf{u}(\beta,a)\cdot\mathbf{p}(\mathbf{y};\mathbf{W}) - v(\beta,a;\mathbf{y}). \tag{189}$$

Then for $\mathbf{y} \in \mathfrak{D} \setminus \mathfrak{H}_{\mathscr{R}}$, we have $f_{\beta,a}(\mathbf{y}) \neq 0$, and hence, we define

$$f = (f_{\beta,a})_{\beta\in\mathbf{V},a\in\{0,1\}} : \mathfrak{D} \setminus \mathfrak{H}_{\mathscr{R}} \to (\mathbb{R}\setminus\{0\})^{2r}. \tag{190}$$

Therefore, for $\mathbf{y} \in \mathfrak{D} \setminus \mathfrak{H}_{\mathscr{R}}$, the point $\mathbf{p}(\mathbf{y};\mathbf{W})$ is a vertex if and only if $f(\mathbf{y})$ is an element of the connected component $(0,\infty)^{2r}$. Since each $f_{\beta,a}$ is continuous and hence $f(\mathfrak{D}^{(\nu)})$ is connected, we see that for a fixed $\mathbf{W} \in \mathscr{W}$, the point $\mathbf{p}(\mathbf{y};\mathbf{W})$ is always a vertex, or never a vertex, on $\mathfrak{D}^{(\nu)}$. Thus,

$$\mathscr{W}_{\mathbf{m}} := \xi_{\mathbf{y}}^{-1}(\mathrm{Vert}(\mathcal{P}_{\mathbf{m},\mathbf{y}})) \subset \mathscr{W} \qquad (\mathbf{y} \in \mathfrak{D}^{(\nu)}) \tag{191}$$

has one-to-one correspondence with $\mathrm{Vert}(\mathcal{P}_{\mathbf{m},\mathbf{y}})$ and is independent of $\mathbf{y}$ on $\mathfrak{D}^{(\nu)}$.

(Step 3.) Now we prove the claim announced just before (188). First, we show that the condition

$$\mathbf{p}(\mathbf{y};\mathbf{W}) \in \mathcal{H}_{\beta,a_\beta}(\mathbf{y}) \tag{192}$$

for some $\beta = \alpha_l \in \mathbf{V} \cap \Psi$ and $a_\beta \in \{0,1\}$ implies $\mathbf{y} \in \mathfrak{H}_{\mathscr{R}}$. For $\mathbf{x} = \mathbf{p}(\mathbf{y};\mathbf{W})$, condition (192) is equivalent to

$$\sum_{\alpha\in\Delta_+\setminus\Psi} x_\alpha \langle \alpha^\vee, \lambda_l\rangle = \langle \mathbf{y}, \lambda_l\rangle + m_l - a_{\alpha_l}. \tag{193}$$

From (179) and (193), we have an overdetermined system with the $|\mathbf{V}\setminus\Psi| = k$ variables x_β for $\beta \in \mathbf{V}\setminus\Psi$ and the $|(\Psi\setminus\mathbf{V}) \cup \{\alpha_l\}| = (k+1)$ equations

$$\sum_{\beta\in\mathbf{V}\setminus\Psi} x_\beta \langle \beta^\vee, \lambda_j\rangle = \langle \mathbf{y}, \lambda_j\rangle + c_j, \tag{194}$$

for $j \in J \cup \{l\}$, where

$$c_j = m_j - a_{\alpha_j} - \sum_{\gamma \in \Delta_+ \setminus (\Psi \cup \mathbf{V})} a_\gamma \langle \gamma^\vee, \lambda_j \rangle \in \mathbb{Z}. \tag{195}$$

Hence, we have

$$\begin{pmatrix} x_{\beta_{i_1}} & \cdots & x_{\beta_{i_k}} & -1 \end{pmatrix} \begin{pmatrix} \langle \beta_{i_1}^\vee, \lambda_{j_1} \rangle & \cdots & \langle \beta_{i_1}^\vee, \lambda_{j_k} \rangle & \langle \beta_{i_1}^\vee, \lambda_l \rangle \\ \vdots & \ddots & \vdots & \vdots \\ \langle \beta_{i_k}^\vee, \lambda_{j_1} \rangle & \cdots & \langle \beta_{i_k}^\vee, \lambda_{j_k} \rangle & \langle \beta_{i_k}^\vee, \lambda_l \rangle \\ \langle \mathbf{y}, \lambda_{j_1} \rangle + c_{j_1} & \cdots & \langle \mathbf{y}, \lambda_{j_k} \rangle + c_{j_k} & \langle \mathbf{y}, \lambda_l \rangle + c_l \end{pmatrix} = \begin{pmatrix} 0 \cdots 0 \end{pmatrix}, \tag{196}$$

where we have put $I = \{i_1, \ldots, i_k\}$ and $J = \{j_1, \ldots, j_k\}$. As the consistency for these equations, we get

$$H(\mathbf{y}; \{\beta_i^\vee\}_{i \in I}, \{\lambda_j\}_{j \in J \cup \{l\}}, \{c_j\}_{j \in J \cup \{l\}}) = 0. \tag{197}$$

By direct substitution, we see that each element of

$$\{\beta^\vee - q\}_{\beta \in \mathbf{V} \setminus \{\alpha_l\}} \cup \{-q\}, \quad q = \sum_{j \in J \cup \{l\}} c_j \alpha_j^\vee \tag{198}$$

satisfies (197), while $\alpha_l^\vee - q$ does not. In fact, if $\mathbf{y} = -q$ or $\mathbf{y} = \beta^\vee - q$ ($\beta \in (\mathbf{V} \cap \Psi) \setminus \{\alpha_l\}$), then the last row of the matrix is $(0, \ldots, 0)$, and if $\mathbf{y} = \beta_{i_p}^\vee - q$ ($\beta_{i_p} \in \mathbf{V} \setminus \Psi$), then the last row is equal to the p-th row, and hence (197) follows, while if $\mathbf{y} = \alpha_l^\vee - q$, then the last row of the matrix is $(0, \ldots, 0, 1)$ and hence

$$H(\mathbf{y}; \{\beta_i^\vee\}_{i \in I}, \{\lambda_j\}_{j \in J \cup \{l\}}, \{c_j\}_{j \in J \cup \{l\}}) = \det(\langle \beta_i^\vee, \lambda_j \rangle)_{j \in J}^{i \in I} \neq 0, \tag{199}$$

because of (180). By (181), we see that $\mathbf{V} \setminus \{\alpha_l\} \subset \Delta$ is a linearly independent subset and hence $(\mathbf{V} \setminus \{\alpha_l\}) \in \mathscr{R}$. It follows that (197) represents the hyperplane $\mathfrak{H}_{\mathbf{V}^\vee \setminus \{\alpha_l^\vee\}} - q \subset \mathfrak{H}_{\mathscr{R}}$. Therefore, (192) implies $\mathbf{y} \in \mathfrak{H}_{\mathscr{R}}$.

Similarly, we see that the condition $\mathbf{p}(\mathbf{y}; \mathbf{W}) \in \mathcal{H}_{\beta, a_\beta}(\mathbf{y})$ for some $\beta = \beta_l \in \mathbf{V} \setminus \Psi$ and $a_\beta \in \{0, 1\}$ yields a hyperplane contained in $\mathfrak{H}_{\mathscr{R}}$ defined by

$$H(\mathbf{y}; \{\beta_i^\vee\}_{i \in I \setminus \{l\}}, \{\lambda_j\}_{j \in J}, \{d_j\}_{j \in J}) = 0, \tag{200}$$

which passes through r points in general position

$$\{\beta^\vee - q\}_{\beta \in \mathbf{V} \setminus \{\beta_l\}} \cup \{-q\}, \tag{201}$$

where

$$q = \sum_{j \in J} d_j \alpha_j^\vee, \tag{202}$$

$$d_j = m_j - a_{\alpha_j} - \sum_{\gamma \in \Delta_+ \setminus (\Psi \cup (\mathbf{V} \setminus \{\beta_i\}))} a_\gamma \langle \gamma^\vee, \lambda_j \rangle \in \mathbb{Z}. \tag{203}$$

This completes the proof of our claim.

(Step 4.) We have checked that on $\mathfrak{D}^{(\nu)}$, the vertices $\mathrm{Vert}(\mathcal{P}_{\mathbf{m},\mathbf{y}})$ neither increase nor decrease and are indexed by $\mathscr{W}_{\mathbf{m}}$. By numbering $\mathscr{W}_{\mathbf{m}}$ as $\{\mathbf{W}_1, \mathbf{W}_2, \ldots\}$, we denote $\mathbf{p}_i(\mathbf{y}) = \mathbf{p}(\mathbf{y}; \mathbf{W}_i)$. We see that on $\mathfrak{D}^{(\nu)}$, the polytopes $\mathcal{P}_{\mathbf{m},\mathbf{y}}$ keep $(n-r)$-dimensional or empty because each vertex is determined by unique $(n-r)$ hyperplanes. Assume that $\mathcal{P}_{\mathbf{m},\mathbf{y}}$ is not empty. Next, we will show that the face poset structure of $\mathcal{P}_{\mathbf{m},\mathbf{y}}$ is independent of $\mathbf{y}$ on $\mathfrak{D}^{(\nu)}$.

Fix $\mathbf{y}_0 \in \mathfrak{D}^{(\nu)}$. Consider a face $\mathcal{F}(\mathbf{y}_0)$ of $\mathcal{P}_{\mathbf{m},\mathbf{y}_0}$ and let

$$\mathrm{Vert}(\mathcal{P}_{\mathbf{m},\mathbf{y}_0}) \cap \mathcal{F}(\mathbf{y}_0) = \{\mathbf{p}_{i_1}(\mathbf{y}_0), \ldots, \mathbf{p}_{i_h}(\mathbf{y}_0)\}. \tag{204}$$

Then by Proposition 2, there exists a subset $\mathcal{J}_0 = \{(\gamma_1, a_{\gamma_1}), (\gamma_2, a_{\gamma_2}), \ldots\} \subset \Delta_+ \times \{0,1\}$ such that $|\mathcal{J}_0| = n - r - \dim \mathcal{F}(\mathbf{y}_0)$ and

$$\{\mathbf{p}_{i_1}(\mathbf{y}_0), \ldots, \mathbf{p}_{i_h}(\mathbf{y}_0)\} \subset \mathcal{F}(\mathbf{y}_0) = \mathcal{P}_{\mathbf{m},\mathbf{y}_0} \cap \bigcap_{(\gamma,a_\gamma) \in \mathcal{J}_0} \mathcal{H}_{\gamma,a_\gamma}(\mathbf{y}_0). \tag{205}$$

By the definition of $\mathbf{p}_{i_1}(\mathbf{y}_0), \ldots, \mathbf{p}_{i_h}(\mathbf{y}_0)$ (see (178)), we find that for $(\gamma, a_\gamma) \in \Delta_+ \times \{0,1\}$, the condition

$$\{\mathbf{p}_{i_1}(\mathbf{y}_0), \ldots, \mathbf{p}_{i_h}(\mathbf{y}_0)\} \subset \mathcal{H}_{\gamma,a_\gamma}(\mathbf{y}_0) \tag{206}$$

is equivalent to

$$\gamma \in \bigcap_{j=1}^{h} (\Delta_+ \setminus \mathbf{V}_{i_j}), \qquad a_\gamma = (\mathbf{A}_{i_1})_\gamma = \cdots = (\mathbf{A}_{i_h})_\gamma, \tag{207}$$

where $\mathbf{W}_{i_j} = (\mathbf{V}_{i_j}, \mathbf{A}_{i_j}) = \xi_{\mathbf{y}_0}^{-1}(\mathbf{p}_{i_j}(\mathbf{y}_0))$. Hence, each $(\gamma, a_\gamma) \in \mathcal{J}_0$ satisfies (207). Assume that there exists a pair $(\gamma', a'_{\gamma'}) \notin \mathcal{J}_0$ satisfying (207). Then

$$\{\mathbf{p}_{i_1}(\mathbf{y}_0), \ldots, \mathbf{p}_{i_h}(\mathbf{y}_0)\} \subset \bigcap_{(\gamma,a_\gamma) \in \mathcal{J}'_0} \mathcal{H}_{\gamma,a_\gamma}(\mathbf{y}_0), \tag{208}$$

where $\mathcal{J}'_0 = \mathcal{J}_0 \cup \{(\gamma', a'_{\gamma'})\}$. Hence, by (28), we have

$$\mathcal{F}(\mathbf{y}_0) = \mathrm{Conv}\{\mathbf{p}_{i_1}(\mathbf{y}_0), \ldots, \mathbf{p}_{i_h}(\mathbf{y}_0)\} \subset \bigcap_{(\gamma,a_\gamma) \in \mathcal{J}'_0} \mathcal{H}_{\gamma,a_\gamma}(\mathbf{y}_0), \tag{209}$$

and in particular,

$$\mathbf{p}_{i_1}(\mathbf{y}_0) \in \bigcap_{(\gamma,a_\gamma)\in\mathcal{J}_0'} \mathcal{H}_{\gamma,a_\gamma}(\mathbf{y}_0). \tag{210}$$

Since $(n-r)$ hyperplanes on which $\mathbf{p}_{i_1}(\mathbf{y}_0)$ lies are uniquely determined and their intersection consists of only $\mathbf{p}_{i_1}(\mathbf{y}_0)$, their normal vectors $\{\mathbf{u}(\gamma,a_\gamma)\}_{(\gamma,a_\gamma)\in\mathcal{J}_0'}$ must be linearly independent. It follows from (209) that $\dim\mathcal{F}(\mathbf{y}_0) \leq n-r-|\mathcal{J}_0'| < n-r-|\mathcal{J}_0| = \dim\mathcal{F}(\mathbf{y}_0)$, which contradicts. Hence, (207) is also a sufficient condition for $(\gamma,a_\gamma)\in\mathcal{J}_0$.

By definition, we have

$$\{\mathbf{p}_{i_1}(\mathbf{y}),\ldots,\mathbf{p}_{i_h}(\mathbf{y})\} \subset \bigcap_{(\gamma,a_\gamma)\in\mathcal{J}_0} \mathcal{H}_{\gamma,a_\gamma}(\mathbf{y}), \tag{211}$$

for all $\mathbf{y}\in\mathfrak{D}^{(\nu)}$. Define

$$\mathcal{F}(\mathbf{y}) = \mathcal{P}_{\mathbf{m},\mathbf{y}} \cap \bigcap_{(\gamma,a_\gamma)\in\mathcal{J}_0} \mathcal{H}_{\gamma,a_\gamma}(\mathbf{y}). \tag{212}$$

Then $\{\mathbf{p}_{i_1}(\mathbf{y}),\ldots,\mathbf{p}_{i_h}(\mathbf{y})\} \subset \mathcal{F}(\mathbf{y})$ and by Proposition 1, we see that $\mathcal{F}(\mathbf{y})$ is a face. Fix another $\mathbf{y}_1\in\mathfrak{D}^{(\nu)}$. Then by the argument at the beginning of this step, there exists a subset $\mathcal{J}_1\subset\Delta_+\times\{0,1\}$ such that $|\mathcal{J}_1| = n-r-\dim\mathcal{F}(\mathbf{y}_1)$ and

$$\begin{aligned}\operatorname{Vert}(\mathcal{P}_{\mathbf{m},\mathbf{y}_1})\cap\mathcal{F}(\mathbf{y}_1) &= \{\mathbf{p}_{i_1}(\mathbf{y}_1),\ldots,\mathbf{p}_{i_h}(\mathbf{y}_1),\mathbf{p}_{i_{h+1}}(\mathbf{y}_1),\ldots,\mathbf{p}_{i_{h'}}(\mathbf{y}_1)\}\\ &\subset \mathcal{F}(\mathbf{y}_1) = \mathcal{P}_{\mathbf{m},\mathbf{y}_1}\cap \bigcap_{(\gamma,a_\gamma)\in\mathcal{J}_1} \mathcal{H}_{\gamma,a_\gamma}(\mathbf{y}_1),\end{aligned} \tag{213}$$

with $h'\geq h$ (because all of $\mathbf{p}_{i_1}(\mathbf{y}_1),\ldots,\mathbf{p}_{i_h}(\mathbf{y}_1)$ are vertices of $\mathcal{F}(\mathbf{y}_1)$, while so far, we cannot exclude the possibility of the existence of other vertices on $\mathcal{F}(\mathbf{y}_1)$). Since each $(\gamma,a_\gamma)\in\mathcal{J}_1$ satisfies

$$\gamma \in \bigcap_{j=1}^{h'}(\Delta_+\setminus\mathbf{V}_{i_j}), \qquad a_\gamma = (\mathbf{A}_{i_1})_\gamma = \cdots = (\mathbf{A}_{i_{h'}})_\gamma, \tag{214}$$

which is equal to or stronger than condition (207), we have $\mathcal{J}_1\subset\mathcal{J}_0$. On the other hand, by comparing (212) and (213), we see that each $(\gamma,a_\gamma)\in\mathcal{J}_0$ satisfies (214). As shown in the previous paragraph, condition (214) is sufficient for $(\gamma,a_\gamma)\in\mathcal{J}_1$, which implies $\mathcal{J}_0=\mathcal{J}_1$ and hence $\dim\mathcal{F}(\mathbf{y}_1)=\dim\mathcal{F}(\mathbf{y}_0)$. If $h'>h$, then (214) implies

$$\{\mathbf{p}_{i_1}(\mathbf{y}_0),\ldots,\mathbf{p}_{i_h}(\mathbf{y}_0),\mathbf{p}_{i_{h+1}}(\mathbf{y}_0),\ldots,\mathbf{p}_{i_{h'}}(\mathbf{y}_0)\}\subset\mathcal{F}(\mathbf{y}_0), \tag{215}$$

which contradicts to (204) and hence $h'=h$. Therefore, for all $\mathbf{y}\in\mathfrak{D}^{(\nu)}$, we see that all faces of $\mathcal{P}_{\mathbf{m},\mathbf{y}}$ are determined at $\mathbf{y}_0$ and are described in the form (212), and we have

$$\operatorname{Vert}(\mathcal{P}_{\mathbf{m},\mathbf{y}})\cap\mathcal{F}(\mathbf{y}) = \{\mathbf{p}_{i_1}(\mathbf{y}),\ldots,\mathbf{p}_{i_h}(\mathbf{y})\}. \tag{216}$$

Assume that $\mathcal{F}'(\mathbf{y}_0) \subset \mathcal{F}(\mathbf{y}_0)$ for faces $\mathcal{F}'(\mathbf{y}_0), \mathcal{F}(\mathbf{y}_0)$ of $\mathcal{P}_{\mathbf{m},\mathbf{y}_0}$. Then by (28), it is equivalent to

$$\mathrm{Vert}(\mathcal{P}_{\mathbf{m},\mathbf{y}_0}) \cap \mathcal{F}'(\mathbf{y}_0) \subset \mathrm{Vert}(\mathcal{P}_{\mathbf{m},\mathbf{y}_0}) \cap \mathcal{F}(\mathbf{y}_0). \tag{217}$$

By applying $\xi_{\mathbf{y}_0}^{-1}$, we obtain an equivalent condition independent of $\mathbf{y}$ and hence $\mathcal{F}'(\mathbf{y}_1) \subset \mathcal{F}(\mathbf{y}_1)$. Therefore, the face poset structure is indeed independent of $\mathbf{y}$ on $\mathfrak{D}^{(\nu)}$.

(Step 5.) By Theorem 1, we have a triangulation of $\mathcal{P}_{\mathbf{m},\mathbf{y}}$ with $(n-r)$-dimensional simplexes $\sigma_{l,\mathbf{m},\mathbf{y}}$ as

$$\mathcal{P}_{\mathbf{m},\mathbf{y}} = \bigcup_{l=1}^{L(\mathbf{m},\mathbf{y})} \sigma_{l,\mathbf{m},\mathbf{y}}, \tag{218}$$

where $L(\mathbf{m},\mathbf{y})$ is the number of the simplexes. From the previous step and by Remark 1, we see that this triangulation does not depend on $\mathbf{y}$ up to the order of simplexes, that is,

$$\{\mathcal{I}(1,\mathbf{m},\mathbf{y}), \ldots, \mathcal{I}(L(\mathbf{m},\mathbf{y}),\mathbf{m},\mathbf{y})\} \tag{219}$$

is independent of $\mathbf{y}$, where $\mathcal{I}(l,\mathbf{m},\mathbf{y})$ is the set of all indices of the vertices of $\sigma_{l,\mathbf{m},\mathbf{y}}$. Reordering $\sigma_{l,\mathbf{m},\mathbf{y}}$ with respect to l if necessary, we assume that each $\mathcal{I}(l,\mathbf{m}) = \mathcal{I}(l,\mathbf{m},\mathbf{y})$ is independent of $\mathbf{y}$ on $\mathfrak{D}^{(\nu)}$. Note that $|\mathcal{I}(l,\mathbf{m})| = n-r+1$. Let $L(\mathbf{m}) = L(\mathbf{m},\mathbf{y})$ if $\mathcal{P}_{\mathbf{m},\mathbf{y}}$ is not empty, and $L(\mathbf{m}) = 0$ otherwise.

By (218) and Lemma 4, we find that the integral in the third expression of (126) is

$$(n-r)! \sum_{l=1}^{L(\mathbf{m})} \mathrm{Vol}(\sigma_{l,\mathbf{m},\mathbf{y}}) \sum_{i \in \mathcal{I}(l,\mathbf{m})} \frac{e^{\mathbf{t}^* \cdot \mathbf{p}_i(\mathbf{y})}}{\prod_{\substack{j \in \mathcal{I}(l,\mathbf{m}) \\ j \neq i}} \mathbf{t}^* \cdot (\mathbf{p}_i(\mathbf{y}) - \mathbf{p}_j(\mathbf{y}))}, \tag{220}$$

where $\mathbf{t}^* = (t_\alpha^*)$ with $t_\alpha^* = t_\alpha - \sum_{i=1}^r t_{\alpha_i} \langle \alpha^\vee, \lambda_i \rangle$. We see that $\mathrm{Vol}(\sigma_{l,\mathbf{m},\mathbf{y}})$ is a polynomial function in $\mathbf{y} = (y_i)_{i=1}^r$ with rational coefficients and its total degree at most $n-r$ on $\mathfrak{D}^{(\nu)}$ due to (34), because in Step 1, we have shown that $\mathbf{p}_i(\mathbf{y})$ depends on $\mathbf{y}$ affine linearly. Therefore, from (126), we have the generating function

$$F(\mathbf{t},\mathbf{y};\Delta) = \left(\prod_{\alpha \in \Delta_+} \frac{t_\alpha}{e^{t_\alpha} - 1} \right) \sum_{j=1}^{(n-r+1)\sum_{\mathbf{m}} L(\mathbf{m})} \frac{f_j(\mathbf{y}) e^{h_j(\mathbf{t},\mathbf{y})}}{g_j(\mathbf{t},\mathbf{y})}, \tag{221}$$

which is valid for all $\mathbf{y} \in \mathfrak{D}^{(\nu)}$, where $f_j \in \mathbb{Q}[\mathbf{y}]$ with its total degree at most $n-r$ and $g_j \in \mathbb{Z}[\mathbf{t},\mathbf{y}]$, $h_j \in \mathbb{Q}[\mathbf{t},\mathbf{y}]$ are of the form

$$g_j(\mathbf{t},\mathbf{y}) = \sum_{\alpha \in \Delta_+} (\langle \phi_\alpha, \mathbf{y} \rangle + c_\alpha) t_\alpha, \qquad \phi_\alpha \in P, \quad c_\alpha \in \mathbb{Z}, \tag{222}$$

$$h_j(\mathbf{t},\mathbf{y}) = \sum_{\alpha\in\Delta_+} (\langle\varphi_\alpha,\mathbf{y}\rangle + d_\alpha)t_\alpha, \quad \varphi_\alpha \in P\otimes\mathbb{Q}, \quad d_\alpha \in \mathbb{Q}. \tag{223}$$

We see from (221) that $F(\mathbf{t},\mathbf{y};\Delta)$ is meromorphically continued from $\mathbb{C}^n \times \mathfrak{D}^{(\nu)}$ to the whole space $\mathbb{C}^n \times (\mathbb{C}\otimes V)$, and that $P(\mathbf{k},\mathbf{y};\Delta)$ is analytically continued to a polynomial function in $\mathbf{y} = (y_i)_{i=1}^r$ with rational coefficients and its total degree at most $|\mathbf{k}| + n - r$ by (124) and Lemma 5. □

Theorem 14. *The function $P(\mathbf{k},\mathbf{y};\Delta)$ is not real analytic in $\mathbf{y}$ on V unless $P(\mathbf{k},\mathbf{y};\Delta)$ is a constant.*

Proof. By Theorem 12, we see that for $\mathbf{k}\in\mathbb{N}_0^n$, $P(\mathbf{k},\mathbf{y};\Delta)$ is a periodic function in $\mathbf{y}$ with its periods $\Psi^\vee$, while by Theorem 13, $P(\mathbf{k},\mathbf{y};\Delta)$ is a polynomial function in $\mathbf{y}$ on some open region. Therefore, such a polynomial expression cannot be extended to the whole space unless $P(\mathbf{k},\mathbf{y};\Delta)$ is a constant. This implies that there are some points on $\mathfrak{H}_{\mathscr{R}}$, at which $P(\mathbf{k},\mathbf{y};\Delta)$ is not real analytic. □

The polynomials $B_{\mathbf{k}}^{(\nu)}(\mathbf{y};\Delta)$ may be regarded as (root-system theoretic) generalizations of Bernoulli polynomials. For instance, they possess the following property.

Theorem 15. *Assume that Δ is an irreducible root system and is not of type A_1. For $\mathbf{k}\in\mathbb{N}_0^n$, $\mathbf{y}\in\partial\mathfrak{D}^{(\nu)}$ and $\mathbf{y}'\in\partial\mathfrak{D}^{(\nu')}$ with $\mathbf{y}\equiv\mathbf{y}' \pmod{Q^\vee}$, we have*

$$B_{\mathbf{k}}^{(\nu)}(\mathbf{y};\Delta) = B_{\mathbf{k}}^{(\nu')}(\mathbf{y}';\Delta). \tag{224}$$

Proof. If $\mathbf{y} = \mathbf{y}'$, then the result follows from the continuity proved in Lemma 8. If $\mathbf{y}\neq\mathbf{y}'$, but $\mathbf{y}\equiv\mathbf{y}' \pmod{Q^\vee}$, we also use the periodicity. □

This theorem also holds in the A_1 case with $k\neq 1$ and can be regarded as a multiple analogue of the formula for the classical Bernoulli polynomials

$$B_k(0) = B_k(1), \tag{225}$$

for $k\neq 1$. Moreover, the formula

$$B_k(1-y) = (-1)^k B_k(y) \tag{226}$$

is well known. In the rest of this section, we will show the results analogous to the above formula for $B_{\mathbf{k}}^{(\nu)}(\mathbf{y};\Delta)$ (Theorem 16) and its vector-valued version (Theorem 18). The latter gives a finite-dimensional representation of Weyl groups. In this framework, (226) can be interpreted as an action of the Weyl group of type A_1 (Example 1). These results will not be used in this chapter, but we insert this topic because of its own interest.

Lemma 13. *Fix* $w \in \mathrm{Aut}(\Delta)$ *and* $\nu \in \mathfrak{J}$. *Then there exist unique* $q_{w,\nu} \in Q^\vee$ *and* $\kappa \in \mathfrak{J}$ *such that*

$$\tau(q_{w,\nu})w\mathfrak{D}^{(\nu)} = \mathfrak{D}^{(\kappa)}. \tag{227}$$

Thus, $\mathrm{Aut}(\Delta)$ *acts on* $\mathfrak{J}$ *as* $w(\nu) = \kappa$. *Moreover,* $q_{v,w(\nu)} + vq_{w,\nu} = q_{vw,\nu}$ *for* $v, w \in \mathrm{Aut}(\Delta)$.

Proof. It can be easily seen from the definition (167) that $\mathfrak{H}_{\mathscr{R}}$ is $\widehat{W}$ invariant. Therefore, $\widehat{W}$ acts on $V \setminus \mathfrak{H}_{\mathscr{R}}$ as homeomorphisms, and a connected component is mapped to another one.

Fix $\mathbf{y} \in w\mathfrak{D}^{(\nu)}$. There exists a unique $q \in Q^\vee$ such that $0 \leq \langle \tau(q)\mathbf{y}, \lambda_j \rangle < 1$ for $1 \leq j \leq r$, that is, $q = -\sum_{j=1}^r a_j \alpha_j^\vee \in Q^\vee$, where $a_j = [\langle \mathbf{y}, \lambda_j \rangle]$ is the integer part of $\langle \mathbf{y}, \lambda_j \rangle$. Denote this q by $q_{w,\nu}$. Then $\tau(q_{w,\nu})\mathbf{y} \in \mathfrak{D}^{(\kappa)}$ for some $\kappa \in \mathfrak{J}$ and thus $\tau(q_{w,\nu})w\mathfrak{D}^{(\nu)} = \mathfrak{D}^{(\kappa)}$.

Let $w, w' \in \mathrm{Aut}(\Delta)$. Assume

$$\tau(q)w\mathfrak{D}^{(\nu)} = \mathfrak{D}^{(\kappa)}, \quad \tau(q')w'\mathfrak{D}^{(\kappa)} = \mathfrak{D}^{(\kappa')}, \quad \tau(q'')w'w\mathfrak{D}^{(\nu)} = \mathfrak{D}^{(\kappa'')}, \tag{228}$$

for $q, q', q'' \in Q^\vee$ and $\nu, \kappa, \kappa', \kappa'' \in \mathfrak{J}$. Then we have

$$\begin{aligned} \mathfrak{D}^{(\kappa')} &= \tau(q')w'\tau(q)w\mathfrak{D}^{(\nu)} \\ &= \tau(q' + w'q)w'w\mathfrak{D}^{(\nu)}. \end{aligned} \tag{229}$$

By the uniqueness, we have $q' + w'q = q''$ and $\kappa' = \kappa''$. □

Theorem 16. *For* $w \in \mathrm{Aut}(\Delta)$,

$$B_{\mathbf{k}}^{(\nu)}(\tau(q_{w^{-1},w(\nu)})w^{-1}\mathbf{y}; \Delta) = \Bigl(\prod_{\alpha \in \Delta_w} (-1)^{-k_\alpha}\Bigr) B_{w\mathbf{k}}^{(w(\nu))}(\mathbf{y}; \Delta). \tag{230}$$

Proof. By Theorems 11 and 12, we have

$$\begin{aligned} P(\mathbf{k}, \tau(q_{w^{-1},w(\nu)})w^{-1}\mathbf{y}; \Delta) &= P(\mathbf{k}, w^{-1}\mathbf{y}; \Delta) \\ &= \Bigl(\prod_{\alpha \in \Delta_{w^{-1}}} (-1)^{-k_{w^{-1}\alpha}}\Bigr) P(w\mathbf{k}, \mathbf{y}; \Delta) \\ &= \Bigl(\prod_{\alpha \in \Delta_w} (-1)^{-k_\alpha}\Bigr) P(w\mathbf{k}, \mathbf{y}; \Delta), \end{aligned} \tag{231}$$

where we have used $w^{-1}\Delta_{w^{-1}} = -\Delta_w$. By (227) in Lemma 13 with replacing w, ν by w^{-1}, $w(\nu)$, respectively, we have

$$\tau(q_{w^{-1},w(\nu)})w^{-1}\mathfrak{D}^{(w(\nu))} = \mathfrak{D}^{(\nu)}. \tag{232}$$

Hence, $\mathbf{y} \in \mathfrak{D}^{(w(\nu))}$ implies $\tau(q_{w^{-1},w(\nu)})w^{-1}\mathbf{y} \in \mathfrak{D}^{(\nu)}$. Therefore, we obtain

$$P(\mathbf{k}, \tau(q_{w^{-1},w(\nu)})w^{-1}\mathbf{y}; \Delta) = B_{\mathbf{k}}^{(\nu)}(\tau(q_{w^{-1},w(\nu)})w^{-1}\mathbf{y}; \Delta), \tag{233}$$

$$P(\mathbf{k}, \mathbf{y}; \Delta) = B_{\mathbf{k}}^{(w(\nu))}(\mathbf{y}; \Delta), \tag{234}$$

by Theorem 13. The theorem of identity implies (230). □

Let $\mathfrak{P}$ be the $\mathbb{Q}$-vector space of all vector-valued polynomial functions of the form $f = (f_\nu)_{\nu\in\mathfrak{J}} : V \to \mathbb{R}^{|\mathfrak{J}|}$ with $f_\nu \in \mathbb{Q}[\mathbf{y}]$. We define a linear map $\phi(w) : \mathfrak{P} \to \mathfrak{P}$ for $w \in \mathrm{Aut}(\Delta)$ by

$$(\phi(w)f)_\nu(\mathbf{y}) = f_{w^{-1}(\nu)}(\tau(q_{w^{-1},\nu})w^{-1}\mathbf{y}). \tag{235}$$

Theorem 17. *The pair* $(\phi, \mathfrak{P})$ *is a representation of* $\mathrm{Aut}(\Delta)$.

Proof. For $\nu, w \in \mathrm{Aut}(\Delta)$, we have

$$\begin{aligned}(\phi(v)\phi(w)f)_\nu(\mathbf{y}) &= (\phi(w)f)_{v^{-1}(\nu)}(\tau(q_{v^{-1},\nu})v^{-1}\mathbf{y}) \\ &= f_{w^{-1}v^{-1}(\nu)}(\tau(q_{w^{-1},v^{-1}(\nu)})w^{-1}\tau(q_{v^{-1},\nu})v^{-1}\mathbf{y}) \\ &= f_{(vw)^{-1}(\nu)}(\tau(q_{w^{-1},v^{-1}(\nu)} + w^{-1}q_{v^{-1},\nu})(vw)^{-1}\mathbf{y}).\end{aligned} \tag{236}$$

Since by Lemma 13 we have $q_{w^{-1},v^{-1}(\nu)} + w^{-1}q_{v^{-1},\nu} = q_{(vw)^{-1},\nu}$, we obtain $\phi(vw) = \phi(v)\phi(w)$. □

Define $\mathbf{B}_{\mathbf{k}}^{(\nu)} \in \mathfrak{P}$ for $\mathbf{k} \in \mathbb{N}_0^n$ and $\nu \in \mathfrak{J}$ by

$$(\mathbf{B}_{\mathbf{k}}^{(\nu)})_\kappa(\mathbf{y}) = (\mathbf{B}_{\mathbf{k}}^{(\nu)})_\kappa(\mathbf{y}; \Delta) = \begin{cases} B_{\mathbf{k}}^{(\nu)}(\mathbf{y}; \Delta), & \text{if } \nu = \kappa, \\ 0, & \text{otherwise}, \end{cases} \tag{237}$$

and let

$$\mathfrak{B}_{\overline{(\mathbf{k},\nu)}} = \sum_{(\mathbf{k}',\nu')\in\overline{(\mathbf{k},\nu)}} \mathbb{Q}\,\mathbf{B}_{\mathbf{k}'}^{(\nu')} \subset \mathfrak{P}, \tag{238}$$

where $\overline{(\mathbf{k}, \nu)}$ is an element of the orbit space $(\mathbb{N}_0^n \times \mathfrak{J})/\mathrm{Aut}(\Delta)$.

Theorem 18. *The vector subspace* $\mathfrak{B}_{\overline{(\mathbf{k},\nu)}}$ *is a finite-dimensional* $\mathrm{Aut}(\Delta)$*-invariant subspace, and the action is*

$$\phi(w)\mathbf{B}_{\mathbf{k}}^{(\nu)} = \Big(\prod_{\alpha\in\Delta_w}(-1)^{-k_\alpha}\Big)\mathbf{B}_{w\mathbf{k}}^{(w(\nu))}. \tag{239}$$

Proof. If $\kappa = w(\nu)$, then we have

$$(\phi(w)\mathbf{B}_{\mathbf{k}}^{(\nu)})_\kappa(\mathbf{y};\Delta) = B_{\mathbf{k}}^{(\nu)}(\tau(q_{w^{-1},w(\nu)})w^{-1}\mathbf{y};\Delta)$$
$$= \Big(\prod_{\alpha\in\Delta_w}(-1)^{-k_\alpha}\Big)B_{w\mathbf{k}}^{(w(\nu))}(\mathbf{y};\Delta), \tag{240}$$

by Theorem 16 and otherwise

$$(\phi(w)\mathbf{B}_{\mathbf{k}}^{(\nu)})_\kappa(\mathbf{y}) = 0. \tag{241}$$

Thus, we obtain (239). □

For the representation in the A_2 case, see (259) and (260).

11.7 Examples

Example 1. The set of positive roots of type A_1 consists of only one root α_1. Hence, we have $\Delta_+ = \Psi = \{\alpha_1\}$ and $2\rho^\vee = \alpha_1^\vee$. We set $t = t_{\alpha_1}$ and $y = \langle \mathbf{y}, \lambda_1\rangle$. Then by Theorem 7, we obtain the generating function $F(\mathbf{t},\mathbf{y};A_1)$ as

$$F(\mathbf{t},\mathbf{y};A_1) = \frac{t}{e^t-1}e^{t\{y\}}. \tag{242}$$

Since $\mathfrak{D} = \{\mathbf{y} \mid 0 \le y \le 1\}$ and $\mathscr{R} = \{\emptyset\}$, we have $\mathfrak{H}_{\mathscr{R}} = Q^\vee$, and hence, $\mathfrak{D}\setminus\mathfrak{H}_{\mathscr{R}}$ consists of only one connected component $\mathfrak{D}\setminus\mathfrak{H}_{\mathscr{R}} = \mathring{\mathfrak{D}} = \mathfrak{D}^{(1)}$. Therefore, we have

$$F^{(1)}(\mathbf{t},\mathbf{y};A_1) = \frac{te^{ty}}{e^t-1} \tag{243}$$

for $\mathbf{y}\in\mathfrak{D}^{(1)}$, which coincides with the generating function of the classical Bernoulli polynomials. Since $\mathfrak{J} = \{1\}$, we see that $\mathbf{B}_{\mathbf{k}}^{(1)}(\mathbf{y};A_1)$ consists of only one component, that is, the classical Bernoulli polynomial $B_k(y)$. The group $\mathrm{Aut}(\Delta)$ is $\{\mathrm{id},\sigma_1\}$. For $\mathbf{y}\in\mathfrak{D}^{(1)}$, we have $\sigma_1\mathbf{y} = -\mathbf{y}$ and hence $\tau(\alpha_1^\vee)\sigma_1\mathbf{y} = \alpha_1^\vee - \mathbf{y} \in \mathfrak{D}^{(1)}$ due to $0 < \langle \alpha_1^\vee - \mathbf{y}, \lambda_1\rangle < 1$, which implies that the action on $\mathbb{Q}[\mathbf{y}]$ is given by $\big(\phi(\sigma_1)f\big)(\mathbf{y}) = f(\alpha_1^\vee - \mathbf{y})$. Therefore, we have the well-known property

$$(\phi(\sigma_1)\mathbf{B}_{\mathbf{k}}^{(1)})(\mathbf{y};A_1) = B_k(1-y) = (-1)^k B_k(y) = (-1)^k\mathbf{B}_{\mathbf{k}}^{(1)}(\mathbf{y};A_1). \tag{244}$$

Example 2. In the root system of type A_2, we have $\Delta_+ = \{\alpha_1,\alpha_2,\alpha_1+\alpha_2\}$ and $\Psi = \{\alpha_1,\alpha_2\}$. Then $\mathscr{R} = \Delta_+$, and from Lemma 12, we have

$$\mathfrak{H}_{\mathscr{R}} = (\mathbb{R}\alpha_1^\vee + \mathbb{Z}\alpha_2^\vee)\cup(\mathbb{R}\alpha_2^\vee + \mathbb{Z}\alpha_1^\vee)\cup(\mathbb{R}(\alpha_1^\vee+\alpha_2^\vee)+\mathbb{Z}\alpha_1^\vee). \tag{245}$$

We see that $\mathfrak{D} \setminus \mathfrak{H}_{\mathscr{R}} = \mathfrak{D}^{(1)} \coprod \mathfrak{D}^{(2)}$, where

$$\mathfrak{D}^{(1)} = \{\mathbf{y} \mid 0 < y_2 < y_1 < 1\}, \tag{246}$$

$$\mathfrak{D}^{(2)} = \{\mathbf{y} \mid 0 < y_1 < y_2 < 1\}, \tag{247}$$

with $y_1 = \langle \mathbf{y}, \lambda_1 \rangle$ and $y_2 = \langle \mathbf{y}, \lambda_2 \rangle$. Let $t_1 = t_{\alpha_1}$, $t_2 = t_{\alpha_2}$, and $t_3 = t_{\alpha_1+\alpha_2}$. For $\mathbf{y} \in \mathfrak{D}^{(1)}$, the vertices of the polytopes $\mathcal{P}_{m_1,m_2,\mathbf{y}}$ in (127) are given by

$$\begin{aligned} 0, y_2 &\qquad \text{for } \mathcal{P}_{0,0,\mathbf{y}}, \\ y_2, y_1 &\qquad \text{for } \mathcal{P}_{0,1,\mathbf{y}}, \\ y_1, 1 &\qquad \text{for } \mathcal{P}_{1,1,\mathbf{y}}. \end{aligned} \tag{248}$$

Then by Theorem 7 and Lemma 4, we have

$$\begin{aligned} &F^{(1)}(\mathbf{t}, \mathbf{y}; A_2) \\ &= \frac{t_1 t_2 t_3 e^{t_1 y_1 + t_2 y_2}}{(e^{t_1} - 1)(e^{t_2} - 1)(e^{t_3} - 1)} \\ &\quad \times \Bigg(y_2 \frac{e^{(t_3 - t_1 - t_2) y_2} - 1}{(t_3 - t_1 - t_2) y_2} + e^{t_2}(y_1 - y_2) \frac{(e^{(t_3 - t_1 - t_2) y_1} - e^{(t_3 - t_1 - t_2) y_2})}{(t_3 - t_1 - t_2)(y_1 - y_2)} \\ &\qquad + e^{t_1 + t_2}(1 - y_1) \frac{(e^{t_3 - t_1 - t_2} - e^{(t_3 - t_1 - t_2) y_1})}{(t_3 - t_1 - t_2)(1 - y_1)} \Bigg) \\ &= \frac{t_1 t_2 t_3 e^{t_1 y_1 + t_2 y_2}}{(e^{t_1} - 1)(e^{t_2} - 1)(e^{t_3} - 1)(t_3 - t_1 - t_2)} \big(e^{(t_3 - t_1 - t_2) y_2} - 1 \\ &\qquad + e^{t_2}(e^{(t_3 - t_1 - t_2) y_1} - e^{(t_3 - t_1 - t_2) y_2}) + e^{t_1 + t_2}(e^{t_3 - t_1 - t_2} - e^{(t_3 - t_1 - t_2) y_1}) \big). \end{aligned} \tag{249}$$

Hence, by the Taylor expansion of (249), we have

$$\begin{aligned} B^{(1)}_{2,2,2}(\mathbf{y}; A_2) = {} & \frac{1}{3780} + \frac{1}{45}(y_1 y_2 - y_1^2 - y_2^2) + \frac{1}{18}(3 y_1 y_2^2 - 3 y_1^2 y_2 + 2 y_1^3) \\ & + \frac{1}{9}(-2 y_1 y_2^3 - 3 y_1^2 y_2^2 + 4 y_1^3 y_2 - 2 y_1^4 + y_2^4) \\ & + \frac{1}{30}(-5 y_1 y_2^4 + 10 y_1^2 y_2^3 + 10 y_1^3 y_2^2 - 15 y_1^4 y_2 + 6 y_1^5) \\ & + \frac{1}{30}(6 y_1 y_2^5 - 5 y_1^2 y_2^4 - 5 y_1^4 y_2^2 + 6 y_1^5 y_2 - 2 y_1^6 - 2 y_2^6). \end{aligned} \tag{250}$$

Table 11.1 Bernoulli numbers for A_2

(k_1,k_2,k_3)	$B_{(k_1,k_2,k_3)}(A_2)$	(k_1,k_2,k_3)	$B_{(k_1,k_2,k_3)}(A_2)$
(0, 0, 0)	1	(0, 1, 1)	−1/12
(0, 1, 3)	1/120	(0, 2, 2)	1/180
(0, 2, 4)	−1/630	(0, 3, 1)	1/120
(0, 3, 3)	−1/840	(0, 4, 2)	−1/630
(0, 4, 4)	1/2100	(1, 0, 1)	−1/12
(1, 0, 3)	1/120	(1, 1, 0)	1/12
(1, 1, 2)	1/180	(1, 1, 4)	−1/630
(1, 2, 1)	−1/180	(1, 2, 3)	1/5040
(1, 3, 0)	−1/120	(1, 3, 2)	−1/5040
(1, 3, 4)	1/12600	(1, 4, 1)	1/630
(1, 4, 3)	−1/12600	(2, 0, 2)	1/180
(2, 0, 4)	−1/630	(2, 1, 1)	−1/180
(2, 1, 3)	1/5040	(2, 2, 0)	1/180
(2, 2, 2)	1/3780	(2, 2, 4)	−1/18900
(2, 3, 1)	1/5040	(2, 4, 0)	−1/630
(2, 4, 2)	−1/18900	(2, 4, 4)	1/103950
(3, 0, 1)	1/120	(3, 0, 3)	−1/840
(3, 1, 0)	−1/120	(3, 1, 2)	−1/5040
(3, 1, 4)	1/12600	(3, 2, 1)	1/5040
(3, 3, 0)	1/840	(3, 3, 4)	−1/277200
(3, 4, 1)	−1/12600	(3, 4, 3)	1/277200
(4, 0, 2)	−1/630	(4, 0, 4)	1/2100
(4, 1, 1)	1/630	(4, 1, 3)	−1/12600
(4, 2, 0)	−1/630	(4, 2, 2)	−1/18900
(4, 2, 4)	1/103950	(4, 3, 1)	−1/12600
(4, 3, 3)	1/277200	(4, 4, 0)	1/2100
(4, 4, 2)	1/103950	(4, 4, 4)	−19/13513500

Similarly, we can calculate $B^{(2)}_{2,2,2}(\mathbf{y}; A_2)$ for $\mathbf{y} \in \mathfrak{D}^{(2)}$, which coincides with (250) with y_1 and y_2 exchanged. In the case $\mathbf{y} = 0$, from Lemma 8, we have

$$\begin{aligned} F(\mathbf{t}; A_2) &= \frac{t_1 t_2 t_3 e^{t_1+t_2}(e^{t_3-t_1-t_2}-1)}{(e^{t_1}-1)(e^{t_2}-1)(e^{t_3}-1)(t_3-t_1-t_2)} \\ &= 1 + \frac{1}{12}(t_1t_2 - t_1t_3 - t_2t_3) + \frac{1}{360}(t_1t_2t_3^2 - t_1^2t_2t_3 - t_1t_2^2t_3) \\ &\quad + \frac{1}{720}(t_1^2t_2^2 + t_1^2t_3^2 + t_2^2t_3^2) + \frac{1}{30240}t_1^2t_2^2t_3^2 + \cdots, \end{aligned} \tag{251}$$

by letting $\mathbf{y} \to 0$ in (249). See Table 11.1 for explicit forms of $B_{(k_1,k_2,k_3)}(A_2)$ with $k_1, k_2, k_3 \le 4$. Note that $B_{(k_1,k_2,k_3)}(A_2) = 0$ for (k_1, k_2, k_3) which are not in the table.

By Theorem 8, we recover Mordell's formula [27]:

$$\zeta_2(2,2,2;A_2) = (-1)^3 \frac{(2\pi\sqrt{-1})^6}{3!} \frac{B_{(2,2,2)}(A_2)}{(2!)^3}$$
$$= (-1)^3 \frac{(2\pi\sqrt{-1})^6}{3!} \frac{1}{30240} = \frac{\pi^6}{2835}. \tag{252}$$

We also discuss the action of $\mathrm{Aut}(\Delta)$. Note that $\mathrm{Aut}(\Delta)$ is generated by $\{\sigma_1, \sigma_2, \omega\}$ where ω is a unique element of Ω such that $\omega \neq \mathrm{id}$, and hence $\omega\alpha_1 = \alpha_2$, $\omega\lambda_1 = \lambda_2$ and $\omega^2 = \mathrm{id}$. Also note that $\alpha_1 = 2\lambda_1 - \lambda_2$ and $\alpha_2 = 2\lambda_2 - \lambda_1$. For $\mathbf{y} \in \mathfrak{D}^{(1)}$, we have

$$\langle \alpha_1^\vee + \sigma_1 \mathbf{y}, \lambda_1 \rangle = 1 + \langle \mathbf{y}, \sigma_1\lambda_1 \rangle = 1 + \langle \mathbf{y}, \lambda_1 - \alpha_1 \rangle = 1 - y_1 + y_2,$$
$$\langle \alpha_1^\vee + \sigma_1 \mathbf{y}, \lambda_2 \rangle = \langle \mathbf{y}, \sigma_1\lambda_2 \rangle = \langle \mathbf{y}, \lambda_2 \rangle = y_2, \tag{253}$$

which implies

$$0 < \langle \alpha_1^\vee + \sigma_1 \mathbf{y}, \lambda_2 \rangle < \langle \alpha_1^\vee + \sigma_1 \mathbf{y}, \lambda_1 \rangle < 1. \tag{254}$$

Therefore, we have $\tau(\alpha_1^\vee)\sigma_1\mathfrak{D}^{(1)} = \mathfrak{D}^{(1)}$ and in a similar way $\sigma_1\mathfrak{D}^{(2)} = \mathfrak{D}^{(2)}$, and so on. Thus, from (235), we see that for $f = (f_1, f_2)$ with $f_1, f_2 \in \mathbb{Q}[\mathbf{y}]$, the action of $\mathrm{Aut}(\Delta)$ is given by

$$(\phi(\sigma_1)f)(\mathbf{y}) = \bigl(f_1(\alpha_1^\vee + \sigma_1\mathbf{y}), f_2(\sigma_1\mathbf{y})\bigr),$$
$$(\phi(\sigma_2)f)(\mathbf{y}) = \bigl(f_1(\sigma_2\mathbf{y}), f_2(\alpha_2^\vee + \sigma_2\mathbf{y})\bigr),$$
$$(\phi(\omega)f)(\mathbf{y}) = \bigl(f_2(\omega\mathbf{y}), f_1(\omega\mathbf{y})\bigr), \tag{255}$$

or in terms of coordinates,

$$(\phi(\sigma_1)f)(y_1, y_2) = \bigl(f_1(1 - y_1 + y_2, y_2), f_2(-y_1 + y_2, y_2)\bigr),$$
$$(\phi(\sigma_2)f)(y_1, y_2) = \bigl(f_1(y_1, y_1 - y_2), f_2(y_1, 1 + y_1 - y_2)\bigr),$$
$$(\phi(\omega)f)(y_1, y_2) = \bigl(f_2(y_2, y_1), f_1(y_2, y_1)\bigr). \tag{256}$$

Then for

$$\mathbf{B}^{(1)}_{2,2,2}(\mathbf{y}; A_2) = (B^{(1)}_{2,2,2}(\mathbf{y}; A_2), 0), \tag{257}$$

$$\mathbf{B}^{(2)}_{2,2,2}(\mathbf{y}; A_2) = (0, B^{(2)}_{2,2,2}(\mathbf{y}; A_2)), \tag{258}$$

we have

$$\phi(\sigma_1)\mathbf{B}^{(1)}_{2,2,2} = \mathbf{B}^{(1)}_{2,2,2},$$
$$\phi(\sigma_2)\mathbf{B}^{(1)}_{2,2,2} = \mathbf{B}^{(1)}_{2,2,2},$$
$$\phi(\omega)\mathbf{B}^{(1)}_{2,2,2} = \mathbf{B}^{(2)}_{2,2,2}. \tag{259}$$

More generally, we can show

$$\begin{aligned}\phi(\sigma_1)\mathbf{B}^{(1)}_{k_1,k_2,k_3} &= (-1)^{-k_1}\mathbf{B}^{(1)}_{k_1,k_3,k_2},\\ \phi(\sigma_2)\mathbf{B}^{(1)}_{k_1,k_2,k_3} &= (-1)^{-k_2}\mathbf{B}^{(1)}_{k_3,k_2,k_1},\\ \phi(\omega)\mathbf{B}^{(1)}_{k_1,k_2,k_3} &= \mathbf{B}^{(2)}_{k_2,k_1,k_3}.\end{aligned} \tag{260}$$

Example 3. The set of positive roots of type $C_2(\simeq B_2)$ consists of $\alpha_1, \alpha_2, 2\alpha_1+\alpha_2$, and $\alpha_1+\alpha_2$. Let $t_1 = t_{\alpha_1}$, $t_2 = t_{\alpha_2}$, $t_3 = t_{2\alpha_1+\alpha_2}$, and $t_4 = t_{\alpha_1+\alpha_2}$. The vertices $(x_{2\alpha_1+\alpha_2}, x_{\alpha_1+\alpha_2})$ of the polytopes $\mathcal{P}_{m_1,m_2}$ in (129) are given by

$$\begin{aligned}&(0,0),(0,1/2),(1,0) && \text{for } \mathcal{P}_{1,1},\\ &(0,1),(0,1/2),(1,0) && \text{for } \mathcal{P}_{1,2},\\ &(0,1),(1,1/2),(1,0) && \text{for } \mathcal{P}_{2,2},\\ &(0,1),(1,1/2),(1,1) && \text{for } \mathcal{P}_{2,3}.\end{aligned} \tag{261}$$

Then by Lemmas 4 and 5, we obtain

$$\begin{aligned}F(\mathbf{t};C_2) &= \left(\prod_{j=1}^{4}\frac{t_j}{\mathrm{e}^{t_j}-1}\right)G(\mathbf{t};C_2)\\ &= 1+\frac{1}{2880}(2t_1t_2t_3^2-4t_1t_2t_4^2-2t_1t_2^2t_3+4t_1t_2^2t_4-4t_1t_3t_4^2-4t_1t_3^2t_4\\ &\qquad +t_1^2t_2t_3-4t_1^2t_2t_4-4t_1^2t_3t_4-t_2t_3t_4^2-2t_2t_3^2t_4-2t_2^2t_3t_4)\\ &\quad +\frac{1}{241920}(3t_1t_2t_3^2t_4^2-3t_1t_2^2t_3t_4^2-3t_1^2t_2t_3^2t_4-3t_1^2t_2^2t_3t_4+2t_1^2t_2^2t_3^2\\ &\qquad +8t_1^2t_2^2t_4^2+8t_1^2t_3^2t_4^2+2t_2^2t_3^2t_4^2)\\ &\quad +\frac{1}{9676800}t_1^2t_2^2t_3^2t_4^2+\cdots,\end{aligned} \tag{262}$$

where

$$\begin{aligned}G(\mathbf{t};C_2) &= \frac{\mathrm{e}^{t_3}}{(t_1+t_2-t_3)(t_1-2t_3+t_4)}-\frac{2\mathrm{e}^{\frac{t_1}{2}+\frac{t_4}{2}}}{(t_1+2t_2-t_4)(t_1-2t_3+t_4)}\\ &\quad +\frac{\mathrm{e}^{t_1+t_2}}{(t_1+t_2-t_3)(t_1+2t_2-t_4)}-\frac{\mathrm{e}^{t_2+t_3}}{(t_2+t_3-t_4)(t_1-2t_3+t_4)}\\ &\quad +\frac{\mathrm{e}^{t_4}}{(t_1+2t_2-t_4)(t_2+t_3-t_4)}+\frac{2\mathrm{e}^{\frac{t_1}{2}+t_2+\frac{t_4}{2}}}{(t_1+2t_2-t_4)(t_1-2t_3+t_4)}\end{aligned}$$

$$+\frac{2e^{\frac{t_1}{2}+t_3+\frac{t_4}{2}}}{(t_1+2t_2-t_4)(t_1-2t_3+t_4)}+\frac{e^{t_1+t_2+t_3}}{(t_2+t_3-t_4)(t_1+2t_2-t_4)}$$
$$-\frac{e^{t_1+t_4}}{(t_2+t_3-t_4)(t_1-2t_3+t_4)}+\frac{e^{t_3+t_4}}{(t_1+t_2-t_3)(t_1+2t_2-t_4)}$$
$$-\frac{2e^{\frac{t_1}{2}+t_2+t_3+\frac{t_4}{2}}}{(t_1+2t_2-t_4)(t_1-2t_3+t_4)}+\frac{e^{t_1+t_2+t_4}}{(t_1+t_2-t_3)(t_1-2t_3+t_4)}. \quad (263)$$

See Tables 11.2–11.4 for explicit forms of $B_{(k_1,k_2,k_3,k_4)}(C_2)$ with $k_1,k_2,k_3,k_4 \leq 4$. Note that $B_{(k_1,k_2,k_3,k_4)}(C_2)=0$ for (k_1,k_2,k_3,k_4) which are not in the table. Using these tables, we obtain

$$\zeta_2(2,2,2,2;C_2)=(-1)^4\frac{(2\pi\sqrt{-1})^8}{2!\cdot 2^2}\frac{B_{(2,2,2,2)}(C_2)}{(2!)^4}$$
$$=(-1)^4\frac{(2\pi\sqrt{-1})^8}{2!\cdot 2^2}\frac{1}{9676800}=\frac{\pi^8}{302400}, \quad (264)$$

and so

$$\zeta_W(2;C_2)=6^2\frac{\pi^8}{302400}=\frac{\pi^8}{8400}. \quad (265)$$

Example 4. The set of positive roots of type A_3 consists of $\alpha_1,\alpha_2,\alpha_3,\alpha_1+\alpha_2,\alpha_2+\alpha_3$, and $\alpha_1+\alpha_2+\alpha_3$. Let $t_1=t_{\alpha_1}$, $t_2=t_{\alpha_2}$, $t_3=t_{\alpha_3}$, $t_4=t_{\alpha_1+\alpha_2}$, $t_5=t_{\alpha_2+\alpha_3}$, and $t_6=t_{\alpha_1+\alpha_2+\alpha_3}$. The vertices $(x_{\alpha_1+\alpha_2},x_{\alpha_2+\alpha_3},x_{\alpha_1+\alpha_2+\alpha_3})$ of the polytopes $\mathcal{P}_{m_1,m_2,m_3}$ in (129) are given by

$$\begin{aligned}
&(0,0,0),(0,0,1),(0,1,0),(1,0,0) && \text{for } \mathcal{P}_{1,1,1},\\
&(0,0,1),(0,1,0),(1,0,0),(1,1,0) && \text{for } \mathcal{P}_{1,2,1},\\
&(0,0,1),(0,1,0),(0,1,1),(1,1,0) && \text{for } \mathcal{P}_{1,2,2},\\
&(0,0,1),(1,0,0),(1,0,1),(1,1,0) && \text{for } \mathcal{P}_{2,2,1},\\
&(0,0,1),(0,1,1),(1,0,1),(1,1,0) && \text{for } \mathcal{P}_{2,2,2},\\
&(0,1,1),(1,0,1),(1,1,0),(1,1,1) && \text{for } \mathcal{P}_{2,3,2}.
\end{aligned} \quad (266)$$

Then by Lemmas 4 and 5, we obtain

$$F(\mathbf{t};A_3)=\left(\prod_{j=1}^{6}\frac{t_j}{e^{t_j}-1}\right)G(\mathbf{t};A_3)$$
$$=1+\cdots+\frac{23}{435891456000}t_1^2t_2^2t_3^2t_4^2t_5^2t_6^2+\cdots, \quad (267)$$

Table 11.2 Bernoulli numbers for C_2

(k_1, k_2, k_3, k_4)	$B_{(k_1,k_2,k_3,k_4)}(C_2)$	(k_1, k_2, k_3, k_4)	$B_{(k_1,k_2,k_3,k_4)}(C_2)$
$(0, 0, 0, 0)$	1	$(0, 1, 1, 2)$	$-1/1440$
$(0, 1, 1, 4)$	$1/20160$	$(0, 1, 2, 1)$	$-1/720$
$(0, 1, 2, 3)$	$1/20160$	$(0, 1, 3, 2)$	$1/10080$
$(0, 1, 3, 4)$	$-1/134400$	$(0, 1, 4, 1)$	$1/2520$
$(0, 1, 4, 3)$	$-1/67200$	$(0, 2, 1, 1)$	$-1/720$
$(0, 2, 1, 3)$	$1/20160$	$(0, 2, 2, 2)$	$1/15120$
$(0, 2, 2, 4)$	$-1/201600$	$(0, 2, 3, 1)$	$1/5040$
$(0, 2, 3, 3)$	$-1/134400$	$(0, 2, 4, 2)$	$-1/50400$
$(0, 2, 4, 4)$	$1/665280$	$(0, 3, 1, 2)$	$1/10080$
$(0, 3, 1, 4)$	$-1/134400$	$(0, 3, 2, 1)$	$1/5040$
$(0, 3, 2, 3)$	$-1/134400$	$(0, 3, 3, 2)$	$-1/67200$
$(0, 3, 3, 4)$	$1/887040$	$(0, 3, 4, 1)$	$-1/16800$
$(0, 3, 4, 3)$	$1/443520$	$(0, 4, 1, 1)$	$1/2520$
$(0, 4, 1, 3)$	$-1/67200$	$(0, 4, 2, 2)$	$-1/50400$
$(0, 4, 2, 4)$	$1/665280$	$(0, 4, 3, 1)$	$-1/16800$
$(0, 4, 3, 3)$	$1/443520$	$(0, 4, 4, 2)$	$1/166320$
$(0, 4, 4, 4)$	$-691/1513512000$	$(1, 0, 1, 2)$	$-1/360$
$(1, 0, 1, 4)$	$1/1260$	$(1, 0, 2, 1)$	$-1/360$
$(1, 0, 2, 3)$	$1/2520$	$(1, 0, 3, 2)$	$1/2520$
$(1, 0, 3, 4)$	$-1/8400$	$(1, 0, 4, 1)$	$1/1260$
$(1, 0, 4, 3)$	$-1/8400$	$(1, 1, 0, 2)$	$1/360$
$(1, 1, 0, 4)$	$-1/1260$	$(1, 1, 2, 0)$	$1/720$
$(1, 1, 2, 2)$	$1/20160$	$(1, 1, 2, 4)$	$-29/2419200$
$(1, 1, 3, 3)$	$1/268800$	$(1, 1, 4, 0)$	$-1/2520$
$(1, 1, 4, 2)$	$-1/86400$	$(1, 1, 4, 4)$	$1/403200$
$(1, 2, 0, 1)$	$1/360$	$(1, 2, 0, 3)$	$-1/2520$
$(1, 2, 1, 0)$	$-1/720$	$(1, 2, 1, 2)$	$-1/20160$
$(1, 2, 1, 4)$	$29/2419200$	$(1, 2, 3, 0)$	$1/5040$
$(1, 2, 3, 2)$	$1/403200$	$(1, 2, 3, 4)$	$-1/1267200$
$(1, 2, 4, 1)$	$-1/151200$	$(1, 2, 4, 3)$	$1/2217600$
$(1, 3, 0, 2)$	$-1/2520$	$(1, 3, 0, 4)$	$1/8400$
$(1, 3, 1, 3)$	$-1/268800$	$(1, 3, 2, 0)$	$-1/5040$
$(1, 3, 2, 2)$	$-1/403200$	$(1, 3, 2, 4)$	$1/1267200$
$(1, 3, 4, 0)$	$1/16800$	$(1, 3, 4, 2)$	$1/2217600$
$(1, 3, 4, 4)$	$-479/4036032000$	$(1, 4, 0, 1)$	$-1/1260$
$(1, 4, 0, 3)$	$1/8400$	$(1, 4, 1, 0)$	$1/2520$
$(1, 4, 1, 2)$	$1/86400$	$(1, 4, 1, 4)$	$-1/403200$
$(1, 4, 2, 1)$	$1/151200$	$(1, 4, 2, 3)$	$-1/2217600$
$(1, 4, 3, 0)$	$-1/16800$	$(1, 4, 3, 2)$	$-1/2217600$
$(1, 4, 3, 4)$	$479/4036032000$		

Table 11.3 Bernoulli numbers for C_2

(k_1,k_2,k_3,k_4)	$B_{(k_1,k_2,k_3,k_4)}(C_2)$	(k_1,k_2,k_3,k_4)	$B_{(k_1,k_2,k_3,k_4)}(C_2)$
(2, 0, 1, 1)	−1/360	(2, 0, 1, 3)	1/2520
(2, 0, 2, 2)	1/3780	(2, 0, 2, 4)	−1/12600
(2, 0, 3, 1)	1/2520	(2, 0, 3, 3)	−1/16800
(2, 0, 4, 2)	−1/12600	(2, 0, 4, 4)	1/41580
(2, 1, 0, 1)	−1/360	(2, 1, 0, 3)	1/2520
(2, 1, 1, 0)	1/1440	(2, 1, 1, 4)	−23/4838400
(2, 1, 2, 1)	−1/20160	(2, 1, 2, 3)	1/806400
(2, 1, 3, 0)	−1/10080	(2, 1, 3, 4)	1/2534400
(2, 1, 4, 1)	1/86400	(2, 1, 4, 3)	−1/2217600
(2, 2, 0, 2)	1/3780	(2, 2, 0, 4)	−1/12600
(2, 2, 1, 1)	−1/20160	(2, 2, 1, 3)	1/806400
(2, 2, 2, 0)	1/15120	(2, 2, 2, 2)	1/604800
(2, 2, 2, 4)	−1/3801600	(2, 2, 3, 1)	1/403200
(2, 2, 4, 0)	−1/50400	(2, 2, 4, 2)	−1/3326400
(2, 2, 4, 4)	1/20697600	(2, 3, 0, 1)	1/2520
(2, 3, 0, 3)	−1/16800	(2, 3, 1, 0)	−1/10080
(2, 3, 1, 4)	1/2534400	(2, 3, 2, 1)	1/403200
(2, 3, 3, 0)	1/67200	(2, 3, 3, 4)	−373/16144128000
(2, 3, 4, 1)	−1/2217600	(2, 3, 4, 3)	53/4036032000
(2, 4, 0, 2)	−1/12600	(2, 4, 0, 4)	1/41580
(2, 4, 1, 1)	1/86400	(2, 4, 1, 3)	−1/2217600
(2, 4, 2, 0)	−1/50400	(2, 4, 2, 2)	−1/3326400
(2, 4, 2, 4)	1/20697600	(2, 4, 3, 1)	−1/2217600
(2, 4, 3, 3)	53/4036032000	(2, 4, 4, 0)	1/166320
(2, 4, 4, 2)	53/1513512000	(2, 4, 4, 4)	−1/172972800
(3, 0, 1, 2)	1/2520	(3, 0, 1, 4)	−1/8400
(3, 0, 2, 1)	1/2520	(3, 0, 2, 3)	−1/16800
(3, 0, 3, 2)	−1/16800	(3, 0, 3, 4)	1/55440
(3, 0, 4, 1)	−1/8400	(3, 0, 4, 3)	1/55440
(3, 1, 0, 2)	−1/2520	(3, 1, 0, 4)	1/8400
(3, 1, 2, 0)	−1/20160	(3, 1, 2, 2)	−1/806400
(3, 1, 2, 4)	1/5068800	(3, 1, 3, 1)	−1/268800
(3, 1, 4, 0)	1/67200	(3, 1, 4, 2)	1/2217600
(3, 1, 4, 4)	−797/16144128000	(3, 2, 0, 1)	−1/2520
(3, 2, 0, 3)	1/16800	(3, 2, 1, 0)	1/20160
(3, 2, 1, 2)	1/806400	(3, 2, 1, 4)	−1/5068800
(3, 2, 3, 0)	−1/134400	(3, 2, 3, 4)	373/32288256000
(3, 2, 4, 1)	1/2217600	(3, 2, 4, 3)	−53/4036032000
(3, 3, 0, 2)	1/16800	(3, 3, 0, 4)	−1/55440
(3, 3, 1, 1)	1/268800	(3, 3, 2, 0)	1/134400
(3, 3, 2, 4)	−373/32288256000	(3, 3, 4, 0)	−1/443520
(3, 3, 4, 2)	−53/4036032000	(3, 3, 4, 4)	1/461260800
(3, 4, 0, 1)	1/8400	(3, 4, 0, 3)	−1/55440
(3, 4, 1, 0)	−1/67200	(3, 4, 1, 2)	−1/2217600
(3, 4, 1, 4)	797/16144128000	(3, 4, 2, 1)	−1/2217600
(3, 4, 2, 3)	53/4036032000	(3, 4, 3, 0)	1/443520
(3, 4, 3, 2)	53/4036032000	(3, 4, 3, 4)	−1/461260800

Table 11.4 Bernoulli numbers for C_2

(k_1, k_2, k_3, k_4)	$B_{(k_1,k_2,k_3,k_4)}(C_2)$	(k_1, k_2, k_3, k_4)	$B_{(k_1,k_2,k_3,k_4)}(C_2)$
(4, 0, 1, 1)	1/1260	(4, 0, 1, 3)	−1/8400
(4, 0, 2, 2)	−1/12600	(4, 0, 2, 4)	1/41580
(4, 0, 3, 1)	−1/8400	(4, 0, 3, 3)	1/55440
(4, 0, 4, 2)	1/41580	(4, 0, 4, 4)	−691/94594500
(4, 1, 0, 1)	1/1260	(4, 1, 0, 3)	−1/8400
(4, 1, 1, 0)	−1/20160	(4, 1, 1, 2)	23/4838400
(4, 1, 2, 1)	29/2419200	(4, 1, 2, 3)	−1/5068800
(4, 1, 3, 0)	1/134400	(4, 1, 3, 2)	−1/2534400
(4, 1, 4, 1)	−1/403200	(4, 1, 4, 3)	797/16144128000
(4, 2, 0, 2)	−1/12600	(4, 2, 0, 4)	1/41580
(4, 2, 1, 1)	29/2419200	(4, 2, 1, 3)	−1/5068800
(4, 2, 2, 0)	−1/201600	(4, 2, 2, 2)	−1/3801600
(4, 2, 2, 4)	373/24216192000	(4, 2, 3, 1)	−1/1267200
(4, 2, 3, 3)	373/32288256000	(4, 2, 4, 0)	1/665280
(4, 2, 4, 2)	1/20697600	(4, 2, 4, 4)	−1/345945600
(4, 3, 0, 1)	−1/8400	(4, 3, 0, 3)	1/55440
(4, 3, 1, 0)	1/134400	(4, 3, 1, 2)	−1/2534400
(4, 3, 2, 1)	−1/1267200	(4, 3, 2, 3)	373/32288256000
(4, 3, 3, 0)	−1/887040	(4, 3, 3, 2)	373/16144128000
(4, 3, 4, 1)	479/4036032000	(4, 3, 4, 3)	−1/461260800
(4, 4, 0, 2)	1/41580	(4, 4, 0, 4)	−691/94594500
(4, 4, 1, 1)	−1/403200	(4, 4, 1, 3)	797/16144128000
(4, 4, 2, 0)	1/665280	(4, 4, 2, 2)	1/20697600
(4, 4, 2, 4)	−1/345945600	(4, 4, 3, 1)	479/4036032000
(4, 4, 3, 3)	−1/461260800	(4, 4, 4, 0)	−691/1513512000
(4, 4, 4, 2)	−1/172972800	(4, 4, 4, 4)	479/1372250880000

where

$$
\begin{aligned}
G(\mathbf{t}; A_3)
&= \frac{e^{t_3+t_4}}{(t_1+t_2-t_4)(t_1-t_3-t_4+t_5)(t_3+t_4-t_6)} - \frac{e^{t_1+t_5}}{(t_2+t_3-t_5)(t_1-t_3-t_4+t_5)(t_1+t_5-t_6)}\\
&- \frac{e^{t_6}}{(t_1+t_2+t_3-t_6)(t_3+t_4-t_6)(t_1+t_5-t_6)} + \frac{e^{t_1+t_2+t_3}}{(t_1+t_2-t_4)(t_2+t_3-t_5)(t_1+t_2+t_3-t_6)}\\
&- \frac{e^{t_4+t_5}}{(t_1+t_2-t_4)(t_2+t_3-t_5)(t_2-t_4-t_5+t_6)} - \frac{e^{t_2+t_3+t_4}}{(t_2+t_3-t_5)(t_1-t_3-t_4+t_5)(t_3+t_4-t_6)}\\
&+ \frac{e^{t_1+t_2+t_5}}{(t_1+t_2-t_4)(t_1-t_3-t_4+t_5)(t_1+t_5-t_6)} + \frac{e^{t_2+t_6}}{(t_3+t_4-t_6)(t_1+t_5-t_6)(t_2-t_4-t_5+t_6)}\\
&+ \frac{e^{t_3+t_4+t_5}}{(t_1+t_2-t_4)(t_3+t_4-t_6)(t_2-t_4-t_5+t_6)} - \frac{e^{t_5+t_6}}{(t_2+t_3-t_5)(t_1+t_2+t_3-t_6)(t_3+t_4-t_6)}\\
&+ \frac{e^{t_1+t_2+t_3+t_5}}{(t_1+t_2-t_4)(t_1+t_2+t_3-t_6)(t_1+t_5-t_6)} - \frac{e^{t_2+t_3+t_6}}{(t_2+t_3-t_5)(t_1+t_5-t_6)(t_2-t_4-t_5+t_6)}
\end{aligned}
$$

$$
\begin{aligned}
&+\frac{e^{t_1+t_4+t_5}}{(t_2+t_3-t_5)(t_1+t_5-t_6)(t_2-t_4-t_5+t_6)}-\frac{e^{t_4+t_6}}{(t_1+t_2-t_4)(t_1+t_2+t_3-t_6)(t_1+t_5-t_6)}\\
&+\frac{e^{t_1+t_2+t_3+t_4}}{(t_2+t_3-t_5)(t_1+t_2+t_3-t_6)(t_3+t_4-t_6)}-\frac{e^{t_1+t_2+t_6}}{(t_1+t_2-t_4)(t_3+t_4-t_6)(t_2-t_4-t_5+t_6)}\\
&-\frac{e^{t_1+t_3+t_4+t_5}}{(t_1+t_5-t_6)(t_2-t_4-t_5+t_6)(t_3+t_4-t_6)}-\frac{e^{t_3+t_4+t_6}}{(t_1+t_2-t_4)(t_1-t_3-t_4+t_5)(t_1+t_5-t_6)}\\
&+\frac{e^{t_1+t_5+t_6}}{(t_2+t_3-t_5)(t_1-t_3-t_4+t_5)(t_3+t_4-t_6)}+\frac{e^{t_1+t_2+t_3+t_6}}{(t_1+t_2-t_4)(t_2+t_3-t_5)(t_2-t_4-t_5+t_6)}\\
&-\frac{e^{t_4+t_5+t_6}}{(t_1+t_2-t_4)(t_2+t_3-t_5)(t_1+t_2+t_3-t_6)}+\frac{e^{t_1+t_2+t_3+t_4+t_5}}{(t_1+t_2+t_3-t_6)(t_3+t_4-t_6)(t_1+t_5-t_6)}\\
&+\frac{e^{t_2+t_3+t_4+t_6}}{(t_2+t_3-t_5)(t_1-t_3-t_4+t_5)(t_1+t_5-t_6)}-\frac{e^{t_1+t_2+t_5+t_6}}{(t_1+t_2-t_4)(t_1-t_3-t_4+t_5)(t_3+t_4-t_6)}.
\end{aligned}
\tag{268}
$$

Therefore, we obtain

$$
\zeta_3(2,2,2,2,2,2;A_3)=(-1)^6\frac{(2\pi\sqrt{-1})^{12}}{4!}\frac{23}{435891456000}=\frac{23}{2554051500}\pi^{12},
\tag{269}
$$

which implies a formula of Gunnells–Sczech [6]:

$$
\zeta_W(2;A_3)=12^2\frac{23}{2554051500}\pi^{12}=\frac{92}{70945875}\pi^{12}.
\tag{270}
$$

In higher rank root systems, generating functions are more involved, since the polytopes are not simplicial any longer. For instance, we have the generating function of type G_2, A_4, B_3, and C_3 with 1010 terms, 5040 terms, 19908 terms, and 20916 terms, respectively, by use of triangulation. In [17], we will improve Theorem 7 and will give more compact forms of the generating functions $F(\mathbf{t},\mathbf{y};\Delta)$, which do not depend on simplicial decompositions. As a result, the numbers of terms in the above generating functions reduce to 15, 125, 68, and 68, respectively (as for the G_2 case, see [18]). In fact, the number for A_n is $(n+1)^{n-1}$, which coincides with the number of all trees on $\{1,\ldots,n+1\}$. See [17] for the details.

Example 5. In Theorem 5, we have already given general forms of functional relations among zeta-functions of root systems. In previous examples, we observed generating functions and special values in several cases, but here, we treat examples of explicit functional relations which can be deduced from the general forms. First, consider the A_2 case (see Example 2). Set

$$
\Delta_+=\Delta_+(A_2)=\{\alpha_1,\alpha_2,\alpha_1+\alpha_2\},
$$

and $\mathbf{y}=0$, $\mathbf{s}=(2,s,2)$ for $s\in\mathbb{C}$ with $\Re s>1$, $I=\{2\}$, that is, $\Delta_{I+}=\{\alpha_2\}$. Then, from (110), we can write the left-hand side of (115) as

$$S(\mathbf{s}, \mathbf{y}; I; \Delta) = \sum_{m,n=1}^{\infty} \frac{1}{m^2 n^s (m+n)^2} + \sum_{\substack{m,n=1 \\ m \neq n}}^{\infty} \frac{1}{m^2 n^s (-m+n)^2}$$

$$= 2\zeta_2(2, s, 2; A_2) + \zeta_2(2, 2, s; A_2).$$

On the other hand, the right-hand side of (115) is

$$\left(\frac{(2\pi\sqrt{-1})^2}{2!} \right)^2 \sum_{m=1}^{\infty} \frac{1}{m^s} \int_0^1 \mathbf{e}(-mx) L_2(x, 0) L_2(-x, 0) \mathrm{d}x$$

$$= \left(\frac{(2\pi\sqrt{-1})^2}{2!} \right)^2 \sum_{m=1}^{\infty} \frac{1}{m^s} \int_0^1 \mathbf{e}(-mx) B_2(x) B_2(1-x) \mathrm{d}x,$$

by using (121). From well-known properties of Bernoulli polynomials, we can calculate the above integral (for details, see Nakamura [28]) and can recover from (115) the formula

$$2\zeta_2(2, s, 2; A_2) + \zeta_2(2, 2, s; A_2) = 4\zeta(2)\zeta(s+2) - 6\zeta(s+4), \tag{271}$$

proved in [36] (see also [28]). The function $\zeta_2(\mathbf{s}; A_2)$ can be continued meromorphically to the whole space $\mathbb{C}^3$ [20], so (271) holds for any $s \in \mathbb{C}$ except for singularities on the both sides. In particular, when $s = 2$, we obtain (252). Similarly, we can treat the $C_2(\simeq B_2)$ case and give some functional relations from (115) by combining the meromorphic continuation of $\zeta_2(\mathbf{s}; C_2)$ which has been shown in [21].

Acknowledgements The authors express their sincere gratitude to the organizers and the editors who gave the occasion of a talk at the conference and the publication of a paper in the present proceedings. The authors are also indebted to the referee for useful comments and suggestions.

References

1. T. M. Apostol, *Introduction to Analytic Number Theory*, Springer, 1976.
2. N. Bourbaki, *Groupes et Algèbres de Lie, Chapitres 4, 5 et 6*, Hermann, Paris, 1968.
3. D. M. Bradley, Partition identities for the multiple zeta function, in 'Zeta Functions, Topology and Quantum Physics', Developments in Mathematics Vol. 14, T. Aoki et al. (eds.), Springer, New York, 2005, pp. 19–29.
4. M. Brion and M. Vergne, Lattice points in simple polytopes, J. Amer. Math. Soc. **10** (1997), 371–392.
5. D. Essouabri, Singularités des séries de Dirichlet associées à des polynômes de plusieurs variables et applications à la théorie analytique des nombres, Thése, Univ. Henri Poincaré - Nancy I, 1995.
6. P. E. Gunnells and R. Sczech, Evaluation of Dedekind sums, Eisenstein cocycles, and special values of L-functions, Duke Math. J. **118** (2003), 229–260.

7. T. Hibi, *Algebraic combinatorics on convex polytopes*, Carslaw Publications, Australia, 1992.
8. J. E. Humphreys, *Introduction to Lie algebras and representation theory*, Graduate Texts in Mathematics, Vol. 9, Springer-Verlag, New York-Berlin, 1972.
9. J. E. Humphreys, *Reflection groups and Coxeter groups*, Cambridge University Press, Cambridge, 1990.
10. V. G. Kac, *Infinite-dimensional Lie algebras*, 3rd ed., Cambridge University Press, Cambridge, 1990.
11. Y. Komori, K. Matsumoto and H. Tsumura, Zeta-functions of root systems, in 'The Conference on L-functions', L. Weng and M. Kaneko (eds.), World Scientific, 2007, pp. 115–140.
12. Y. Komori, K. Matsumoto and H. Tsumura, Zeta and L-functions and Bernoulli polynomials of root systems, Proc. Japan Acad. Series A **84** (2008), 57–62.
13. Y. Komori, K. Matsumoto and H. Tsumura, On multiple Bernoulli polynomials and multiple L-functions of root systems, in 'Analytic Number Theory and Related Areas' (Kyoto, 2007), RIMS Kokyuroku **1665** (2009), 114–126.
14. Y. Komori, K. Matsumoto and H. Tsumura, Functional relations for zeta-functions of root systems, in 'Number Theory: Dreaming in Dreams - Proceedings of the 5th China-Japan Seminar', T. Aoki, S. Kanemitsu and J. -Y. Liu (eds.), Ser. on Number Theory and its Appl. Vol. 6, World Scientific, 2010, pp. 135–183.
15. Y. Komori, K. Matsumoto and H. Tsumura, An introduction to the theory of zeta-functions of root systems, in 'Algebraic and Analytic Aspects of Zeta Functions and L-functions', G. Bhowmik, K. Matsumoto and H. Tsumura (eds.), MSJ Memoirs, Vol. 21, Mathematical Society of Japan, 2010, pp. 115–140.
16. Y. Komori, K. Matsumoto and H. Tsumura, On Witten multiple zeta-functions associated with semisimple Lie algebras II, J. Math. Soc. Japan, **62** (2010), 355–394.
17. Y. Komori, K. Matsumoto and H. Tsumura, On multiple Bernoulli polynomials and multiple L-functions of root systems, Proc. London Math. Soc. **100** (2010), 303–347.
18. Y. Komori, K. Matsumoto and H. Tsumura, On Witten multiple zeta-functions associated with semisimple Lie algebras IV, Glasgow Math. J. **53** (2011), 185–206.
19. I. G. Macdonald, *Symmetric functions and Hall polynomials*, 2nd ed., Oxford Univ. Press, Oxford, 1995.
20. K. Matsumoto, On the analytic continuation of various multiple zeta-functions, in 'Number Theory for the Millennium II, Proc. Millennial Conference on Number Theory', M. A. Bennett et al. (eds.), A K Peters, 2002, pp. 417–440.
21. K. Matsumoto, On Mordell-Tornheim and other multiple zeta-functions, in 'Proceedings of the Session in analytic number theory and Diophantine equations' (Bonn, January-June 2002), D. R. Heath-Brown and B. Z. Moroz (eds.), Bonner Mathematische Schriften Nr. 360, Bonn 2003, n.25, 17pp.
22. K. Matsumoto, T. Nakamura, H. Ochiai and H. Tsumura, On value-relations, functional relations and singularities of Mordell-Tornheim and related triple zeta-functions, Acta Arith. **132** (2008), 99–125.
23. K. Matsumoto, T. Nakamura and H. Tsumura, Functional relations and special values of Mordell-Tornheim triple zeta and L-functions, Proc. Amer. Math. Soc. **136** (2008), 2135–2145.
24. K. Matsumoto and H. Tsumura, On Witten multiple zeta-functions associated with semisimple Lie algebras I, Ann. Inst. Fourier **56** (2006), 1457–1504.
25. K. Matsumoto and H. Tsumura, Functional relations for various multiple zeta-functions, in 'Analytic Number Theory' (Kyoto, 2005), RIMS Kokyuroku **1512** (2006), 179–190.
26. K. Matsumoto and H. Tsumura, Functional relations among certain double polylogarithms and their character analogues, Šiauliai Math. Sem. **3** (**11**) (2008), 189–205.
27. L. J. Mordell, On the evaluation of some multiple series, J. London Math. Soc. **33** (1958), 368–371.
28. T. Nakamura, A functional relation for the Tornheim double zeta function, Acta Arith. **125** (2006), 257–263.
29. M. V. Subbarao and R. Sitaramachandrarao, On some infinite series of L. J. Mordell and their analogues, Pacific J. Math. **119** (1985), 245–255.

30. R. P. Stanley, Decompositions of rational convex polytopes, in ‘Combinatorial mathematics, optimal designs and their applications’ (Proc. Sympos. Combin. Math. and Optimal Design, Colorado State Univ., Fort Collins, Colo., 1978), Ann. Discrete Math. **6** (1980), 333–342.
31. A. Szenes, Iterated residues and multiple Bernoulli polynomials, Internat. Math. Res. Notices **18** (1998), 937–958.
32. A. Szenes, Residue formula for rational trigonometric sums, Duke Math. J. **118** (2003)189–228.
33. L. Tornheim, Harmonic double series, Amer. J. Math. **72** (1950), 303–314.
34. H. Tsumura, On Witten’s type of zeta values attached to $SO(5)$, Arch. Math. (Basel) **84** (2004), 147–152.
35. H. Tsumura, On some functional relations between Mordell-Tornheim double L-functions and Dirichlet L-functions, J. Number Theory **120** (2006), 161–178.
36. H. Tsumura, On functional relations between the Mordell-Tornheim double zeta functions and the Riemann zeta function, Math. Proc. Cambridge Philos. Soc. **142** (2007), 395–405.
37. E. T. Whittaker and G. N. Watson, *A Course of Modern Analysis*, 4th ed., Cambridge University Press (1958).
38. E. Witten, On quantum gauge theories in two dimensions, Comm. Math. Phys. **141** (1991), 153–209.
39. D. Zagier, Values of zeta functions and their applications, in ‘First European Congress of Mathematics’ Vol. II, A. Joseph et al. (eds.), Progr. Math. Vol. 120, Birkhäuser, 1994, pp. 497–512.
40. G. M. Ziegler, *Lectures on polytopes*, Springer-Verlag, New York, 1995.

Chapter 12
A Pseudo Twin Primes Theorem

Alex V. Kontorovich

Abstract Selberg identified the "parity" barrier that sieves alone cannot distinguish between integers having an even or odd number of factors. We give here a short and self-contained demonstration of parity breaking using bilinear forms, modeled on the twin primes conjecture.

Keywords Exponential sums • Bilinear forms • Piatetski-Shapiro primes

12.1 Introduction

The twin prime conjecture states that there are infinitely many primes p such that $p + 2$ is also prime. A refined version of this conjecture is that $\pi_2(x)$, the number of prime twins lying below a level x, satisfies

$$\pi_2(x) \sim C \frac{x}{\log^2 x},$$

as $x \to \infty$, where $C \approx 1.32032\ldots$ an arithmetic constant.

The best result towards the twin prime conjecture is Chen's [Che73], stating that there are infinitely many primes p for which $p + 2$ is either itself prime or the product of two primes. This statement is a quintessential exhibition of the "parity" barrier identified by Selberg that sieve methods alone cannot distinguish between sets having an even or odd number of factors. Vinogradov's resolution [Vin37] of

A.V. Kontorovich (✉)
Yale University, Mathematics Department, PO Box 208283, New Haven, CT 06520-8283, USA
e-mail: alex.kontorovich@yale.edu

D. Bump et al. (eds.), *Multiple Dirichlet Series, L-functions and Automorphic Forms*, Progress in Mathematics 300, DOI 10.1007/978-0-8176-8334-4_12,

the ternary Goldbach problem introduced the idea that estimating certain bilinear forms can sometimes break this barrier, and there have since been many impressive instances of this phenomenon; see, e.g., [FI98, HB01].

In this note, we aim to illustrate parity breaking in a simple, self-contained example. Consider an analogue of the twin prime conjecture where instead of intersecting two copies of the primes, we intersect one copy of the primes with a set which analytically mimics the primes. For $x > 2$ let

$$\mathrm{iL}(x) \sim x \log x$$

denote the inverse to the logarithmic integral function

$$\mathrm{Li}(x) := \int_2^x \frac{dt}{\log t}.$$

Definition 1. Let $\widehat{\pi}(x)$ denote the number of primes $p \leq x$ such that $p = \lfloor \mathrm{iL}(n) \rfloor$ for some integer n.

Here $\lfloor \cdot \rfloor$ is the floor function, returning the largest integer not exceeding its argument. Our main goal is to demonstrate.

Theorem 1. *As $x \to \infty$,*

$$\widehat{\pi}(x) \sim \frac{x}{\log^2 x}.$$

Notice that the constant above is 1, that is, there is no arithmetic interference. This theorem follows also from the work of Leitmann [Lei77]; both his proof and ours essentially mimic Piatetski–Shapiro's theorem [Pu53]. Our aim is to give a short derivation of this statement from scratch.

12.1.1 Outline

In Sect. 12.2, we give bounds for exponential sums of linear and bilinear type; these are used in the sequel. We devote Sect. 12.3 to reducing Theorem 1 to an estimate for exponential sums over primes. The latter are treated in Sect. 12.4 by Vaughan's identity, relying on the bounds of Sect. 12.2 to establish Theorem 1.

12.2 Estimates for Linear and Bilinear Sums

In this section, we develop preliminary bounds of linear and bilinear type, which are used in the sequel. We require first the following two well-known estimates due originally to Weyl [Wey21] and van der Corput [vdC21, vdC22]; see, e.g., Theorem 2.2 and Lemma 2.5 of [GK91].

Lemma 1 (van der Corput). *Suppose f has two continuous derivatives and for $0 < c < C$, we have $c\Delta \le f'' \le C\Delta$ on $[N, 2N]$. Then*

$$\sum_{N<n\le N_1\le 2N} e(f(n)) \ll_{C,c} N\Delta^{1/2} + \Delta^{-1/2}.$$

This is proved by truncating Poisson summation, comparing the sum to the integral, and integrating by parts two times.

Lemma 2 (Weyl, van der Corput). *Let $z_k \in \mathbb{C}$ be any complex numbers, $k = K+1, \ldots, 2K$. Then for any $Q \le K$,*

$$\left|\sum_{K<k\le 2K} z_k\right|^2 \le \frac{K+Q}{Q}\sum_{|q|<Q}\left(1-\frac{|q|}{Q}\right)\sum_{K<k,k+q\le 2K} z_k\bar{z}_{k+q}.$$

To prove this, shift the interval by q and average the contributions over $|q| < Q$.

12.2.1 Estimating Type I Sums

We use Lemma 1 to prove

Lemma 3. *For any integer h and $\ell \ge 1$,*

$$\sum_{N<n\le N_1\le 2N} e(h\,\mathrm{Li}(n\ell)) \ll \begin{cases} N & \text{if } h = 0 \\ (N|h|\ell)^{\frac{1}{2}}\log(N\ell) & \text{otherwise.} \end{cases}$$

Remark 1. Here as throughout, the implied constant is absolute unless otherwise specified.

Proof. Let Ξ denote the sum in question. The trivial estimate is N. Assume without loss of generality $h > 0$. Apply Lemma 1 with $f(n) = h\,\mathrm{Li}(n\ell)$, taking $\Delta = \frac{h\ell}{N\log^2(N\ell)}$. Thus,

$$\Xi \ll N\left(\frac{h\ell}{N\log^2(N\ell)}\right)^{1/2} + \left(\frac{N\log^2(N\ell)}{h\ell}\right)^{1/2} \ll (Nh\ell)^{1/2}\log(N\ell),$$

so we are done. □

12.2.2 Estimating Type II Sums

We first require the following estimate.

Lemma 4. *For positive integers h, k, q, and $L \geq 10$, let*

$$S_0(q;k) := \sum_{L<\ell\leq 2L} e\Big(h\big[\operatorname{Li}(\ell k) - \operatorname{Li}(\ell(k+q))\big]\Big). \tag{1}$$

Then

$$S_0(q;k) \ll (Lhq)^{1/2},$$

where the implied constant is absolute, that is, independent of k.

Proof. We again apply Lemma 1, this time choosing for the function $f(x) = h(\operatorname{Li}(xk) - \operatorname{Li}(x(k+q)))$. Then for $L < x \leq 2L$,

$$f''(x) = h\left(\frac{-k}{x\log^2(xk)} + \frac{k+q}{x\log^2(x(k+q))}\right) = hq\frac{\log(k'x)-2}{x\log(k'x)} \asymp \frac{hq}{L},$$

for some $k' \in [k, k+q)$ by the mean value theorem in k. Thus, we can take $\Delta = \frac{hq}{L}$ and

$$S_0 \ll L\left(\frac{hq}{L}\right)^{1/2} + \left(\frac{L}{hq}\right)^{1/2} \ll (Lhq)^{1/2},$$

as desired. □

With this estimate in hand, we control Type II sums as follows (see also [GK91, Lemma 4.13]).

Lemma 5. *Let $\alpha(\ell)$ and $\beta(k)$ be sequences of complex numbers supported in $(L, 2L]$ and $(K, 2K]$, respectively, and suppose that*

$$\sum_{\ell} |\alpha(\ell)|^2 \ll L\log^{2A} L \quad \text{and} \quad \sum_{k} |\beta(k)|^2 \ll K\log^{2B} K. \tag{2}$$

Then

$$\sum_{L<\ell\leq 2L}\ \sum_{K<k\leq 2K} \alpha(\ell)\beta(k)e(h\operatorname{Li}(\ell k)) \ll KL^{5/6}h^{1/6}\log^A L\ \log^B K. \tag{3}$$

The implied constant in (3) *depends only on those in* (2).

Proof. Let S denote the sum on the left-hand side. By Cauchy–Schwartz,

$$|S|^2 \ll \left(\sum_{\ell}|\alpha(\ell)|^2\right)\sum_{\ell}\left|\sum_{k}\beta(k)e(h\operatorname{Li}(\ell k))\right|^2.$$

Let $Q \leq K$ be a parameter to be chosen later. Using Lemma 2 and (2), we get

$$|S|^2 \ll L \log^{2A} L \, \frac{K+Q}{Q} \sum_{|q|<Q} \left(1 - \frac{|q|}{Q}\right)$$

$$\times \sum_{\ell} \sum_{K<k, k+q \leq 2K} \beta(k)\bar{\beta}(k+q)\, e\Big(h\big[\operatorname{Li}(\ell k) - \operatorname{Li}(\ell(k+q))\big]\Big)$$

$$\ll L \log^{2A} L \frac{K}{Q} \sum_{1 \leq |q| < Q} \sum_{k} |\beta(k)\bar{\beta}(k+q)||S_0(q;k)| + \frac{K^2L^2}{Q} \log^{2A} L \log^{2B} K,$$

where S_0 is defined by (1).

Using Cauchy's inequality, $|x\bar{y}| \leq \frac{1}{2}(|x|^2 + |y|^2)$, and the fact that $|S_0(q;k)| = |S_0(-q;k+q)|$, we get

$$|S|^2 \ll \frac{K^2L^2}{Q} \log^{2A} L \ \log^{2B} K + \frac{LK}{Q} \log^{2A} L \sum_{k} |\beta(k)|^2 \sum_{1 \leq q < Q} |S_0(q;k)|.$$

From Lemma 4, we have the estimate

$$\frac{1}{Q} \sum_{1 \leq q < Q} |S_0(q;k)| \ll (LhQ)^{1/2},$$

so we finally see that

$$|S|^2 \ll \frac{K^2L^2}{Q} \log^{2A} L \ \log^{2B} K + L^{3/2} K^2 \log^{2A} L \ \log^{2B} K Q^{1/2} h^{1/2}.$$

The choice $Q = \lfloor L^{1/3} h^{-1/3} \rfloor$ gives the desired result. □

12.3 Reduction to Exponential Sums

In this section, we reduce the statement of Theorem 1 to a certain exponential sum over primes. We follow standard methods; see, e.g., [GK91, HB83], which we include here for completeness. If $p = \lfloor \operatorname{iL}(n) \rfloor$, then $p \leq \operatorname{iL}(n) < p + 1$ or, equivalently, $\operatorname{Li}(p) \leq n < \operatorname{Li}(p+1)$. The existence of an integer in the interval $[\operatorname{Li}(p), \operatorname{Li}(p+1))$ is indicated by the value $\lfloor \operatorname{Li}(p+1) \rfloor - \lfloor \operatorname{Li}(p) \rfloor$, so we have

$$\widehat{\pi}(x) = \sum_{p \leq x} \Big(\lfloor \operatorname{Li}(p+1) \rfloor - \lfloor \operatorname{Li}(p) \rfloor \Big).$$

Write $\lfloor\theta\rfloor = \theta - \psi(\theta) - \frac{1}{2}$, where ψ is the shifted fractional part

$$\psi(\theta) := \{\theta\} - \frac{1}{2} \in \left[-\frac{1}{2}, \frac{1}{2}\right).$$

So we have

$$\widehat{\pi}(x) = \sum_{p\leq x}\left[\operatorname{Li}(p+1) - \operatorname{Li}(p)\right] + \sum_{p\leq x}\left[\psi(\operatorname{Li}(p)) - \psi(\operatorname{Li}(p+1))\right].$$

Since $\operatorname{Li}'(x) = \frac{1}{\log x}$, we use the Taylor expansion:

$$\operatorname{Li}(p+1) = \operatorname{Li}(p) + \frac{1}{\log p} + O\left(\frac{1}{p\log^2 p}\right)$$

to get:

$$\widehat{\pi}(x) = \sum_{p\leq x}\frac{1}{\log p} + \sum_{p\leq x}\left[\psi(\operatorname{Li}(p)) - \psi(\operatorname{Li}(p+1))\right] + O(1).$$

By partial summation and a crude form of the prime number theorem,

$$\begin{aligned}\sum_{p\leq x}\frac{1}{\log p} &= \int_2^x \frac{\mathrm{d}\pi(t)}{\log t} = \frac{\pi(x)}{\log x} + O\left(\int_2^x \frac{\pi(t)}{t\log^2 t}\mathrm{d}t\right)\\ &= \frac{x}{\log^2 x} + O\left(\frac{x}{\log^3 x}\right).\end{aligned}$$

Therefore, to prove Theorem 1, it suffices to show that

$$\sum_{p\leq x}\left[\psi(\operatorname{Li}(p)) - \psi(\operatorname{Li}(p+1))\right] \ll \frac{x}{\log^3 x}. \tag{4}$$

Equivalently, split the sum into dyadic segments and apply partial summation to reduce (4) to the statement that for any $N < N_1 \leq 2N$,

$$\Sigma := \sum_{N<n\leq N_1\leq 2N} \Lambda(n)\left[\psi(\operatorname{Li}(n)) - \psi(\operatorname{Li}(n+1))\right] \ll \frac{N}{\log^2 N}, \tag{5}$$

with $N \ll x$. Here Λ is the von Mangoldt function:

$$\Lambda(n) = \begin{cases}\log p & \text{if } n = p^k \text{ is a prime power}\\ 0 & \text{otherwise.}\end{cases}$$

The truncated Fourier series of ψ is

$$\psi(\theta) = \sum_{0<|h|\le H} c_h\, e(\theta h) + O(g(\theta, H)), \tag{6}$$

where $e(x) = \mathrm{e}^{2\pi i x}$, $c_h = \frac{1}{2\pi i h}$, and

$$g(\theta, H) = \min\left(1, \frac{1}{H\|\theta\|}\right).$$

Here $\|\cdot\|$ is the distance to the nearest integer. In the above, H is a parameter which we will choose later, eventually setting

$$H = \log^4 N. \tag{7}$$

The function g has Fourier expansion

$$g(\theta, H) = \sum_{h\in\mathbb{Z}} a_h\, e(\theta h),$$

in which

$$a_h \ll \min\left(\frac{\log 2H}{H}, \frac{H}{|h|^2}\right). \tag{8}$$

Using (6), write the sum in (5) as $\Sigma = \Sigma_1 + O(\Sigma_2)$ where

$$\Sigma_1 := \sum_n \Lambda(n) \sum_{0<|h|\le H} c_h \Big[e(h\,\mathrm{Li}(n)) - e(h\,\mathrm{Li}(n+1))\Big]$$

and

$$\Sigma_2 := \sum_{n\sim N} \Lambda(n)\Big[g(\mathrm{Li}(n), H) + g(\mathrm{Li}(n+1), H)\Big].$$

We first dispose of Σ_2. Using positivity of g, the bound (8), and Lemma 3, we have

$$\begin{aligned}
\Sigma_2 &\ll \log N \sum_{n\sim N} g(\mathrm{Li}(n), H) \ll \log N \sum_{h\in\mathbb{Z}} |a_h| \left|\sum_{n\sim N} e(\mathrm{Li}(n)h)\right| \\
&\ll \log N \left[\frac{\log 2H}{H} N + \sum_{h\neq 0} \frac{H}{|h|^2} (N|h|)^{1/2} \log N\right] \\
&\ll (\log N)^2 \Big(N/H + N^{1/2} H\Big).
\end{aligned}$$

This bound is acceptable for (5) on setting H according to (7).

Next, we massage Σ_1. On writing $\phi_h(x) = 1 - e(h(\mathrm{Li}(x+1) - \mathrm{Li}(x))$, we see by partial summation that

$$\begin{aligned}
\Sigma_1 &\ll \sum_{1\le h\le H} h^{-1} \left| \sum_{N<n\le N_1} \Lambda(n)\phi_h(n)e(h\,\mathrm{Li}(n)) \right| \\
&\ll \sum_{1\le h\le H} h^{-1} \left| \phi_h(N_1) \sum_{N<n\le N_1} \Lambda(n)e(h\,\mathrm{Li}(n)) \right| \\
&\quad + \int_N^{N_1} \sum_{1\le h\le H} h^{-1} \left| \frac{\partial\phi_h(x)}{\partial x} \sum_{N<n\le x} \Lambda(n)e(h\,\mathrm{Li}(n)) \right| \mathrm{d}x \\
&\ll \frac{1}{\log N} \max_{N_2\le 2N} \sum_{1\le h\le H} \left| \sum_{N<n\le N_2} \Lambda(n)e(h\,\mathrm{Li}(n)) \right|.
\end{aligned}$$

Here we used the bounds

$$\phi_h(x) \ll h(\mathrm{Li}(x+1) - \mathrm{Li}(x)) \ll \frac{h}{\log N}$$

and

$$\frac{\partial\phi_h(x)}{\partial x} \ll h\left(\frac{1}{\log(x+1)} - \frac{1}{\log(x)}\right) \ll \frac{h}{N\log^2 N}$$

for $N \le x \le 2N$. We have thus reduced Theorem 1 to the statement that for all $N < N_2 \le 2N$,

$$S := \sum_{0<h\le H} \left| \sum_{N<n\le N_2\le 2N} \Lambda(n)e(h\,\mathrm{Li}(n)) \right| \ll \frac{N}{\log N}. \tag{9}$$

We establish this fact in the next section.

12.4 Proof of Theorem 1

Our goal in this section is to demonstrate (9), thereby establishing Theorem 1. We will actually prove more; instead of a log savings, we will save a power:

Theorem 2. *For S defined in* (9) *and any $\epsilon > 0$, we have*

$$S \ll_\epsilon N^{21/22+\epsilon}, \qquad \text{as} \qquad N \to \infty.$$

Fix u and v, parameters to be chosen later, and let $F(s) = \sum_{1\le n\le v} \Lambda(n)n^{-s}$ and $M(s) = \sum_{1\le n\le u} \mu(n)n^{-s}$, where μ is the Möbius function:

$$\mu(n) = \begin{cases} (-1)^k & \text{if } n \text{ is the product of } k \text{ distinct primes} \\ 0 & \text{if } n \text{ is not square-free.} \end{cases}$$

The functions F and M are the truncated Dirichlet polynomials of the functions $-\zeta'/\zeta$ and $1/\zeta$, respectively, where $\zeta(s)$ is the Riemann zeta function. Notice, for instance, that

$$\frac{\zeta'}{\zeta}(s) + F(s) = -\sum_{n>v} \Lambda(n)n^{-s}.$$

Comparing the Dirichlet coefficients on both sides of the identity

$$\frac{\zeta'}{\zeta} + F = \left(\frac{\zeta'}{\zeta} + F\right)(1 - \zeta M) + \zeta' M + \zeta F M$$

gives for $n > v$:

$$-\Lambda(n) = -\sum_{\substack{k\ell=n\\ k>v,\ell>u}} \Lambda(k)\sum_{\substack{d|\ell\\ d>u}} \mu(d) - \sum_{\substack{k\ell=n\\ \ell\le u}} \log k\ \mu(\ell) + \sum_{\substack{k\ell m=n\\ \ell\le v,m\le u}} 1\cdot\Lambda(\ell)\mu(m), \quad (10)$$

This formula is originally due to Vaughan [Vau77] (see also [GK91, Lemma 4.12]). Assume for now that $v \le N$ (we will eventually set u and v to be slightly less than $\sqrt{N}$). Multiply the above identity by $e(h\,\mathrm{Li}(n))$ and sum over n:

$$\begin{aligned}\sum_{N<n\le N_2\le 2N} \Lambda(n)e(h\,\mathrm{Li}(n)) &= \sum_{u<\ell\le N_2/v}\ \sum_{\substack{N/\ell\le k\le N_2/\ell\\ v<k}} \Lambda(k)a(\ell)e(h\,\mathrm{Li}(k\ell)) \\ &\quad + \sum_{\ell\le u}\ \sum_{N/\ell\le k\le N_2/\ell} \mu(\ell)\log k\ e(h\,\mathrm{Li}(k\ell)) \\ &\quad - \sum_{r\le uv}\ \sum_{N/r\le k\le N_2/r} b(r)e(h\,\mathrm{Li}(kr)) \\ &= S_1 + S_2 - S_3,\end{aligned}$$

where

$$a(\ell) = \sum_{\substack{d|\ell\\ d>u}} \mu(d), \text{ and } b(r) = \sum_{\substack{\ell m=r\\ \ell\le v,m\le u}} \Lambda(\ell)\mu(m).$$

It is the bilinear nature of the above identity which we exploit, forgetting the arithmetic nature of the coefficients a, b, μ, and Λ and just treating them as arbitrary. The savings then comes from the matrix norm of $\{e(h\,\mathrm{Li}(kl))\}_{k,\ell}$. This is achieved as follows.

Notice that $|a(\ell)|$ is at most $d(\ell)$, the number of divisors of ℓ, and similarly $|b(r)| \le \sum_{d|r} \Lambda(d) = \log r$, so we have the estimates

$$\sum_{L<\ell\le 2L} |a(\ell)|^2 \ll L \log^3 L, \text{ and } \sum_{R<r\le 2R} |b(r)|^2 \ll R \log^2 R.$$

It now suffices to show that $\sum_{0<h<H} |S_i| \ll_\epsilon N^{21/22+\epsilon}$ for each $i = 1, 2, 3$ by choosing u and v appropriately. We treat the sums of S_i individually in the next three subsections.

12.4.1 The Sum S_2

Let $G(x) := \sum_{k\le x} e(h\,\mathrm{Li}(k\ell))$. By Lemma 3, $G(x) \ll (xh\ell)^{\frac{1}{2}} \log(x\ell)$, so by partial integration, we get

$$\begin{aligned} S_2 &= \sum_{\ell\le u} \mu(\ell) \sum_{N/\ell\le k\le N_2/\ell} \log k\; e(h\,\mathrm{Li}(k\ell)) \\ &\ll \sum_{\ell\le u} \left| \int_{N/\ell}^{N_2/\ell} \log x \,\mathrm{d}G(x) \right| \\ &\ll \sum_{\ell\le u} \left(\sqrt{Nh} \log^2 N + \int_{N/\ell}^{N_2/\ell} \frac{1}{x} \sqrt{xh\ell} \log(x\ell) \mathrm{d}x \right) \\ &\ll \sqrt{Nhu} \log^2 N. \end{aligned}$$

Thus, $\sum_{1\le h<H} |S_2| \ll_\epsilon N^{21/22+\epsilon}$ (as desired) on taking $u = N^{5/11}$ and recalling (7).

12.4.2 The Sum S_1

Rewrite S_1 and split it into $\ll \log^2 N$ sums of the form:

$$\begin{aligned} S_1 &= \sum_{\substack{N\le k\ell\le N_2 \\ v<k, u<\ell}} \alpha(k)\beta(\ell) e(h\,\mathrm{Li}(k\ell)) \\ &\ll \log^2 N \sum_{L<\ell\le 2L} \sum_{\substack{K<k\le 2K \\ N<k\ell\le N_2}} \alpha(k)\beta(\ell) e(h\,\mathrm{Li}(k\ell)). \end{aligned}$$

The roles of k and ℓ are essentially symmetric (allowing α and β to be either Λ or a affects only powers of log and not the final estimate), and taking $v = u$, we may arrange it so $N^{5/11} \le K \le N^{1/2} \le L \le N^{6/11}$.

Now using Lemma 5, we find that:

$$\begin{aligned} S_1 &\ll \log^2 N \left(KL^{5/6}h^{1/6}\log^2 L \ \log^2 K\right) \\ &\ll \log^6 N \left(N^{21/22}h^{1/6}\right). \end{aligned}$$

Thus, $\sum_h |S_1| \ll_\epsilon N^{21/22+\epsilon}$, as desired.

12.4.3 The Sum S_3

Recall S_3 and break it according to

$$\begin{aligned} S_3 &= \sum_{r \le uv} b(r) \sum_{N/r \le k \le N_2/r} e(h\,\mathrm{Li}(kr)) \\ &= \sum_{r \le u} + \sum_{u < r \le uv} \\ &= S_4 + S_5. \end{aligned}$$

We treat S_4 exactly as S_2, getting $S_4 \ll (Nh)^{1/2} \log N(u \log u)$, which is clearly sufficiently small.

For S_5, the analysis is identical to that of S_1 and gives the same estimate, so we are done.

Acknowledgments The author wishes to thank Peter Sarnak for suggesting this problem, and generously lending of his time. Thanks also to Dorian Goldfeld and Patrick Gallagher for enlightening conversations, and to Tim Browning, Gautam Chinta, Steven J. Miller, and the referee for corrections to an earlier draft.

References

[Che73] Jing Run Chen. On the representation of a larger even integer as the sum of a prime and the product of at most two primes. *Sci. Sinica*, 16:157–176, 1973.

[FI98] John Friedlander and Henryk Iwaniec. The polynomial $X^2 + Y^4$ captures its primes. *Ann. of Math. (2)*, 148(3):945–1040, 1998.

[GK91] S.W. Graham and G. Kolesnik. *Van der Corput's Method of Exponential Sums*, volume 126. London Math. Soc., Lecture Notes, 1991.

[HB83] D. R. Heath-Brown. The Pjateckiĭ-Šapiro prime number theorem. *J Number Theory*, 16:242–266, 1983.

[HB01] D. R. Heath-Brown. Primes represented by $x^3 + 2y^3$. *Acta Math.*, 186(1):1–84, 2001.

[Lei77] D. Leitmann. The distribution of prime numbers in sequences of the form $[f(n)]$. *Proc. London Math. Soc.*, 35(3):448–462, 1977.

[Pu53] I. I. Pjateckiĭ-Šapiro. On the distribution of prime numbers in sequences of the form $[f(n)]$. *Mat. Sb.*, 33:559–566, 1953.

[Vau77] Robert-C. Vaughan. Sommes trigonométriques sur les nombres premiers. *C. R. Acad. Sci. Paris Sér. A-B*, 285(16):A981–A983, 1977.

[vdC21] J. G. van der Corput. Zahlentheoretische Abschätzungen. *Math. Ann.*, 84(1–2):53–79, 1921.

[vdC22] J. G. van der Corput. Verschärfung der Abschätzung beim Teilerproblem. *Math. Ann.*, 87(1–2):39–65, 1922.

[Vin37] I. M. Vinogradov. Representation of an odd number as a sum of three primes. *Dokl. Akad. Nauk SSSR*, 15:291–294, 1937.

[Wey21] H. Weyl. Zur Abschätzung von $\zeta(1 + it)$. *Math. Z.*, 10:88–101, 1921.

Chapter 13
Principal Series Representations of Metaplectic Groups Over Local Fields

Peter J. McNamara

Abstract Let G be a split reductive algebraic group over a non-archimedean local field. We study the representation theory of a central extension $\widetilde{G}$ of G by a cyclic group of order n, under some mild tameness assumptions on n. In particular, we focus our attention on the development of the theory of principal series representations for $\widetilde{G}$ and its applications to the study of Hecke algebras via a Satake isomorphism.

Keywords Principal series representations • Metaplectic group • Satake isomorphism

13.1 Introduction

Let F be a non-archimedean local field with ring of integers O_{F} and assume that G is a split reductive group over F that arises by base extension from a smooth reductive group scheme $\mathbf{G}$ over O_{F}. Let n be a positive integer such that $2n$ is coprime to the residue characteristic of F and that $F^{\times}$ contains $2n$ distinct $2n$th roots of unity. The object of this chapter is to study the principal series representations of a group $\widetilde{G}$ which arises as a central extension of (the F-points of) G by the cyclic group μ_n of order n. This means that there is an exact sequence of topological groups

$$1 \to \mu_n \to \widetilde{G} \to G \to 1$$

with the kernel μ_n lying inside the centre of $\widetilde{G}$.

P.J. McNamara (✉)
Department of Mathematics, Stanford University, Stanford, CA 94305-2125, USA
e-mail: petermc@math.stanford.edu

D. Bump et al. (eds.), *Multiple Dirichlet Series, L-functions and Automorphic Forms*, Progress in Mathematics 300, DOI 10.1007/978-0-8176-8334-4_13,

This metaplectic group $\widetilde{G}$ is a locally compact, totally disconnected topological group. Following the example of reductive case, we study the simplest family of representations, namely those which are induced from the inverse image in $\widetilde{G}$ of a Borel subgroup of G. Such representations have been studied in the literature for particular classes of groups. Kazhdan and Patterson [KP84] have a detailed study in the case of $G = GL_n$, while Savin [Sav04] has considered the case of G simply laced and simply connected. The double cover of a general simply connected algebraic group has also been considered by Loke and Savin [LS10]. This chapter, in developing a theory in greater generality borrows heavily on the results and arguments from the above-mentioned papers, as often the existing arguments in the literature can be generalised to the case of an arbitrary reductive group. In another direction, we mention the work of Weissman [Wei09] where the representation theory of metaplectic non-split tori is studied.

This chapter is intended to be partly expository, and partly an extension of the work mentioned above, redone to hold in slightly greater generality. We begin by giving an overview of the construction of the metaplectic group. In order to carry out our construction in the desired generality, we are forced to use a significant amount of the theory of these extensions in the semisimple simply connected case, after which we proceed to the general reductive case along the lines of the approach in [FL10].

Our study of the representation theory begins by focusing on the metaplectic torus and its representations, which govern a large part of the following theory. This metaplectic torus is no longer abelian, but its irreducible representations are finite dimensional by a version of the Stone–von Neumann theorem.

Following this, we construct the principal series representations for a metaplectic group by the familiar method of inducing from a Borel subgroup. The theory of Jacquet modules and intertwining operators between such representations is developed in this generality.

We then study the Hecke algebras of anti-genuine compactly supported locally constant functions invariant on the left and the right with respect to an open compact subgroup. The two cases of interest to us are when this compact subgroup is taken to be the maximal compact subgroup $K = \mathbf{G}(\mathcal{O}_F)$ or an Iwahori subgroup. In the former case, we present a metaplectic version of the Satake isomorphism in Theorem 10, while in the Iwahori case, we give a presentation of the corresponding Hecke algebra in terms of generators and relations, following Savin [Sav88, Sav04].

With the study of the structure of these Hecke algebras, we propose a combinatorial definition of the dual group to a metaplectic group which extends the notion of a dual group to a reductive group. This dual group is always a reductive group, so unlike in the reductive case, a metaplectic group cannot be recovered from its dual group. We hope that this notion of a dual group will prove to be useful in order to bring the study of metaplectic groups under the umbrella covered by the Langlands functoriality conjectures. It is worth noting that the root datum for the dual group has also appeared in [Lus93, Sect. 2.2.5], [FL10] and [Rei].

It is believed to be possible to develop this theory while working under the weaker assumption that F only contains n nth roots of unity, though in order to achieve this, a large amount of extra complications in formulae is necessary.

The author would like to thank Ben Brubaker and Omer Offen for useful conversations.

13.2 Preliminaries

As mentioned in the introduction, F will denote a non-archimedean local field, O_F is its valuation ring, and k its residue field. Let q be the cardinality of k. We choose once and for all a uniformising element $\varpi \in O_F$ (so $O_F/\varpi O_F = k$).

Let $\mathbf{G}$ be a split reductive group scheme over O_F. Throughout, our practice will be to use boldface letters to denote the group scheme and roman letters to denote the corresponding group of F-points. Let $\mathbf{B}$ be a Borel subgroup of $\mathbf{G}$ with unipotent radical $\mathbf{U}$, and let $\mathbf{T}$ be a maximal split torus contained in $\mathbf{B}$. We let $Y = \mathrm{Hom}(\mathbb{G}_m, \mathbf{T})$ be the group of cocharacters of $\mathbf{T}$.

Our object of study is not G itself but a central extension of $\widetilde{G}$ by the group μ_n of nth roots of unity. A construction of $\widetilde{G}$ is given in Sect. 13.3. For any subgroup H of G, we will use the notation $\widetilde{H}$ to denote the inverse image of H in $\widetilde{G}$; it is a central extension of H by μ_n. The topology on G induces the structure of a topological group on $\tilde{G}$. We will use p to denote the natural projection map $\widetilde{G} \to G$.

We consistently phrase our results in terms of coroots and cocharacters. Given a coroot α, there is an associated morphism of group schemes $\varphi_\alpha : \mathbf{SL_2} \longrightarrow \mathbf{G}$. Accordingly, we define elements of G by $w_\alpha = \varphi_\alpha((\begin{smallmatrix} 0 & 1 \\ -1 & 0 \end{smallmatrix}))$ and $e_\alpha(x) = \varphi_\alpha((\begin{smallmatrix} 1 & x \\ 0 & 1 \end{smallmatrix}))$. We will show in Sect. 13.4 that the Weyl group and unipotent subgroups of G split in the central extension $\widetilde{G}$ (in the latter case canonically) and use the same notation for the corresponding lifts to $\widetilde{G}$. For any $x \in F$ and $\lambda \in Y$, we will denote the image in T of x under λ by x^λ.

We fix a positive integer n which shall be the degree of our cover. The assumption on n we work under is that $2n|q-1$. This implies the condition that F contains $2n$ $2n$th roots of unity that was mentioned in the introduction.

We require some knowledge of the existence and properties of the Hilbert symbol, which we shall now recap. These results concerning the Hilbert symbol are well known, a reference for this material may be given by Serre's book [Ser62].

The Hilbert symbol is a bilinear map $(\cdot,\cdot) : F^\times \times F^\times \to \mu_n$ such that

$$(s,t)(t,s) = 1 = (t,-t) = (t,1-t).$$

Due to our assumptions on n, we can calculate the Hilbert symbol via the equation

$$(s,t) = \overline{\left((-1)^{v(s)v(t)}\frac{s^{v(t)}}{t^{v(s)}}\right)}^{\frac{q-1}{n}}$$

where the bar indicates to take the image in the residue field and v is the valuation in F. Particular special cases that we will make liberal use of throughout this chapter without further comment are the identities $(-1, x) = 1$ and $(\varpi^a, \varpi^b) = 1$, both of which require $q - 1$ to be divisible by $2n$, as opposed to only being divisible by n.

A representation (π, V) of $\widetilde{G}$ is a vector space V equipped with a group homomorphism $\pi : \widetilde{G} \longrightarrow \mathrm{Aut}\,(V)$. We say that (π, V) is *smooth* if the stabiliser of every vector contains an open subgroup, and *admissible* if for every open compact subgroup K, the subspace V^K of vectors fixed by K is finite dimensional.

Fix a faithful character $\epsilon : \mu_n \longrightarrow \mathbb{C}^\times$. We will only have cause to consider representations of $\tilde{G}$ in which the central μ_n acts by ϵ. (If μ_n did not act faithfully on an irreducible representation, then this representation would factor through a smaller cover of G.) Such representations will be called *genuine*.

We always will assume our representations to be genuine, smooth and admissible, and denote the category of such representations by $\mathrm{Rep}\,(\widetilde{G})$.

For any subgroup B of a group A, we use $C_A(B)$ (respectively $N_A(B)$) to denote the centraliser (respectively normaliser) of B in A.

13.3 Construction of the Extension

This section is devoted to showing the existence of the central extension $\widetilde{G}$ that we will be studying.

The cocharacter group Y comes equipped with a natural action of the Weyl group W. Let $B : Y \times Y \longrightarrow \mathbb{Z}$ be a W-invariant symmetric bilinear form on Y such that $Q(\alpha) := B(\alpha, \alpha)/2 \in \mathbb{Z}$ for all coroots α. (There will never be any possible confusion between this use of the symbol B and its use for a Borel subgroup.) Associated to such a B, we will construct an appropriate central extension.

We begin by recalling the following result of Brylinski and Deligne [BD01].

Proposition 1. *Suppose that* $\mathbf{G}$ *is semisimple and simply connected. The category of central extensions of* $\mathbf{G}$ *by* $\mathbf{K}_2$ *as sheaves on the big Zariski site* $Spec(F)_{Zar}$ *is equivalent to the category of integer valued Weyl group invariant quadratic forms on* Y*, where the only morphisms in the latter category are the identity morphisms.*

Upon taking F-points, this yields a central extension

$$1 \to K_2(F) \to E \to G \to 1.$$

Since F is assumed to be a local non-archimedean field containing n nth roots of unity, there is a surjection $K_2(F) \to \mu_n$ given by the Hilbert symbol and hence we obtain our central extension

$$1 \to \mu_n \to \widetilde{G} \to G \to 1.$$

In particular, if we consider the case where $\mathbf{G} = \mathbf{SL_2}$, then we have defined a map of abelian groups $\xi : \mathbb{Z} \longrightarrow H^2(SL_2(F), \mu_n)$.

Theorem 1. *For any split reductive group G, with our assumptions on F, n and B as above, there exists a central extension $\widetilde{G}$ of G by μ_n, such that for each coroot α, the pullback of the central extension under ϕ_α to a central extension of $SL_2(F)$ is incarnated by the cohomology class of $\xi(Q(\alpha))$, and when restricted to a central extension of T, the commutator $[\cdot,\cdot] : T \times T \longrightarrow \mu_n$ is given by*

$$[x^\lambda, y^\mu] = (x, y)^{B(\lambda,\mu)}$$

for all $x, y \in F^\times$ and $\lambda, \mu \in Y$.

A more explicit form of the commutator formula is possible, and we discuss it now since it will be of use to us later. Pick a basis $e_1, \ldots, e_r$ of Y. This induces an explicit isomorphism $(F^\times)^r \simeq T$, namely $(t_1, \ldots, t_r) \mapsto t_1^{e_1} \ldots t_r^{e_r}$. Via this isomorphism, suppose that $s = (s_1, \ldots, s_r)$ and $t = (t_1, \ldots, t_r)$ are two elements of T. Then their commutator is given by the formula

$$[s, t] = \prod_{i,j} (s_i, t_j)^{b_{ij}} = \prod_i \left(s_i, \prod_j t_j^{b_{ij}} \right), \tag{1}$$

where the integers b_{ij} are defined in terms of the bilinear form B by

$$B\left(\sum_i x_i e_i, \sum_j y_j e_j \right) = \sum_{i,j} b_{ij} x_i y_j .$$

We first discuss Theorem 1 in the case where $\mathbf{G}$ is semisimple and simply connected. In this case, our desired central extensions of G by μ_n are well known to exist. Steinberg [Ste62] computed the universal central extension of G, and Matsumoto [Mat69] computed the kernel of this extension, showing it to be equal to the group $K_2(F)$ except when G is of symplectic type, in which case $K_2(F)$ is canonically a quotient of this kernel. In fact, the adjoint group always acts by conjugation on the universal central extension, and the group of coinvariants is always $K_2(F)$. More recently, Brylinski and Deligne [BD01] have generalised this construction, proving the result we quoted above as Proposition 1.

Theorem 1 is thus known in this semisimple simply connected case. The desired commutator relation upon restriction to the torus is proved in [BD01, Proposition 3.14].

It is most important to us that this central extension is both derived from a solution to a universal problem (so that any automorphism of G lifts to an automorphism of the extension), and that this extension has no automorphisms.

We now discuss Theorem 1 in the case where $\mathbf{G} = \mathbf{T}$ is a torus. In this case, the construction of [BD01] does not produce all natural central extensions. For example, even in the case of $\mathbf{GL}_1$ over a field of Laurent series, the work of [AdCK89] produces a central extension whose commutator is the Hilbert symbol, whereas K_2-based methods only produce the square of this extension. The author does not know how to generalise this construction to the case of mixed characteristic, so we shall proceed in the following ad hoc manner, making use of the assumption that $\mu_{2n} \subset F$. We will write $(\cdot,\cdot)_{2n}$ for the Hilbert symbol with values in μ_{2n} and reserve $(\cdot,\cdot)$ for the Hilbert symbol with values in μ_n.

We can associate to $2B$ a quadratic form Q and consider the corresponding central extension

$$1 \to \mu_{2n} \to E \to T \to 1.$$

This extension may be either constructed from the work of [BD01], or we may construct it explicitly from the 2-cocycle

$$\sigma(s,t) = \prod_{i \le j} (s_i, t_j)_{2n}^{q_{ij}}$$

where $Q(\sum_i y_i e_i) = \sum_{i \le j} q_{ij} y_i y_j$.

This central extension has commutator

$$[x^\lambda, y^\mu] = (x,y)_{2n}^{2B(\lambda,\mu)} = (x,y)^{B(\lambda,\mu)}.$$

We will realise $\widetilde{T}$ as an index 2 subgroup of E.

Since the commutator takes values only in the subgroup μ_n of μ_{2n}, when we quotient out by μ_n, we obtain a central extension

$$1 \to \mu_2 \to E' \to T \to 1,$$

where the group E' is abelian.

The central extension splits over $\mathbf{T}(O_{\mathrm{F}})$ since the Hilbert symbol is trivial on $O_{\mathrm{F}}^\times \times O_{\mathrm{F}}^\times$.

Now choose a splitting of $\mathbf{T}(O_{\mathrm{F}})$ and quotient E' by its image. We arrive at a central extension

$$1 \to \mu_n \to E'' \to T/\mathbf{T}(O_{\mathrm{F}}) \to 1.$$

This is a short exact sequence of abelian groups with the last term free. Hence, it splits, so we get a surjection $E \twoheadrightarrow \mu_2$. Taking the kernel of this surjection as $\widetilde{T}$, we get our desired central extension of T by μ_n.

We now complete the proof of Theorem 1 for an arbitrary split reductive $\mathbf{G}$ by a reduction to the semisimple simply connected case and the torus case. This argument follows the proof of [FL10, Proposition 1]. For our purposes here, it will be most convenient for us to reinterpret a central extension of a group H by a group A as a group morphism $H \to BA$, where BA is the (Milnor) classifying space of A. We caution the reader that $f : H \longrightarrow BA$ is not a homomorphism of topological groups but instead is only a group homomorphism up to homotopy. For example, the two maps $H \times H \to BA$ given by $(f \times f) \circ m$ and $m \circ f$, where m is the multiplication, are not equal but are homotopic.

We briefly indicate how this translation works. By definition, a morphism $H \to BA$ is an A-torsor over H. Since A is abelian, the multiplication map from $A \times A$ to A is a group homomorphism, and this induces a group structure on BA. The extra structure of a group homomorphism on the map $H \to BA$ is what yields the datum of a group structure on the total space of the torsor represented by this map. It is this group structure on this total space which is the central extension of H by A.

Let G^{sc} be the simply connected cover of the derived group of G, and let T^{sc} be the inverse image of T in G^{sc}. Suppose that we have two central extensions $G^{\text{sc}} \to B\mu_n$ and $T \to B\mu_n$ which are isomorphic upon restriction to T^{sc}. Suppose furthermore that the extension $G^{\text{sc}} \to B\mu_n$ is invariant under the conjugation action of T on G^{sc} and the trivial T-action on μ_n. To this set of data, we construct a central extension $G \to B\mu_n$.

Consider the semidirect product $G^{\text{sc}} \rtimes T$ where T is acting on G^{sc} by conjugation. The datum we have of group morphisms $T \to B\mu_n$ and $G^{\text{sc}} \to B\mu_n$ with the latter being T-equivariant is equivalent to that of a group morphism $G^{\text{sc}} \rtimes T \to B\mu_n$.

Now consider the multiplication map from $G^{\text{sc}} \rtimes T$ to G. This is a surjective group homomorphism with kernel isomorphic to T^{sc} embedded inside $G^{\text{sc}} \rtimes T$ via $t \mapsto (t, t^{-1})$.

If we assume that the restrictions of $G^{\text{sc}} \to B\mu_n$ and $T \to B\mu_n$ to T^{sc} are assumed isomorphic, the restriction of $G^{\text{sc}} \rtimes T \to B\mu_n$ is a trivial central extension. Choosing a trivialisation, we may thus factor $G^{\text{sc}} \rtimes T \to B\mu_n$ through the quotient G of $G^{\text{sc}} \rtimes T$ by T^{sc} and thus we get our desired extension $G \to B\mu_n$.

Now from our knowledge of the semisimple simply connected case and the torus case, associated to B, we may construct our desired central extensions of G^{sc} and T by μ_n. It remains to show that these two extensions agree upon restriction to T^{sc} and that $G^{\text{sc}} \to B\mu_n$ is invariant under the conjugation action of T.

The former property may be seen by an analogous argument to the one appearing in the proof of the torus case: both extensions split over $\mathbf{T}^{\text{sc}}(O_{\text{F}})$ and have the same commutator, so their difference in $H^2(T, \mu_n)$ is abelian, splitting over $\mathbf{T}^{\text{sc}}(O_{\text{F}})$, hence trivial. The latter property holds since the extension by K_2 is a canonically constructed object that has no automorphisms. Consequently, the action of T on G^{sc} by conjugation extends to an action on the cover $\widetilde{G}^{\text{sc}}$ that is trivial when restricted to the central μ_n. This completes our proof of Theorem 1.

If we restrict ourselves to the case where SL_2, then we can be very explicit about our extension. Following Kubota [Kub69], we have the following formula for $\sigma \in H^2(SL_2(F), \mu_n)$ such that multiplication in $\widetilde{G}$ is given by $(g_1, \zeta_1)(g_2, \zeta_2) = (g_1 g_2, \zeta_1 \zeta_2 \sigma(g_1, g_2))$:

$$\sigma(g, h) = \left(\frac{x(gh)}{x(g)}, \frac{x(gh)}{x(h)} \right)^{Q(\alpha)}, \tag{2}$$

where for $g = \left(\begin{smallmatrix} a & b \\ c & d \end{smallmatrix}\right) \in SL_2(F)$, we define $x(g) = c$ unless $c = 0$ in which case $x(g) = d$. In this formula, α is a simple coroot.

13.4 Splitting Properties

A subgroup H of G is said to be split by the central extension if we have an isomorphism $p^{-1}(H) \simeq \mu_n \times H$ that commutes with the projection maps to H. In this section, we shall show that the unipotent subgroups and the maximal compact subgroup K are split in $\widetilde{G}$. We also discuss splittings of discrete subgroups corresponding to the coroot lattice and the Weyl group.

Proposition 2. *Any unipotent subgroup U of G is split canonically by the central extension $\widetilde{G}$.*

Proof. This result is proved in greater generality in [MW94, Appendix 1]. Since we are concerned only with the case where $(n, q) = 1$, a simple proof can be given. The assumptions on n ensure that the map $U \to U$ given by $u \mapsto u^n$ is bijective. If $u \in U$, write $u = u_1^n$ and let $\tilde{u_1}$ be any lift of u_1 to $\widetilde{G}$. Then define $s(u) = (\tilde{u_1})^n$. This is well defined, invariant under conjugation and is the proposed section determining the splitting.

So it suffices to show that s is a group homomorphism. For U abelian, this is trivial. In general, U is solvable, write U' for the quotient group $U/[U, U]$, and by induction, we may assume that s is a homomorphism when restricted to $[U, U]$. We now form the quotient $\widetilde{U}' = \widetilde{U}/s([U, U])$. Suppose $u_1, u_2 \in U$. Form $\zeta = s(u_1)s(u_2)s(u_1u_2)^{-1}$. A priori, we have $\zeta \in \mu_n$. Projecting into $\widetilde{U}'$ and using the abelian case of the proposition imply that $\zeta \in s([U, U])$ and thus we have $\zeta = 1$ so s is a homomorphism as desired. □

For the corresponding result for the maximal compact subgroup, the splitting is no longer unique, and in order for the splitting to exist, it is essential that n is coprime to the residue characteristic.

Theorem 2. *[BD01, Sect. 10.7], [Moo68, Lemma 11.3] The extension $\widetilde{G}$ of G splits over the maximal compact subgroup $K = \mathbf{G}(O_F)$.*

Proof. Let K_1 denote the kernel of the surjection $\mathbf{G}(O_F) \to \mathbf{G}(k)$. Then K_1 is a pro-p group, hence has trivial cohomology with coefficients in μ_n. By the Lyndon–Hochschild–Serre spectral sequence, we thus have an isomorphism $H^2(K, \mu_n) \simeq H^2(\mathbf{G}(k), \mu_n)$. Let $\mathbf{M}$ be the normaliser of $\mathbf{T}$ in $\mathbf{G}$. Since the index of $\mathbf{M}(k)$ in $\mathbf{G}(k)$ is coprime to n, we know that the map $H^2(\mathbf{G}(k), \mu_n) \to H^2(\mathbf{M}(k), \mu_n)$ is injective. $\mathbf{T}(k)$ can be considered as the k points of the group scheme of $(q-1)$th roots of unity in $\mathbf{T}$, which is etale, so lifts uniquely into $\mathbf{T}(O_F)$. The group generated by this lift together with the elements $w_\alpha \in K$ form a lift of $\mathbf{M}(k)$ into K. Thus, it suffices to show that our central extension is trivial when restricted to this lift of $\mathbf{M}(k)$ in K. However, we have explicit knowledge of the central extension in terms of a 2-cocycle on M thanks to [Mat69, Lemme 6.5] (the non-simply connected case works similarly), allowing us to complete the proof in this manner. □

When needed, we will denote by κ^* the lifting of K to $\widetilde{G}$. This lifting κ^* is not unique, being well defined only up only to a homomorphism from K to μ_n.

In [KP84], a canonical choice is made in the case of $G = GL_n$; however, this failure of uniqueness shall not be of concern to us.

Just as in the case of $G = SL_2$ when we were able to provide an explicit formula for the cocycle σ, again in this case we are able to provide an explicit formula for the splitting κ, following Kubota [Kub69]. Writing s for the section $g \mapsto (g, 1)$, and κ^* as $\kappa^*(k) = s(k)\kappa(k)$, we have

$$\kappa\begin{pmatrix} a & b \\ c & d \end{pmatrix} = \begin{cases} (c,d)^{Q(\alpha)} & \text{if } 0 < |c| < 1 \\ 1 & \text{otherwise,} \end{cases} \tag{3}$$

for all $\left(\begin{smallmatrix} a & b \\ c & d \end{smallmatrix}\right) \in SL_2(O_F)$, where again α is the simple coroot.

The group Y, considered as a subgroup of T by the injection $\lambda \mapsto \varpi^\lambda$, is trivially split in our central extension since it is a free abelian group whose cover is abelian (here we require the fact that $\mu_{2n} \subset F$ to conclude that the Hilbert symbols appearing in the commutator vanish).

Let W_0 be the subgroup of G generated by all elements of the form w_α for α a coroot. This group W_0 is a finite cover of the Weyl group W. It is split in our extension since it is a subgroup of K. Let $W_{a,0}$ be the corresponding cover of the affine Weyl group, that is, the group generated by Y and W_0. We can see this subgroup of G is also split in our extension by following through the construction of [Mat69, Lemme 6.5], noting that the Weyl group action on Y preserves its splitting. For any $w \in W_{a,0}$, we will identify w with its image under this splitting in $\widetilde{G}$, and we will denote this splitting by s when necessary.

13.5 Heisenberg Group Representations

A Heisenberg group is defined to be any two-step nilpotent subgroup, that is, H is a Heisenberg group if its commutator subgroup is central. The metaplectic torus $\widetilde{T}$ is an example of such a group, since the commutator subgroup $[\widetilde{T}, \widetilde{T}]$ is contained in the central μ_n. The representation theory of Heisenberg groups is well understood; in particular, we will make use of the following version of the Stone–von Neumann theorem, compare, for example, with [Wei09, Theorem 3.1].

Theorem 3. *Let H be a Heisenberg group with centre Z such that H/Z is finite and let χ be a character of Z. Suppose that* $\ker(\chi) \cap [H, H] = \{1\}$. *Then up to isomorphism, there is a unique irreducible representation of H with central character χ. It can be constructed as follows: Let A be a maximal abelian subgroup of H and let χ' be any extension of χ to A. Then inducing this representation from A to H produces the desired representation.*

Proof. Let π be an irreducible H-representation with central character χ. Since H/Z is finite, π is finite dimensional. Considering π as an A-representation, this implies that it has a one-dimensional quotient χ' which must be an extension of the character χ to A.

By Frobenius reciprocity, there is thus a non-trivial H-morphism from π to $\mathrm{Ind}_A^H \chi'$. To conclude that π is isomorphic to $\mathrm{Ind}_A^H \chi'$, we need to prove that this induced representation is irreducible.

Since $\mathrm{Ind}_A^H \chi'$ is generated by a non-zero function supported on A, to prove irreducibility, it suffices to show that any H-invariant subspace contains such a function. So suppose $f \neq 0$ is in an H-invariant subspace M. Then by translating by an element of H, we may assume that the support of f contains A. Now suppose that there exists $h \in H \setminus A$ such that $f(h) \neq 0$. Then since A is a maximal abelian subgroup of H, there exists $a \in A$ such that $[h, a] \neq 1$. Now consider $(a - \epsilon([a,h])\chi'(a))f \in M$. It has strictly smaller support than f (since it vanishes at h) and is non-zero on A. So continual application of this method will, by finite dimensionality, produce a non-zero function in M supported on A and thus we have proved the irreducibility of $\mathrm{Ind}_A^H \chi'$.

To finish the proof of the theorem, we need to show that if we have two different extensions χ_1 and χ_2 of χ to A, then after inducing to H, we get isomorphic representations. Given two such extensions, $\chi_1\chi_2^{-1}$ is a character of A/Z, and we extend it to a character of H/Z. Our assumption that $\ker(\chi)\cap[H,H] = \{1\}$ implies that the pairing $\langle\cdot,\cdot\rangle : H/Z \times H/Z \longrightarrow \mathbb{C}^\times$ given by $\langle h_1, h_2\rangle = \chi([h_1,h_2])$ is nondegenerate and hence that every character of H/Z is of the form $h \mapsto \chi([h,x])$ for some $x \in H$. Hence, there exists $x \in H$ such that $\chi_1\chi_2^{-1}(a) = \chi([a,x])$ for all $a \in A$. This implies that the characters χ_1 and χ_2 conjugate by x under the conjugation action of H on A. Hence, when induced to representations of H, the induced representations are isomorphic. □

Corollary 1. *Genuine representations of $\widetilde{T}$ are parametrised by characters of $Z(\widetilde{T})$.*

Proof. $\widetilde{T}$ is a Heisenberg group so we only need to check that the conditions of the above theorem are satisfied. The condition that $\widetilde{T}/Z(\widetilde{T})$ is finite is satisfied since T^n (the subgroup of nth powers in T) is central and $(F^\times)^n$ is of finite index in $F^\times$. The condition that $[H,H]\cap\ker(\chi) = \{1\}$ is satisfied for genuine characters χ, since $[\widetilde{T},\widetilde{T}] \subset \mu_n$ and χ is faithful on μ_n. Hence, we may apply Theorem 3 to obtain our desired corollary. □

We now produce explicitly a choice of maximal abelian subgroup of $\widetilde{T}$ that we can use later on for our convenience.

Lemma 1. *The group $C_{\widetilde{T}}(\widetilde{T} \cap K)$ is a maximal abelian subgroup of $\widetilde{T}$.*

Proof. Since $\widetilde{T} \cap K$ is abelian, it is clear that any maximal abelian subgroup of $\widetilde{T}$ containing $\widetilde{T} \cap K$ is contained in $C_{\widetilde{T}}(\widetilde{T} \cap K)$. So it suffices to prove that $C_{\widetilde{T}}(\widetilde{T} \cap K)$ is abelian.

Recall the basis $e_1, \ldots, e_r$ of Y which was used to introduce coordinates on T and the coefficients b_{ij} of the bilinear form B introduced at the beginning of Sect. 13.3. We need to make use of the equation (1), which we reproduce here for convenience:

$$[s,t] = \prod_{i,j} (s_i, t_j)^{b_{ij}} = \prod_i \left(s_i, \prod_j t_j^{b_{ij}} \right). \tag{4}$$

Thus, the condition for t to be in $C_{\widetilde{T}}(\widetilde{T} \cap K)$ is that $\prod_j t_j^{b_{ij}}$ has valuation divisible by n for all i. Now suppose that s and t are elements of $C_{\widetilde{T}}(\widetilde{T} \cap K)$. Let x_i and y_i be the valuations of s_i and t_i, respectively. Then, we have that (s_i, t_j) is equal to $((-1)^{x_i y_j} s_i^{y_j} / t_j^{x_i})^{\frac{q-1}{n}}$ after reduction modulo ϖ. Hence, we may compute the commutator

$$[s,t] = \left(\prod_{i,j} (-1)^{x_i y_j} \prod_i s_i^{\sum_j y_j b_{ij}} \prod_j t_j^{\sum_i x_i b_{ij}} \right)^{\frac{q-1}{n}}.$$

Since we are assuming that $2n|q-1$, the power of -1 which appears in this product is even. By assumption on s and t, all exponents of s_i and t_j are divisible by n. So the whole product is a $q-1$-th power, so after reduction modulo ϖ becomes 1. Hence, s and t commute, so $C_{\widetilde{T}}(\widetilde{T} \cap K)$ is abelian, as required. □

We will use H to denote this maximal abelian subgroup $C_{\widetilde{T}}(\widetilde{T} \cap K)$.

13.6 Principal Series Representations

We begin by studying the class of representations that will be our main object of study. Let (π, V) be a genuine, smooth admissible representation of $\widetilde{T}$. The group $\widetilde{B}$ contains U as a normal subgroup with quotient naturally isomorphic to $\widetilde{T}$. Via this quotient, we consider (π, V) as a representation of $\widetilde{B}$ on which U acts trivially.

Definition 1. For (π, V) a smooth representation of $\widetilde{T}$, we define the (normalised) induced representation $I(V)$ as follows:

The space of $I(V)$ is the space of all locally constant functions $f : \widetilde{G} \longrightarrow V$ such that

$$f(bg) = \delta^{1/2}(b)\pi(b)f(g)$$

for all $b \in \widetilde{B}$ and $g \in \widetilde{G}$ where δ is the modular quasicharacter of $\widetilde{B}$ and we are considering (π, V) as a representation of $\widetilde{B}$. The action of $\widetilde{G}$ on $I(V)$ is given by right translation. In this way, we define an induction functor

$$I : \mathrm{Rep}\,(\widetilde{T}) \longrightarrow \mathrm{Rep}\,(\widetilde{G}).$$

Suppose now that χ is a genuine character of $Z(\widetilde{T})$. We denote by $i(\chi) = (\pi_\chi, V_\chi)$ a representative of the corresponding isomorphism class of irreducible representations of $\widetilde{T}$ with central character χ. By the considerations in the above section, $i(\chi)$ is finite dimensional. We will write $I(\chi)$ for the corresponding induced representation $I(i(\chi))$ of $\widetilde{G}$. Such representations $I(\chi)$ will be called *principal series representations*.

We now define a family of principal series representations, called *unramified*, that will be of principal interest to us.

Definition 2. A genuine character χ of $Z(\widetilde{T})$ is said to be unramified if it has an extension to H that is trivial on $\widetilde{T} \cap K$. We use the same adjective unramified for the corresponding representation $I(\chi)$ of $\widetilde{G}$.

Lemma 2. *An unramified principal series representation $I(\chi)$ has a one-dimensional space of K-fixed vectors.*

Proof. Suppose $f \in I(\chi)^K$. Let $g = f(1) \in i(\chi)$. By the Iwasawa decomposition $G = UAK$, we may write any $g \in \widetilde{G}$ as $g = uak$ with $u \in U$, $a \in \widetilde{T}$ and $k \in K$. Then we have $f(g) = f(uak) = \pi_\chi(a)g$.

The element a is well defined up to right multiplication by an element $\eta \in \widetilde{T} \cap K$. This induces (the only) compatibility condition, we thus require that $f(g) = \pi_\chi(a\eta)g = \pi_\chi(a)\pi_\chi(\eta)g$. Thus, we have that $f \mapsto f(1)$ is an isomorphism from $I(\chi)^K$ to $i(\chi)^{\widetilde{T}\cap K}$.

If $g \in i(\chi)^{\widetilde{T}\cap K}$, then for all $t \in \widetilde{T}$ and $\eta \in \widetilde{T} \cap K$ we have

$$g(t) = g(t\eta) = [t,\eta]g(\eta t) = [t,\eta]\chi(\eta)g(t).$$

Since χ is unramified, $\chi(\eta) = 1$, so either $[t,\eta] = 1$ or $g(t) = 0$. The function $g \in i(\chi)$ is determined by its restriction to a set of coset representatives of $H\backslash\widetilde{T}$. By the definition of H we know that $[t,\eta] = 1$ for all $\eta \in \widetilde{T} \cap K$ if and only if $t \in H$. Thus, we have shown that $i(\chi)^{\widetilde{T}\cap K}$ is one dimensional, proving the lemma. □

A K-fixed vector in such a representation will be called *spherical*.

We now define an action of the Weyl group W on principal series representations.

The group $\widetilde{M}$ acts on $\widetilde{T}$ by conjugation, and hence acts on $\text{Rep}(\widetilde{T})$. Explicitly, write $c_m : \widetilde{T} \longrightarrow \widetilde{T}$ for the operation of conjugation by $m \in \widetilde{M}$ on $\widetilde{T}$. Then for (π, V), the action of $m \in \widetilde{M}$ is defined by $(\pi, V)^m = (\pi^m, V^m)$ where $V^m = V$ and π^m is the composition $\pi \circ c_m : \widetilde{T} \longrightarrow \text{Aut}(V)$.

Unfortunately, when we restrict this action to $\widetilde{T}$, we do not obtain the identity. However, we may still define an action of the Weyl group on $\text{Rep}(\widetilde{T})$. Recall from the discussion at the end of Sect. 13.4, that the group W_0 lifts to $\widetilde{M}$. In this realisation of W_0, the kernel of the surjection $W_0 \to W$ lies in $Z(\widetilde{T})$. Since the conjugation action of $Z(\widetilde{T})$ is trivial, we are able to define an action of W on $\text{Rep}(\widetilde{T})$ by first restricting the action of $\widetilde{M}$ to an action of W_0, which then induces a well-defined action of W on $\text{Rep}(\widetilde{T})$.

In a similar but simpler manner, one may define an action of W on the space of characters of $Z(\widetilde{T})$. These two actions are compatible in the sense that $i(\chi^w) = i(\chi)^w$ for all characters χ and $w \in W$.

To proceed, we require the theory of the Jacquet functor, and the results regarding the composition of the Jacquet functor with the induction functor I. These results are all contained in [BZ77].

Definition 3. The Jacquet functor J from Rep $(\widetilde{G})$ to Rep $(\widetilde{T})$ is defined to be the functor of U-coinvariants.

Explicitly, for an object V in Rep $(\widetilde{G})$, $J(V)$ is defined to be the largest quotient of V on which U acts trivially, that is, we quotient out by the submodule generated by all elements of the form $\pi(u)v - v$ with $u \in U$ and $v \in V$. Since $\widetilde{T}$ normalises U (as U has a unique splitting, conjugation by $\widetilde{T}$ must preserve this splitting), the action of $\widetilde{T}$ on V induces an action on $J(V)$, so the image of J is indeed in Rep $(\widetilde{T})$.

The work in [BZ77] is in sufficient generality to cover our circumstances. In particular, we have the following two propositions.

Proposition 3. *[BZ77, Proposition 1.9(a)] The Jacquet functor is exact.*

Proposition 4. *[BZ77, Proposition 1.9(b)] The Jacquet functor J is left adjoint to the induction functor I. That is, there exists a natural isomorphism*

$$Hom_{\widetilde{T}}(J(V), W) \cong Hom_{\widetilde{G}}(V, I(W)).$$

The main result of [BZ77] is their Theorem 5.2, from which we derive the following important corollary.

Corollary 2. *The composition factors of the Jacquet module $J(I(\chi))$ are given by $i(\chi^w)$ as w runs over W.*

Proof. The derivation of this corollary follows in exactly the same manner as in the reductive case. In the notation of Bernstein and Zelevinsky [BZ77], we apply their Theorem 5.2 with $\mathbf{G} = \widetilde{G}$, $\mathbf{Q} = \mathbf{P} = \widetilde{B}$, $\mathbf{N} = \mathbf{M} = \widetilde{T}$, $\mathbf{V} = \mathbf{U} = U$ and $\psi = \theta = 1$. □

We say that a character χ of $Z(\widetilde{T})$ is regular if $\chi^w \neq \chi$ for all $w \neq 1$.

Proposition 5. *If χ is regular, then $J(I(\chi))$ is semisimple.*

Proof. Decompose the $\widetilde{T}$-module $J(I(\chi))$ into $Z(\widetilde{T})$-eigenspaces—this must be a semisimple decomposition since we are dealing with commuting operators on a finite-dimensional space. As χ is assumed to be regular, these eigenvalues of $Z(\widetilde{T})$ are all distinct. This shows that the filtration from Corollary 2 splits as $\widetilde{T}$-modules, so we are done. □

13.7 Intertwining Operators

We start with these results on the spaces of morphisms between various principal series representations.

Theorem 4. *1. For two characters χ_1 and χ_2 of $Z(\widetilde{T})$, we have*

$$Hom_{\widetilde{G}}(I(\chi_1), I(\chi_2)) = 0$$

unless there exists $w \in W$ such that $\chi_2 = \chi_1^w$.
2. Suppose that χ is regular. Then for all $w \in W$ we have

$$\dim Hom_{\widetilde{G}}(I(\chi), I(\chi^w)) = 1.$$

Proof. Since J is left adjoint to I, we have

$$\mathrm{Hom}_{\widetilde{G}}(I(\chi_1), I(\chi_2)) = \mathrm{Hom}_{\widetilde{T}}(J(I(\chi_1)), i(\chi_2)).$$

Our knowledge of the description of the composition series of $J(I(\chi_1))$ from Corollary 2 and Proposition 5 completes the proof. □

This section will be dedicated to the explicit construction and analysis of these spaces $\mathrm{Hom}_{\widetilde{G}}(I(\chi), I(\chi^w))$. Elements in these spaces are referred to as intertwining operators.

Suppose $s \in \mathbb{C}$. Associated to s is a one-dimensional representation δ^s of $\widetilde{B}$ given by raising the modular quasicharacter δ to the sth power. Accordingly, given any representation V of $\widetilde{T}$, we define a family I_s of representations of $\widetilde{G}$ by

$$I_s(V) = I(V \otimes \delta^s).$$

For each $V \in \mathrm{Rep}(\widetilde{T})$, this family of representations is a trivialisable vector bundle over $\mathbb{C}$. We choose a trivialisation as follows:

To each $f \in I(V) = I_0(V)$ and $s \in C$, we define the element $f_s \in I_s(V)$ by

$$f_s(bk) = \delta(b)^s f(bk) \tag{5}$$

for any $b \in \widetilde{B}$ and $k \in K$. It is easily checked that this is well defined, the claim $f_s \in I_s(V)$ is true and that $s \mapsto f_s$ does define a section.

Our strategy for constructing intertwining operators is as follows: We shall first construct intertwining operators via an integral representation that is only absolutely convergent on a cone in the set of all possible characters χ. We then make use of the trivialising section we have just constructed to meromorphically continue these intertwining operators to all $I(\chi)$.

For any finite-dimensional $\widetilde{T}$ representation (π, V), and any coroot α, we define the α-radius $r_\alpha(V)$ to be the maximum absolute value of an eigenvalue of the operator $\pi(\varpi^\alpha)$ on V. This turns out to be independent of the choice of uniformiser ϖ since $T(O_F)$ is compact.

For $w \in W$ and such a finite-dimensional representation (π, V), the intertwining operator $T_w : I(V) \longrightarrow I(V^w)$ is defined by the integral

$$(T_w f)(g) = \int_{U_w} f(w^{-1}ug)\mathrm{d}u. \tag{6}$$

whenever this is absolutely convergent.

To check that T_w does indeed map $I(V)$ into $I(V^w)$ is a simple calculation. Note that the underlying vector spaces of V and V^w are equal as per the definition of the W-action on such representations in Sect. 13.6.

Lemma 3. *Suppose that $w_1, w_2 \in W$ are such that $\ell(w_1w_2) = \ell(w_1)+\ell(w_2)$. Then $T_{w_1w_2} = T_{w_1}T_{w_2}$, whenever their defining integrals are absolutely convergent.*

Proof. This result is a simple application of Fubini's theorem. □

Let us now restrict ourselves to a study of the case where $w = w_\alpha$ is the simple reflection corresponding to the simple coroot α.

Theorem 5. *The defining integral (6) for the intertwining operator T_{w_α} is absolutely convergent for $r_\alpha(V) < 1$.*

Proof. In SL_2, we have the following identity:

$$\begin{pmatrix} 0 & -1 \\ 1 & 0 \end{pmatrix}\begin{pmatrix} 1 & x \\ 0 & 1 \end{pmatrix} = \begin{pmatrix} 1/x & -1 \\ 0 & x \end{pmatrix}\begin{pmatrix} 1 & 0 \\ 1/x & 1 \end{pmatrix}.$$

We apply the morphism ϕ_α to interpret this as an identity in G. This equation lifts to $\widetilde{G}$ as the relevant Kubota cocycles are trivial. We are thus able to write

$$\begin{aligned}(T_{w_\alpha}f)(g) &= \int_F f(w_\alpha^{-1}e_\alpha(x)g)\mathrm{d}x \\ &= \int_F f(e_\alpha(-1/x)x^\alpha e_{-\alpha}(1/x)g)\mathrm{d}x \\ &= \int_F \delta^{1/2}(x^\alpha)\pi(x^\alpha)f(e_{-\alpha}(1/x)g)\mathrm{d}x.\end{aligned}$$

In the above, $e_\gamma(x)$ is the canonical lift from G to $\widetilde{G}$ of the one-dimensional unipotent subgroup corresponding to the coroot γ, as defined in Sect. 13.2.

Since f is locally constant, there exists a positive number N such that for $|x| \geq N$ we have $f(e_{-\alpha}(1/x)g) = f(g)$. Now we shall break up our integral over F into a sum of two integrals, the first over $|x| < N$ and the second over $|x| \geq N$. The first integral is an integral over a compact set so is automatically convergent. We will now study the second integral in greater detail.

Note that $\varpi^{n\alpha}$ is central in $\widetilde{T}$. We may assume without loss of generality that $f(g) \in V$ is an eigenvector of $\pi(\varpi^{n\alpha})$ with corresponding eigenvalue $(q^{-1}x_\alpha)^n$. Then our second integral becomes

$$\int_{|x|\geq N}(\delta^{1/2}\pi)(x^{\alpha})f(e_{-\alpha}(1/x)g)\mathrm{d}x$$
$$=\left(\int_{m\leq v(x)<m+n}(\delta^{1/2}\pi)(x^{\alpha})f(g)\right)\left(\sum_{i=0}^{\infty}x_{\alpha}^{\mathrm{in}}\right). \tag{7}$$

This is absolutely convergent if and only if $|x_\alpha| < 1$, proving the theorem. □

For ease of exposition, we shall now restrict ourselves to the case of intertwining operators from $I(\chi)$ to $I(\chi^w)$. Under this restriction, the complex numbers x_α are essentially well defined, in that different choices of eigenvectors will only change them by an nth root of unity. We may pick any such eigenvector to define the x_α, any subsequent formulae will be independent of such choices.

Define a renormalised version of the intertwining operator by

$$\widetilde{T}_w = \prod_{\substack{\alpha>0\\ w\alpha<0}}(1-x_a^n)T_w. \tag{8}$$

Lemma 4. *The collection of renormalised intertwining operators $\widetilde{T}_{w_\alpha}$ satisfies the braid relations.*

Proof. Since we know that the unnormalised intertwining operators T_{w_α} satisfy the braid relations, to check this lemma, it suffices to check that $c(w_1w_2, x) = c(w_1, w_2x)c(w_2, x)$ where $c(w, x)$ is the renormalising coefficient in (8). This is a triviality. □

We are now in a position to analytically continue the intertwining operators $\widetilde{T}_w$.

If λ is the eigenvalue of $\pi(\varpi^{n\alpha})$ on V, then λq^{-s} is the eigenvalue of $\pi(\varpi^{n\alpha})$ on $V\otimes\delta^s$. Then by (7), $(\widetilde{T}_{w_\alpha}f_s)(g)$ is a polynomial in λq^{-s}, so in particular is a holomorphic function in s. Recall that the section f_s is as defined in (5).

For $\Re(s)$ sufficiently large, the defining integral for $\widetilde{T}_{w_\alpha}f_s$ is absolutely convergent. Thus we can define $\widetilde{T}_{w_\alpha}f_s$ for all $s\in\mathbb{C}$ by analytic continuation. In particular, for all V, we have now defined

$$\widetilde{T}_{w_\alpha}: I(V)\longrightarrow I(V^{w_\alpha})$$

and since the maps $\widetilde{T}_{w_\alpha}$ satisfy the braid relations, we have also defined

$$\widetilde{T}_w: I(V)\longrightarrow I(V^w)$$

for all $w\in W$.

Now let us suppose that $V = i(\chi)$ is an irreducible unramified representation of $\widetilde{T}$. By Lemma 2, $I(V)$ contains a K-fixed vector. Let ϕ_K be such a vector for $I(V)$ and ϕ_K^w be such a vector for $I(V^w)$. We normalise these spherical functions such that

$(\phi_K(1_{\widetilde{G}}))(1_{\widetilde{T}}) = 1$. The spherical vectors ϕ_K and ϕ_K^w are related by $\widetilde{T}_w$ in a manner given by the following theorem. The integer n_α is defined to be $\frac{n}{(n,Q(\alpha))}$ where the notation $(\cdot,\cdot)$ here is that of the greatest common divisor.

Theorem 6. *[McN11, Theorem 6.5]*

$$\widetilde{T}_w\phi_K = \prod_{\alpha\in\Phi_w}\left(1-q^{-1}x_\alpha^{n_\alpha}\right)\frac{1-x_\alpha^n}{1-x_\alpha^{n_\alpha}}\phi_K^w.$$

Proof. The proof in [McN11] is for the case of G semisimple and simply connected, so we need to show how to reduce to this case. First, we note that $T_w\phi_K$ is a priori K-fixed, so by Lemma 2, it suffices to calculate the integral

$$I_\chi = \left(\int_U \phi_K(w^{-1}u)du\right)(1_{\widetilde{T}}).$$

Consider the natural map from the corresponding simply connected semisimple group G_{ss}^{sc} to G. We can pullback the central extension $\widetilde{G}$ of G to a central extension of G_{ss}^{sc} and thus consider the corresponding group H_{ss}^{sc}. The character χ of H can be extended to a character χ' of H_{ss}^{sc}. In calculating I_χ, only group elements in the image of G_{ss}^{sc} occur and we see that the calculation is the same as for the corresponding integral $I_{\chi'}$. In this way, this theorem is reduced to the semisimple, simply connected case. □

Corollary 3. *For generic χ (so on a Zariski open subset of such characters), the intertwining operator $\widetilde{T}_w$ induces an isomorphism $I(\chi) \simeq I(\chi^w)$.*

Proof. The functor $\widetilde{T}_w$ restricts to a morphism from $I(\chi)^K$ to $I(\chi^w)^K$. These two spaces are one dimensional, so we have an isomorphism as long as $\widetilde{T}_w\phi_K$ is non-zero. The Corollary now follows immediately from Theorem 6. □

At this point, we have developed the theory as far as is necessary for the purposes of the Satake isomorphism. Following the works of Casselman [Cas], Kazhdan–Patterson [KP84] and Rodier [Rod81], one could push this line of thought further to produce stronger results on the composition series of principal series representations, though we shall not do this here.

13.8 Whittaker Functions

In this section, we consider (π, V) a spherical genuine admissible representation of $\widetilde{G}$. Let ψ be a character of U such that the restriction of ψ to each one-dimensional subgroup U_α for α a simple coroot is non-trivial.

Let $\mathcal{W}$ denote the space of smooth functions $f : \widetilde{G} \longrightarrow \mathbb{C}$ such that $f(\zeta ng) = \zeta\psi(n)f(g)$ for $\zeta \in \mu_r$ and $n \in N$. Then a Whittaker model for (π, V) is defined to be a $\widetilde{G}$-morphism from V to $\mathcal{W}$. A Whittaker function is any non-zero spherical vector in a Whittaker model. It thus is a function $W_\chi : \widetilde{G} \longrightarrow \mathbb{C}$ satisfying

$$W_\chi(\zeta ngk) = \zeta\psi(n)W_\chi(g) \quad \text{for } \zeta \in \mu_r, n \in N, g \in G, k \in K. \tag{9}$$

Define the twisted Jacquet functor J_ψ from $\mathrm{Rep}\,(\widetilde{G})$ to $\mathrm{Vect}_{\mathbb{C}}$ by $J_\psi(V) = V/V_\psi(U)$, where $V_\psi(U)$ is the subspace of V generated by the vectors $\pi(u)v - \psi(u)v$ for all $u \in U$ and $v \in V$. There is a natural bijection between $J_\psi(V)$ and the vector space of Whittaker models of V.

Theorem 5.2 of [BZ77] can be used to compute the dimension of the space of Whittaker functions in the same manner as it was used to compute the composition series of a Jacquet module of an induced representation.

Theorem 7. *The dimension of the space of Whittaker functions for a principal series representation $I(\chi)$ is $|\widetilde{T}/H|$.*

We apply [BZ77, Theorem 5.2] with $\mathbf{G} = \widetilde{G}, \mathbf{P} = \widetilde{B}, \mathbf{M} = \widetilde{T}, \mathbf{U} = \mathbf{Q} = \mathbf{V} = U, \mathbf{N} = 1, \theta = 1$ and ψ non-trivial as above. Of the glued functors that appear in the composition series of $J_\psi(I(\chi))$ via [BZ77, Theorem 5.2], only one is non-zero, and it is the forgetful functor from $\mathrm{Rep}\,(\widetilde{T})$ to $\mathrm{Vect}_{\mathbb{C}}$.

If f is a spherical vector in $I(\chi)$, then we can construct a Whittaker function as the integral

$$W(g) = \int_U f(w_0ug)\psi(u)\mathrm{d}u.$$

Technically speaking, this is a $i(\chi)$-valued function, so to obtain a $\mathbb{C}$-valued Whittaker function, we should compose with a functional on $i(\chi)$. Such a choice is made in [McN11] where a complex-valued Whittaker function is evaluated. In fact, in [McN11], a basis for the space of Whittaker functions is computed together with the production of an explicit formula for $W(t)$ with $t \in \widetilde{T}$ in the case where $G = SL_n$. Note that by (9) and the Iwasawa decomposition, W is completely determined by its restriction to $\widetilde{T}$. There is an alternative method of Chinta and Offen [CO] for calculating these metaplectic Whittaker functions. Their method more closely follows the lines of the original work of Casselman and Shalika [CS80], again working in type A.

13.9 The Spherical Hecke Algebra

We call a complex-valued function f on $\widetilde{G}$ anti-genuine if, for all $\zeta \in \mu_n$ and $g \in \widetilde{G}$, we have $f(\zeta g) = \zeta^{-1}f(g)$. This notion is of use to us since we are only studying genuine representations of $\widetilde{G}$. If we decompose the algebra $C_c^\infty(\widetilde{G})$ of smooth compactly supported functions on $\widetilde{G}$ into a direct sum of eigenspaces under the action of μ_n, then only the anti-genuine functions act non-trivially on a genuine representation of $\widetilde{G}$. We now define and study a version of the spherical Hecke algebra for the metaplectic group.

Considering K as a subgroup of $\widetilde{G}$ via κ^*, let $\mathcal{H}(\widetilde{G}, K)$ denote the algebra of K-bi-invariant anti-genuine compactly supported smooth (locally constant) complex-valued functions. In other words, a compactly supported smooth function $f : \widetilde{G} \longrightarrow \mathbb{C}$ is in $\mathcal{H}(\widetilde{G}, K)$ if and only if $f(\zeta k_1 g k_2) = \zeta^{-1} f(g)$ for all $\zeta \in \mu_r$, $g \in G$ and $k_1, k_2 \in K$. The algebra structure is given by convolution; for $f_1, f_2 \in \mathcal{H}(\widetilde{G}, K)$, we define

$$(f_1 f_2)(g) = \int_{\widetilde{G}} f_1(h) f_2(h^{-1} g) \mathrm{d}h,$$

where the Haar measure on $\widetilde{G}$ is normalised such that $K \times \mu_n$ has measure 1.

We have the following two results about the structure of $\mathcal{H}(\widetilde{G}, K)$. In the case of $G = GL_n$, these appear in [KP86].

Theorem 8. *$\mathcal{H}(\widetilde{G}, K)$ is commutative.*

We will not prove this in this section, but instead note that it follows immediately from the Satake isomorphism, Theorem 10.

Theorem 9. *The support of $\mathcal{H}(\widetilde{G}, K)$ is given by $\mu_n KHK$.*

Proof. The Cartan decomposition $G = KTK$ implies that every (K, K) double coset of G contains a representative of the form ϖ^λ, and this decomposition clearly lifts to $\widetilde{G}$. So it suffices to find the set of λ for which the double coset $\mu_n K \varpi^\lambda K$ supports a function in $\mathcal{H}(\widetilde{G}, K)$.

Fix λ, and let K^λ denote the subgroup $K \cap \varpi^{-\lambda} K \varpi^\lambda$ of G. We define a function $\phi^\lambda : K^\lambda \longrightarrow \mu_n$ as follows. For $k \in K^\lambda$ there exists a unique $k' \in K$ such that $k\varpi^\lambda = \varpi^\lambda k'$. We lift this identity into $\widetilde{G}$ using our choice of splitting of K, and define $\phi^\lambda(k)$ by $k\varpi^\lambda = \phi(k)\varpi^\lambda k'$.

It is straightforward to check that ϕ^λ is a group homomorphism. Furthermore, there is a function in $\mathcal{H}(\widetilde{G}, K)$ supported on $\mu_n K \varpi^\lambda K$ if and only if the homomorphism ϕ^λ is trivial.

The normal subgroup $K_1 \cap K^\lambda$ of K^λ is a pro-p group; hence, the homomorphism ϕ^λ is trivial when restricted to this subgroup.

There is a canonicial isomorphism $K^\lambda/(K_1 \cap K^\lambda) \simeq \mathbf{P}(k)$ for some parabolic subgroup $\mathbf{P}$ of $\mathbf{G}$. The above shows that ϕ^λ factors to a homomorphism from $\mathbf{P}(k)$ to μ_n. The group $\mathbf{P}(k)$ is generated by $\mathbf{T}(k)$ and unipotent elements. Since ϕ^λ is necessarily trivial on any unipotent element, it is completely determined by its restriction to $\mathbf{T}(k)$.

We know that the restriction of ϕ^λ to $\mathbf{T}(O_\mathrm{F})$ is trivial if and only if $\varpi^\lambda \in H$, by the definition of H. This completes our proof. □

13.10 Satake Isomorphism

The approach we shall take in presenting the Satake isomorphism was learnt by the author from a lecture of Kazhdan in the reductive case, and differs from that which is generally considered as, for example, in [Gro98]. First, we define a free abelian group Λ which shall be of fundamental importance for the remainder of this chapter. Let

$$\Lambda = \{\lambda \in Y \mid s(\varpi^{\lambda}) \in H\} = \{x \in Y \mid B(x,y) \in n\mathbb{Z}\ \forall\ y \in Y\}.$$

The equivalence of the two given presentations is a consequence of the commutator formula (1). This group Λ is also naturally isomorphic to the abelian group $H/(\widetilde{T} \cap K \times \mu_n)$, and carries an action of the Weyl group, inherited from the action of W on T.

The aim of this section is to prove the following.

Theorem 10 (Satake Isomorphism). *Let $\mathbb{C}[\Lambda]$ denote the group algebra of Λ. Then there is a natural isomorphism between the spherical Hecke algebra $\mathcal{H}(\widetilde{G},K)$ and the W-invariant subalgebra, $\mathbb{C}[\Lambda]^W$.*

Let Z_Λ denote the complex affine variety $\mathrm{Hom}(\Lambda,\mathbb{C}^\times)$ and Γ_Λ be the ring of regular functions on Z_Λ. We shall first define a homomorphism from $\mathcal{H}(\widetilde{G},K)$ to Γ_Λ.

To any $\chi \in Z_\Lambda$, there is an associated genuine unramified principal series representation $I(\chi) = (\pi_\chi, V_\chi)$ of $\widetilde{G}$. By Lemma 2, this representation has the property that $\dim V_\chi^K = 1$, and thus V_χ^K is a one-dimensional representation of $\mathcal{H}(\widetilde{G},K)$. We again use $\pi_\chi : \mathcal{H}(\widetilde{G},K) \longrightarrow \mathrm{End}(V_\chi^K) \cong \mathbb{C}$ to denote this representation.

From this representation, we obtain a ring homomorphism $S : \mathcal{H}(\widetilde{G},K) \longrightarrow \Gamma_\Lambda$ given by $Sf(\chi) = \pi_\chi(f)$. This is the Satake map. A priori, the image of this map lies in the set of functions from Z_λ to $\mathbb{C}$, though it will follow from the results proven below that the image lies in the ring of regular functions on Z_λ.

For any abelian group Λ, there is a canonical isomorphism between Γ_Λ and the group ring of Λ (which is actually the same as given above, if we can take $\widetilde{G} = \Lambda$ in the definition of the Satake map).

Let us identify Γ_Λ with $\mathbb{C}[\Lambda]$ via this isomorphism. Using this, we will from now assume that S has image in $\mathbb{C}[\Lambda]$.

Lemma 5. *We have the following formula for the Satake map $S : \mathcal{H}(\widetilde{G},K) \longrightarrow \mathbb{C}[\Lambda]$:*

$$(Sf)(\lambda) = \delta^{1/2}(\varpi^{\lambda}) \int_U f(\varpi^{\lambda} u)du. \tag{10}$$

Proof. We begin by unfolding of the integral definition of the action of f on the spherical vector ϕ_K. From this we get

$$\begin{aligned}\pi_\chi(f)\phi_K &= \int_{\widetilde{G}} f(g)\pi_\chi(g)\phi_K \mathrm{d}g \\ &= \int_K \int_{\widetilde{B}} f(bk)\pi_\chi(bk)\phi_K d_L b \,\mathrm{d}k \\ &= \int_{\widetilde{B}} f(b)\pi_\chi(b)\phi_K d_L b \\ &= \int_{\widetilde{T}} \int_U f(tu)\pi_\chi(tu)\phi_K \mathrm{d}u \,\mathrm{d}t \\ &= \int_{\widetilde{T}} \left(\delta^{1/2}(t)\int_U f(tu)\mathrm{d}u\right)(\delta^{-1/2}\pi_\chi)(t)\phi_K \mathrm{d}t.\end{aligned}$$

It was shown in the proof of Lemma 2 that for $t \in \widetilde{T}$, $(\delta^{-1/2}\pi_\chi)(t)\phi_K = \chi(t)\phi_K$ if $t \in H$ and is zero otherwise. Thus, we may restrict our integral over $\widetilde{T}$ to an integral over H. Since the integrand is invariant under $\widetilde{T} \cap K \times \mu_n$, we obtain the following sum over Λ:

$$\pi_\chi(f) = \sum_{\lambda\in\Lambda} \delta^{1/2}(\varpi^\lambda)\int_U f(\varpi^\lambda u)\mathrm{d}u\,\chi(\varpi^\lambda).$$

Under the isomorphism $\Gamma_\Lambda \simeq \mathbb{C}[\Lambda]$, this gives us (10) as required. □

Lemma 6. *The image of the Satake map lies in* Γ_Λ^W.

Proof. By Corollary 3, we have, for generic χ, an isomorphism between $I(\chi)^K$ and $I(\chi^w)^K$. Thus, the image of the Satake map is W-invariant. To complete the proof, it remains to show that the image of S consists of regular functions on Z_λ (or equivalently that $(Sf)(\lambda)$ is non-zero for only finitely many λ). For this we use the integral expression from Lemma 5. To see this, we need to remark that any $f \in \mathcal{H}(\widetilde{G}, K)$ is compactly supported and used [BT72, Proposition 4.4.4(i)]. □

Theorem 11. *The Satake map S gives an isomorphism between $\mathcal{H}(\widetilde{G}, K)$ and $\mathbb{C}[\Lambda]^W$.*

Proof. For dominant $\lambda \in \Lambda$, we define basis elements c_λ and d_λ of $\mathcal{H}(\widetilde{G}, K)$ and $\mathbb{C}[\Lambda]^W$, respectively.

Let c_λ be the function in $\mathcal{H}(\widetilde{G}, K)$ that is supported on $\mu_n K\varpi^\lambda K$ and takes the value 1 at $s(\varpi^\lambda)$. That the set of all such c_λ form a basis of $\mathcal{H}(\widetilde{G}, K)$ is known from Theorem 9.

Let $d_\lambda \in \mathbb{C}[\Lambda]^W$ be the characteristic function of the orbit $W\lambda$.

Write $Sc_\lambda = \sum_\mu a_{\lambda\mu}d_\mu$. We shall show that $a_{\lambda\lambda} \neq 0$ and that $a_{\lambda\mu} = 0$ unless $\mu \leq \lambda$, which suffices to prove that S is bijective. Since we already know that S is a homomorphism, this is sufficient to prove our theorem.

To show that $a_{\lambda\lambda} \neq 0$, we must calculate $Sc_\lambda(\lambda)$. Notice that for $u \in U$, we have $\varpi^\lambda u \in K\varpi^\lambda K$ if and only if $u \in K$ so in the calculation of the integral (10), the integrand is non-zero only on $K \cap U$, where it takes the value 1; hence, the integral is non-zero, so $a_{\lambda\lambda} \neq 0$ as desired.

To show that $a_{\lambda\mu} = 0$ unless $\mu \leq \lambda$, we again look at calculating $Sc_\lambda(\mu)$ via the integral (10). We again appeal to a result from the structure theory of reductive groups over local fields [BT72, Proposition 4.4.4(i)] to say that $\varpi^\mu U \cap K\varpi^\lambda K = 0$ unless $\mu \leq \lambda$, which immediately gives us our desired vanishing result, so we are done. □

13.11 The Dual Group to a Metaplectic Group

Motivated by the Satake isomorphism in the previous section, we will now give a combinatorial definition of a dual group to a metaplectic group. This group $\widetilde{G}^\vee$ will be a split reductive group, so to define it, it will suffice to give a root datum (X, Φ, X', Φ').

We use Δ to denote the set of all coroots. Throughout this section, lower-case Greek letters will be used to denote coroots. If α is a simple coroot, recall that the integer n_α is defined to be the quotient $n_\alpha = \frac{n}{(n,Q(\alpha))}$ (where $(\cdot,\cdot)$ here is used to denote the greatest common divisor).

We define a root datum (X, Φ, X', Φ') by

$$\begin{aligned} X &= \Lambda, \\ \Phi &= \{n_\alpha \alpha \mid \alpha \in \Delta\}, \\ X' &= \mathrm{Hom}\,(\Lambda, \mathbb{Z}) \subset \mathrm{Hom}\,(\mathbf{T}, \mathbb{G}_m) \otimes \mathbb{Q}, \\ \Phi' &= \{n_\alpha^{-1}\alpha^\vee \mid \alpha \in \Delta\}, \end{aligned}$$

and we define the dual group $\widetilde{G}^\vee$ of $\widetilde{G}$ to be the reductive group associated to this root datum.

Theorem 12. *The quadruple (X, Φ, X', Φ') defines a root datum.*

Proof. To check that Φ and Φ' are stable under the Weyl group is straightforward. For example, if $w\alpha = \beta$, then $Q(\alpha) = Q(\beta)$ so $wn_\alpha\alpha = n_\beta\beta$. The only part involving significant work is to check that $\Phi \subset X$ and $\Phi' \subset X'$.

To check that $\Phi \subset X$, it suffices to show that for all $\alpha \in \Delta$ and $y \in Y$ we have that $B(\alpha, y)$ is divisible by $Q(\alpha)$.

Consider the set $L_y = y + \mathbb{Z}\alpha$. It is a w_α stable subset of Y. There are two possibilities, either L_y contains z which is fixed by w_α or L_α contains z such that $w_\alpha z = z + \alpha$.

In the former case, consider the $\mathbb{Q}$-subspace of $Y \otimes \mathbb{Q}$ spanned by z and α. On this subspace, we have $Q(m\alpha + nz) = Am^2 + Bn^2 + Cmn$ for some $A, B, C \in \mathbb{Q}$. Since Q is invariant under w_α, we must have that $C = 0$. Then $B(\alpha, z) = 0$, so since $Q(\alpha)$ divides $B(\alpha, \alpha)$, it must divide $B(\alpha, y)$.

In the latter case, we calculate that $B(\alpha, z) = -Q(\alpha)$, so proceed as in the former case, so we are done.

We now show that $\Phi' \subset X'$.

Firstly, we use the fact that $B(\alpha, \alpha) = 2Q(\alpha)$ to conclude that

$$n_\alpha \mathbb{Z}\alpha \subset X \cap \mathbb{Q}\alpha \subset \frac{n_\alpha}{2}\mathbb{Z}\alpha.$$

Now consider some $\beta \in X$, and let $M_\beta = (\beta + \mathbb{Q}\alpha) \cap X$. A priori, there are three options.

The first is that there exists $\gamma \in M_\beta$ such that $w_\alpha \gamma = \gamma$, which implies $\langle \alpha^\vee, \gamma \rangle = 0$.

The second is that there exists $\gamma \in M_\beta$ such that $w_\alpha \gamma = \gamma + n_\alpha \alpha$ which implies $\langle \alpha^\vee, \gamma \rangle = n_\alpha$.

In the third potential case, we would have $\gamma \in M_\beta$ such that $w_\alpha \gamma = \gamma + \frac{n_\alpha}{2}\alpha$. For this to occur, we would require that $2|n_\alpha$, so in this case $B(\gamma, \alpha) \notin n\mathbb{Z}$. However, this last statement implies that $\gamma \notin X$, which cannot occur.

Thus, since we know that $\beta = \gamma + \frac{kn_\alpha}{2}\alpha$ for some integer k, we obtain that $\langle \beta, \alpha^\vee \rangle \in n_\alpha \mathbb{Z}$. This shows that $n_\alpha^{-1}\alpha^\vee \in \operatorname{Hom}(X, \mathbb{Z}) = X'$, as required. □

Thus, we have a root datum, so defining $\widetilde{G}^\vee$ as the split reductive group corresponding to this root datum is well defined.

As a consequence, we may consider the Satake isomorphism to be the existence of a natural isomorphism

$$\mathcal{H}(\widetilde{G}, K) \cong \mathbb{C}[\Lambda]^W \cong K_0(\mathrm{Rep}(\widetilde{G}^\vee)) \otimes \mathbb{C}.$$

13.12 Iwahori–Hecke Algebra

There is an alternative Hecke algebra associated to the group $\widetilde{G}$, defined in the same fashion as the spherical Hecke algebra $\mathcal{H}(\widetilde{G}, K)$, but considering a standard Iwahori subgroup I (defined to be the inverse image of $\mathbf{B}(k)$ under the surjection $K \to \mathbf{G}(k)$) in place of the hyperspecial maximal compact subgroup K. We will denote this Hecke algebra by $\mathcal{H}(\widetilde{G}, I)$; it is the algebra of anti-genuine I-biinvariant compactly supported locally constant functions on $\widetilde{G}$.

Let J denote the normaliser in $\widetilde{G}$ of $\widetilde{T} \cap K$.

Theorem 13. *The support of the algebra $\mathcal{H}(\widetilde{G}, I)$ is IJI.*

Proof. We also use the decomposition $G = IMI$. Suppose that $t \in M$ and $t \notin J$. Then there exists $k \in \widetilde{T} \cap K$ such that $tkt^{-1} \notin \widetilde{T} \cap K$. Since we are assuming $t \in WT$, we have that $p(t) \in T \cap K$. Thus $tkt^{-1} = \zeta k'$ for some $k' \in \widetilde{T} \cap K$ and $\zeta \in \mu_n$ with $\zeta \neq 1$. Hence any $f \in \mathcal{H}(\widetilde{G}, I)$ has $f(t) = 0$ so we have proved that the support of $\mathcal{H}(\widetilde{G}, I)$ lies in IJI.

For the reverse implication, we need to show that if $t \in J$ then there exists $f \in \mathcal{H}(\widetilde{G}, I)$ with $f(t) \neq 0$. To do this, we need to show that whenever $p(i_1 t i_2) = p(t)$ for $i_1, i_2 \in I$, then $i_1 t i_2 = t$.

Let I_p denote the maximal pro-p subgroup of I (it is the inverse image of $\mathbf{U}(k)$ under the projection $K \to \mathbf{G}(k)$). The torus $\mathbf{T}(k)$ over the residue field lifts to I and every element of I can be uniquely written as a product of an element of $\mathbf{T}(k)$ with an element of I_p.

After projection to G, $i_2 \in I \cap t^{-1}It$. Write $i_2 = j_1 j_2$ with $j_1 \in \mathbf{T}(k)$ and $j_2 \in I_p$. Then $ti_2t^{-1} = tj_1t^{-1}tj_2t^{-1}$. We have $tj_1t^{-1} \in \mathbf{T}(k)$ because t normalises $\widetilde{T} \cap K$ and $\mathbf{T}(k)$ consists of all elements of order $q-1$ in this group. Since tj_2t^{-1} topologically generates a pro-p group, it must be that $tj_2t^{-1} \in I_p$ since it is a priori in $\tilde{I}$ which also has a unique maximal pro-p subgroup. Thus, $ti_2t^{-1} = tj_1t^{-1}tj_2t^{-1} \in I$, so we are done. □

Let W_a denote inverse image of Λ under the projection from the affine Weyl group to Q. Then there is an isomorphism $I\backslash IJI/I \simeq W_a$. As a corollary of the above theorem, we are able to exhibit a basis for $\mathcal{H}(\widetilde{G}, I)$. For any $w \in W$, we are able to exhibit a choice of a lifting $w \in \widetilde{G}$ which is an element of our embedding $W_0 \hookrightarrow \widetilde{G}$ from the discussion at the end of Sect. 13.4. There is an embedding $\Lambda \subset W_a$ and $W \subset W_a$. Using these inclusions, we identify elements of λ as elements of W_a and for each simple coroot α denote by $s_\alpha \in W_a$ the corresponding simple reflection.

Corollary 4. *For each $w \in W_a$, then there is a function T_w in $\mathcal{H}(\widetilde{G}, I)$, supported on $\mu_n IwI$ and taking the value 1 at w. Then the collection of these T_w for $w \in W_a$, forms a $\mathbb{C}$ basis for the algebra $\mathcal{H}(\widetilde{G}, I)$.*

It is possible to write down a system of generators and relations for the algebra $\mathcal{H}(\widetilde{G}, I)$. The following is a corrected version of [Sav88, Proposition 3.1.2]. The change is in the definition of Savin's integer m, which has been replaced by n_α (although $m = n_\alpha$ in a large number of cases, in general they are not even equal in the rank one case).

Let Λ^+ denote the set of dominant elements of Λ and Δ denote the set of simple coroots.

Theorem 14. *The following relations hold in $\mathcal{H}(\widetilde{G}, I)$:*

1. *$T_\lambda T_\mu = T_{\lambda+\mu}$ for $\lambda, \mu \in \Lambda^+$.*
2. *If $s_\alpha \lambda = \lambda$ for $\alpha \in \Delta$ and $\lambda \in \Lambda^+$ then T_{s_α} and T_λ commute.*
3. *If $\lambda \in \Lambda^+$ and $\langle \alpha^\vee, \lambda \rangle = n_\alpha$ then*
$$T_\lambda T_{s_\alpha}^{-1} T_\lambda T_{s_\alpha}^{-1} = q^{n_\alpha - 1} T_{2\lambda - n_\alpha \alpha}.$$
4. *If $\lambda \in \Lambda^+$ and $\langle \alpha, \lambda \rangle = 2n_\alpha$ then*
$$T_\lambda T_{s_\alpha}^{-1} T_\lambda T_{s_\alpha}^{-1} = q^{2n_\alpha - 1} T_{2\lambda - 2n_\alpha \alpha} + (q-1) q^{n_\alpha - 1} T_{2\lambda - n_\alpha \alpha} T_{s_\alpha}^{-1}.$$
5. *$(T_{s_\alpha} - q)(T_{s_\alpha} + 1) = 0$ for $\alpha \in \Delta$.*
6. *For $w_1, w_2 \in W$ with $\ell(w_1 w_2) = \ell(w_1) + \ell(w_2)$ we have $T_{w_1 w_2} = T_{w_1} T_{w_2}$.*

Proof. The proof given by Savin [Sav88] is applicable here and correct until we reduce to a rank one calculation in proving parts 3 and 4. We will present this rank one calculation here. It does not appear in [Sav88] and is the source of the inaccurate statement in [Sav88, Proposition 3.1.2]. To carry out this computation, we will be making use of the explicit formulae for the 2-cocyle and the splitting given in equations (2) and (3), respectively.

In the rank one case, the statements of parts 3 and 4 simplify to the following, where s is the sole reflection in the Weyl group:

(3') If $\lambda = n_\alpha \alpha/2 \in \Lambda^+$ then

$$T_\lambda T_s^{-1} T_\lambda = q^{n_\alpha - 1} T_s$$

(4') If $\lambda = n_\alpha \alpha \in \Lambda^+$ then

$$T_\lambda T_s^{-1} T_\lambda = q^{2n_\alpha - 1} T_s + (q-1) q^{n_\alpha - 1} T_\lambda.$$

We know from Savin's proof that $T_{s_i}^{-1} T_\lambda = T_{s_i \lambda}$ and so need to calculate the product $T_\lambda T_{s\lambda}$. In particular, we need to calculate $T_\lambda T_{s\lambda}(\varpi^\lambda)$, which is the task we shall accomplish.

Let us first consider the case where our rank one group is $G = SL_2$. We may thus write $\varpi^\lambda = \begin{pmatrix} \varpi^l & 0 \\ 0 & \varpi^{-l} \end{pmatrix}$ for some integer l (actually $l = \langle \alpha^\vee, \lambda \rangle / 2$). Note that $2lQ(\alpha)$ is divisible by n, which will have the consequence that all powers of Hilbert symbols that appear will be ± 1, a feature we will exploit, simplifying our expressions by freely inverting such symbols on a whim.

We write $T_\lambda T_{s\lambda}(\varpi^\lambda)$ as an integral over $\widetilde{G}/\mu_n \simeq G$.

$$T_\lambda T_{s\lambda}(\varpi^\lambda) = \int_G T_\lambda(h) T_{s\lambda}(h^{-1}\varpi^\lambda) \mathrm{d}h.$$

This integrand is non-zero when $h^{-1} \in I\varpi^{-\lambda} I \cap I s \varpi^{\lambda} I \varpi^{-\lambda}$. We shall work modulo I on the left. Thus we have

$$h^{-1} = \begin{pmatrix} 0 & -\varpi^{-l} \\ \varpi^l & 0 \end{pmatrix} \begin{pmatrix} a & b \\ c & d \end{pmatrix} \begin{pmatrix} \varpi^{-l} & 0 \\ 0 & \varpi^l \end{pmatrix},$$

where $\left(\begin{smallmatrix} a & b \\ c & d \end{smallmatrix}\right) \in I$. The condition for $h^{-1} \in I\varpi^{-\lambda} I$ is equivalent to $c = \varpi^l u$ for some $u \in \mathcal{O}_F^\times$.

We calculate $h = \begin{pmatrix} b\varpi^{2l} & d \\ -a & -u\varpi^{-l} \end{pmatrix}$ and $s(h)^{-1} = s(h^{-1})$.

For $i_1 = \begin{pmatrix} -u & -d\varpi^l \\ 0 & -u^{-1} \end{pmatrix}$ and $i_2 = \left(\begin{smallmatrix} 1 & 0 \\ -au^{-1}\varpi^l & 1 \end{smallmatrix}\right)$ we have $i_1 h i_2 = \varpi^\lambda$, $\kappa(i_1) = \kappa(i_2) = 1$ and $\sigma(i_1 h, i_2)\sigma(i_1, h) = (au, \varpi^{lQ(\alpha)})$.

For $i_3 = \left(\begin{smallmatrix} d & -b \\ -u\varpi^l & a \end{smallmatrix}\right)$, we have $h^{-1}\varpi^\lambda i_3 = s\varpi^\lambda$, $\kappa(i_3) = (a, \varpi^\lambda)$ and $\sigma(h^{-1}, \varpi^\lambda)\sigma(h^{-1}\varpi^\lambda, i_3) = (a, \varpi^{lQ(\alpha)})$.

Thus, overall, our integrand $T_\lambda(h)T_{s\lambda}(h^{-1}\varpi^\lambda)$ is equal to $(au, \varpi^{lQ(\alpha)})$ where it is supported. Hence, the integral $T_\lambda T_{s\lambda}(\varpi^\lambda)$ is equal zero if n does not divide $lQ(\alpha)$ and the value of the appropriate volume, namely $q^{2n_\alpha-1}$, otherwise. To complete the proof in the SL_2 case, we need to note that in case 3, $2l = n_\alpha$ and thus n does not divide $lQ(\alpha)$ as can be seen by looking at 2-adic valuations. In case 4, $l = n_\alpha$ so n trivially divides $lQ(\alpha)$.

Now we turn to the case of $G = PGL_2$. We write $\varpi^\lambda = \left(\begin{smallmatrix}\varpi^l & 0\\ 0 & 1\end{smallmatrix}\right)$. In order to have our integrand $T_\lambda(h)T_{s\lambda}(h^{-1}\varpi^\lambda)$ non-zero, by consideration of the valuation of the determinant, we must have that l is even. This immediately proves our result when l is odd. For in case 4, for PGL_2 we have $l = 2n_\alpha$, so if l is odd, we must be in case 3.

If $\lambda \in \Lambda^+$ is such that $\langle\alpha^\vee, \lambda\rangle = 2n_\alpha$ and $\lambda/2 \in \Lambda^+$, then the (4) is a formal consequence of (3) and (5). Thus, we may reduce to the case where λ is a minimal non-zero element of Λ^+. This implies that l divides n, so in particular, n is even.

Let E be an unramified quadratic extension of F. Consider the natural map $SL_2(E) \to PGL_2(E)$ and restrict this to the preimage of $PGL_2(F)$. Note that ϖ^λ and all elements of the Iwahori subgroup I lie in the image of this map. Accordingly, we will be able to make use of the above calculation for $SL_2(E)$.

Since n is even, $\frac{q+1}{2} \equiv 1 \pmod n$. For $s, t \in E$ with $s^2, t^2 \in F$, we thus have the following identity of Hilbert symbols:

$$
\begin{aligned}
(s^2, t^2)_F^{Q(\alpha/2)} &= \overline{\left((-1)^{v(s)v(t)}\frac{s^{v(t)}}{t^{v(s)}}\right)}^{\frac{q-1}{n}Q(\alpha)} = \overline{\left((-1)^{v(s)v(t)}\frac{s^{v(t)}}{t^{v(s)}}\right)}^{\frac{q^2-1}{n}\frac{Q(\alpha)}{2}} \\
&= (s,t)_E^{Q(\alpha)/2}.
\end{aligned}
$$

We interpret this in the following manner: Since our central extensions are determined by their restriction to maximal tori, this shows that the pullback of the extension of $PGL_2(F)$ to its inverse image in $SL_2(E)$ is the same as the restriction of the central extension on $SL_2(E)$ corresponding to the quadratic form Q' defined by $Q' = Q/2$. We are able to push forward a cocycle on SL_2 to a cocycle on PGL_2 since the centre of SL_2 remains central when lifted to $\widetilde{SL}_2$.

As a result of this relationship between the covers of $PGL_2(F)$ and $SL_2(E)$, we are able to use the SL_2 calculations above for the proof in the PGL_2 case. We have $\langle\alpha^\vee, \lambda\rangle = l$ and $T_\lambda(h)T_{s\lambda}(h^{-1}\varpi^\lambda)$ is non-zero if and only if n divides $lQ(\alpha)/2$.

In case 3, $l = n_\alpha$. Since n is known to be even, it does not divide $lQ(\alpha)/2$ by the same 2-adic argument as in the SL_2 case.

In case 4, $l = 2n_\alpha$ and in this case n trivially divides $lQ(\alpha)/2$.

This completes our calculation and so, combined with the work in [Sav88], completes the proof. □

There is a stronger statement, giving a presentation for the Hecke algebra $\mathcal{H}(\widetilde{G}, I)$.

Theorem 15. *[Sav88] The set of relations presented in Theorem 14 provides a complete set of relations for the algebra $\mathcal{H}(\widetilde{G}, I)$.*

Proof. The proof of Savin [Sav88] of this theorem goes through without change. □

13.13 Further Work with the Dual Group

Corollary 5 ([Sav88]). *Suppose that $\widetilde{G}$ and $\widetilde{H}$ are two metaplectic groups with isomorphic dual groups $\widetilde{G}^\vee \cong \widetilde{H}^\vee$ and Iwahori subgroups $I^{\widetilde{G}}$ and $I^{\widetilde{H}}$, respectively. Then, there is an isomorphism of Iwahori–Hecke algebras:*

$$\mathcal{H}_\epsilon\left(\widetilde{G}, I^{\widetilde{G}}\right) \cong \mathcal{H}_\epsilon\left(\widetilde{H}, I^{\widetilde{H}}\right).$$

Proof. This is an immediate consequence of the description of these Hecke algebras in terms of generators and relations in Theorem 15. To see this explicitly, we rewrite the relations without any occurrences of n_α in the exponents of q by defining new variables $U_s = T_s$ and $U_\lambda = q^{-\langle \rho^\vee, \lambda \rangle} T_\lambda$. □

To the data of a metaplectic cover of a split group (i.e., the group G, the quadratic form Q and the degree of the cover n), let us propose to define the L-group of $\widetilde{G}$ to be the complex reductive group $\widetilde{G}^\vee(\mathbb{C})$. We hope that this definition will provide a way to bring the study of the metaplectic groups into the paradigm that is the Langlands functoriality conjectures.

The above corollary together with the metaplectic Satake isomorphism provides a starting point for correspondences between local representations with an Iwahori-fixed or spherical vector, respectively. In the spherical case, we have the following.

Proposition 6. *Suppose $\widetilde{G}$ and $\widetilde{H}$ are two metaplectic (possibly reductive) groups with a continuous homomorphism ${}^L\widetilde{G} \to {}^L\widetilde{H}$. Then there is a natural correspondence from spherical representations of $\widetilde{G}$ to spherical representations of $\widetilde{H}$.*

Proof. The homomorphism ${}^L\widetilde{G} \to {}^L\widetilde{H}$ defines a functor $\mathrm{Rep}\,({}^L H) \to \mathrm{Rep}\,({}^L G)$. Taking Grothendieck groups and using the Satake isomorphism, we obtain a natural morphism of spherical Hecke algebras $\mathcal{H}(\widetilde{H}, K) \to \mathcal{H}(\widetilde{G}, K)$, hence a map between representations of these spherical Hecke algebras, and thus a correspondence of representations from spherical representations of $\widetilde{G}$ to spherical representations of $\widetilde{H}$. □

We end this chapter with a short discussion of a categorified version of the metaplectic Satake isomorphism due to Finkelberg and Lysenko [FL10]. Suppose that F is a field of Laurent series $F = k((t))$ over a field k with some mild assumption on the characteristic of k not being too small. Corresponding to $\widetilde{G}$, there is a central extension of the loop group $\mathbf{G}(F)$ by $\mathbb{G}_m(k)$ as group ind-schemes over k. This central extension splits over $\mathbf{G}(O_\mathrm{F})$, so we obtain a $\mathbb{G}_m$ torsor over the affine Grassmannian $Gr = \mathbf{G}(F)/\mathbf{G}(O_{\mathrm{F}})$ (as an ind-scheme over k). The group $K = \mathbf{G}(O_{\mathrm{F}})$ acts on the total space of this torsor E° by left multiplication and the group μ_n acts by multiplication fibrewise. Again choose a faithful character ϵ of $\mu_n(k)$. Consider ϵ as a representation of $\pi_1(\mathbb{G}_m)$ and let L^ϵ be the corresponding one-dimensional local system on $\mathbb{G}_m$. One considers the category of perverse sheaves on E° which are K- and $(\mathbb{G}_m, L^\epsilon)$-equivariant. Finkelberg and Lysenko give this category the structure of a tensor category and show that it is equivalent to the category of representations of a reductive algebraic group. They construct explicitly the root system of this group and it can be seen to be the same as the root system constructed above for the group we denoted $\widetilde{G}^\vee$.

References

[AdCK89] E. Arbarello, C. de Concini and V. G. Kac. The infinite wedge representation and the reciprocity law for algebraic curves In *Theta functions—Bowdoin 1987, Part 1 (Brunswick, ME, 1987)*, volume 49 of *Proc. Sympos. Pure Math.*, pages 171–190. Amer. Math. Soc., Providence, RI, 1989.

[BD01] J-L. Brylinski and P. Deligne. Central extensions of reductive groups by $\mathbf{K}_2$. *Publ. Math. Inst. Hautes Études Sci.*, (94):5–85, 2001.

[BT72] F. Bruhat and J. Tits. Groupes réductifs sur un corps local. *Inst. Hautes Études Sci. Publ. Math.*, (41):5–251, 1972.

[BZ77] I. N. Bernstein and A. V. Zelevinsky. Induced representations of reductive p-adic groups. I. *Ann. Sci. École Norm. Sup. (4)*, 10(4):441–472, 1977.

[Cas] W. Casselman. Introduction to the theory of admissible representations of p-adic reductive groups. *preprint*. http://www.math.ubc.ca/~cass/research.html.

[CS80] W. Casselman and J. Shalika. The unramified principal series of p-adic groups. II. The Whittaker function. *Compositio Math.*, 41(2):207–231, 1980.

[CO] G. Chinta and O. Offen. A metaplectic Casselman-Shalika formula for GL_r. *preprint*.

[FL10] M. Finkelberg and S. Lysenko. Twisted geometric Satake equivalence. *J. Inst. Math. Jussieu*, 9(4):719–739, 2010.

[Gro98] B. H. Gross. On the Satake isomorphism. In *Galois representations in arithmetic algebraic geometry (Durham, 1996)*, volume 254 of *London Math. Soc. Lecture Note Ser.*, pages 223–237. Cambridge Univ. Press, Cambridge, 1998.

[KP84] D. A. Kazhdan and S. J. Patterson. Metaplectic forms. *Inst. Hautes Études Sci. Publ. Math.*, (59):35–142, 1984.

[KP86] D. A. Kazhdan and S. J. Patterson. Towards a generalized Shimura correspondence. *Adv. in Math.*, 60(2):161–234, 1986.

[Kub69] T. Kubota. *On automorphic functions and the reciprocity law in a number field.* Lectures in Mathematics, Department of Mathematics, Kyoto University, No. 2. Kinokuniya Book-Store Co. Ltd., Tokyo, 1969.

[LS10] H. Y. Loke and G. Savin. Modular forms on nonlinear double covers of algebraic groups. *Trans. Amer. Math. Soc.*, 362(9):4901–4920, 2010.

[Lus93] G. Lusztig. *Introduction to quantum groups*, Birkhäuser Boston Inc., Boston, MA, 1993.

[Mat69] H. Matsumoto. Sur les sous-groupes arithmétiques des groupes semi-simples déployés. *Ann. Sci. École Norm. Sup. (4)*, 2:1–62, 1969.

[McN11] P. J. McNamara. Metaplectic Whittaker functions and crystal bases. *Duke Math J.*, 156(1):1–31, 2011.

[Moo68] C. C. Moore. Group extensions of p-adic and adelic linear groups. *Inst. Hautes Études Sci. Publ. Math.*, (35):157–222, 1968.
[MW94] C. Mœglin and J-L. Waldspurger. *Décomposition spectrale et séries d'Eisenstein*, volume 113 of *Progress in Mathematics*. Birkhäuser Verlag, Basel, 1994. Une paraphrase de l'Écriture.
[Rei] R. Reich. Twisted geometric Satake equivalence via gerbes on the factorizable grassmannian *preprint*, arXiv:1012.5782.
[Rod81] F. Rodier. Décomposition de la série principale des groupes réductifs p-adiques. In *Noncommutative harmonic analysis and Lie groups (Marseille, 1980)*, volume 880 of *Lecture Notes in Math.*, pages 408–424. Springer, Berlin, 1981.
[Sav88] G. Savin. Local Shimura correspondence. *Math. Ann.*, 280(2):185–190, 1988.
[Sav04] G. Savin. On unramified representations of covering groups. *J. Reine Angew. Math.*, 566:111–134, 2004.
[Ser62] J-P. Serre. *Corps locaux*. Publications de l'Institut de Mathématique de l'Université de Nancago, VIII. Actualités Sci. Indust., No. 1296. Hermann, Paris, 1962.
[Ste62] Robert Steinberg. Générateurs, relations et revêtements de groupes algébriques. In *Colloq. Théorie des Groupes Algébriques (Bruxelles, 1962)*, pages 113–127. Librairie Universitaire, Louvain, 1962.
[Wei09] M. H. Weissman. Metaplectic tori over local fields. *Pacific J. Math.*, 241(1):169–200, 2009.

Chapter 14
Excerpt from an Unwritten Letter

S.J. Patterson

Abstract This note is concerned with the representation constructed by Chinta and Gunnells in (Constructing Weyl group multiple Dirichlet series, J. Amer. Math. Soc. 23, 2010, 189–215), a representation of the Weyl group of an irreducible root system on an infinite-dimensional algebra over a base field. Chinta and Gunnells in (Constructing Weyl group multiple Dirichlet series, J. Amer. Math. Soc. 23, 2010, 189–215). The first group of remarks is that this result can, at least, in principle, be constructed and understood from the point of view of the representation theory of local metaplectic groups. The original proof is by means of generators, relations and computer algebra, and so a representation-theoretical proof makes the construction and verification more "natural."

The second group of remarks concerns the application of this local theorem to the global problem of determining the Fourier–Whittaker coefficients of metaplectic theta functions and the closely related problem of the distribution of the values of Gauss sums and their generalizations. These applications are still very preliminary, but the prospects are encouraging.

Keywords Chinta–Gunnells averaging method • Metaplectic Whittaker function

This should have been part of a joint letter to Ben Brubaker, Dan Bump, Gautam Chinta, Sol Friedberg and Jeff Hoffstein. The purpose was to formulate with a broad brush how I understand the papers of Chinta and Gunnells and of Chinta and Offen.

We shall need a number of results of H. Matsumoto which I shall recall here. Let F be a non-archimedean local field containing the nth roots of 1. Let Φ be a simple reduced root system which corresponds to a simple complex Lie algebra $\mathfrak{g}$. From this, one constructs an algebraic group (Chevalley group) G (over $\mathbb{Z}$) such that

S.J. Patterson (✉)
Mathematisches Institut, Bunsenstr. 3-5, 37073 GÖTTINGEN, Germany
e-mail: sjp@uni-math.gwdg.de

D. Bump et al. (eds.), *Multiple Dirichlet Series, L-functions and Automorphic Forms*, Progress in Mathematics 300, DOI 10.1007/978-0-8176-8334-4_14,

the Lie algebra of $G(\mathbb{C})$ is $\mathfrak{g}$. There is no need to worry about the finer points here. We fix a set of positive roots Φ_+. We denote by $\Phi_{\mathbb{R}}$ the real vector space in which Φ is embedded. As usual, it is acted upon by the Weyl group W of Φ, and there is an invariant positive definite inner product $(\cdot,\cdot)$ on $\Phi_{\mathbb{R}}$.

For each $\alpha \in \Phi_+$ there is a morphism $h_\alpha : \mathrm{SL}_2 \to G$. The theory of Chevalley groups builds G out of these maps. It is useful to denote by $\check{\alpha}(x)$ the image under h_α of the diagonal matrix with entries x and x^{-1}. As we are not concerned here with changes of the field, we write G for $G(F)$, etc., when there should be no danger of confusion. The element $\check{\alpha}(x)$ lies in the distinguished Cartan subgroup H of G. Recall that the roots originate as elements of $\mathrm{Hom}(H, F^\times)$. We then have $\beta(\check{\alpha}(x)) = x^{M_\Phi(\alpha,\beta)}$, where $M_\Phi(\alpha,\beta) = \frac{2(\alpha,\beta)}{(\beta,\beta)}$ is the appropriate entry of the Cartan matrix of Φ.

Let N denote the unipotent group of G generated by the images under the h_α of the unipotent upper-triangular matrices in SL_2. Let e be a nontrivial additive character of F. Let Φ_0 be the set of simple roots in Φ_+. Let $r : \Phi_0 \to F$. Then, we can construct a character e_r of N with $e_r\left(h_\alpha\left(\left(\begin{smallmatrix} cc1 & x \\ 0 & 1 \end{smallmatrix}\right)\right)\right) = e(r(\alpha)x)$.

Matsumoto showed how to glue the metaplectic extensions of the $\mathrm{SL}_2(F)$ into a global one of G. We recall what we need here; the details of the construction itself are not relevant at the moment. Let $(\cdot,\cdot)$ be the n-th order Hilbert symbol in F. Then for each $\alpha \in \Phi$ with α a long root, Kubota's construction [13] gives a metaplectic cover of $\mathrm{SL}_2(F)$ associated with $(\cdot,\cdot)$, and we consider it as a cover of $h_\alpha(\mathrm{SL}_2(F))$; if there are two different lengths of root, then for a short root we use instead $(\cdot,\cdot)^m$ where $m = 2$ for the cases B_*, C_*, F_4 and $m = 3$ for G_2. Matsumoto shows that there is a metaplectic extension $\tilde{G}$ of G whose restriction to the image of h_α is the given one, [18, Theorems 5.10 and 8.2]. One can give a cocycle on a Zariski open subset of $G \times G$, but it is not of much use. What is relevant is that the commutator of (lifts of) elements of H is an invariant and is known. The commutator of $\check{\alpha}(x)$ and $\check{\beta}(y)$ is $(x,y)^{M_\Phi(\alpha,\beta)m(\alpha)}$ where $m(\alpha) = 1$ if α is long and has the value above if α is short should there be two different lengths. Note that $m(\alpha)(\alpha,\alpha)$ takes on the same value for all roots. There is no need to normalize the inner product. See also [20].

The centre of G is finite and the same clearly is true of $\tilde{G}$. We fix a homomorphism $\varepsilon : \mu_n(F) \to \mathbb{C}^\times$. It makes life easier, but it is not absolutely necessary to assume that ε is injective. Since 2 and 3 are primes, the centre of $\tilde{H}$ is generated by $\mu_n(F)$ and the images of $\check{\alpha}(x)$ with $x \in F^{\times n/\gcd(m(\alpha),n)}$ where $m(\alpha)$ are as above. We denote the centre of $\tilde{H}$ by $\tilde{H}_Z$, and we denote by $\tilde{H}_*$ a maximal abelian subgroup.

Now, we can introduce the manifold $\Omega(\varepsilon)$ of quasicharacters of $\tilde{H}_Z$ which restrict to ε on $\mu_n(F)$. We can extend any $\omega \in \Omega(\varepsilon)$, in several ways, to $\tilde{H}_*$. From this extension, we can form the induced representation of $\tilde{H}$ which depends only on ω and is irreducible. Let μ be the square root of the modulus function of H acting on N. Then, we can induce $\omega_*\mu$ on $\tilde{H}_*\tilde{N}$ as usual to $\tilde{G}$. Denote the resulting representation by $V(\omega)$. We can regard $V(\omega)$ as a fibre of a holomorphic vector bundle over $\Omega(\varepsilon)$.

We can now introduce as usual the intertwining operators, but care is called for. Let $M \subset G$ be the normalizer of H (monomial group). In the usual dispensation of things, the Weyl group is defined to be M/H. It is *not* a subgroup of G, but there is a covering group of W which is; the order of the covering is $2^{\text{rank of } \Phi}$—for details, see [18, Sect. 6]. It is this fact that seems to have given rise to many of the problems when -1 is not an nth power.

For our purposes, we have to replace the usual Weyl group with $\tilde{M}/\tilde{H}_Z$. This is also a covering group of the Weyl group. One could use the lift of the image of Matsumoto's group in it for the discussion of intertwining operators. Generally, one defines the intertwining operator $I_w(\omega) : V(\omega) \to V({}^w\omega)$ for $w \in \tilde{M}$ as a $\tilde{G}$-morphisms where ${}^w\omega(h) = \omega(w^{-1}hw)$. For $h \in \tilde{H}_Z$, one has that ${}^w\omega(h)$ depends only on the class of w in the (metaplectic) Weyl group, but the extension to $\tilde{H}_*$ depends on w in $\tilde{M}/\tilde{H}_*$. For $h \in \tilde{H}_* \subset \tilde{M}$ one has that $I_h(\omega_*)$ is $(\omega_*\mu)(h)^{-1}$ times the identity.

One has that generically $I_{w_1}({}^{w_2}\omega)I_{w_2}(\omega)$ is a multiple of $I_{w_1w_2}(\omega)$. The word "generically" needs an explanation. For a simple root α, let $\omega_\alpha(x) = \omega(\check{\alpha}(x))$ for $x \in F^{\times n/\gcd(n,m_\alpha)}$. Then, all of the functions are rational on each connected component of $\Omega(\epsilon)$ in certain roots of the $(x_\alpha)_{\alpha\in\Phi_0}$ where $x_\alpha = \omega_\alpha(\pi^{\times n/\gcd(n,m_\alpha)})$ and π is a uniformizer of F - see [1]. (The fact that x_α depends on the choice of π will cause us no problems.) We understand the word "generically" to mean that the result holds true on a Zariski open subset of the $(x_\alpha)_{\alpha\in\Phi_0}$.

The multiple above is 1 if the length of w_1w_2 is the sum of the lengths of w_1 and w_2. Otherwise, it is a product of L or gamma functions (depending on the language one uses) and monomials. The calculations for SL_2 are straightforward, and one can reduce the calculation in the general case to this one using standard techniques as in [1, 6, 7].

One can use the conclusions of this calculation to investigate the reducibility of the $V(\omega)$. Again, the techniques were established around 1970—see [6] for the classical case and [15, Corollary 2.2.7] for the case where -1 is a nth power. One can, as in the standard case, renormalize the intertwining operators by forming certain multiples $\tilde{I}_w(\omega)$ for which $\tilde{I}_{w_1}({}^{w_2}\omega)\tilde{I}_{w_2}(\omega)$ is an monomial times $\tilde{I}_{w_1w_2}(\omega)$.

Now, we turn to the Whittaker models. We introduced the e_r above, and we shall assume that e_r is nondegenerate in the sense that $r(\alpha) \neq 0$ for all $\alpha \in \Phi_0$. The space of Whittaker models $\mathrm{Wh}(\omega, e_r)$ is the subspace of the dual space of $V(\omega)$ of linear forms λ with $\lambda(nv) = e_r(n)\lambda(v)$ for $n \in N$. Generically, this space is of dimension $[H : H_Z]^{\frac{1}{2}}$. We form the dual maps ${}^t\tilde{I}_w(\omega) : \mathrm{Wh}({}^w\omega, e_r) \to \mathrm{Wh}(\omega, e_r)$ which can be represented by certain matrices.

We should note here that $e_r(hnh^{-1}) = e_{r'}(n)$ where $r'(\alpha) = \alpha(h)r(\alpha)$, and we say r and r' are in the same class in this case. It is easy to compare the $\mathrm{Wh}(\omega, e_r)$ for different r's in the same class.

One can construct a basis of $\mathrm{Wh}(\omega, e_r)$ at least for the (x_α) in an open set of $\Omega(\varepsilon)$ as integrals

$$v \mapsto \int_{N^\star} v(w^{-1}n)\bar{e}_r(n)\mathrm{d}n$$

where $N^\star$ is the standard lift of N and w runs through elements of $\tilde{M}$ of maximal length in the Weyl group taken modulo $\tilde{H}_*$. This gives a more precise meaning as to how the ${}^t\tilde{I}_w(\omega) : \mathrm{Wh}({}^w\omega, e_r) \to \mathrm{Wh}(\omega, e_r)$ can be represented as matrices. Let us now fix a basis $\{\lambda_1(\omega, r), \ldots, \lambda_N(\omega, r)\}$ of $\mathrm{Wh}(\omega, e_r)$ such as one of the type described above. We shall assume that this basis is rational in ω in the sense above. This is the case with the functionals those given above. Such a basis can be specialized to a basis almost everywhere on $\Omega(\epsilon)$. Define hr so that $(hr)(\alpha) = \alpha(h)r(\alpha)$ $(\alpha \in \Phi_0)$ leads to a matrix representation of the action of ${}^t\tilde{I}_w(\omega)$. It can be written as

$$ {}^t\tilde{I}_w(\omega)(\lambda_i({}^w\omega) = \sum_j \tau_{ij}(\omega, r)\lambda_j(\omega) $$

As usual, the coefficients $\tau_{ij}(\omega, r)$ will be rational. Now, for $\lambda \in \mathrm{Wh}(\omega, e_r)$, the linear form $\lambda \circ h : v \mapsto \lambda(hv)$ belongs to $\mathrm{Wh}(\omega, e_{hr})$. One can give a fairly general formula for the coefficients in terms of the generalized gamma functions (which, one should recall, are defined first as generalized functions):

$$ \Gamma_n(\chi, re_o) = \int_{F^{\times n}} \chi(x)e_o(rx)|x|^{-1}\mathrm{d}x $$

where χ is a quasicharacter on $F^{\times n}$ and $r \neq 0$, [10, Sect. 1]. It is elementary to express these functions in terms of the gamma functions of Tate's thesis, the $\Gamma_1(\chi', re_o)$. By means of generalized functions (or otherwise), one can reduce the calculation of the matrix coefficients to the case of SL_2. If we assume also that all the ω_α are unramified in the sense that $\omega_\alpha(x) = 1$ if $|x|_F = 1$ (where $|\cdot|_F$ is the norm on F), then the calculation become very explicit. This was done in the case of GL_r in [12]. We obtain a functional equation of Jacquet's type, [11, Sect. 3].

This construction yields a representation of $\tilde{M}$ on $\bigoplus_{w \in M/H_*} \mathrm{Wh}({}^w\omega, e_r)$. It should be in essence the Chinta–Gunnells representation. Actually, their theorem is a very general one, but it seems as if one can deduce it from this more special one using Bernstein's techniques which have been developed in this context by Banks [1]. This would give a more transparent proof of their theorem which makes use of computer algebra (see [8, Proof of Theorem 3.2]) and thereby only the straightforward formal properties of monomials and Gauss sums. Moreover, as Matsumoto's construction does not need the assumption that -1 be an nth power, we see that a version of the Chinta–Gunnells representation has to exist in all cases[1]. It is an important consequence of [8] that their theorem can dissociated from the theory of Chevalley groups and that it may be considered as a construction from Φ without the intervention of the Chevalley group or its covers. One can recover their full theorem from this version using the techniques of [1].

---

[1]The condition that -1 be an nth power is used in [8, Sect. 3] in the formula $\gamma(j)\gamma(-j) = q^{-1}$ for $j \not\equiv 0 \pmod{n}$ (their notations). This holds true if $(-1, \pi)_{n,F} = 1$. If $(-1, \pi)_{n,F} = -1$, we can replace $\gamma(j)$ by $i^{j^2}\gamma(j)$ with $i^2 = -1$. From this, one can derive a representation of the same type as [8, Theorem 3.2] when $(-1, \pi)_{n,F} = -1$.

One can use these considerations to determine Whittaker functions and spherical functions. The techniques for doing so go back to Casselman and Shalika [6, 7]. Under the restriction that -1 be an nth power, they have been used by various authors [5, 8, 9, 15–17] where they are augmented by either the use using the asymptotics of Whittaker functions and the Gidinkin–Karpelevich formula and Jacquet's functional equation or with the Hecke algebra (cf. [19]). There are other approaches, as for example the direct one of [3]. All these formulae are under the assumption about -1, but again there seems to be nothing essential here. One should note that one should take into account the different classes of maximal compact open subgroups; for the classification, see [2].

In fact, Chinta and Gunnells do not determine the Whittaker functions but rather the coefficients of the multiple Dirichlet series which is done inductively [8, Theorem 3.5]. It is this formula that I want to discuss here and in connection with it to point out what are for me the two most serious open problems at the present time, both of which seem to me to be not extremely difficult. They both concern the global case.

Let k be a global field containing the nth roots of 1. I shall assume that the multiple Weyl Dirichlet series exist in this case. The purpose of the arguments above was to persuade you the restriction on -1 is unnecessary. The theory is well established in the case A_1, and the arguments of [8] allow one to extend the definition to the other simple root systems. Although they should be associated with forms on the cover of the Chevalley group, the argument of [8] shows that it is a question only about root systems. The connection with metaplectic forms has been established in several cases, but this will not be relevant here. Let S be a finite set of places containing all those v where $|n|_v \neq 1$. We shall assume that the ring of S-integers R is a principal ideal domain. We let $k_S = \prod_{v \in S}$, and we let e be a nontrivial additive character on k_S trivial on R. We can, and shall, assume that $\{x \in k : e|xR = 1\} = R$. Let me denote the coefficients of the metaplectic multiple Weyl series by $g_\Phi(\mathbf{r}, \epsilon, \mathbf{c})$—this you prefer to write this as $H(c)$. Here, $\mathbf{r} : \Phi_+ \to R$. For a number of purposes, it is convenient to regard $\mathbf{r}$ as an element of $\Phi_Z \otimes_{\mathbb{Z}} R$.

The function $g_\Phi(\mathbf{r}, \epsilon, \mathbf{c})$ is, in principle, known—by your work—through the property of twisted multiplicativity and a list, still somewhat implicit in the general case, of values depending on $\mathrm{ord}(r(\alpha))$ and $\mathrm{ord}(c(\alpha))$. This is like defining the Gauss sum $g(r, \varepsilon, c) = \sum_{x \pmod{c}} \left(\frac{x}{c}\right) e(rx/c)$ by demanding that it have the twisted multiplicativity property and that if π is a uniformizer of F, $r_o \in R^\times$

$$g(r_o\pi^\rho, \epsilon, \pi^\gamma) = \begin{cases} 0 & \text{if } \gamma > \rho + 1 \\ q^{\gamma-1} g(r_o, \epsilon^\gamma, \pi) & \text{if } \gamma = \rho + 1, \gamma \not\equiv 0 \pmod{n} \\ -q^{\gamma-1} & \text{if } \gamma = \rho + 1, \gamma \equiv 0 \pmod{n} \\ 0 & \text{if } \gamma \leq \rho, \gamma \not\equiv 0 \pmod{n} \\ q^\gamma - q^{\gamma-1} & \text{if } 0 < \gamma \leq \rho, \gamma \equiv 0 \pmod{n} \\ 1 & \text{if } 0 = \gamma \end{cases}$$

where q is the cardinality of the residue class field. This would clearly not be the correct approach. As far as I can see, neither the method of crystal bases nor that of Chinta–Gunnells leads quickly to a unified formula. It is perhaps an act of faith to believe that there is one, but I think that one can subscribe to it for the moment. It is, from my point of view, a central problem to give a good global description of g_Φ.

Let us consider the case A_1. Let N be the modulus function (norm) on k_S with respect to the additive Haar measure. We define the Gauss sum $g(r,\epsilon,c)$ in R as usual. We write

$$\psi^o(r,\epsilon,\eta,s) = \sum_{c\sim\eta} g(r,\epsilon,c)N(c)^{-s}$$

which converges in $\mathrm{Re}(s) > \frac{3}{2}$. Here, we write $\sim$ to indicate that two elements lie in the same coset of $k_S^\times/k_S^{\times n}$. The sum is taken modulo $R^{\times n}$. The ^{o} indicates that we shall not worry about the L or zeta factors needed to produce the most elegant functional equation and to limit the poles to nonspurious ones. This function has an analytic continuation to the entire plane as a meromorphic function. There is at most one pole in $\mathrm{Re}(s) > 1$; it is at $s = 1 + \frac{1}{n}$. We denote the residue there by $\rho^o(r,\epsilon,\eta)$.

Let π be a prime in R and let S' be the union of S and the valuation associated with π. Let $q = \mathrm{N}(\pi)$. Then (see [12, Prop. II.3.2]),

$$\psi_S^o(r_o\pi^m,\epsilon,\eta,s) = \psi_{S'}^o(r_o\pi^m,\epsilon,\eta,s)\frac{(1-q^{n-ns-1})-(1-q^{-1})q^{(n-ns)([\frac{m}{n}]+1)}}{1-q^{n-ns}}$$
$$+\ \psi_{S'}^o(r_o\pi^{-m-2},\epsilon,\eta\pi^{-m-1},s)g(r_o,\epsilon^{m+1},\pi)q^{(m+1)(1-s)-s}\epsilon(\eta,\pi)^{m+1}$$

or, what is the same (with $(m)_n$ the least nonnegative residue of $m \pmod n$)

$$\psi_S^o(r_o\pi^m,\epsilon,\eta,s) = \frac{1-q^{n-ns-1}}{1-q^{n-ns}}\Bigg\{\psi_{S'}^o(r_o\pi^m,\epsilon,\eta,s)$$
$$-\,q^{(m+1)(1-s)}\Bigg\{\psi_{S'}^o(r_o\pi^m,\epsilon,\eta,s)\frac{(1-q^{-1})q^{(1-s)(n-(m)_n)}}{1-q^{n-ns-1}}$$
$$-\psi_{S'}^o(r_o\pi^{-m-2},\epsilon,\eta\pi^{-m-1},s)\frac{g(r_o,\epsilon^{m+1},\pi)q^{-s}\epsilon(\eta,\pi)^{m+1}}{1-q^{n-ns-1}}(1-q^{n-ns})\Bigg\}\Bigg\}$$

The second term of this equation reflects the action of a reflection through the Chinta–Gunnells representation on ψ^o. This means that we can surmise the form of the general case. The two terms may be considered as giving the asymptotics in m. The major pole of $\psi_S^o(r\pi^m,\epsilon,\eta,s)$, should it exist, is at $s = 1 + \frac{1}{n}$ and the Periodicity Theorem asserts that the residue there is periodic in m with period n. This means that the residue of the term corresponding to the nontrivial element of "the Weyl group" is regular at this point. From this, we obtain certain relations between the residues—see, for example, [10, Theorem 1.9]. These are, for m with $0 \le m \le n-2$,

$$\rho(r_o\pi^m,\epsilon,\eta)$$
$$= N(\pi)^{-\frac{m+1}{n}} g(r_o,\epsilon^{m+1},\pi)\epsilon((-\eta,\pi))^{m+1}\rho(r_o\pi^{n-m-2},\epsilon,\eta\pi^{-m-1}) \qquad (\star)$$

and for $m = n-1$

$$\rho(r_o\pi^{n-1},\epsilon,\eta) = 0 \qquad (\star\star)$$

We have written here $\rho_S(r,\epsilon,\eta)$ for $\rho_S^o(r,\epsilon,\eta)\zeta_k(2)$. Together with the Periodicity Theorem, these determine ρ essentially completely if $n = 2$ or 3. If $n > 3$, then they do not suffice, and one of the major problems is to understand the function ρ better. It seems unrealistic to expect an explicit formula. C. Eckhardt and I proposed a partial formula in the case $n = 4$ and $k = \mathbb{Q}(\sqrt{-1})$. Beyond this case, one can only make conjectures, and as I tried to explain at the conference for Dorian Goldfeld's 60th birthday, all the ρ's seem to have a common transcendental factor (meaning a number given in terms of transcendental functions).[2] This was first proposed, in the case $n = 6, k = \mathbb{Q}(\mu_3)$, by G. Wellhausen in his thesis [21]. Again, following the ideas of Wellhausen, it seems plausible that, divided by this factor, the coefficients are as small as $(\star)$ allows them to be. More precisely, if the transcendental factor is T, then it appears that all the $(\rho(r,\epsilon,\eta)/T)^n$ lie in an algebraic number field of finite degree over $\mathbb{Q}$. Moreover, it seems plausible that for $\varepsilon > 0$, we have $\rho(r,\epsilon,\eta) = \mathcal{O}(\mathrm{N}(r)^{\varepsilon})$ and perhaps for r free of nth powers even that $\rho(r,\epsilon,\eta) = \mathcal{O}(N(r)^{-\frac{1}{2n}+\varepsilon})$. Finally, there are indications that the norm of the denominator of $(\rho(\mu_o\beta,\cdot)/T)^n\mathrm{N}(\beta)$ is $\mathcal{O}(\mathrm{N}(\beta)^{\varepsilon})$. This latter suggestion is that the height be "as small as possible."

Now, let us meditate on the general case. The second general question that seems to me to be need investigation is to determine when a situation analogous to that with $n = 2$ and $n = 3$. We assume we can construct

$$\psi^o_{\Phi,S}(\mathbf{r},\eta,\mathbf{s}) = \sum_{\mathbf{c}\sim\eta} g_{\Phi}(\mathbf{r},\epsilon,\mathbf{c})N(\mathbf{c})^{-\mathbf{s}}$$

where $c \in H^+(R)$ taken modulo $H(R^{\times n})$ and $H^+(R)$ is the semigroup of $H(k_S)$ generated by the $\check{\alpha}(x)$ with $x \in R - \{0\}$ and $\alpha \in \Phi_+$. Further, $H(R^{\times n})$ is the subgroup of $H(k_S)$ generated by the $\check{\alpha}(x)$ with $x \in R^{\times n}$. Next, $\mathbf{s} \in \Phi_Z \otimes \mathbb{C}$. We can write any such $\mathbf{s}$ uniquely as $\sum_{\alpha\in\Phi_0} s(\alpha)\alpha$, and so we can also think of

[2]The precise nature of T is unclear. In the case of the cubic theta function and in Wellhausen's conjectures in the case $n = 6$, we find a factor $(2\pi)^{1-1/n}\Gamma(1/n)$. If this were also the case when $n = 4$, the constant of [10, pp. 240,251] which was numerically estimated as 0.14742376— note that a digit was omitted on p. 240—could be $(\frac{1}{4}(2\pi)^{\frac{3}{4}}\gamma(1/4))^2$ This is numerically 0.1475425748.... This is close but not close enough in view of the accuracy of the calculations of [10]. A much better estimate is $\frac{(2\pi)^3}{128\Gamma(1/4)^2}$ which is numerically 0.1474237606.... This is very puzzling.

$\mathbf{s}$ as a function from Φ_0 to $\mathbb{C}$. The norm N induces a map from $H(k_S)$ to $H^o(\mathbb{R})$ the connected component of the identity of $H(\mathbb{R})$, and we can interpret $\mathbf{s}$ as a quasicharacter of $H^o(\mathbb{R})$ which we write as $y \mapsto y^{\mathbf{s}}$. For w in the Weyl group, we can define $w(\mathbf{s})$ by $\mathrm{N}(w_1^{-1}xw_1)^{\mathbf{s}} = \mathrm{N}(x)^{w(\mathbf{s})}$ where w_1 is a representative of w. Finally, $\eta \in H(k_S)$, and we write $\mathbf{c} \sim \eta$ when $\mathbf{c} \in \eta H_Z(k_S)$ where $H_Z(k_S)$ is the group covered by $\tilde{H}_Z(k_S)$. Then, what Chinta and Gunnells have proved can be reformulated as follows. Let π be a prime in R. Let $\mathbf{f} \in \Phi_Z$ which we can write uniquely as $\sum_{\alpha\in\Phi_0} f(\alpha)\alpha$. We suppose for convenience that S and e are such that $\{y \in k : e|yk_S = 1\}$ of e is R. Then, $\mathbf{r}_o$ is a function defined on Φ_0 with values in R. Let $\mathbf{r}_o$ be such that $\mathbf{r}_o(\alpha)$ is not divisible by π for any α. The expression $\mathbf{r}_o\pi^{\mathbf{f}}$ is defined by component-wise multiplication on Φ_0. Then, Chinta and Gunnells have shown that $\psi^o_{\Phi,S}(\mathbf{r}_0\pi^{\mathbf{f}}, \eta, \mathbf{s})$ can, if all the $\mathbf{f}(\alpha)$ are dominant, be written as,

$$\sum_{w\in W} F(\mathbf{r}_0, w, \eta, \mathbf{s})\mathrm{N}(\pi^{\mathbf{f}})^{w(\mathbf{s}-\rho)+\rho}$$

and is 0 otherwise. Here, $\rho = \frac{1}{2}\sum_{\alpha\in\Phi_+}\alpha$ as before but now interpreted as a map from Φ_0 to $\mathbb{C}$. The use of ψ^o instead of a normalized function, leads, as above, to some extraneous products of zeta or L-factors which we shall not discuss here. Apart from them, $F(\mathbf{r}_0, I, \eta, \mathbf{s})$ will be $\psi^o_{\Phi,S'}(\mathbf{r}, \eta, \mathbf{s})$, and the other $F(\mathbf{r}_0, w, \eta, \mathbf{s})$ are the consequence of the action of the Chinta–Gunnells representation on some $\psi^o_{\Phi,S'}(\mathbf{r}_o\pi^{\star}, \star, \star, \mathbf{s})$.

Since we can approach the major singularity $(\mathbf{s}, \check{\alpha}) = 1+\gcd(n, m_\alpha)/n$ $(\alpha \in \Phi_0)$ by first selecting out one α and taking the corresponding residue, we see the residues are invariant when the αth component is multiplied with a $n/\gcd(n, m_\alpha)$th power; see [4]. This is the corresponding version of the Periodicity Theorem in this context. In the sum over the "Weyl group," this means that at the major singularity all the terms vanish except for the one corresponding to the identity. Consequently, the residues satisfy a system of linear equations. These reflect, as we see, the action of the intertwining operators, and just as with them, it should follow that all these relations follow from the ones for simple reflections. We should have a set of relations parametrized by Φ_0 essentially of the same shape as $(\star)$ above. Without worrying now about the finer points of the definition of the residue, for example, about the zeta functions needed for the functional equation, we can formulate the putative generalization of $(\star)$ as follows. Write $\rho_o(\mathbf{r}, \epsilon, \eta)$ for the residue. Then, we might expect for each $\alpha \in \Phi_0$

$$\rho_\Phi(\mathbf{r_o}\pi^{\mathbf{f}}, \epsilon, \eta) = \mathrm{N}(\pi)^{-(f(\alpha)+1)\gcd(n,m(\alpha))/n} g(\mathbf{r_o}(\alpha), \epsilon^{-(f(\alpha)+1)m(\alpha)}, \pi)$$
$$\epsilon(-\eta(\alpha), \pi)^{(f(\alpha)+1)m(\alpha)} \rho_\Phi\left(\mathbf{r_o}\pi^{s_\alpha(\mathbf{f})-2(\rho-s_\alpha(\rho))}, \epsilon, \eta\pi^{-\frac{1}{2}(\mathbf{f}-s_\alpha(\mathbf{f}))+(\rho-s_\alpha(\rho))}\right) \qquad (\dagger)$$

when $f(\alpha) < n/\gcd(n, m(\alpha))$. Moreover, we would expect if for some $\alpha \in \Phi_o$ we have $(\mathbf{f} + \rho)(\alpha) \equiv 0 \pmod{n/\gcd(n, m(\alpha))}$, then

$$\rho_\Phi(\mathbf{r_o}\pi^{\mathbf{f}}, \epsilon, \eta) = 0 \qquad (\ddagger)$$

If we accept these formulae, then we can attempt to use them to determine the $\rho_\Phi(\mathbf{r}, \epsilon, \eta)$ in terms of finitely many values as happens in the case of A_1 when $n = 2$ and $n = 3$. In the cases of the classical root systems, then one can carry out the calculation without too much trouble, but the results depend on various congruences, and I shall not go into details here. The most useful consequence would seem to be that the rank 2 cases can be used to garner information about the case $n = 5$. The case (with $k = \mathbb{Q}(\mu_5)$) is the only one that seems within the range of numerical investigation at the present time and which has not yet been investigated. In the case of A_* the argument leading to [12, Theorem II.2.5] shows that "uniqueness almost everywhere" leads to "uniqueness everywhere". This is based on the special role that Whittaker models play in the case of general linear groups, and there is no parallel to them in general. Consequently, neither the existence nor uniqueness statements can be asserted. Whether they are true or not is an open question.

There is an additional technique which one could potentially make use of this context. If we select out one simple root from the Dynkin diagram and construct the residue in $s(\alpha)$ at the remaining ones, then we have an Eisenstein series in one variable. It would be interesting to investigate the coefficients of the Whittaker functions. This idea was proposed in the context of cuspidal representations by Langlands in [14] and has been developed much further by F. Shahidi in a series of papers. The optimist would hope that they are Dirichlet series of Hecke type, or quadratic ones of Rankin–Selberg type, or triple products of Garrett type in the branches of the Dynkin diagram left after excising the simple root selected. The cases which would be most interesting at the outset would be A_2, A_3 (with the middle root excised), D_4 (with the central root excised) and G_2. Nothing concrete is known here. To investigate these questions, one should probably attempt to understand the Chinta–Gunnells representation in more detail, especially their Theorem (3.5) and its variants.

References

1. W. Banks, A corollary to Bernstein's theorem and Whittaker functions on the metaplectic group, Math.Research Letters 5(1998) 781–790.
2. K. S. Brown, Buildings, Springer-Verlag, 1989.
3. B. Brubaker, D. Bump, S. Friedberg, Weyl group multiple Dirichlet series, Eisenstein series and crystal bases, Ann. of Math. 173(2011)1081–1120.
4. B. Brubaker, D. Bump, S. Friedberg, J. Hoffstein, Coefficients of the n-fold theta function and Weyl group multiple Dirichlet series, In: Contributions in Analytic and Algebraic Number Theory, Festschrift for S. J. Patterson, Edd. V. Blomer, P. Mihailescu, Springer Proceedings in Mathematics, Springer-Verlag, 2012,83–95.
5. W. Banks, D. Bump, D. Lieman, Whittaker–Fourier coefficients of metaplectic Eisenstein series, Compositio Math. 135(2)(2003)153–178.
6. W. Casselman, J. Shalika The unramified principal series of p-adic groups, I, The spherical function, Compositio Math. 40(1980) 387–406.
7. W. Casselman, J. Shalika The unramified principal series of p-adic groups, II, The Whittaker function, Compositio Math. 41(1980)207–231.

8. G. Chinta, P. Gunnells, Constructing Weyl group multiple Dirichlet series, J. Amer. Math. Soc.23(2010)189–215.
9. G. Chinta, O. Offen, A metaplectic Casselman–Shalika formula for GL_r, Amer. J. Math. (to appear)
10. C. Eckhardt, S. J. Patterson, On the Fourier coefficients of biquadratic theta series, Proc. London Math.Soc. (3)64(1992)225–264.
11. H. Jacquet, Fonctions de Whittaker associées aux groupes de Chevalley, Bull.Soc.Math.France 95(1967) 243–309.
12. D. A. Kazhdan, S. J. Patterson, Metaplectic forms, Publ. Math. IHES 59(1984)35–142.
13. T. Kubota, Automorphic forms and the reciprocity law in a number field, Kyoto University, 1969.
14. R.P. Langlands, Euler Products, Yale Mathematical Monographs 1, Yale U.Press, 1971.
15. P. J. McNamara, Whittaker functions on metaplectic groups, Ph.D.Thesis, MIT, 2010.
16. P. J. McNamara, Principal series representations of metaplectic groups over local field.
17. P. J. McNamara, Metaplectic Whittaker Functions and Crystal Bases. Duke Math Journal 156(1)(2011)1–31.
18. H. Matsumoto, Sur les sous-groupes arithmétiques des groupes semi–simples déployés, Ann. scient. Éc. Norm. Sup. 2(1969)1–62.
19. G. Savin, Local Shimura correspondence, Math. Ann. 280(1988)185–190.
20. G. Savin, On unramified representations of covering groups, J. reine angew. Math.566(2004)113–134.
21. G. Wellhausen, Fourierkoeffizienten von Thetafunktionen sechster Ordnung, Dissertation, Göttingen, 1996, available at: http://webdoc.sub.gwdg.de/ebook/e/1999/mathgoe/preprint/mg.96.15.dvi.Z.

Chapter 15
Two-Dimensional Adelic Analysis and Cuspidal Automorphic Representations of GL(2)

Masatoshi Suzuki

Abstract Two-dimensional adelic objects were introduced by I. Fesenko in his study of the Hasse zeta function associated to a regular model $\mathcal{E}$ of the elliptic curve E. The Hasse–Weil L-function $L(E,s)$ of E appears in the denominator of the Hasse zeta function of $\mathcal{E}$. The two-dimensional adelic analysis predicts that the integrand h of the boundary term of the two-dimensional zeta integral attached to $\mathcal{E}$ is mean-periodic. The mean-periodicity of h implies the meromorphic continuation and the functional equation of $L(E,s)$. On the other hand, if E is modular, several nice analytic properties of $L(E,s)$, in particular the analytic continuation and the functional equation, are obtained by the theory of the cuspidal automorphic representation of $GL(2)$ over the ordinary ring of adele (one dimensional adelic object). In this chapter, we try to relate in analytic way the theory of two-dimensional adelic object to the theory of cuspidal automorphic representation of $GL(2)$ over the one-dimensional adelic object, under the assumption that E is modular. In the first approximation, they are dual each other.

Keywords L-functions of elliptic curves • Two-dimensional adelic analysis • Mean-periodic functions • Cuspidal automorphic representations of GL(2)

15.1 Introduction

Let $X \to \operatorname{Spec}\mathbb{Z}$ be a scheme separated and of finite type. The Hasse zeta function of X is defined by the Euler product

$$\zeta_X(s) = \prod_{x \in X_0} (1 - |\kappa(x)|^{-s})^{-1},$$

M. Suzuki (✉)
Department of Mathematics, Tokyo Institute of Technology, 2-12-1 Ookayama, Meguro-ku, Tokyo 152-8551, Japan
e-mail: msuzuki@math.titech.ac.jp

D. Bump et al. (eds.), *Multiple Dirichlet Series, L-functions and Automorphic Forms*, Progress in Mathematics 300, DOI 10.1007/978-0-8176-8334-4_15,

where X_0 is the set of all closed points x of X with residue field $\kappa(x)$ of cardinality $|\kappa(x)| < \infty$. For a number field k with the ring of integers $\mathcal{O}_k$ the Hasse zeta function of the affine scheme $\operatorname{Spec}\mathcal{O}_k$ is the Dedekind zeta function $\zeta_k(s) = \prod_{\mathfrak{p}\subset\mathcal{O}_k}(1-|\mathcal{O}_k/\mathfrak{p}|^{-s})^{-1}$. It is conjectured that $\zeta_X(s)$ has several nice analytic properties such as a meromorphic continuation and a functional equation. However, very little is established when the dimension of X is larger than one.

From now on, we concentrate on characteristic zero case. If the dimension of X is one, the Hasse zeta function $\zeta_X(s)$ is essentially the Dedekind zeta function $\zeta_k(s)$. Due to the celebrated work of Iwasawa and Tate, the analytic properties of $\zeta_k(s)$ are obtained by the Fourier analysis on the adeles $\mathbb{A}_k$. The completed Dedekind zeta function $\widehat{\zeta}_k(s)$ is defined by multiplying $\zeta_k(s)$ with a finite product of Γ-factors. It has the integral representation

$$\widehat{\zeta}_k(s) = \int_{\mathbb{A}_k^\times} f(x)|x|^s \mathrm{d}\mu_{\mathbb{A}_k^\times}(x) =: \zeta_k(f,s),$$

where f is an appropriate Schwartz–Bruhat function on $\mathbb{A}_k$ and $|\ |$ is a module on the ideles $\mathbb{A}_k^\times$ of k. On the other hand, $\zeta_k(f,s)$ is written as

$$\zeta_k(f,s) = \xi(f,s) + \xi(\hat{f}, 1-s) + \omega(f,s)$$

on $\Re(s) > 1$, where $\hat{f}$ is the Fourier transform of f on $\mathbb{A}_k$. Here $\xi(f,s)$ is an entire function given by an integral converging absolutely for every $s \in \mathbb{C}$, and the term $\omega(f,s)$ which is called as the *boundary term* in [4–6, 15] is expressed as the integral

$$\omega(f,s) = \int_0^1 h_f(x)x^s \frac{\mathrm{d}x}{x}$$

for some function h_f on $(0,1)$. Hence, the meromorphic continuation and the functional equation for $\widehat{\zeta}_k(s)$ are equivalent to the meromorphic continuation and the functional equation for $\omega(f,s)$. In this sense, the analytic properties of the function $h_f(x)$ are crucial for a better understanding of the boundary term $\omega(f,s)$. Fourier analysis and analytic duality on $k \subset \mathbb{A}_k$ leads to

$$h_f(x) = -\mu\left(\mathbb{A}_k^1/k^\times\right)\left(f(0) - x^{-1}\hat{f}(0)\right).$$

As a consequence, the boundary term $\omega(f,s)$ is a rational function of s invariant with respect to $f \mapsto \hat{f}$ and $s \mapsto (1-s)$. Thus, $\widehat{\zeta}_k(s)$ admits a meromorphic continuation to $\mathbb{C}$ and satisfies a functional equation with respect to $s \mapsto (1-s)$.

Let E an elliptic curve over k and let $\mathcal{E} \to B = \operatorname{Spec}\mathcal{O}_k$ be a regular model of E over k. Then the description of geometry of models in [10, Theorems 3.7, 4.35 in Chap. 9 and Sect. 10.2.1 in Chap. 10] implies that

$$\zeta_{\mathcal{E}}(s) = n_{\mathcal{E}}(s)\zeta_E(s) \quad \text{with} \quad \zeta_E(s) = \frac{\zeta_k(s)\zeta_k(s-1)}{L(E,s)} \tag{1}$$

on $\Re(s) > 2$, where $L(E,s)$ is the Hasse–Weil L-function of E. Here, $n_{\mathcal{E}}(s)$ is the product of zeta functions of affine lines over finite extension $\kappa(b_j)$ of the residue fields $\kappa(b)$, where b are places of bad reduction of E:

$$n_{\mathcal{E}}(s) = \prod_{j=1}^{J} \left(1 - |\kappa(b_j)|^{1-s}\right)^{-1} \tag{2}$$

where J is the number of singular fibres of $\mathcal{E} \to B$ (see [5, Sect. 7.3]).

The automorphic conjecture for E/k asserts that there exists a cuspidal automorphic representation π_E of $\mathrm{GL}_2(\mathbb{A}_k)$ such that

$$L(E,s) = L(\pi_E, s - 1/2).$$

Then the general theory of L-function $L(\pi,s)$ attached to a cuspidal automorphic representation π of $\mathrm{GL}_2(\mathbb{A}_k)$ implies the analytic continuation and functional equation of $L(E,s)$ via $L(\pi,s)$. The analytic properties of $L(\pi,s)$ are obtained by extending the Iwasawa–Tate theory from the commutative group $\mathrm{GL}_1(\mathbb{A}_k)$ to the noncommutative group $\mathrm{GL}_2(\mathbb{A}_k)$. In this story, the theory of noncommutative group $\mathrm{GL}_2(\mathbb{A}_k)$ relates to $\zeta_{\mathcal{E}}(s)$ via the modularity conjecture and the *L-function* $L(E,s)$ *of* E.

In contrast with the above story, I. Fesenko proposed another way to study $\zeta_{\mathcal{E}}(s)$ in [4–6] by using a commutative group associated with two-dimensional adeles. The ordinary ring of adeles $\mathbb{A}_k$ is regarded as a one-dimensional object in the sense that it is associated to the one-dimensional scheme $\operatorname{Spec} \mathcal{O}_k$. He introduced the two-dimensional adelic space $\mathbf{A}_{\mathcal{E}}$ associated to the two-dimensional scheme $\mathcal{E}$ and established a theory of translation invariant measure and integrals on its subring $\mathbb{A}_{\mathcal{E},S} \prec \mathbf{A}_{\mathcal{E}}$, where S is a set of curves on $\mathcal{E}$ consisting of finitely many horizontal curves and all vertical fibers. Using a measure theory on the two-dimensional adelic space, he defined the zeta integral

$$\zeta_{\mathcal{E},S}(f,s) = \int_{T_{\mathcal{E},S}} f(t)|t|^s \mathrm{d}\mu(t),$$

where f is a generalized Schwartz–Bruhat function on $\mathbb{A}_{\mathcal{E},S} \times \mathbb{A}_{\mathcal{E},S}$, $T_{\mathcal{E},S}$ is certain subgroup of $\mathbb{A}_{\mathcal{E},S}^{\times} \times \mathbb{A}_{\mathcal{E},S}^{\times}$, $|\ |$ is a module function on $T_{\mathcal{E},S}$, and $\mathrm{d}\mu$ is a measure on $T_{\mathcal{E},S}$ (see [5, Sect. 5]). The zeta integral $\zeta_{\mathcal{E},S}(f,s)$ converges absolutely for $\Re(s) > 2$. If the test function f_0 is well-chosen, the zeta integral $\zeta_{\mathcal{E},S}(f_0,s)$ equals

$$\zeta_{\mathcal{E},S}(f_0,s) = \prod_{\text{finite}} \widehat{\zeta}_{k_i}(s/2)^2 \cdot c_{\mathcal{E}}^{1-s} \cdot \zeta_{\mathcal{E}}(s)^2, \tag{3}$$

where k_i is an extension of k determined by each horizontal curve in S and $c_{\mathcal{E}}$ is a positive real number determined by $\mathcal{E}$. On the other hand, as well as the Iwasawa–Tate theory, the zeta integral $\zeta_{\mathcal{E},S}(f,s)$ decomposes into three parts

$$\zeta_{\mathcal{E},S}(f,s) = \xi(f,s) + \xi(\hat{f}, 2-s) + \omega(f,s)$$

on $\Re(s) > 2$, where $\xi(f,s)$ is an entire function and $\hat{f}$ is the Fourier transform of f on $\mathbb{A}_{\mathcal{E},S} \times \mathbb{A}_{\mathcal{E},S}$. The third term $\omega(f,s)$ is called the boundary term of the zeta integral $\zeta_{\mathcal{E},S}(f,s)$. Hence, the meromorphic continuation of the boundary term $\omega(f_0,s)$ attached to f_0 in (3) implies the meromorphic continuation of the Hasse zeta function $\zeta_{\mathcal{E}}(s)$. Therefore, if the meromorphic continuation of the boundary term $\omega(f,s)$ is proved without proving the (difficult) modularity property, for example, by using Fourier analysis and analytic duality on the two-dimensional adelic space, it leads to the meromorphic continuation of the L-function $L(E,s)$ without the modularity!

One possible approach for the meromorphic continuation of $\omega(f,s)$ is proposed by the theory of mean-periodic functions as follows. (see [5, Sect. 7], and see also [15]. For the general theory of mean-periodic functions, see Kahane [8], Schwartz [13], or references of [15]). As in the Iwasawa–Tate theory, the boundary term is written as

$$\omega(f,s) = \int_0^1 h_f(x) \cdot x^s \frac{\mathrm{d}x}{x} = \int_0^\infty h_f(\mathrm{e}^{-t}) \cdot \mathrm{e}^{-st} \mathrm{d}t \tag{4}$$

for some function h_f on $(0,1)$, namely, the boundary term is expressed as the Laplace transform of $h_f(\mathrm{e}^{-t})$. Standing on this fact, the boundary term is connected with the theory of mean-periodic functions.

Let $\mathfrak{X}$ be a locally convex separated topological $\mathbb{C}$-vector space consisting of complex valued functions on $\mathbb{R}_+^\times = (0,\infty)$. It has the natural representation τ of $\mathbb{R}_+^\times$ as $(\tau_a F)(x) = F(x/a)$ for every $F \in \mathfrak{X}$. For a function $F \in \mathfrak{X}$, we denote by $\mathcal{T}(F)$ the closure of the set $\{\tau_a F \mid a \in \mathbb{R}_+^\times\}$ with respect to the topology on $\mathfrak{X}$. A function $F \in \mathfrak{X}$ is called *mean-periodic* if the closure $\mathcal{T}(F)$ is not equal to the whole space $\mathfrak{X}$. Using the representation τ, the convolution $F * \varphi$ of $F \in \mathfrak{X}$ and $\varphi \in \mathfrak{X}^\vee$ is defined by

$$(F * \varphi)(x) = \left\langle \tau_x \check{F}, \varphi \right\rangle,$$

where $\check{F}(x) = F(x^{-1})$ and $\langle \cdot, \cdot \rangle$ is the pairing on $\mathfrak{X} \times \mathfrak{X}^\vee$. The mean-periodicity condition $\mathcal{T}(F) \neq \mathfrak{X}$ of F is equivalent to the assertion that the space

$$\mathcal{T}(F)^\perp := \{\varphi \in X^\vee \mid G * \varphi = 0 \text{ for all } G \in \mathcal{T}(F)\}$$

of annihilators of $\mathcal{T}(F)$ with respect to the convolution is nontrivial. The general theory of mean-periodic functions assert that the Laplace transform of $F(e^{-t})$ (the Mellin transform of $F(x)$) extends meromorphically to the whole complex plane whenever the function F is mean-periodic.

Now we suppose that the function h_{f_0} determined by (3) and (4) belongs to the space $\mathfrak{X}$. Then the conjectural mean-periodicity of h_{f_0} implies the meromorphic continuation of the Hasse zeta function $\zeta_{\mathcal{E}}(s)$ [5,6], and the mean-periodicity of h_{f_0} is equivalent to $\mathcal{T}(h_{f_0})^{\perp} \neq \{0\}$. Hence, it is important to understand the space of annihilators $\mathcal{T}(h_{f_0})^{\perp}$ attached to h_{f_0} in order to prove the mean-periodicity of h_{f_0} without any assumption.

In this paper, we describe the space of annihilators $\mathcal{T}(h_{f_0})^{\perp}$ in the case of $k = \mathbb{Q}$ by using the cuspidal automorphic representation of $\mathrm{GL}_2(\mathbb{A}_{\mathbb{Q}})$ whose existence follows from the modularity of $E/\mathbb{Q}$ (Theorems 3.1 and 3.2). The restriction $k = \mathbb{Q}$ is settled for the simplicity; see Remark 3.1. Such descriptions of $\mathcal{T}(h_{f_0})^{\perp}$ suggest some duality between the commutative theory of two-dimensional adeles $\mathbf{A}_{\mathcal{E}}$, $\mathbb{A}_{\mathcal{E},S}$ and the noncommutative theory (GL_2-theory) of one-dimensional adele $\mathbb{A}_{\mathbb{Q}}$. To study the space of annihilators $\mathcal{T}(H_{\mathcal{E}})^{\perp}$, we use the theory of C. Soulé [14] and A. Deitmar [3] which is a generalization of the work of A. Connes [1] (see also the book [2]) about the spectral interpretation of the zeros of Hecke L-functions.

In Sect. 15.2, we review the theory of Connes after R. Meyer [11, 12], and review the theory of mean-periodic function involving the Mellin–Carleman (Laplace–Carleman) transform. In addition, we review several facts on the boundary term $\omega(f_0, s)$ according to [15]. Then, in Sect. 15.3, we state the results (Theorems 3.1 and 3.2), and we prove them in Sect. 15.4. In the final section, we comment on the space of annihilators leave the setting of elliptic curves.

15.2 Preliminaries

At the first stage of Connes' work [1], certain weighted Hilbert space together with some artificial parameter is introduced, and a natural generalization of such weighted Hilbert space is used in works of Soulé [14] and Deitmar [3] about the spectral interpretation of the zeros of an automorphic L-function attached to a cuspidal automorphic representation. However, such weighted Hilbert space is not useful for the theory of mean-periodic functions due to the artificial parameter. In particular, if we use one of their weighted Hilbert spaces, the off-line zeros of the L-function do not appear in the spectrum. This difficulty is overcome by the work of Meyer [11, 12] by introducing the strong Schwartz space (which is named by Connes–Marcolli [2]) and dismissing the Hilbert space setting.

15.2.1 Connes' Spectral Interpretation of Zeros

In this part, we review Connes' work about the spectral interpretation of the zeros of the Riemann zeta function after the work of Meyer. For detailed and rigorous description, see [1] and [12].

Let $S(\mathbb{R})$ be the Schwartz space on $\mathbb{R}$ which consists of smooth functions on $\mathbb{R}$ satisfying

$$\|f\|_{m,n} = \sup_{x\in\mathbb{R}} |x^m f^{(n)}(x)| < \infty$$

for all nonnegative integer m and n. It is a Fréchet space over the complex numbers with the topology induced from the family of seminorms $\| \ \|_{m,n}$. Let us define the Schwartz space $S(\mathbb{R}_+^\times)$ on $\mathbb{R}_+^\times$ and its topology via the homeomorphism

$$S(\mathbb{R}) \to S(\mathbb{R}_+^\times);\ f(t) \mapsto f(-\log x),$$

where $t = -\log x$. The *strong Schwartz space* $\mathbf{S}(\mathbb{R}_+^\times)$ is defined by

$$\mathbf{S}(\mathbb{R}_+^\times) := \bigcap_{\beta\in\mathbb{R}} \Big\{ f : \mathbb{R}_+^\times \to \mathbb{C}, \Big[x \mapsto x^{-\beta} f(x)\Big] \in S(\mathbb{R}_+^\times) \Big\}. \tag{5}$$

One of the family of seminorms on $\mathbf{S}(\mathbb{R}_+^\times)$ defining its topology is given by

$$\|f\|_{m,n} = \sup_{x\in\mathbb{R}_+^\times} |x^m f^{(n)}(x)| \tag{6}$$

for every integers m and nonnegative integers n. The strong Schwartz space $\mathbf{S}(\mathbb{R}_+^\times)$ is a Fréchet space over the complex numbers where the family of seminorms defining its topology is given in (6). This space is closed under the multiplication by a complex number and the pointwise addition and multiplication [12].

Let $S(\mathbb{R})_0$ be the subspace of $S(\mathbb{R})$ consisting of all even functions $\phi \in S(\mathbb{R})$ satisfying $\phi(0) = \hat{\phi}(0) = 0$, where $\hat{\phi}$ is the Fourier transform of ϕ on $\mathbb{R}$. For a function $\phi \in S(\mathbb{R})_0$, we define the function $\mathfrak{E}(\phi)$ on the positive real line $\mathbb{R}_+^\times$ by the formula

$$\mathfrak{E}(\phi)(x) = \sum_{n=1}^{\infty} \phi(nx). \tag{7}$$

Then we find that $\mathfrak{E}(\phi)$ is of rapid decay as $x \to +\infty$. In addition, we have the reciprocal formula $\mathfrak{E}(\phi)(x) = x^{-1}\mathfrak{E}(\hat{\phi})(x^{-1})$ by the Poisson summation formula. Hence, $\mathfrak{E}(\phi)(x)$ is also of rapid decay as $x \to 0^+$, since the subspace $S(\mathbb{R})_0$ is closed under the Fourier transform. Hence, the function $\mathfrak{E}(\phi)$ belongs to the strong Schwartz space $\mathbf{S}(\mathbb{R}_+^\times)$ for every $\phi \in S(\mathbb{R})_0$. In other words, $\mathfrak{E}$ is a map from the subspace $S(\mathbb{R})_0$ into the strong Schwartz space. Denote by $\mathcal{V} \subset \mathbf{S}(\mathbb{R}_+^\times)$ the range of the map $\mathfrak{E} : S(\mathbb{R})_0 \hookrightarrow \mathbf{S}(\mathbb{R}_+^\times)$:

$$\mathcal{V} = \left\{ \Phi \in \mathbf{S}(\mathbb{R}_+^\times) \mid \Phi(x) = \mathfrak{E}(\phi) \text{ for some } \phi \in S(\mathbb{R})_0 \right\}.$$

Then the "discrete part" of the "orthogonal complement"

$$\mathcal{V}^\perp = \{\psi \in \mathbf{S}(\mathbb{R}_+^\times)^\vee \mid \langle \mathfrak{E}(\phi), \psi \rangle = 0 \text{ for all } \phi \in S(\mathbb{R})_0 \} \subset \mathbf{S}(\mathbb{R}_+^\times)^\vee$$

is spanned by functions

$$\{x^{\rho}(\log x)^k \mid \zeta(\rho) = 0,\ 0 < \Re(\rho) < 1,\ 0 \le k < m_\rho\},$$

where $\mathbf{S}(\mathbb{R}_+^\times)^\vee$ is the dual space of $\mathbf{S}(\mathbb{R}_+^\times)$ and m_ρ is the multiplicity of the zero ρ. Roughly, it is shown as follows. Let D be the $\mathbb{R}_+^\times$-invariant operator on $\mathbf{S}(\mathbb{R}_+^\times)$ defined by $(Df)(x) = xf'(x)$, and let $D^\vee$ be the adjoint operator of D on $\mathbf{S}(\mathbb{R}_+^\times)^\vee$ defined by $\langle f, D^\vee\varphi\rangle = \langle Df, \varphi\rangle$ $(f \in \mathbf{S}(\mathbb{R}_+^\times))$. Then all eigenvalues of $D^\vee$ are $s \in \mathbb{C}$ with eigenvector x^s, and each eigenspace is spanned by $\{\psi_{s,k}(x) = x^s(\log x)^k \mid 0 \le k < \infty\}$. We have

$$\begin{aligned}\langle \mathfrak{E}(\phi), \psi_{s,k}\rangle &= \int_0^\infty \mathfrak{E}(\phi)(x)x^s(\log x)^k \frac{\mathrm{d}x}{x}\\ &= \frac{\mathrm{d}^k}{\mathrm{d}s^k}\int_0^\infty \mathfrak{E}(\phi)(x)\,x^s\frac{\mathrm{d}x}{x} = \frac{\mathrm{d}^k}{\mathrm{d}s^k}\big(\zeta(s)\Phi(s)\big),\end{aligned}$$

where Φ is the Mellin transform of ϕ. Hence, the condition that $\langle \mathfrak{E}(\phi), \psi_{s,k}\rangle = 0$ for every $\phi \in S(\mathbb{R})_0$ implies $\zeta(s) = 0$ and $k < m_s$. The trivial zeros are excluded by using the functional equation of $\zeta(s)\Phi(s)$. (In the case of the large space $S(\mathbb{R}_+^\times) \supset \mathbf{S}(\mathbb{R}_+^\times)$, all eigenvalues of $D^\vee$ on $S(\mathbb{R}_+^\times)$ are $i\gamma \in i\mathbb{R}$ with eigenvector $x^{i\gamma}$).

15.2.2 Mean-Periodicity on the Strong Schwartz Space

In this part, we review the theory of mean-periodic functions attached to the strong Schwartz space $\mathbf{S}(\mathbb{R}_+^\times)$ according to Sect. 2 of [15]. For the general theory of mean-periodic functions, see Kahane [8], Schwartz [13], or references of [15].

Let $\mathbf{S}(\mathbb{R}_+^\times)^\vee$ be the dual space of the strong Schwartz space $\mathbf{S}(\mathbb{R}_+^\times)$ with the weak $*$-topology, and denote by $\langle\cdot,\cdot\rangle$ the pairing on $\mathbf{S}(\mathbb{R}_+^\times) \times \mathbf{S}(\mathbb{R}_+^\times)^\vee$, namely, $\langle f, \varphi\rangle = \varphi(f)$ for $f \in \mathbf{S}(\mathbb{R}_+^\times)$ and $\varphi \in \mathbf{S}(\mathbb{R}_+^\times)^\vee$. Using the (multiplicative) representation

$$(\tau_x f)(y) = f(y/x) \quad (x \in \mathbb{R}_+^\times),$$

of the multiplicative group $\mathbb{R}_+^\times$ on $\mathbf{S}(\mathbb{R}_+^\times)$, define the (multiplicative) convolution $f * \varphi$ for $f \in \mathbf{S}(\mathbb{R}_+^\times)$ and $\varphi \in \mathbf{S}(\mathbb{R}_+^\times)^\vee$ by

$$(f * \varphi)(x) = \langle \tau_x \check{f}, \varphi\rangle \quad (x \in \mathbb{R}_+^\times),$$

where $\check{f}(x) := f(x^{-1})$. In addition, define the dual representation $\tau^\vee$ on $\mathbf{S}(\mathbb{R}_+^\times)^\vee$ by

$$\langle f, \tau_x^\vee \varphi\rangle := \langle \tau_x f, \varphi\rangle.$$

For a $\mathbb{C}$-vector space V, we identify the bidual space $(V^{\vee})^{\vee}$ (the dual space of $V^{\vee}$ with respect to the weak $*$-topology on $V^{\vee}$) with V in the following way. For a continuous linear functional F on $V^{\vee}$ with respect to its weak $*$-topology, there exists $v \in V$ such that $F(v^{\vee}) = v^{\vee}(v)$ for every $v^{\vee} \in V^{\vee}$. Therefore, we do not distinguish the pairing on $(V^{\vee})^{\vee} \times V^{\vee}$ from the pairing on $V \times V^{\vee}$.

Definition 2.1. Let $\mathfrak{X} = \mathbf{S}(\mathbb{R}_{+}^{\times})^{\vee}$. An element $x \in \mathfrak{X}$ is said to be $\mathfrak{X}$-mean-periodic if there exists a nontrivial element $x^{\vee}$ in $\mathfrak{X}^{\vee}$ satisfying $x * x^{\vee} = 0$.

For $x \in \mathfrak{X} = \mathbf{S}(\mathbb{R}_{+}^{\times})^{\vee}$, we denote by $\mathcal{T}(x)$ the closure of the $\mathbb{C}$-vector space spanned by $\{\tau_g^{\vee}(x), g \in \mathbb{R}_{+}^{\times}\}$. Then the Hahn–Banach theorem implies the following equivalence of the $\mathfrak{X}$-mean-periodicity.

Proposition 2.1. *Let $\mathfrak{X} = \mathbf{S}(\mathbb{R}_{+}^{\times})^{\vee}$. An element $x \in \mathfrak{X}$ is $\mathfrak{X}$-mean-periodic if and only if the closure $\mathcal{T}(x)$ is not equal to the whole space $\mathfrak{X}$.*

Let $L^1_{\mathrm{loc,poly}}(\mathbb{R}_{+}^{\times})$ be the space of locally integrable functions on $\mathbb{R}_{+}^{\times}$ satisfying

$$h(x) = \begin{cases} O(x^a) & \text{as } x \to +\infty, \\ O(x^{-a}) & \text{as } x \to 0^+ \end{cases}$$

for some real number $a \geq 0$. Each $h \in L^1_{\mathrm{loc,poly}}(\mathbb{R}_{+}^{\times})$ gives rise to a distribution $\varphi_h \in \mathbf{S}(\mathbb{R}_{+}^{\times})^{\vee}$ defined by

$$\langle f, \varphi_h \rangle = \int_0^{+\infty} f(x)h(x)\frac{\mathrm{d}x}{x}, \quad \forall f \in \mathbf{S}(\mathbb{R}_{+}^{\times}).$$

If there is no confusion, we denote φ_h by h itself, and use the notations $\langle f, h \rangle = \langle f, \varphi_h \rangle$ and $h(x) \in \mathbf{S}(\mathbb{R}_{+}^{\times})^{\vee}$. Under this convention, we have

$$x^{\lambda} \log^k(x) \in L^1_{\mathrm{loc,poly}}(\mathbb{R}_{+}^{\times}) \subset \mathbf{S}(\mathbb{R}_{+}^{\times})^{\vee}$$

for every $(k, \lambda) \in \mathbb{Z}_{\geq 0} \times \mathbb{C}$. Moreover, for every $h \in L^1_{\mathrm{loc,poly}}(\mathbb{R}_{+}^{\times})$, the convolution $f * \varphi_h$ with $f \in \mathbf{S}(\mathbb{R}_{+}^{\times})$ coincides with the ordinary convolution on functions on $\mathbb{R}_{+}^{\times}$, namely,

$$(f * h)(x) = \langle \tau_x \check{f}, f \rangle = \int_0^{+\infty} f(x/y)h(y)\frac{\mathrm{d}y}{y} = \int_0^{+\infty} f(y)h(x/y)\frac{\mathrm{d}y}{y}.$$

For a function $h \in L^1_{\mathrm{loc,poly}}(\mathbb{R}_{+}^{\times})$, we set

$$h^+(x) := \begin{cases} 0 & \text{if } x \geq 1, \\ h(x) & \text{otherwise} \end{cases} \qquad h^-(x) := \begin{cases} h(x) & \text{if } x \geq 1, \\ 0 & \text{otherwise.} \end{cases}$$

Clearly both h^+ and h^- belong to $L^1_{\mathrm{loc,poly}}(\mathbb{R}_{+}^{\times})$ if h belongs to $L^1_{\mathrm{loc,poly}}(\mathbb{R}_{+}^{\times})$.

Lemma 2.1. *Let $h \in L^1_{\mathrm{loc,poly}}(\mathbb{R}_+^\times)$. Suppose that $f * h = 0$ for some nontrivial $f \in \mathbf{S}(\mathbb{R}_+^\times)$. Then the Mellin transforms*

$$\mathsf{M}(f * h^{\pm})(s) = \int_0^{+\infty} (f * h^{\pm})(x) x^s \frac{\mathrm{d}x}{x}$$

are entire functions on $\mathbb{C}$.

Definition 2.2. Let $h \in L^1_{\mathrm{loc,poly}}(\mathbb{R}_+^\times)$. Suppose that $f * h = 0$ for some nontrivial $f \in \mathbf{S}(\mathbb{R}_+^\times)$. Then the *Mellin–Carleman transform* $\mathsf{M}C(h)(s)$ of $h(x)$ is defined by

$$\mathsf{M}C(h)(s) := \frac{\mathsf{M}(f * h^+)(s)}{\mathsf{M}(f)(s)} = -\frac{\mathsf{M}(f * h^-)(s)}{\mathsf{M}(f)(s)}.$$

The Mellin–Carleman transform $\mathsf{M}C(h)$ does not depend on the particular choice of nontrivial f satisfying $f * h = 0$. By Lemma 2.1 we have

Proposition 2.2. *Let $h \in L^1_{\mathrm{loc,poly}}(\mathbb{R}_+^\times) \subset \mathbf{S}(\mathbb{R}_+^\times)^\vee$. Suppose that h is $\mathbf{S}(\mathbb{R}_+^\times)^\vee$-mean-periodic, in other words, suppose that $f * h = 0$ for some nontrivial $f \in \mathbf{S}(\mathbb{R}_+^\times)$. Then the Mellin–Carleman transform $\mathsf{M}C(h)(s)$ of $h(x)$ is a meromorphic function on $\mathbb{C}$.*

The Mellin–Carleman transform $\mathsf{M}C(h)(s)$ of $h(x)$ is *not* a generalization of the Mellin transform of h but is a generalization of the half Mellin transform

$$\int_0^1 h(x) x^s \frac{\mathrm{d}x}{x}.$$

See Sect. 2 of [15] for more details.

15.2.3 Boundary Term $\omega(f_0, s)$

In this part, we review several formulas for the boundary term $\omega(f_0, s)$ attached to the elliptic surface $\mathcal{E} \to \operatorname{Spec}\mathbb{Z}$ according to Sect. 5 of [15]. Let E be an elliptic curve over $\mathbb{Q}$ with conductor q_E. Then the completed L-function $\Lambda(E, s)$ is defined by

$$\Lambda(E, s) := q_E^{s/2}(2\pi)^{-s}\Gamma(s)L(E, s),$$

where $L(E, s)$ is the Hasse–Weil L-function appearing in (1). It is conjectured that $\Lambda(E, s)$ is continued to an entire function and satisfies the functional equation $\Lambda(E, s) = \omega_E \Lambda(E, 2 - s)$ for some sign $\omega_E \in \{\pm 1\}$. By (1), the meromorphic continuation and the functional equation of $\Lambda(E, s)$ implies the meromorphic continuation and the functional equation of $\zeta_{\mathcal{E}}(s)$. Moreover such nice analytic properties of $\Lambda(E, s)$ imply the mean-periodicity of the boundary term $\omega(f_0, s)$ as follows.

Theorem 2.1. *Let E be an elliptic curve over $\mathbb{Q}$ and let $\mathcal{E} \to \operatorname{Spec}\mathbb{Z}$ be its regular model. Assume that $\Lambda(E,s)$ is continued meromorphically to $\mathbb{C}$ with finitely many poles and satisfies the functional equation*

$$\Lambda(E,s)^2 = \Lambda(E,2-s)^2.$$

Then the function

$$h_{\mathcal{E}}(x) := f_{\mathcal{E}}(x) - x^{-1} f_{\mathcal{E}}(x^{-1}) \tag{8}$$

with

$$f_{\mathcal{E}}(x) = \frac{1}{2\pi i}\int_{(c)} \Lambda(s/2+1/4)^2 c_{\mathcal{E}}^{-s-1/2} \zeta_{\mathcal{E}}(s+1/2)^2 x^{-s} ds \quad (c>1) \tag{9}$$

belongs to $\mathbf{S}(\mathbb{R}_+^\times)^\vee$, where $c_{\mathcal{E}}$ is a positive real constant determined by the singular fiber of $\mathcal{E}$ [5, Sect. 5]. *Moreover, $h_{\mathcal{E}}$ is $\mathbf{S}(\mathbb{R}_+^\times)^\vee$-mean-periodic and has the expansion*

$$h_{\mathcal{E}}(x) = \lim_{T\to\infty} \sum_{\Im(\lambda)\le T} \sum_{m=1}^{m_\lambda} C_m(\lambda)\frac{(-1)^{m-1}}{(m-1)!} x^{-\lambda}(\log x)^{m-1},$$

where λ are poles of $\Lambda(s/2+1/4)^2 c_{\mathcal{E}}^{-s-1/2}\zeta_{\mathcal{E}}(s+1/2)^2$ of multiplicity m_λ and $C_m(\lambda)$ are constants determined by the principal part at $s=\lambda$;

$$\Lambda(s/2+1/4)^2 c_{\mathcal{E}}^{-s-1/2}\zeta_{\mathcal{E}}(s+1/2)^2 = \sum_{m=1}^{m_\lambda}\frac{C_m(\lambda)}{(s-\lambda)^m} + O(1) \quad \text{when} \quad s\to\lambda,$$

and the sum over λ converges uniformly on every compact subset of $\mathbb{R}_+^\times$.

Proof. See Sect. 5 of [15]. □

Therefore, the mean-periodicity of $h_{\mathcal{E}}(x)$ and the meromorphic continuation of $\Lambda(E,s)^2$ are equivalent to each other in the first approximation.

Remark 2.1. Let S be the set of curves on $\mathcal{E}$ consisting of one horizontal curve which is the image of the zero section of $\mathcal{E} \to \operatorname{Spec}\mathbb{Z}$ and all vertical fibers of $\mathcal{E} \to \operatorname{Spec}\mathbb{Z}$. Then we have

$$\begin{aligned}\zeta_{\mathcal{E},S}(f_0,s) &= \Lambda(s/2)^2 c_{\mathcal{E}}^{-s}\zeta_{\mathcal{E}}(s)^2\\ &= \int_1^\infty x^{-1/2} f_{\mathcal{E}}(x)x^s\frac{dx}{x} + \int_1^\infty x^{-1/2} f_{\mathcal{E}}(x)x^{2-s}\frac{dx}{x} + \int_0^1 x^{-1/2}h_{\mathcal{E}}(x)x^s\frac{dx}{x}.\end{aligned} \tag{10}$$

Hence, the function $h_{f_0}(x)$ in the introduction is $x^{-1/2}h_{\mathcal{E}}(x)$.

As mentioned in the introduction, it is desirable to prove the mean-periodicity of $h_{\mathcal{E}}(x)$ without the meromorphic continuation of $\Lambda(E,s)$ which is a consequence of the modularity of E.

15.3 Statement of Results

Throughout this section we denote by $\mathbb{A}$ the adele ring $\mathbb{A}_{\mathbb{Q}}$ of $\mathbb{Q}$. At first, we settle the following basic assumption.

Basic assumption and notations. *Suppose that $E/\mathbb{Q}$ is modular. We denote by (π, V_π) the corresponding cuspidal automorphic representation in $L^2(GL_2(\mathbb{Q})\backslash GL_2(\mathbb{A}), \mathbf{1})$, where $\mathbf{1}$ is the trivial central character.*

Needless to say, the modularity of $E/\mathbb{Q}$ is now a theorem by the famous work of Wiles et al. However it is not proved for a general number field k. We emphasize this assumption for the future study of this direction.

For the function $h_{\mathcal{E}}$ of (8), we have

$$\{x^{-\lambda}(\log x)^k\} \subset \mathcal{T}(h_{\mathcal{E}}) \subset \mathbf{S}(\mathbb{R}_+^\times)^\vee, \tag{11}$$

where λ extends over all common poles of Mellin–Carleman transforms $\mathsf{MC}(\varphi)(s)$ with $h_{\mathcal{E}} * \varphi = 0$ that are just all poles of $\mathsf{MC}(h_{\mathcal{E}})$. By (1) and (10), the poles of $\mathsf{MC}(h_{\mathcal{E}})$ coincide with the zeros of $L(E,s)$, ignoring the cancellations in (1). Recall the construction of Sect. 15.2.1:

$$\begin{aligned} \{\mathfrak{E}(\phi)\ ;\ \phi \in S(\mathbb{R})_0\} \subset \mathcal{V} \subset \mathbf{S}(\mathbb{R}_+^\times), \\ \{x^{\rho}(\log x)^k\} \subset \mathcal{V}^\perp \subset \mathbf{S}(\mathbb{R}_+^\times)^\vee, \end{aligned} \tag{12}$$

where ρ extends over all nontrivial zeros of $\zeta(s)$ that are also common zeros of the Mellin transforms of $\mathsf{M}(\mathfrak{E}(\phi))$ with $\phi \in S(\mathbb{R})_0$. Comparing (11) and the second line of (12), we expect that the space of annihilators $\mathcal{T}(h_{\mathcal{E}})^\perp$ is an analogue of $\mathcal{V}$ so that the set of common zeros of the Mellin transforms $\mathsf{M}(\psi)$ with $\psi \in \mathcal{T}(h_{\mathcal{E}})^\perp$ is the set of all zeros of $\Lambda(E,s)$.

15.3.1 Construction on the Positive Real Line

In this part, we construct the space of annihilators $\mathcal{T}(h_{\mathcal{E}})^\perp$ attached to the function $h_{\mathcal{E}}$ in (8) by using $\mathrm{GL}_2(\mathbb{A})$-theory of Soulé [14] which is an extension of the original theory of Connes [1], and using the Godement–Jacquet zeta integrals [7].

Let $M = \mathrm{Mat}_2$ and $G = \mathrm{GL}_2$. Let $||: G_{\mathbb{A}} \to \mathbb{R}_+^\times$ be the module map given by $|g| = |\det g|_{\mathbb{A}}$. Let f_π be an admissible matrix coefficient of the cuspidal automorphic representation (π, V_π) on $L^2(G_{\mathbb{Q}}\backslash G_{\mathbb{A}}) = L^2(G_{\mathbb{Q}}\backslash G_{\mathbb{A}}, \mathbf{1})$, namely,

$$f_\pi(g) = \int_{Z_{\mathbb{A}} G_{\mathbb{Q}} \backslash G_{\mathbb{A}}} \varphi_1(hg)\varphi_2(h) \mathrm{d}h,$$

for some $\varphi_1, \varphi_2 \in \mathcal{A}_0(G) \cap V_\pi$, where $\mathcal{A}_0(G) \subset L^2(G_{\mathbb{Q}} \backslash G_{\mathbb{A}})$ is the space of cuspidal automorphic forms with central character $\mathbf{1}$. Let ϕ be a Schwartz–Bruhat function on $M_{\mathbb{A}}$. For a positive real number x, we set $G_x = \{g \in G_{\mathbb{A}} \mid |g| = x\}$. Define a complex valued function $\mathfrak{E}(\phi, f_\pi)$ on $\mathbb{R}_+^\times$ by

$$\mathfrak{E}(\phi, f_\pi)(x) = \int_{G_x} \phi(g) f_\pi(g) \mathrm{d}g \quad (x \in \mathbb{R}_+^\times) \tag{13}$$

as an analogue of (7). Then

1. The integral (13) converges absolutely.
2. For a given integer $N > 0$, there exists a positive constant $C = C_N$ such that

$$|\mathfrak{E}(\phi, f_\pi)(x)| \le C x^{-N} \tag{14}$$

   for every $x \in \mathbb{R}_+^\times$.
3. We have the functional equation

$$\mathfrak{E}(\phi, f_\pi)(x) = x^{-2} \mathfrak{E}(\hat{\phi}, \check{f}_\pi)(x^{-1}), \tag{15}$$

   where $\hat{\phi}$ is the Fourier transform of ϕ and $\check{f}_\pi(g) = f_\pi(g^{-1})$.

In addition, we find that

$$\mathfrak{E}(\phi, f_\pi)(x) = \sum_{\xi \in G_{\mathbb{Q}}} \int_{G_{\mathbb{Q}} \backslash G_1 \times G_{\mathbb{Q}} \backslash G_1} \varphi_1(g)\varphi_2(h) \int_{\mathbb{A}^1/\mathbb{Q}^\times} \phi(h^{-1} a \tau_x \xi g) \mathrm{d}a \, \mathrm{d}h \, \mathrm{d}g. \tag{16}$$

A function $\phi \in S(M_{\mathbb{A}})$ is called *gaussian* if every archimedean component has the form $P(m)\exp(-a|m|^2)$ $(m \in M(F_v))$ for some polynomial function P on $M(F_v)$ and a positive real number $a > 0$. Put

$$S(\pi) = \{(\phi, f_\pi) \mid \phi \in S(M_{\mathbb{A}}) : \text{gaussian}, \ f_\pi\text{: admissible coefficient of } \pi\}.$$

Then (14) and (15) show that $\mathfrak{E}$ is a map from $S(\pi)$ into $\mathbf{S}(\mathbb{R}_+^\times)$:

$$\mathfrak{E} : S(\pi) \to \mathbf{S}(\mathbb{R}_+^\times); \ (\phi, f_\pi) \mapsto \mathfrak{E}(\phi, f_\pi).$$

We denote by $\mathcal{V}_\pi \subset \mathbf{S}(\mathbb{R}_+^\times)$ the image of this map. Using the function

$$w_0(x) = \frac{1}{2\pi i} \int_{(c)} \Gamma(s/4)^2 \cdot \frac{(c_{\mathcal{E}}/q_E)^s}{n_{\mathcal{E}}(s)^2} \cdot s^4(s-2)^4 \cdot (s-1)^2 \cdot x^{-s} \mathrm{d}s, \tag{17}$$

we define the space $\mathcal{W}_\pi$ by

$$\mathcal{W}_\pi = w_0 * \mathcal{V}_\pi * \mathcal{V}_\pi = \mathrm{span}_{\mathbb{C}}\{w_0 * v_1 * v_2 \mid v_i \in \mathcal{V}_\pi\}. \tag{18}$$

Then the space $\mathcal{W}_\pi$ is a subspace of $\mathbf{S}(\mathbb{R}_+^\times)$, since $w_0 \in \mathbf{S}(\mathbb{R}_+^\times)$ and $\mathbf{S}(\mathbb{R}_+^\times)$ is closed under the multiplicative convolution. For $h \in \mathbf{S}(\mathbb{R}_+^\times)$, we define

$$\mathcal{T}(h)^\perp = \{\, g \in \mathbf{S}(\mathbb{R}_+^\times) \mid g * \tau = 0 \text{ for all } \tau \in \mathcal{T}(h)\}$$

and

$$\mathcal{W}_\pi^\perp = \{\, \varphi \in \mathbf{S}(\mathbb{R}_+)^\vee \mid w * \varphi = 0 \text{ for all } w \in \mathcal{W}_\pi \,\}.$$

Theorem 3.1. *Let E be an elliptic curve over $\mathbb{Q}$ and let $\mathcal{E} \to \mathrm{Spec}\,\mathbb{Z}$ be its regular model. Let $h_\mathcal{E}$ be the function in* (8)*. Then we have*

$$\mathcal{T}(h_\mathcal{E})^\perp \supset \mathcal{W}_\pi \quad \textit{and} \quad \mathcal{T}(h_\mathcal{E}) \subset \mathcal{W}_\pi^\perp.$$

Hence, $\mathcal{W}_\pi \neq \{0\}$ means $\mathbf{S}(\mathbb{R}_+^\times)^\vee$-mean-periodicity of $h_\mathcal{E}(x)$. Further, the equality

$$\mathcal{T}(h_\mathcal{E})^\perp = \mathcal{W}_\pi \quad \textit{or} \quad \mathcal{T}(h_\mathcal{E}) = \mathcal{W}_\pi^\perp$$

implies the absence of cancellations of zeros between

$$(s-1)\widehat{\zeta}(s/2)\widehat{\zeta}(s)\widehat{\zeta}(s-1) \quad \textit{and} \quad n_\mathcal{E}(s)^{-1}\Lambda(E,s).$$

15.3.2 Adelic Construction

In this part, we consider the adelic version of the construction in the previous subsection according to Deitmar [3]. Let $S(M_\mathbb{A})_0$ be the space of all $\phi \in S(M_\mathbb{A})$ such that ϕ and $\hat{\phi}$ send $\{g \in M_\mathbb{A} \mid \det(g) = 0\} = M_\mathbb{A} \setminus G_\mathbb{A}$ to zero. For $\phi \in S(M_\mathbb{A})_0$, we define functions $\mathfrak{E}(\phi)$ and $\hat{\mathfrak{E}}(\phi)$ on $G_\mathbb{A}$ by

$$\begin{aligned}\mathfrak{E}(\phi)(g) &= \sum_{\gamma \in M_\mathbb{Q}} \phi(\gamma g) = \sum_{\gamma \in G_\mathbb{Q}} \phi(\gamma g), \\ \hat{\mathfrak{E}}(\phi)(g) &= \sum_{\gamma \in M_\mathbb{Q}} \phi(g\gamma) = \sum_{\gamma \in G_\mathbb{Q}} \phi(g\gamma).\end{aligned} \tag{19}$$

Then for every $\phi \in S(M_{\mathbb{A}})_0$, we have:

1. The sums $\mathfrak{E}(\phi)$ and $\hat{\mathfrak{E}}(\phi)$ converge locally uniformly in g with all derivatives.
2. For any $N > 0$ there exists $C > 0$ such that

$$|\mathfrak{E}(\phi)(g)|,\ |\hat{\mathfrak{E}}(\phi)(g)| \leq C\min\bigl(|g|, |g|^{-1}\bigr)^N. \tag{20}$$

3. For $g \in G_{\mathbb{A}}$ we have the functional equation

$$\mathfrak{E}(\phi)(g) = |g|^{-2}\hat{\mathfrak{E}}(\hat{\phi})(g^{-1}). \tag{21}$$

Hence, $\mathfrak{E}(\phi)$ belongs to the strong Schwartz space

$$\mathbf{S}(G_{\mathbb{Q}}\backslash G_{\mathbb{A}}) = \bigcap_{\beta\in\mathbb{R}} |\ |^{\beta} S(G_{\mathbb{Q}}\backslash G_{\mathbb{A}}).$$

Let $G_{\mathbb{A}}^1$ be the kernel of the module map $g \mapsto |g|$. Fix a splitting $\text{ß} : \mathbb{R}_+^{\times} \to G_{\mathbb{A}}$ of the exact sequence $1 \to G_{\mathbb{A}}^1 \to G_{\mathbb{A}} \to 1$ such that $(\mathrm{id}, \text{ß}) : G_{\mathbb{A}}^1 \times \mathbb{R}_+^{\times} \to G_{\mathbb{A}}$ is an isomorphism. We denote by R the image of splitting ß. Let $\varphi_\pi \in V_\pi \subset L^2(G_{\mathbb{Q}}\backslash G_{\mathbb{A}}^1) \simeq L^2(\mathsf{R}G_{\mathbb{Q}}\backslash G_{\mathbb{A}})$ be a vector $\varphi_\pi = \otimes_v \varphi_{\pi,v}$ such that $\varphi_{\pi,v}$ is a normalized class one vector for almost all places. Further, we assume that φ_π is smooth and $\varphi_\pi(1) \neq 0$.

We define

$$\mathcal{W}_\pi = \mathrm{span}_{\mathbb{C}}\bigl\{(w_0 \circ \text{ß}) * (\mathfrak{E}(\phi_1)\cdot\varphi_\pi) * (\mathfrak{E}(\phi_2)\cdot\varphi_\pi) \bigm| \phi_i \in S(M_{\mathbb{A}})\bigr\} \subset \mathbf{S}(G_{\mathbb{Q}}\backslash G_{\mathbb{A}}),$$

where w_0 is the function in (17), $(\mathfrak{E}(\phi)\cdot\varphi_\pi)(x) = \mathfrak{E}(\phi)(x)\cdot\varphi_\pi(x)$ and $*$ is the convolution on $G_{\mathbb{Q}}\backslash G_{\mathbb{A}}$ via the right regular representation R, and

$$\mathcal{W}_\pi^{\perp} = \{\eta \in \mathbf{S}(G_{\mathbb{Q}}\backslash G_{\mathbb{A}})^{\vee} \mid w * \eta \equiv 0,\ \forall w \in \mathcal{W}_\pi\}.$$

For $\eta \in \mathbf{S}(G_{\mathbb{Q}}\backslash G_{\mathbb{A}})^{\vee}$ we define

$$\mathcal{T}(\eta) = \mathrm{span}_{\mathbb{C}}\{R^{\vee}(g)\eta \mid g \in G_{\mathbb{A}}\},$$

where $R^{\vee}$ is the transpose of the right regular representation of $G_{\mathbb{A}}$ on $\mathbf{S}(G_{\mathbb{Q}}\backslash G_{\mathbb{A}})$ with respect to the pairing $\langle\ ,\ \rangle$ of $\mathbf{S}(G_{\mathbb{Q}}\backslash G_{\mathbb{A}})$ and $\mathbf{S}(G_{\mathbb{Q}}\backslash G_{\mathbb{A}})^{\vee}$.

Theorem 3.2. *Let $h_{\mathcal{E}}$ be the function on $\mathbb{R}_+^{\times}$ associated to the Hasse zeta function $\zeta_{\mathcal{E}}(s)^2$ as in Theorem 2.1. Under the above notations, we have*

$$\mathcal{T}(h_{\mathcal{E}} \circ \text{ß}) \subset \mathcal{W}_\pi^{\perp}.$$

The equality

$$\mathcal{T}(h_{\mathcal{E}} \circ \text{ß}) = \mathcal{W}_\pi^{\perp}$$

implies the absence of cancellation of the zeros between

$$(s-1)\widehat{\xi}(s/2)\widehat{\xi}(s)\widehat{\xi}(s-1) \quad \textit{and} \quad n_{\mathcal{E}}(s)^{-1}\Lambda_E(s).$$

Remark 3.1. There are no essential obstructions to extend the above results to a general algebraic number fields k under the assumption for automorphic properties of the L-function of E/k, since Soulé [14] and Deitmar [3] are done for general global fields k and analytic properties of $h_{\mathcal{E}}$ for general algebraic number fields k are similar to the case of the rational number field under automorphic properties of E/k (see Sect. 5 of [15], in particular the series expansion of $h_{\mathcal{E}}$ consisting of $x^{\lambda}(\log x)^k$).

15.4 Proof of Results

15.4.1 Proof of Theorem 3.1

First, we prove the implication $\mathcal{W}_\pi \subset \mathcal{T}(h_{\mathcal{E}})^{\perp}$. It suffices to prove that $w * h_{\mathcal{E}} \equiv 0$ for every $w \in \mathcal{W}_\pi$. By Theorem 2.1 the function $h_{\mathcal{E}}$ is a series consisting of functions $f_{\lambda,k}(x) = x^{-\lambda}(\log x)^k$. For a function $w \in \mathcal{W}_\pi$, we have

$$\begin{aligned} w * f_{\lambda,k}(x) &= \int_0^\infty w(y) f_{\lambda,k}(x/y)\frac{\mathrm{d}x}{x} \\ &= \sum_{j=1}^{k}(-1)^k\binom{k}{j} x^{-\lambda}(\log x)^{k-j}\int_0^\infty w(y)y^{\lambda}(\log y)^j\frac{\mathrm{d}y}{y}. \end{aligned}$$

Here

$$\int_0^\infty w(y)y^{\lambda}(\log y)^j\frac{\mathrm{d}y}{y} = \frac{\mathrm{d}^j}{\mathrm{d}\lambda^j}\int_0^\infty w(y)y^{\lambda}\frac{\mathrm{d}y}{y}.$$

By definition (18) of the space $\mathcal{W}_\pi$, we have

$$\begin{aligned} \int_0^\infty w(y)y^{\lambda}\frac{\mathrm{d}y}{y} &= \int_0^\infty \left(w_0 * \mathfrak{E}(\phi_1, f_\pi) * \mathfrak{E}(\phi_2, f_\pi')\right)(y)y^{\lambda}\frac{\mathrm{d}y}{y} \\ = \int_0^\infty w_0(y)y^{\lambda}\frac{\mathrm{d}y}{y} &\cdot \int_0^\infty \mathfrak{E}(\phi_1, f_\pi)(y)y^{\lambda}\frac{\mathrm{d}y}{y} \cdot \int_0^\infty \mathfrak{E}(\phi_2, f_\pi')(y)y^{\lambda}\frac{\mathrm{d}y}{y} \end{aligned}$$

for some (ϕ_1, f_π), $(\phi_2, f_\pi') \in S(\pi)$. From the construction of $\mathcal{V}_\pi$, we have

$$\int_0^\infty \mathfrak{E}(\phi, f_\pi)(y)y^{\lambda}\frac{\mathrm{d}y}{y} = F_{\phi, f_\pi}(\lambda)L(\pi, \lambda - 1/2) = F_{\phi, f_\pi}(\lambda)L(E, \lambda)$$

where $F_{\phi,f_\pi}(\lambda)$ is an entire function determined by (ϕ, f_π) (see [7, Theorem 13.8] and [14, Sect. 2.5]). The second equality is a consequence of the modularity of $E/\mathbb{Q}$. As a consequence of the above argument, we obtain

$$\int_0^\infty w(y)y^\lambda \frac{\mathrm{d}y}{y} = \Gamma(\lambda/4)^2\lambda^4(\lambda-2)^4(\lambda-1)^2(c_{\mathcal{E}}/q_E)^\lambda \\ \times n_{\mathcal{E}}(\lambda)^{-2}L(E,\lambda)^2F_{\phi_2,f_\pi}(\lambda)F_{\phi_2,f'_\pi}(\lambda), \tag{22}$$

and $w * h_{\mathcal{E}}$ is a series consisting of (22) and its jth derivative with $j \le m_\lambda$. Because $E/\mathbb{Q}$ is modular, $\Lambda(E,s)$ is an entire function. Therefore, the complex number λ appearing in the expansion of $h_{\mathcal{E}}(x)$ is one of the following:

1. $\lambda = 0$ or 2 and $m_\lambda = 4$.
2. $\lambda \ne 1$ is a zero of $\Lambda(E,s)$ with $n_{\mathcal{E}}(\lambda)^{-1} \ne 0$ and $0 \le m_\lambda \le$ the multiplicity of zero of $\Lambda(E,s)^2$ at $s=\lambda$.
3. $\lambda \ne 1$ is a common zero of $\Lambda(E,s)$ and $n_{\mathcal{E}}(s)^{-1}$, and $-2 \le m_\lambda - 2 \le$ the multiplicity of zero of $\Lambda(E,s)^2$ at $s=\lambda$.
4. $\lambda \ne 1$ is a zero of $n_{\mathcal{E}}(s)^{-1}$ with $\Lambda(E,\lambda) \ne 0$, and $m_\lambda = 2$.
5. $\lambda = 1$ and $-2-2J \le m_\lambda - 2 - 2J \le$ the multiplicity of zero of $\Lambda(E,s)^2$ at $s=\lambda$, where J is the number of singular fibers of $\mathcal{E}$ (see (2)).

Hence, $w * h_{\mathcal{E}} \equiv 0$. Because w was arbitrary, we obtain $\mathcal{W}_\pi \subset \mathcal{T}(h_{\mathcal{E}})^\perp$.

The other implication $\mathcal{T}(h_{\mathcal{E}}) \subset \mathcal{W}_\pi^\perp$ is proved by a similar way. The following fact is useful for this direction (see [7, Sect. 13] and [14, Sect. 2.5]); there exists finitely many $(\phi_\alpha, f_{\pi,\alpha}) \in S(\pi)$ such that

$$\sum_\alpha \int_0^\infty \mathfrak{E}(\phi_\alpha, f_{\pi,\alpha})(x)x^s\frac{\mathrm{d}x}{x} = L(\pi, s-1/2).$$

The final assertion for $\mathcal{T}(h_{\mathcal{E}})^\perp = \mathcal{W}_\pi$ is obvious from (22) and (1)–(5). For $\mathcal{T}(h_{\mathcal{E}}) = \mathcal{W}_\pi^\perp$, we note that $\mathcal{W}_\pi^\perp$ consists of $f_{\lambda,k}$ such that λ is a zero of $n_{\mathcal{E}}(s)^{-2}\Lambda(E,s)s^4(s-2)^4(s-1)^2$ and $k \le$ the multiplicity of λ [14]. If $f_{\lambda,k} \in \mathcal{T}(h_{\mathcal{E}})$, then λ is a pole of order $\ge k$ of $\mathsf{MC}(h_{\mathcal{E}})$ by the general theory of mean-periodic function (e.g., [8, Theorem in lecture 4]). Hence, the cancellation cannot occur when $\mathcal{T}(h_{\mathcal{E}}) = \mathcal{W}^\perp$. □

15.4.2 Proof of Theorem 3.2

This is proved similarly to the proof of Theorem 3.1. For $\mathcal{T}(h_{\mathcal{E}} \circ \text{ß}) \subset \mathcal{W}_\pi^\perp$, it is sufficient to prove that $(h_{\mathcal{E}} \circ \text{ß}) * w = 0$ for any $w \in \mathcal{W}_\pi$. By the expansion of $h_{\mathcal{E}}(x)$ in Theorem 2.1, $h_{\mathcal{E}} \circ \text{ß}$ is a series consisting of $f_{\lambda,k} \circ \text{ß}$. We have

$$(f_{\lambda,k} \circ \text{ß}) * w(y) = \sum_{j=0}^{k} (-1)^j \binom{k}{j} |y|^{-\lambda} (\log |y|)^{k-j} \int_{G_{\mathbb{Q}} \backslash G_{\mathbb{A}}} w(x)|x|^{\lambda} (\log |x|)^j \mathrm{d}x.$$

Here

$$\int_{G_{\mathbb{Q}} \backslash G_{\mathbb{A}}} w(x)|x|^{\lambda} (\log |x|)^j \mathrm{d}x = \frac{\mathrm{d}^j}{\mathrm{d}\lambda^j} \int_{G_{\mathbb{Q}} \backslash G_{\mathbb{A}}} w(x)|x|^{\lambda} \mathrm{d}x,$$

and

$$\int_{G_{\mathbb{Q}} \backslash G_{\mathbb{A}}} w(x)|x|^{\lambda} \mathrm{d}x = \int_{\mathbb{R}_+^{\times}} w_0(x) x^{\lambda} \frac{\mathrm{d}x}{x}$$
$$\times \int_{G_{\mathbb{Q}} \backslash G_{\mathbb{A}}} \mathfrak{E}(\phi_1)(x)\varphi_{\pi}(x)|x|^{\lambda} \mathrm{d}x \int_{G_{\mathbb{Q}} \backslash G_{\mathbb{A}}} \mathfrak{E}(\phi_2)(x)\varphi_{\pi}(x)|x|^{\lambda} \mathrm{d}x,$$

since $|\ |^{\lambda}$ is a multiplicative (quasi) character. By Lemma 3.5 of [3],

$$\int_{G_{\mathbb{Q}} \backslash G_{\mathbb{A}}} \mathfrak{E}(\phi)(x)\varphi_{\pi}(x)|x|^{\lambda} \mathrm{d}x = L(\pi, s - 1/2) F_{\phi,\varphi_{\pi}}(s) = L(E, s) F_{\phi,\varphi_{\pi}}(s),$$

where $F_{\phi,\varphi_{\pi}}(s)$ is an entire function. Therefore, $(f_{\lambda,k} \circ \text{ß}) * w(y) = 0$ for each λ, $1 \le k \le m_{\lambda}$ appearing in the expansion of $h_{\mathcal{E}}$, since λ is a zero of $L(E, s)$ or a zero of $\int_{\mathbb{R}_+^{\times}} w_0(x) x^{\lambda} \frac{\mathrm{d}x}{x}$. Hence, $(h_{\mathcal{E}} \circ \text{ß}) * w = 0$ for any $w \in \mathcal{W}_{\pi}$. □

15.5 Note on Mean-Periodicity and Modularity

Recall the discussion of Sect. 15.3. Many elements of the space of annihilators of $\mathcal{T}(h)$ with respect to the convolution satisfied a kind of modular relation (see (15) or (21)). In this final section, we observe that the space of annihilators of $\mathcal{T}(h)$ contains an element satisfying some modular relation in general. It shows a duality of mean-periodicity and certain kind of modularity.

15.5.1 Basic Assumptions

Let X be a locally convex separated topological $\mathbb{C}$-vector space consisting of $\mathbb{C}$-valued functions on $\mathbb{R}_+^{\times}$ satisfying:

(X-1) The Hahn–Banach theorem is available in X.
(X-2) The Mellin transform $\mathsf{M}(g)(s)$ is defined for every element g of the dual space $X^{\vee}$, and $\mathsf{M}(g)(s)$ is an entire function.
(X-3) The spectral synthesis holds in X.

The second condition is settled in order to use the theory of the Mellin–Carleman transform (see Definition 2.2). Recall that a function $h \in X$ is called (X-)mean-periodic if there exists nontrivial $g \in X^\vee$ such that $h * g \equiv 0$. By (X-1), the mean-periodicity of $h \in X$ is equivalent to $\mathcal{T}(h) \neq X$. For a mean-periodic function $h \in X$, the set of all poles of $\mathsf{MC}(h; s)$ is called the *spectrum of* h. We denote by $\mathrm{Spec}(h)$ the set of spectrum of h without multiplicity.

Throughout this section, we study mean-periodic functions $h \in X$ satisfying the following three conditions:

(h-1) $h(x) = -x^{-1}h(x^{-1})$.

(h-2) $\displaystyle\sum_{\lambda \in \mathrm{Spec}(h)} \frac{m_\lambda}{(1+|\lambda|)^2} < \infty$, where m_λ is the multiplicity of $\lambda \in \mathrm{Spec}(h)$.

(h-3) two numbers

$$\Lambda_- := \inf\{\Re(\lambda)\,;\, \lambda \in \mathrm{Spec}(h)\}, \quad \Lambda_+ := \sup\{\Re(\lambda)\,;\, \lambda \in \mathrm{Spec}(h)\}$$

are both finite, namely, $\mathrm{Spec}(h)$ is contained in some vertical strip of finite width.

15.5.2 Symmetry of the Spectrum

Lemma 5.1. *Let $h \in X$ be a mean-periodic function satisfying* (h-1). *Then*

$$\lambda \in \mathrm{Spec}(h) \quad \textit{implies} \quad 1-\lambda \in \mathrm{Spec}(h). \tag{23}$$

Remark 5.1. We do not know whether (23) holds if we exclude (X-3).

Proof. By (X-3), every mean-periodic function $h \in X$ has an expansion

$$h(x) = \sum_{\lambda \in \mathrm{Spec}(h)} \sum_{m=0}^{m_\lambda - 1} a_m(\lambda)\, x^{-\lambda} (\log x)^m \quad (a_m(\lambda) \neq 0), \tag{24}$$

where the meaning of the series on the right-hand side should be considered in the sense of the topology on X. The series expansion (24) and (h-1) implies (23). Moreover,

$$a_m(\lambda) = (-1)^{m+1} a_m(1-\lambda)$$

for every $\lambda \in \mathrm{Spec}(h)$ and $0 \le m < m_\lambda$. □

15.5.3 Modular Type Realization of the Mean-Periodicity

Take a nontrivial $g \in X^\vee$ satisfies $h * g \equiv 0$, and denote by $\widetilde{G}_0(s)$ the Weierstrass product attached to $\mathrm{Spec}(h)$:

$$\widetilde{G}_0(s) = s^{m_0} \prod_{0 \neq \lambda \in \mathrm{Spec}(h)} \left\{ \left(1 - \frac{s}{\lambda}\right) \exp\left(\frac{s}{\lambda}\right) \right\}^{m_\lambda}.$$

By definition of the spectrum, $\mathrm{Spec}(h)$ is a subset of the zeros of the Mellin transform $\mathsf{M}(g;s)$, and $\widetilde{G}_0(s)$ divide $\mathsf{M}(g;s)$ as $\mathsf{M}(g;s) = \widetilde{G}_0(s) \cdot \widetilde{G}_1(s)$ for some entire function $\widetilde{G}_1(s)$. Define

$$B_0 := -\frac{1}{2} \sum_{\lambda \in \mathrm{Spec}(h)} m_\lambda \left(\frac{1}{\lambda} + \frac{1}{1-\lambda} \right)$$

which is well defined by (h-2) and (23), and put

$$G_0(s) := \exp(B_0 s) \cdot \widetilde{G}_0(s). \tag{25}$$

Then $G_0(s)$ satisfies the functional equation

$$G_0(s) = G_0(1-s) \tag{26}$$

by (23) and the correction factor $\exp(B_0 s)$.

In the later, we take $X = C^\infty_{\mathrm{poly}}(\mathbb{R}^\times_+)$ which is the space of smooth functions on $\mathbb{R}^\times_+$ having at most polynomial growth as $x \to 0^+$ and $x \to +\infty$, and satisfies (X-1), (X-2), and (X-3) (see Sect. 2 of [15]).

Theorem 5.1. *Let $X = C^\infty_{\mathrm{poly}}(\mathbb{R}^\times_+)$. Suppose that $h \in X$ is mean-periodic and satisfies* (h-1), (h-2), *and* (h-3). *In addition, suppose that h satisfies one of the following conditions:*

($\star$) *For any $-\infty < a < b < +\infty$, there exist positive numbers $C_{a,b}$, $A_{a,b}$, and $t_{a,b}$ such that*

$$|G_0(\sigma + it)| \leq C_{a,b} |t|^{-1-A_{a,b}}$$

for every $a \leq \sigma \leq b$ and $|t| \geq t_{a,b}$, where $G_0(s)$ is the entire function defined in (25).

($\star\star$) *There exists $\delta > 0$ such that*

$$\sum_{\lambda \in \mathrm{Spec}(h)} \frac{m_\lambda}{(1+|\lambda|)^{2-\delta}} < \infty.$$

Then there exists a nonzero function $g \in X^\vee$ satisfying $h * g = 0$ and

$$g(x^{-1}) = xg(x). \tag{27}$$

Proof. First, we consider the case when ($\star$) holds. Then, by ($\star$), the inverse Mellin transform $g_0(x)$ of $G_0(s)$ is defined by

$$g_0(x) = \frac{1}{2\pi i} \int_{c-i\infty}^{c+i\infty} G_0(s)\, x^{-s} \mathrm{d}s$$

for each $c \in \mathbb{R}$, and $g_0(x)$ is smooth. The functional equation (26) implies $g_0(x^{-1}) = xg_0(x)$. Moreover, by moving the path of integration to the right (resp. left), we have

$$g_0(x) = O(x^{-N}) \quad (\text{resp. } g_0(x) = O(x^{N}))$$

for every $N > 0$. Here the moving the path of integration is justified by ($\star$). Hence $g_0(x)$ belongs to the dual space of $X = C^{\infty}_{\mathrm{poly}}(\mathbb{R}^{\times}_{+})$. Using the expansion (24), we have

$$\begin{aligned}
(h * g_0)(x) &= \int_0^\infty h(x/y) g_0(y) \frac{\mathrm{d}y}{y} \\
&= \sum_{\lambda \in \mathrm{Spec}(h)} \sum_{m=0}^{m_\lambda - 1} \sum_{j=0}^{m} \binom{m}{j} (-1)^{m-j} a_m(\lambda)\, x^{-\lambda} (\log x)^j \int_0^\infty g_0(y) y^{\lambda} (\log y)^{m-j} \frac{\mathrm{d}y}{y} \\
&= \sum_{\lambda \in \mathrm{Spec}(h)} \sum_{m=0}^{m_\lambda - 1} \sum_{j=0}^{m} \binom{m}{j} (-1)^{m-j} a_m(\lambda)\, x^{-\lambda} (\log x)^j G_0^{(m-j)}(\lambda) = 0,
\end{aligned}$$

since $\mathrm{Spec}(h)$ is a subset of zeros of $G_0(s)$. Hence, we have $h * g_0 = 0$. Thus, the desired assertion is obtained by taking $g = g_0$.

Successively, suppose that ($\star\star$) holds. Then ($\star\star$) implies

$$|G_0(s)| \le \exp(C_\varepsilon |s|^{2-\delta+\varepsilon}) \tag{28}$$

for every $\varepsilon > 0$ by the general theory of the Weierstrass product [9, Sect. 4.3]. Define

$$G(s) := \exp(s(s-1)) \cdot G_0(s).$$

Then $G(s)$ satisfies $G(s) = G(1-s)$ and

$$|G(\sigma + it)| \le \exp(\sigma(\sigma - 1) - t^2) \exp(C_\varepsilon |s|^{2-\delta+\varepsilon}) \tag{29}$$

for every $\varepsilon > 0$. Hence, in particular, for any $-\infty < a < b < +\infty$, there exists positive numbers $C_{a,b}$, $A_{a,b}$, and $t_{a,b}$ such that

$$|G(\sigma + it)| \le C_{a,b} |t|^{-1-A_{a,b}}$$

for every $a \le \sigma \le b$ and $|t| \ge t_{a,b}$. Thus, similar to the proof of the first case, we find that the inverse Mellin transform $g(x)$ of $G(s)$ is a required function. □

As in (23), the spectrum of a mean-periodic function h has the symmetry $\lambda \mapsto 1-\lambda$ as a consequence of (h-1) (and (X-3)). However, we especially mention that the symmetry $\lambda \mapsto 1-\lambda$ is *not* a simple consequence of the functional equation $\mathsf{MC}(h)(s) = \mathsf{MC}(h)(1-s)$ of the Mellin–Carleman transform, since condition (h-1) itself does not imply such functional equation. The symmetry of the spectrum is a consequence of the mean-periodicity of h *and* condition (h-1), since these two properties lead to the above functional equation of $\mathsf{MC}(h)(s)$ (cf. Proposition 2.10 and Remark 2.11 of [15]). Therefore, one interpretation of Theorem 5.1 is that the mean-periodicity of $h \in X$ involves a kind of modularity implicitly (together with (h-1)∼(h-3)), and it is realized in the dual space $X^\vee$ as the modular type equality (27). In the next subsection, we try to express equality (27) coming from the mean-periodicity of $h \in X$ as a kind of summation formula (see (33)) in more number theoretical setting.

15.5.4 A Kind of Summation Formula

Suppose that the entire function $G_0(s)$ defined by (25) has the form

$$G_0(s) = \Gamma(h;s)D(h;s) \tag{30}$$

such that

$$D(h;s) = \sum_{n=1}^{\infty} \frac{c_n}{n^s} \quad \text{(converges absolutely for } \Re(s) \gg 0) \tag{31}$$

and $\Gamma(h;s)$ is a Γ-like function which has no zeros in $\Lambda_- \le \Re(s) \le \Lambda_+$, and $D(h;s)$, $\Gamma(h;s)$ are meromorphic functions on $\mathbb{C}$. Then Spec(h) is a subset of all zeros of $D(h;s)$.

For a "nice" test function $\phi \in S(\mathbb{R})$, for example, its Mellin transform $\mathsf{M}(\phi)(s)$ vanishes at each pole of $D(h;s)$, we "define"

$$\mathfrak{E}_h(\phi)(x) := \sum_{n=1}^{\infty} c_n \phi(nx) \tag{32}$$

referring to definition (7). Then $\mathfrak{E}_h(\phi)(x)$ is of rapid decay as $x \to +\infty$, and

$$\mathsf{M}(\mathfrak{E}_h(\phi))(s) = D(h;s)\mathsf{M}(\phi)(s)$$

is an entire function. Since $G_0(s) = G_0(1-s)$, we have

$$D(h;s)\mathsf{M}(\phi)(s) = D(h;1-s)\frac{\Gamma(h;1-s)}{\Gamma(h;s)}\mathsf{M}(\phi)(s).$$

Taking the inverse Mellin transform of both sides, we formally obtain

$$\sum_{n=1}^{\infty} c_n\,\phi(nx) = \frac{1}{x}\sum_{n=1}^{\infty} c_n\,(\mathsf{F}_h\phi)\Big(\frac{n}{x}\Big), \tag{33}$$

where

$$\mathsf{F}_h\phi(x) := \frac{1}{2\pi i}\int_{c-i\infty}^{c+i\infty} \frac{\Gamma(h;1-s)}{\Gamma(h;s)}\mathsf{M}(\phi;s)\,x^{s-1}\mathrm{d}s. \tag{34}$$

Let $S(h)$ be the subspace of $S(\mathbb{R})$ such that:

1. $\phi(x) = \phi(-x)$ for every $\phi \in S(h)$.
2. $\mathsf{M}(\phi)(s)$ vanishes at all poles of $D(h;s)$ for every $\phi \in S(h)$.
3. $S(h)$ is closed under the operator F_h of (34).

If $S(h)$ is not empty and not trivial, $\mathfrak{E}_h(\phi)(x)$ is also of rapid decay as $x \to 0^+$ by (33) for every $\phi \in S(h)$, and hence, $\mathfrak{E}_h(\phi)(x)$ belongs to the dual space of $C^\infty_{\mathrm{poly}}(\mathbb{R}^\times_+)$, and satisfies $h * \mathfrak{E}_h(\phi) = 0$. That is, the space of annihilators of $\mathcal{T}(h)$ contains the image of the map $\mathfrak{E}_h : S(h) \hookrightarrow C^\infty_{\mathrm{poly}}(\mathbb{R}^\times_+)^\vee$.

It is possible to regard the "summation formula" (33) as a trace of equality (27) (cf. Sects. 15.2.1, 15.3.1, and 15.3.2).

Acknowledgments The author thanks Ivan Fesenko for his many helpful comments and questions to this research and also the School of Mathematical Sciences of the University of Nottingham for the hospitality on author's stay in November 2007–February 2008, August 2008. The author had fruitful conversations and discussions with Guillaume Ricotta and Ivan Fesenko during this stay. This work was partially supported by Grant-in-Aid for JSPS Fellows.

References

1. Alain Connes, *Trace formula in noncommutative geometry and the zeros of the Riemann zeta function*, Selecta Math. (N.S.) **5** (1999), no. 1, 29–106. MR MR1694895 (2000i:11133)
2. Alain Connes and Matilde Marcolli, *Noncommutative geometry, quantum fields and motives*, American Mathematical Society Colloquium Publications, vol. 55, American Mathematical Society, Providence, RI, 2008. MR MR2371808 (2009b:58015)
3. Anton Deitmar, *A Polya-Hilbert operator for automorphic L-functions*, Indag. Math. (N.S.) **12** (2001), no. 2, 157–175. MR MR1913639 (2003h:11055)
4. Ivan Fesenko, *Analysis on arithmetic schemes. I*, Doc. Math. (2003), no. Extra Vol., 261–284 (electronic), Kazuya Kato's fiftieth birthday. MR MR2046602 (2005a:11186)
5. Ivan Fesenko, *Adelic approach to the zeta function of arithmetic schemes in dimension two*, Mosc. Math. J. **8** (2008), no. 2, 273–317, 399–400, (also available at author's web page). MR MR2462437
6. Ivan Fesenko, *Analysis on arithmetic schemes. II*, J. K-Theory **5** (2010), no. 3, 437–557. MR 2658047
7. Roger Godement and Hervé Jacquet, *Zeta functions of simple algebras*, Springer-Verlag, Berlin, 1972, Lecture Notes in Mathematics, Vol. 260. MR MR0342495 (49 #7241)

8. Jean-Pierre Kahane, *Lectures on mean periodic functions*, Tata Inst. Fundamental Res., Bombay, 1959.
9. Boris Ya. Levin, *Lectures on entire functions*, Translations of Mathematical Monographs, vol. 150, American Mathematical Society, Providence, RI, 1996, In collaboration with and with a preface by Yu. Lyubarskii, M. Sodin and V. Tkachenko, Translated from the Russian manuscript by Tkachenko. MR MR1400006 (97j:30001)
10. Qing Liu, *Algebraic geometry and arithmetic curves*, Oxford Graduate Texts in Mathematics,, vol. 6, Oxford University Press, Oxford, 2006, Translated from the French by Reinie Erné. MR MR1917232 (2003g:14001)
11. Ralf Meyer, *On a representation of the idele class group related to primes and zeros of L-functions*, Duke Math. J. **127** (2005), no. 3, 519–595. MR MR2132868 (2006e:11128)
12. Ralf Meyer, *A spectral interpretation for the zeros of the Riemann zeta function*, Mathematisches Institut, Georg-August-Universität Göttingen: Seminars Winter Term 2004/2005, Universitätsdrucke Göttingen, Göttingen, 2005, pp. 117–137. MR MR2206883 (2006k:11166)
13. Laurent Schwartz, *Théorie générale des fonctions moyenne-périodiques*, Ann. of Math. (2) **48** (1947), 857–929. MR MR0023948 (9,428c)
14. Cristophe Soulé, *On zeroes of automorphic L-functions*, Dynamical, spectral, and arithmetic zeta functions (San Antonio, TX, 1999), Contemp. Math., vol. 290, Amer. Math. Soc., Providence, RI, 2001, pp. 167–179. MR MR1868475 (2003c:11103)
15. Masatoshi Suzuki, Ivan Fesenko, and Guillaume Ricotta, *Mean-periodicity and zeta functions*, Annales de l'Institut Fourier (to appear) (2012).